Lecture Notes in Mathematics

Edited by A. Dold and B. Eckmann

T0184806

1176

R. R. Bruner · J. P. May
J. E. McClure · M. Steinberger

H_∞ Ring Spectra and their Applications

Springer-Verlag
Berlin Heidelberg New York Tokyo

Authors

Robert R. Bruner
Department of Mathematics, Wayne State University
Detroit, Michigan 48202, USA

J. Peter May
Department of Mathematics, University of Chicago
Chicago, Illinois 60637, USA

James E. McClure
Department of Mathematics, University of Kentucky
Lexington, Kentucky 40506, USA

Mark Steinberger
Department of Mathematics, Northern Illinois University
De Kalb, Illinois 60115, USA

Mathematics Subject Classification (1980): 55N15, 55N20, 55P42, 55P47, 55Q10, 55Q35, 55Q45, 55S05, 55S12, 55S25, 55T15

ISBN 3-540-16434-0 Springer-Verlag Berlin Heidelberg New York Tokyo
ISBN 0-387-16434-0 Springer-Verlag New York Heidelberg Berlin Tokyo

Printing and binding: Beltz Offsetdruck, Hemsbach/Bergstr.
2146/3140-543210

PREFACE

This volume concerns spectra with enriched multiplicative structure. It is a truism that interesting cohomology theories are represented by ring spectra, the product on the spectrum giving rise to the cup products in the theory. Ordinary cohomology with mod p coefficients has Steenrod operations as well as cup products. These correspond to an enriched multiplicative structure on the Eilenberg-MacLane spectrum HZ_p. Atiyah has shown that the Adams operations in KU-theory are related to similar structure on its representing spectrum and tom Dieck and Quillen have considered Steenrod operations in cobordism coming from similar structure on Thom spectra. Kahn, Toda, Milgram, and others have exploited the same kind of structure on the sphere spectrum to construct and study homotopy operations, and Nishida's proof of the nilpotency of the stable stems is also based on this structure on the sphere spectrum.

In all of this work, the spectrum level structure is either implicit or treated in an ad hoc way, although Tsuchiya gave an early formulation of the appropriate notions. Our purpose is to give a thorough study of such structure and its applications. While there is much that is new here, we are also very interested in explaining how the material mentioned above, and other known results, can be rederived and, in many cases, sharpened and generalized in our context.

The starting point of our work is the existence of extended powers of spectra generalizing the extended powers

$$D_jX = E\Sigma_j \ltimes_{\Sigma_j} X^{(j)} = E\Sigma_j \times_{\Sigma_j} X^{(j)}/E\Sigma_j \times_{\Sigma_j} \{*\}$$

of based spaces X. Here Σ_j is the symmetric group on j letters, $E\Sigma_j$ is a contractible space on which Σ_j acts freely, the symbol $\ltimes$ denotes the "half smash product", and $X^{(j)}$ denotes the j-fold smash power of X. This construction and its variants play a fundamental role in homotopy theory. They appear ubiquitously in the study of torsion phenomena.

It will come as no surprise to anyone that extended powers of spectra can be constructed and shown to have all of the good properties present on the space level. However, those familiar with the details of the analysis of smash products of spectra will also not be surprised that there are onerous technical details involved. In working with spectra, the precise construction of smash products is seldom relevant, and I think most workers in the field are perfectly willing to use them without bothering to learn such details. The same attitude should be taken towards extended powers.

With this in mind, we have divided our work into two parts, of which this volume is the first. We here assume given extended powers and structured spectra and show how to exploit them. This part is meant to be accessible to anyone with a standard background in algebraic topology and some vague idea of what the stable category is. (However, we should perhaps insist right at the outset that, in stable homotopy theory, it really is essential to work in a good stable category and not merely to think in terms of cohomology theories on spaces; only in the former do we have such basic tools as cofibration sequences.) All of the technical work, or rather all of it which involves non-standard techniques, is deferred until the second volume.

We begin by summarizing the properties of extended powers of spectra and introducing the kinds of structured ring spectra we shall be studying. An H_∞ ring spectrum is a spectrum E together with suitably related maps $D_jE \to E$ for $j \geq 0$. The notion is analogous to that of an E_∞ space which I took as the starting point of my earlier work in infinite loop space theory. Indeed, H_∞ ring spectra may be viewed as analogs of infinite loop spaces, and we shall also give a notion of H_n ring spectrum such that H_n ring spectra are analogs of n-fold loop spaces. However, it is to be emphasized that this is only an analogy: the present theory is essentially independent of infinite loop space theory. The structure maps of H_∞ ring spectra give rise to homology, homotopy, and cohomology operations. However, for a complete theory of cohomology operations, we shall need the notion of an H_∞^d ring spectrum. These have structural maps $D_j\Sigma^{di}E \to \Sigma^{dji}E$ for $j \geq 0$ and all integers i.

While chapter I is prerequisite to everything else, the blocks II, III, IV-VI, and VII-IX are essentially independent of one another and can be read in any order.

In chapter II, which is primarily expository and makes no claim to originality, I give a number of rather direct applications of the elementary properties of extended powers of spectra. In particular, I reprove Nishida's nilpotency theorems, explain Jones' recent proof of the Kahn-Priddy theorem, and describe the relationship of extended powers to the Singer construction and to theorems of Lin and Gunawardena.

In chapter III, Mark Steinberger introduces homology operations for H_∞ (and for H_n) ring spectra. These are analogs of the by now familiar (Araki-Kudo, Dyer-Lashof) homology operations for iterated loop spaces. He also carries out extensive calculations of these operations in the standard examples. In particular, it turns out that the homology of HZ_p is monogenic with respect to homology operations, a fact which neatly explains many of the familiar splittings of spectra into wedges of Eilenberg-MacLane and Brown-Peterson spectra.

In chapters IV-VI, Bob Bruner introduces homotopy operations for H_∞ ring spectra and gives a thorough analysis of the behavior of the H_∞ ring structure with respect to the Adams spectral sequence and its differentials. As very special cases, he uses this theory to rederive the Hopf invariant one differentials and certain key odd primary differentials due to Toda. The essential point is the relationship between the structure maps $D_pE \to E$ and Steenrod operations in the E_2 term of the Adams spectral sequence. Only a few of the Steenrod operations survive to homotopy operations, and the attaching maps of the spectra D_pS^q naturally give rise to higher differentials on the remaining Steenrod operations. An attractive feature of Bruner's work is his systematic exploitation of a "delayed" Adams spectral sequence originally due to Milgram to keep track of these complex phenomena.

In chapters VII-IX, Jim McClure relates the notion of an H_∞^d ring spectrum to structure on the familiar kinds of spectra used to represent cohomology theories on spaces. For example, he shows that the representing spectrum KU for complex periodic K-theory is an H_∞^2 ring spectrum, that the Atiyah-Bott-Shapiro orientations give rise to an H_∞^2 ring map $MSpin^c \to KU$, and that similar conclusions hold with $d = 8$ in the real case. He then describes a general theory of cohomology operations and discusses its specialization to ordinary theory, K-theory, and cobordism. Finally, he gives a general theory of homology operations and uses the resulting new operations in complex K-theory to compute the K-theory of $QX = \operatorname{colim} \Omega^n\Sigma^n X$ as a functor of X. This is a striking generalization of work of Hodgkin and of Miller and Snaith, who treated the cases $X = S^0$ and $X = RP^n$ by different methods.

Our applications - and I have only mentioned some of the highlights - are by no means exhaustive. Indeed, our examples show that this is necessarily the case. Far from being esoteric objects, the kinds of spectra we study here abound in nature and include most of the familiar examples of ring spectra. Their internal structure is an essential part of the foundations of stable homotopy theory and should be part of the tool kit of anybody working in this area of topology.

There is a single table of contents, bibliography, and index for the volume as a whole, but each chapter has its own introduction; a reading of these will give a much better idea of what the volume really contains. References are generally by name (Lemma 3.1) within chapters and by number (II.3.1) when to results in other chapters. References to "the sequel" or to [Equiv] refer to "Equivariant stable homotopy theory", which will appear shortly in this series; it contains the construction and analysis of extended powers of spectra.

J. Peter May

Feb. 29, 1984

CONTENTS

CHAPTER I

EXTENDED POWERS AND H_∞ RING SPECTRA

by J. P. May

In this introductory chapter, we establish notations to be adhered to throughout and introduce the basic notions we shall be studying. In the first section, we introduce the equivariant half-smash product of a π-space and a π-spectrum, where π is a finite group. In the second, we specialize to obtain the extended powers of spectra. We also catalog various homological and homotopical properties of these constructions for later use. While the arguments needed to make these two sections rigorous are deferred to the sequel (alias [Equiv] or [51]), the claims the reader is asked to accept are all of the form that something utterly trivial on the level of spaces is also true on the level of spectra. The reader willing to accept these claims will have all of the background he needs to follow the arguments in the rest of this volume.

In sections 3 and 4, we define H_∞ ring spectra and H_∞^d ring spectra in terms of maps defined on extended powers. We also discuss various examples and catalog our techniques for producing such structured ring spectra.

§1. Equivariant half-smash products

We must first specify the categories in which we shall work. All spaces are to be compactly generated and weak Hausdorff. Most spaces will be based; $\mathcal{T}$ will denote the category of based spaces.

Throughout this volume, by a spectrum E we shall understand a sequence of based spaces E_i and based homeomorphisms $\tilde{\sigma}_i: E_i \to \Omega E_{i+1}$, the notation σ_i being used for the adjoints $\Sigma E_i \to E_{i+1}$. A map $f: E \to E'$ of spectra is a sequence of based maps $f_i: E_i \to E_i'$ strictly compatible with the given homeomorphisms; f is said to be a weak equivalence if each f_i is a weak equivalence. There results a category of spectra $\mathcal{S}$. There is a cylinder functor $E \wedge I^+$ and a resulting homotopy category $h\mathcal{S}$. The stable category $\overline{h\mathcal{S}}$ is obtained from $h\mathcal{S}$ by adjoining formal inverses to the weak equivalences, and we shall henceforward delete the adjective "weak". $\overline{h\mathcal{S}}$ is equivalent to the other stable categories in the literature, and we shall use standard properties and constructions without further comment. Definitions of virtually all such constructions will appear in the sequel.

Define $h\mathcal{T}$ and $\overline{h\mathcal{T}}$ analogously to $h\mathcal{S}$ and $\overline{h\mathcal{S}}$. For $X \in \mathcal{T}$, define $QX = \operatorname{colim} \Omega^n\Sigma^n X$, the colimit being taken with respect to suspension of maps $S^n \to \Sigma^n X$. Define adjoint functors

$$\Sigma^\infty: \mathcal{T} \to \mathcal{S} \qquad \text{and} \qquad \Omega^\infty: \mathcal{S} \to \mathcal{T}$$

by $\Sigma^\infty X = \{Q\Sigma^i X\}$ and $\Omega^\infty E = E_0$. (This conflicts with the notation used in most of my previous work, where Σ^∞ and Ω^∞ had different meanings and the present Σ^∞ was called Q_∞; the point of the change is that the present Σ^∞ is by now generally recognized to be the most appropriate infinite suspension functor, and the notation Ω^∞ for the underlying infinite loop space functor has an evident mnemonic appeal.) We then have $QX = \Omega^\infty\Sigma^\infty X$, and the inclusion and evaluation maps $\eta: X \to \Omega^n\Sigma^n X$ and $\varepsilon: \Sigma^n\Omega^n Y \to Y$ pass to colimits to give $\eta: X \to \Omega^\infty\Sigma^\infty X$ for a space X and $\varepsilon: \Sigma^\infty\Omega^\infty E \to E$ for a spectrum E. For any homology theory h_*, ε induces the stabilization homomorphism $\tilde{h}_* E_0 \to h_* E$ obtained by passage to colimits from the suspensions associated to the path space fibrations $E_i \to PE_{i+1} \to E_{i+1}$ for $i \geq 0$.

Let π be a finite group, generally supposed embedded as a subgroup of some symmetric group Σ_j. By a based π-space, we understand a left π-space with a basepoint on which π acts trivially. We let $\pi\mathcal{T}$ denote the resulting category. Actually, most results in this section apply to arbitrary compact Lie groups π.

Let W be a free unbased right π-space and form W^+ by adjoining a disjoint basepoint on which π acts trivially. For $X \in \pi\mathcal{T}$, define the "equivariant half-smash product" $W \ltimes_\pi X$ to be $W^+ \wedge_\pi X$, the orbit space of $W \times X/W \times \{*\}$ obtained by identifying $(w\sigma, x)$ and $(w, \sigma x)$ for $w \in W$, $x \in X$, and $\sigma \in \pi$.

In the sequel, we shall generalize this trivial construction to spectra. That is, we shall explain what we mean by a "π-spectrum E" and we shall make sense of "$W \ltimes_\pi E$"; this will give a functor from the category $\pi\mathcal{S}$ of π-spectra to $\mathcal{S}$. For intuition, with $\pi \subset \Sigma_j$, one may think of E as consisting of based π-spaces E_{ji} for $i \geq 0$ together with π-equivariant maps $E_{ji} \wedge S^j \to E_{j(i+1)}$ whose adjoints are homeomorphisms, where π acts on $S^j = S^1 \wedge \ldots \wedge S^1$ by permutations and acts diagonally on $E_{ji} \wedge S^j$.

The reader is cordially invited to try his hand at making sense of $W \ltimes_\pi E$ using nothing but the definitions already on hand. He will quickly find that work is required. The obvious idea of getting a spectrum from the evident sequence of spaces $W \ltimes_\pi E_{ji}$ and maps

$$\Sigma(W \ltimes_\pi E_{ji}) \to W \ltimes_\pi (E_{ji} \wedge S^j) \to W \ltimes_\pi E_{j(i+1)}$$

is utterly worthless, as a moment's reflection on homology makes clear (compare II.5.6 below). The quickest form of the definition, which is not the form best suited for proving things, is set out briefly in VIII §8 below. The skeptic is invited to refer to the detailed constructions and proofs of the sequel. The pragmatist is invited to accept our word that everything one might naively hope to be true about $W \ltimes_\pi E$ is in fact true.

The first and perhaps most basic property of this construction is that it generalizes the stabilization of the space level construction. If X is a based π-space, then $\Sigma^\infty X$ is a π-spectrum in a natural way.

Proposition 1.1. For based π-spaces X, there is a natural isomorphism of spectra

$$W \ltimes_\pi \Sigma^\infty X \cong \Sigma^\infty (W \ltimes_\pi X).$$

The construction enjoys various preservation properties, all of which hold trivially on the space level.

Proposition 1.2 (i) The functor $W \ltimes_\pi (?)$ from $\pi\mathcal{S}$ to $\mathcal{S}$ preserves wedges, pushouts, and all other categorical colimits.
(ii) If X is a based π-space and $E \wedge X$ is given the diagonal π action, then $W \ltimes (E \wedge X) \cong (W \ltimes E) \wedge X$ before passage to orbits over π; if π acts trivially on X

$$W \ltimes_\pi (E \wedge X) \cong (W \ltimes_\pi E) \wedge X$$

(iii) The functor $W \ltimes_\pi (?)$ preserves cofibrations, cofibres, telescopes, and all other homotopy colimits.

Taking $X = I^+$ in (ii), we see that the functor $W \ltimes_\pi (?)$ preserves π-homotopies between maps of π-spectra.

Let F(X,Y) denote the function space of based maps $X \to Y$ and give $F(W^+,Y)$ the π action $(\sigma f)(w) = f(w\sigma)$ for $f: W \to Y$, $\sigma \in \pi$, and $w \in W$. For π-spaces X and spaces Y, we have an obvious adjunction

$$\mathcal{J}(W \ltimes_\pi X, Y) \cong \pi\mathcal{J}(X, F(W^+, Y)).$$

We shall have an analogous spectrum level adjunction

$$\mathcal{S}(W \ltimes_\pi E, D) \cong \pi\mathcal{S}(E, F[W, D])$$

for spectra D and π-spectra E. Since left adjoints preserve colimits, this will imply the first part of the previous result.

Thus the spectrum level equivariant half-smash products can be manipulated just like their simple space level counterparts. This remains true on the calculational level. In particular, we shall make sense of and prove the following result.

Theorem 1.3. If W is a free π-CW complex and E is a CW spectrum with cellular π action, then $W \ltimes_\pi E$ is a CW spectrum with cellular chains

$$C_*(W \ltimes_\pi E) \cong C_*W \otimes_\pi C_*E.$$

Moreover, the following assertions hold.

(i) If D is a π-subcomplex of E, then $W \ltimes_\pi D$ is a subcomplex of $W \ltimes_\pi E$ and

$$(W \ltimes_\pi E)/(W \ltimes_\pi D) = W \ltimes_\pi (E/D).$$

(ii) If W^n is the n-skeleton of W, then $W^{n-1} \ltimes_\pi E$ is a subcomplex of $W^n \ltimes_\pi E$ and

$$(W^n \ltimes_\pi E)/(W^{n-1} \ltimes_\pi E) \simeq [(W^n/\pi)/(W^{n-1}/\pi)] \wedge E.$$

(iii) With the notations of (i) and (ii),

$$W^{n-1} \ltimes_\pi D = (W^n \ltimes_\pi D) \cap (W^{n-1} \ltimes_\pi E) \subset W^n \ltimes_\pi E.$$

The calculation of cellular chains follows from (i)-(iii), the simpler calculation of chains for ordinary smash products, and an analysis of the behavior of the π actions with respect to the equivalences of (ii).

So far we have considered a fixed group, but the construction is also natural in π. Thus let $f:\rho \to \pi$ be a homomorphism and let $g:V \to W$ be f-equivariant in the sense that $g(v\sigma) = g(v)f(\sigma)$ for $v \in V$ and $\sigma \in \rho$, where V is a ρ-space and W is a π-space. For π-spectra E, there is then a natural map

$$g \ltimes 1 \colon V \ltimes_\rho (f^*E) \to W \ltimes_\pi E,$$

where f^*E denotes E regarded as a ρ-spectrum by pullback along f.

For $X \in \pi\mathscr{T}$ and $Y \in \rho\mathscr{T}$, we have an obvious adjunction

$$\pi\mathscr{T}(\pi^+ \wedge_\rho Y, X) \cong \rho\mathscr{T}(Y, f^*X).$$

We shall have an analogous extension of action functor which assigns a π-spectrum $\pi \ltimes_\rho F$ to a ρ-spectrum F and an analogous adjunction

$$\pi\mathscr{L}(\pi \ltimes_\rho F, E) \cong \rho\mathscr{L}(F, f^*E).$$

Moreover, the following result will hold.

<u>Lemma 1.4</u>. With the notations above,

$$W \ltimes_\pi (\pi \ltimes_\rho F) = W \ltimes_\rho F.$$

When $\rho = e$ is the trivial group, $\pi \ltimes F$ is the free π-spectrum generated by a spectrum F. Intuitively, $\pi \ltimes F$ is the wedge of copies of F indexed by the elements of π and given the action of π by permutations. Here the lemma specializes to give

$$W \ltimes_{\pi} (\pi \ltimes F) = W \ltimes F,$$

and the nonequivariant spectrum $W \ltimes F$ is (essentially) just $W^{+} \wedge F$. Note that, with $\rho = e$ and V a point in the discussion above, we obtain a natural map

$$\iota : E \to W \ltimes_{\pi} E$$

depending on a choice of basepoint for W.

For finite groups π and ρ, there are also natural isomorphisms

$$\alpha : (W \ltimes_{\pi} E) \wedge (V \ltimes_{\rho} F) \to (W \times V) \ltimes_{\pi \times \rho} (E \wedge F)$$

and, if $\rho \subset \Sigma_j$,

$$\beta : V \ltimes_{\rho} (W \ltimes_{\pi} E)^{(j)} \to (V \times W^{j}) \ltimes_{\rho \int \pi} E^{(j)}$$

for π-spaces W, π-spectra E, ρ-spaces V, and ρ-spectra F. Here $E^{(j)}$ denotes the j-fold smash power of E and $\rho \int \pi$ is the wreath product, namely $\rho \times \pi^{i}$ with multiplication

$$(\sigma, \mu_1, \ldots, \mu_j)(\tau, \nu_1, \ldots, \nu_j) = (\sigma\tau, \mu_{\tau(1)}\nu_1, \ldots, \mu_{\tau(j)}\nu_j).$$

The various actions are defined in the evident way. These maps will generally be applied in composition with naturality maps of the sort discussed above.

We need one more general map. If E and F are π-spectra and π acts diagonally on $E \wedge F$, there is a natural map

$$\delta : W \ltimes_{\pi} (E \wedge F) \to (W \ltimes_{\pi} E) \wedge (W \ltimes_{\pi} F).$$

All of these maps ι, α, β, and δ are generalizations of their evident space level analogs. That is, when specialized to suspension spectra, they agree under the isomorphisms of Proposition 1.1 with the suspensions of the space level maps. Moreover, all of the natural commutative diagrams relating the space level maps generalize to the spectrum level, at least after passage to the stable category.

§2. Extended powers of spectra

The most important examples of equivariant half-smash products are of the form $W \ltimes_{\pi} E^{(j)}$ for a spectrum E, where $\pi \subset \Sigma_j$ acts on $E^{(j)}$ by permutations. It requires a little work to make sense of this, and the reader is asked to accept from the sequel that one can construct the j-fold smash power as a functor from $\mathcal{S}$ to $\pi\mathcal{S}$ with all the good properties one might naively hope for. The general properties of these extended powers (or j-adic constructions) are thus direct consequences of the assertions of the previous section. The following consequence of Theorem 1.3 is particularly important.

Corollary 2.1. If W is a free π-CW complex and E is a CW spectrum, then $W \ltimes_\pi E^{(j)}$ is a CW-spectrum with

$$C_*(W \ltimes_\pi E^{(j)}) \cong C_*W \otimes_\pi (C_*E)^j.$$

Thus, with field coefficients, $C_*(W \ltimes_\pi E^{(j)})$ is chain homotopy equivalent to $C_*W \otimes_\pi (H_*E)^j$.

Indeed, $C_*(E^{(j)}) \cong (C_*E)^j$ as a π-complex, where $(C_*E)^j$ denotes the j-fold tensor power. This implies the first statement, and the second statement is a standard, and purely algebraic, consequence (e.g. [68,1.1]).

We shall be especially interested in the case when W is contractible. While all such W yield equivalent constructions, for definiteness we restrict attention to $W = E\pi$, the standard functorial and product-preserving contractible π-free CW-complex (e.g. [70,p.31]). For this W, we define

$$D_\pi E = W \ltimes_\pi E^{(j)}.$$

When $\pi = \Sigma_j$, we write $D_\pi E = D_j E$. Since $E\Sigma_1$ is a point, $D_1 E = E$. We adopt the convention that $D_0 E = E^{(0)} = S$ for all spectra E, where S denotes the sphere spectrum $\Sigma^\infty S^0$.

We adopt analogous notations for spaces X. Thus $D_j X = E\Sigma_j \ltimes_{\Sigma_j} X^{(j)}$, $D_1 X = X$, and $D_0 X = S^0$. Since there is a natural isomorphism $\Sigma^\infty(X^{(j)}) \cong (\Sigma^\infty X)^{(j)}$ of π-spectra, Proposition 1.1 implies the following important consistency statement.

Corollary 2.2. For based spaces X, there is a natural isomorphism of spectra

$$D_\pi \Sigma^\infty X \cong \Sigma^\infty D_\pi X.$$

Corollary 2.1 has the following immediate consequence.

Corollary 2.3. With field coefficients,

$$H_* D_\pi E \cong H_*(\pi;(H_*E)^j).$$

In general, we only have a spectral sequence. Since the skeletal filtrations of $E\pi$ and $B\pi$ satisfy $(E\pi)^n/\pi = (B\pi)^n$, part (ii) of Theorem 1.3 gives a filtration of $D_\pi E$ with successive quotients $[(B\pi)^n/(B\pi)^{n-1}] \wedge E^{(j)}$.

Corollary 2.4. For any homology theory k_*, there is a spectral sequence with $E_2 = H_*(\pi;k_*E^{(j)})$ which converges to $k_*(D_\pi E)$.

This implies the following important preservation properties.

Proposition 2.5. Let T be a set of prime numbers.

(i) If $\lambda:E \to E_T$ is a localization of E at T, then $D_\pi(E_T)$ is T-local and $D_\pi\lambda:D_\pi E \to D_\pi(E_T)$ is a localization at T.

(ii) If $\gamma:E \to \hat{E}_T$ is a completion of E at T, then the completion at T of $D_\pi\gamma:D_\pi E \to D_\pi(\hat{E}_T)$ is an equivalence.

Proof. We refer the reader to Bousfield [21] for a nice treatment of localizations and completions of spectra. By application of the previous corollary with $k_* = \pi_*$, we see that $D_\pi(E_T)$ has T-local homotopy groups and is therefore T-local. (Note that there is no purely homological criterion for recognizing when general spectra, as opposed to bounded below spectra, are T-local.) Taking k_* to be ordinary homology with T-local or mod p coefficients, we see that $D_\pi\lambda$ is a Z_T-homology isomorphism and $D_\pi\gamma$ is a Z_p-homology isomorphism for all $p \in T$. The conclusions follow.

Before proceeding, we should make clear that, except where explicitly stated otherwise, we shall be working in the appropriate homotopy categories $\bar{h}\mathcal{J}$ or $\bar{h}\mathcal{S}$ throughout this volume. Maps and commutative diagrams are always to be understood in this sense.

The natural maps discussed at the end of the previous section lead to natural maps

$$\iota_j:E^{(j)} \to D_jE$$

$$\alpha_{j,k}:D_jE\wedge D_kE \to D_{j+k}E$$

$$\beta_{j,k}:D_jD_kE \to D_{jk}E$$

and

$$\delta_j:D_j(E\wedge F) \to D_jE\wedge D_jF .$$

These are compatible with their obvious space level analogs in the sense that the following diagrams commute.

$$\begin{array}{ccc} & \overset{\iota_j}{\nearrow} & D_j\Sigma^\infty X \\ \Sigma^\infty(X^{(j)}) & & \wr\Vert \\ & \underset{\Sigma^\infty\iota_j}{\searrow} & \Sigma^\infty D_j X \end{array} \qquad \begin{array}{ccc} D_j(\Sigma^\infty X\wedge\Sigma^\infty Y) & \xrightarrow{\delta_j} & D_j\Sigma^\infty X\wedge D_j\Sigma^\infty Y \\ \wr\Vert & & \wr\Vert \\ \Sigma^\infty D_j(X\wedge Y) & \xrightarrow{\Sigma^\infty\delta_j} & \Sigma^\infty(D_jX\wedge D_jY) \end{array}$$

$$\begin{array}{ccc} D_j\Sigma^\infty X\wedge D_k\Sigma^\infty X & \xrightarrow{\alpha_{j,k}} & D_{j+k}\Sigma^\infty X \\ \wr\Vert & & \wr\Vert \\ \Sigma^\infty(D_jX\wedge D_kX) & \xrightarrow{\Sigma^\infty\alpha_{j,k}} & \Sigma^\infty D_{j+k}X \end{array} \qquad \begin{array}{ccc} D_jD_k\Sigma^\infty X & \xrightarrow{\beta_{j,k}} & D_{jk}\Sigma^\infty X \\ \wr\Vert & & \wr\Vert \\ \Sigma^\infty D_jD_kX & \xrightarrow{\Sigma^\infty\beta_{j,k}} & \Sigma^\infty D_{jk}X \end{array}$$

These maps will play an essential role in our theory. H_∞ ring spectra will be defined in terms of maps $D_jE \to E$ such that appropriate diagrams commute. Just as the notion of a ring spectrum presupposes the coherent associativity and commutativity of the smash product of spectra in the stable category, so the notion of an H_∞ ring spectrum presupposes various coherence diagrams relating the extended powers.

Before getting to these, we describe the specializations of our transformations when one of j or k is zero or one.

Remarks 2.6. When j or k is zero, the specified transformations specialize to identity maps (this making sense since $D_0E = S$ and S is the unit for the smash product) with one very important exception, namely $\beta_{j,0}: D_jS \to S$. these maps play a special role in our theory, and we shall also write $\xi_j = \beta_{j,0}$. Observe that D_jS^0 is just $B\Sigma_j^+$, the union of $B\Sigma_j$ and a disjoint basepoint 0. We have the discretization map $d: B\Sigma_j^+ \to S^0$ specified by $d(0) = 0$ and $d(x) = 1$ for $x \in B\Sigma_j$, and ξ_j is given explicitly as

$$D_jS = D_j\Sigma^\infty S^0 \cong \Sigma^\infty D_jS^0 \xrightarrow{\Sigma^\infty d} \Sigma^\infty S^0 = S.$$

Remarks 2.7. The transformations ι_1, $\beta_{j,1}$, $\beta_{1,j}$, and δ_1 are all given by identity maps, and

$$\alpha_{1,1} = \iota_2 : E\wedge E \to D_2E.$$

The last equation is generalized in Lemma 2.11 below.

We conclude this section with eight lemmas which summarize the calculus of extended powers of spectra. Even for spaces, such a systematic listing is long overdue, and every one of the diagrams specified will play some role in our theory. The proofs will be given in the sequel, but in all cases the analogous space level assertion is quite easy to check.

Let $\tau: E\wedge F \to F\wedge E$ denote the commutativity isomorphism in $\overline{h}\mathscr{S}$.

<u>Lemma 2.8.</u> $\{\alpha_{j,k}\}$ is a commutative and associative system, in the sense that the following diagrams commute.

$$\begin{array}{ccc} D_jE\wedge D_kE & & \\ {\scriptstyle\tau}\downarrow & \searrow^{\alpha_{j,k}} & \\ & & D_{j+k}E \\ & \nearrow_{\alpha_{k,j}} & \\ D_kE\wedge D_jE & & \end{array} \qquad \text{and} \qquad \begin{CD} D_iE\wedge D_jE\wedge D_kE @>{\alpha_{i,j}\wedge 1}>> D_{i+j}E\wedge D_kE \\ @V{1\wedge\alpha_{j,k}}VV @VV{\alpha_{i+j,k}}V \\ D_iE\wedge D_{j+k}E @>{\alpha_{i,j+k}}>> D_{i+j+k}E \end{CD}$$

Write $\alpha_{i,j,k}$ for the composite in the second diagram, and so on inductively.

<u>Lemma 2.9.</u> $\{\beta_{j,k}\}$ is an associative system, in the sense that the following diagrams commute.

$$\begin{CD} D_iD_jD_kE @>{\beta_{i,j}}>> D_{ij}D_kE \\ @V{D_i\beta_{j,k}}VV @VV{\beta_{ij,k}}V \\ D_iD_{jk}E @>{\beta_{i,jk}}>> D_{ijk}E \end{CD}$$

Write $\beta_{i,j,k}$ for the composite, and so on inductively.

<u>Lemma 2.10.</u> Each δ_j is commutative and associative, in the sense that the following diagrams commute.

$$\begin{CD} D_j(E\wedge F) @>{\delta_j}>> D_jE\wedge D_jF \\ @V{D_j\tau}VV @VV{\tau}V \\ D_j(F\wedge E) @>{\delta_j}>> D_jF\wedge D_jE \end{CD} \qquad \text{and} \qquad \begin{CD} D_j(E\wedge F\wedge G) @>{\delta_j}>> D_j(E\wedge F)\wedge D_jG \\ @V{\delta_j}VV @VV{\delta_j\wedge 1}V \\ D_jE\wedge D_j(F\wedge G) @>{1\wedge\delta_j}>> D_jE\wedge D_jF\wedge D_jG \end{CD}$$

Continue to write δ_j for the composite in the second diagram, and so on inductively.

Our next two lemmas relate the remaining transformations to the ι_j.

<u>Lemma 2.11.</u> The following diagrams commute.

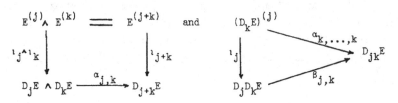

$$\begin{array}{ccc} E^{(j)}\wedge E^{(k)} & = & E^{(j+k)} \\ \downarrow{\scriptstyle \iota_j\wedge\iota_k} & & \downarrow{\scriptstyle \iota_{j+k}} \\ D_jE\wedge D_kE & \xrightarrow{\alpha_{j,k}} & D_{j+k}E \end{array} \qquad \text{and} \qquad \begin{array}{ccc} (D_kE)^{(j)} & \xrightarrow{\alpha_{k,\dots,k}} & D_{jk}E \\ \downarrow{\scriptstyle \iota_j} & \nearrow{\scriptstyle \beta_{j,k}} & \\ D_jD_kE & & \end{array}$$

<u>Lemma 2.12.</u> The following diagram commutes, where ν_j is the evident shuffle isomorphism

$$\begin{array}{ccc} (E\wedge F)^{(j)} & \xrightarrow{\nu_j} & E^{(j)}\wedge F^{(j)} \\ \downarrow{\scriptstyle \iota_j} & & \downarrow{\scriptstyle \iota_j\wedge\iota_j} \\ D_j(E\wedge F) & \xrightarrow{\delta_j} & D_jE\wedge D_jF \end{array}$$

Our last three lemmas of diagrams are a bit more subtle and appear to be new already on the level of spaces.

<u>Lemma 2.13.</u> The following diagram commutes.

$$\begin{array}{ccc} D_iD_kE\wedge D_jD_kE & \xrightarrow{\beta_{i,k}\wedge\beta_{j,k}} & D_{ik}E\wedge D_{jk}E \\ \downarrow{\scriptstyle \alpha_{i,j}} & & \downarrow{\scriptstyle \alpha_{ik,jk}} \\ D_{i+j}D_kE & \xrightarrow{\beta_{i+j,k}} & D_{ik+jk}E \end{array}$$

<u>Lemma 2.14.</u> The following diagrams commute.

$$\begin{array}{ccccc} D_j(E\wedge F)\wedge D_k(E\wedge F) & & \xrightarrow{\alpha_{j,k}} & & D_{j+k}(E\wedge F) \\ \downarrow{\scriptstyle \delta_j\wedge\delta_k} & & & & \downarrow{\scriptstyle \delta_{j+k}} \\ D_jE\wedge D_jF\wedge D_kE\wedge D_kF & \xrightarrow{1\wedge\tau\wedge 1} & D_jE\wedge D_kE\wedge D_jF\wedge D_kF & \xrightarrow{\alpha_{j,k}\wedge\alpha_{j,k}} & D_{j+k}E\wedge D_{j+k}F \end{array}$$

and

$$\begin{array}{ccccc} D_jD_k(E\wedge F) & & \xrightarrow{\beta_{j,k}} & & D_{jk}(E\wedge F) \\ \downarrow{\scriptstyle D_j\delta_k} & & & & \downarrow{\scriptstyle \delta_{jk}} \\ D_j(D_kE\wedge D_kE) & \xrightarrow{\delta_j} & D_jD_kE\wedge D_jD_kF & \xrightarrow{\beta_{j,k}\wedge\beta_{j,k}} & D_{jk}E\wedge D_{jk}F \end{array}$$

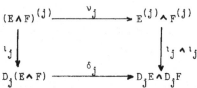

Lemma 2.15. The following diagram commutes.

$$\begin{array}{ccccc} D_i(D_jE\wedge D_kE) & \xrightarrow{\ \delta_i\ } & D_iD_jE\wedge D_iD_kE & \xrightarrow{\ \beta_{i,j}\wedge\beta_{i,k}\ } & D_{ij}E\wedge D_{ik}E \\ \Big\downarrow{\scriptstyle D_i\alpha_{j,k}} & & & & \Big\downarrow{\scriptstyle \alpha_{ij,ik}} \\ D_iD_{j+k}E & & \xrightarrow{\ \beta_{i,j+k}\ } & & D_{ij+ik}E \end{array}$$

When j = k = 1, this diagram specializes to

$$\begin{array}{ccc} D_j(E\wedge E) & \xrightarrow{\ \delta_j\ } & D_jE\wedge D_jE \\ \Big\downarrow{\scriptstyle D_j\iota_2} & & \Big\downarrow{\scriptstyle \alpha_{j,j}} \\ D_jD_2E & \xrightarrow{\ \beta_{j,2}\ } & D_{2j}E \end{array}$$

(On a technical note, all of these coherence diagrams except those of Lemma 2.15 will commute for the extended powers associated to an arbitrary operad; Lemma 2.15 requires restriction to E_∞ operads.)

§3. H_∞ ring spectra

Recall that a (commutative) ring spectrum is a spectrum E together with a unit map $e:S \to E$ and a product map $\phi:E\wedge E \to E$ such that the following diagrams commute (in the stable category, as always).

$$\begin{array}{ccccc} E\wedge S & \xrightarrow{\ 1\wedge e\ } & E\wedge E & \xleftarrow{\ e\wedge 1\ } & S\wedge E \\ & \searrow\!\!\searrow & \Big\downarrow{\scriptstyle\phi} & \swarrow\!\!\swarrow & \\ & & E & & \end{array} \qquad \begin{array}{ccc} E\wedge E\wedge E & \xrightarrow{\ \phi\wedge 1\ } & E\wedge E \\ \Big\downarrow{\scriptstyle 1\wedge\phi} & & \Big\downarrow{\scriptstyle\phi} \\ E\wedge E & \xrightarrow{\ \phi\ } & E \end{array} \qquad \begin{array}{ccc} E\wedge E & & \\ \Big\downarrow{\scriptstyle\tau} & \overset{\phi}{\searrow} & E \\ E\wedge E & \underset{\phi}{\nearrow} & \end{array}$$

In fact, this notion incorporates only a very small part of the full structure generally available.

Definition 3.1. An H_∞ ring spectrum is a spectrum E together with maps $\xi_j:D_j \to E$ for $j \geq 0$ such that ξ_1 is the identity map and the following diagrams commute for $j,k \geq 0$.

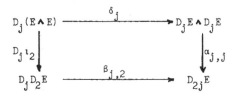

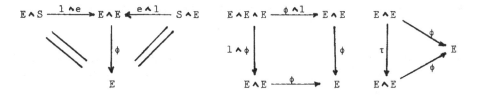

$$\begin{array}{ccccc} D_jE \wedge D_kE & \xrightarrow{\alpha_{j,k}} & & & D_{j+k}E \\ \downarrow{\scriptstyle \xi_j \wedge \xi_k} & & & & \downarrow{\scriptstyle \xi_{j+k}} \\ E \wedge E & \xrightarrow{\iota_2} & D_2E & \xrightarrow{\xi_2} & E \end{array} \quad \text{and} \quad \begin{array}{ccc} D_jD_kE & \xrightarrow{\beta_{j,k}} & D_{jk}E \\ \downarrow{\scriptstyle D_j\xi_k} & & \downarrow{\scriptstyle \xi_{jk}} \\ D_jE & \xrightarrow{\xi_j} & E \end{array}$$

A map $f:E \to F$ between H_∞ ring spectra is an H_∞ ring map if $\xi_j \circ D_jf = f \circ \xi_j$ for $j \geq 0$.

This is a valid sharpening of the notion of a ring spectrum in view of the following consequence of Remarks 2.6 and Lemma 2.8.

Lemma 3.2. With $e = \xi_0:S \to E$ and $\phi = \xi_2 \circ \iota_2:E \wedge E \to E$, an H_∞ ring spectrum is a ring spectrum and an H_∞ ring map is a ring map.

There are various variants and alternative forms of the basic definition that will enter into our work. For a first example, we note the following facts.

Proposition 3.3. Let E be a ring spectrum with maps $\xi_j:D_jE \to E$ such that $\xi_0 = e$, $\xi_1 = 1$, and $\phi = \xi_2\iota_2$. If the first diagram of Definition 3.1 commutes, then ξ_j factors as the composite

$$D_jE = D_jE \wedge S \xrightarrow{1 \wedge e} D_jE \wedge E \xrightarrow{\alpha_{j,1}} D_{j+1}E \xrightarrow{\xi_{j+1}} E .$$

Conversely, if all ξ_j so factor and the second diagram of Definition 1.1 commutes, then the first diagram also commutes and thus E is an H_∞ ring spectrum.

Proof. The first part is an elementary diagram chase. The second part results from Lemmas 2.8 and 2.11 via a rather lengthy diagram chase.

The definition of an H_∞ ring spectrum, together with the formal properties of extended powers, implies the following important closure and consistency properties of the category of H_∞ ring spectra.

Proposition 3.4. The following statements hold, where E and F are H_∞ ring spectra.
(i) With $\xi_j = \beta_{j,0}:D_jS \to S$, the sphere spectrum S is an H_∞ ring spectrum, and $e:S \to E$ is an H_∞ ring map.
(ii) The smash product $E \wedge F$ is an H_∞ ring spectrum with structural maps the composites

$$D_j(E \wedge F) \xrightarrow{\delta_j} D_jE \wedge D_jF \xrightarrow{\xi_j \wedge \xi_j} E \wedge F;$$

the resulting product is the standard one, $(\phi \wedge \phi)(1 \wedge \tau \wedge 1)$.

(iii) The composite $\xi_j \iota_j : E^{(j)} \to E$ is the j-fold iterated product on E and is itself an H_∞ ring map for all j.

Proof. These are elementary diagram chases based respectively on:

(i) Remarks 2.6 and the case $k = 0$ and $E = S$ of Lemmas 2.9 and 2.13.

(ii) Lemmas 2.12 and 2.14.

(iii) Remarks 2.7 and Lemmas 2.9 and 2.11.

In view of Proposition 2.5, we have the following further closure property of the category of H_∞ ring spectra.

Proposition 3.5. If E is an H_∞ ring spectrum, then its localization E_T and completion $\hat{E}_T$ at any set of primes T admit unique H_∞ ring structures such that $\lambda : E \to E_T$ and $\gamma : E \to \hat{E}_T$ are H_∞ ring maps.

Proof. The assertion is obvious in the case of localization. In the case of completion, $\xi_j : D_j \hat{E}_T \to \hat{E}_T$ can and must be defined as the composite

$$D_j \hat{E}_T \xrightarrow{\gamma} (D_j \hat{E}_T)^{\wedge}_T \xrightarrow{((D_j\gamma)^{\wedge}_T)^{-1}} (D_j E)^{\wedge}_T \xrightarrow{(\xi_j)^{\wedge}_T} \hat{E}_T .$$

An easy calculation in ordinary cohomology shows that Eilenberg-MacLane spectra are H_∞ ring spectra.

Proposition 3.6. The Eilenberg-MacLane spectrum HR of a commutative ring R admits a unique H_∞ ring structure, and this structure is functorial in R. If E is a connective H_∞ ring spectrum and $i : E \to H(\pi_0 E)$ is the unique map which induces the identity homomorphism on π_0, then i is an H_∞ ring map.

Proof. Corollary 2.1 implies that $\iota_j : F^{(j)} \to D_j F$ induces an isomorphism in R-cohomology in degree 0 for any connective spectrum F. Moreover, by the Hurewicz theorem and universal coefficients, $H^0(F;R)$ may be identified with $\mathrm{Hom}(\pi_0 F, R)$. Thus we can, and by Proposition 3.4(iii) must, define $\xi_j : D_j HR \to HR$ to be that cohomology class which restricts under ι_j to the j-fold external power of the fundamental class or, equivalently under the identification above, to the j-fold product on R. Similarly, the commutativity of the diagrams in Definition 3.1 is checked by restricting to smash powers and considering cohomology in degree 0. The same argument gives the functoriality. For the last statement, the maps $\xi_j D_j i$ and $i\xi_j$ from $D_j E$ to $H(\pi_0 E)$ are equal because they both restrict under ι_j to the cohomology class given by the iterated product $(\pi_0 E)^j \to \pi_0 E$.

We shall continue to write i for its composite with any map $H(\pi_0 E) \to HR$ induced by a ring homomorphism $\pi_0 E \to R$. We think of such a map $i:E \to HR$ as a counit of E. the composite $ie:S \to HR$ is clearly the unit of HR.

In the rest of this section, we consider the behavior of H_∞ ring spectra with respect to the functors Σ^∞ and Ω^∞. Note first that if E is a ring spectrum, then its unit $e:S \to E$ is determined by the restriction of $e_0:QS^0 \to E_0$ to S^0. If the two resulting basepoints 0 and 1 of E_0 lie in the same component, then e is the trivial map and therefore E is the trivial spectrum.

Definition 3.7. An H_∞ space with zero, or $H_{\infty 0}$ space, is a space X with basepoint 0 together with based maps $\xi_j:D_jX \to X$ for $j \geq 0$ such that the diagrams of Definition 3.1 commute with E replaced by X. Note that $\xi_0:S^0 \to X$ gives X a second basepoint 1. An H_∞ space is a space Y with basepoint 1 together with based maps $E\Sigma_j \times_{\Sigma_j} Y^j \to Y$ for $j \geq 0$ such that the evident analogs of the diagrams of Definition 3.1 commute; $Y^+ = Y \amalg \{0\}$ is then an $H_{\infty 0}$ space.

We remind the reader that we are working up to homotopy (i.e., in $\bar{h}\mathcal{J}$). There is a concomitant notion of a (homotopy associative and commutative) H-space with zero, or H_0-space, given by maps $e:S^0 \to X$ and $\phi:X \wedge X \to X$ such that the diagrams defining a ring spectrum commute with E replaced by X. It is immediately obvious that, mutatis mutandis, Lemma 3.2 and Propositions 3.3-3.5 remain valid for spaces. A commutative ring $R = K(R,0)$ is evidently an $H_{\infty 0}$ space, ξ_j being given by the j-fold product with the $E\Sigma_j$ coordinate ignored.

The isomorphisms $D_j\Sigma^\infty X \cong \Sigma^\infty D_jX$ together with the compatibility of the space and spectrum level transformations ι_j, $\alpha_{j,k}$, and $\beta_{j,k}$ under these isomorphisms have the following immediate consequence.

Proposition 3.8. If X is an $H_{\infty 0}$ space, then $\Sigma^\infty X$ is an H_∞ ring spectrum with structural maps

$$\Sigma^\infty \xi_j : D_j\Sigma^\infty X \cong \Sigma^\infty D_jX \to \Sigma^\infty X.$$

The relationship of Ω^∞ to H_∞ ring structures is a bit more subtle since it is not true that $D_j\Omega^\infty E \cong \Omega^\infty D_jE$. However, the evaluation map $\varepsilon:\Sigma^\infty\Omega^\infty E \to E$ induces

$$D_j\varepsilon:\Sigma^\infty D_j\Omega^\infty E \cong D_j\Sigma^\infty\Omega^\infty E \to D_jE,$$

the adjoint $(\Omega^\infty D_j\varepsilon)\eta$ of which is a natural map

$$\zeta_j:D_j\Omega^\infty E \to \Omega^\infty D_jE \quad \text{or} \quad \zeta_j:D_jE_0 \to (D_jE)_0 .$$

<u>Proposition 3.9</u>. If E is an H_∞ ring spectrum, then E_0 is an $H_{\infty 0}$ space with structural maps

$$(\xi_j)_0 \circ \zeta_j : D_j E_0 \to E_0 .$$

<u>Proof</u>. We must check that the commutativity of the diagrams of Definition 3.1 for E implies their commutativity for E_0. For the first diagram, it is useful to introduce the natural map

$$\zeta : E_0 \wedge F_0 \xrightarrow{\ \eta\ } Q(E_0 \wedge F_0) \cong (\Sigma^\infty E_0 \wedge \Sigma^\infty F_0)_0 \xrightarrow{(\varepsilon \wedge \varepsilon)_0} (E \wedge F)_0$$

for spectra E and F. The relevant diagrams then look as follows

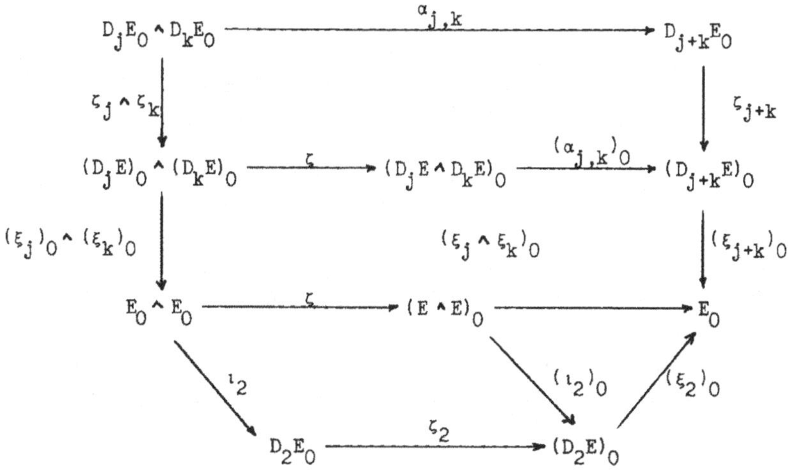

and

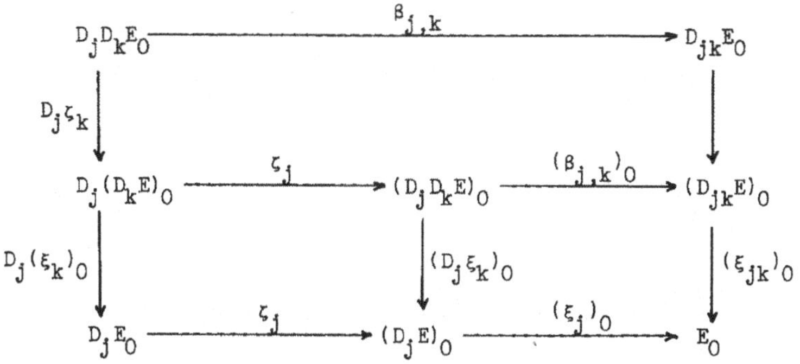

In the upper diagram, $\zeta_2\iota_2 = (\iota_2)_0\zeta$ by the naturality of η and ι_2 and the compatibility of the space and spectrum level maps ι_2. The commutativity of the top rectangles of both diagrams follows similarly, via fairly elaborate chases, from naturality and compatibility diagrams together with the fact that the composite $\varepsilon o \Sigma^\infty\eta:\Sigma^\infty \to \Sigma^\infty\Omega^\infty\Sigma^\infty \to \Sigma^\infty$ is the identity transformation.

The preceding results combine in the following categorical description of the relationship between $H_{\infty 0}$ spaces and H_∞ ring spectra.

Proposition 3.10. If X is an $H_{\infty 0}$ space, then $\eta:X \to \Omega^\infty\Sigma^\infty X$ is a map of $H_{\infty 0}$ spaces. If E is an H_∞ ring spectrum, then $\varepsilon:\Sigma^\infty\Omega^\infty E \to E$ is a map of H_∞ ring spectra. Therefore Σ^∞ and Ω^∞ restrict to an adjoint pair of functors relating the categories of $H_{\infty 0}$ spaces and of H_∞ ring spectra.

The proof consists of easy diagram chases. It follows that if E is an H_∞ ring spectrum, then $\varepsilon_0:QE_0 \to E_0$ is a map of $H_{\infty 0}$ spaces. As we shall explain in the sequel, the significance of this fact is that it implies that the 0th space of an H_∞ ring spectrum is an "H_∞ ring space".

§4. Power operations and H_∞^d ring sprectra

Just as the product of a ring spectrum gives rise to an external product in its represented cohomology theory on spectra and thus to an internal cup product in its represented cohomology theory on spaces, so the structure maps ξ_j of an H_∞ ring spectrum give rise to external and internal extended power operations.

Definitions 4.1. Let E be an H_∞ ring spectrum. For a spectrum Y, define

$$\mathcal{P}_j:E^0Y = [Y,E] \to [D_jY,E] = E^0D_jY$$

by letting $\mathcal{P}_j(h) = \xi_j \circ D_jh$ for $h:Y \to E$. For a based space X, let $\tilde{E}^*X$ denote the reduced cohomology of X and define

$$P_j:\tilde{E}^0X = E^0\Sigma^\infty X \to E^0\Sigma^\infty(B\Sigma_j^+ \wedge X) = \tilde{E}^0(B\Sigma_j^+ \wedge X)$$

by $P_j(h) = (\Sigma^\infty d)^*\mathcal{P}_j(h)$ for $h:\Sigma^\infty X \to E$, where

$$d = 1 \times \Delta:B\Sigma_j^+ \wedge X = E\Sigma_j \ltimes_{\Sigma_j} X \to E\Sigma_j \ltimes_{\Sigma_j} X^{(j)} = D_jX.$$

Of course, the main interest is in the case $j = p$ for a prime p. A number of basic properties of these operations can be read off directly from the definition of an H_∞ ring spectrum, the most important being that $\iota_j^*\mathcal{P}_j(h) = h^j$, where

$h^j \in E^0(Y^{(j)})$ is the external j^{th} power of h, and similarly for the internal operations. McClure will give a systematic study in chapter VIII. While we think of the $\mathcal{P}_j$ as cohomology operations, they can be manipulated to obtain various other kinds of operations. For example, we can define homotopy operations on π_*E parametrized by elements of $E_*D_jS^q$.

Definition 4.2. Let E be an H_∞ ring spectrum. For $\alpha \in E_rD_jS^q$, define $\tilde{\alpha}:\pi_qE \to \pi_rE$ by $\tilde{\alpha}(h) = \alpha/\mathcal{P}_j(h)$ for $h \in \pi_qE$. Explicitly, $\tilde{\alpha}(h)$ is the composite

$$S^r \xrightarrow{\alpha} D_jS^q \wedge E \xrightarrow{\mathcal{P}_j(h)\wedge 1} E \wedge E \xrightarrow{\phi} E.$$

These operations will make a fleeting appearance in our study of nilpotency relations in the next chapter, and Bruner will study them in detail in the case E = S in chapter V. McClure will introduce a related approach to homology operations in chapter VIII.

Returning to Definition 4.1 and replacing Y by Σ^jY for any i, we obtain operations $\mathcal{P}_j:E^{-i}Y \to E^0D_j\Sigma^iY$. A moment's reflection on the Steenrod operations in ordinary cohomology makes clear that we would prefer to have operations $E^{-i}Y \to E^{-ji}D_jY$ for all i. However, the twisting of suspension coordinates which obstructs the equivalence of $D_j\Sigma^iY$ with $\Sigma^{ji}D_jY$ makes clear that the notion of an H_∞ ring spectrum is inadequate for this purpose. For $Y = \Sigma^\infty X$, one can set up a formalism of twisted coefficients to define one's way around the obstruction, but this seems to me to be of little if any use calculationally. Proceeding adjointly, we think of E^iY as $[Y,\Sigma^iE]$ and demand structural maps $\xi_j:D_j\Sigma^iE \to \Sigma^{ji}E$ for all integers i rather than just for i = 0. We can then define extended power operations

$$\mathcal{P}_j:E^iY = [Y,\Sigma^iE] \to [D_jY,\Sigma^{ji}E] = E^{ji}D_jY$$

by letting $\mathcal{P}_j(h) = \xi_j \circ D_jh$ for $h:Y \to \Sigma^iE$; internal operations

$$P_j:\tilde{E}^iX = E^i\Sigma^\infty X \to E^{ji}\ \Sigma^\infty(B\Sigma_j^+ \wedge X) = \tilde{E}^{ji}(B\Sigma_j^+ \wedge X)$$

for spaces are given by $P_j(h) = (\Sigma^\infty d)^*\mathcal{P}_j(h)$, as in Definition 4.1.

In practice, this demands too much. One can usually only obtain maps $\xi_j:D_j\Sigma^{di}E \to \Sigma^{dji}E$ for all j and i and some fixed $d > 0$, often 2 and always a power of 2. In favorable cases, one can use twisted coefficients or restriction to cyclic groups to fill in the missing operations, in a manner to be explained by McClure in chapter VIII. The experts will recall that some such argument was already necessary to define the classical mod p Steenrod operations on odd dimensional classes when $p > 2$.

Definition 4.3. Let d be a positive integer. An H_∞^d ring spectrum is a spectrum E together with maps

$$\xi_{j,i}: D_j \Sigma^{di} E \to \Sigma^{dji} E$$

for all $j \geq 0$ and all integers i such that each $\xi_{1,i}$ is an identity map and the following diagrams commute for all $j \geq 0$, $k \geq 0$, and all integers h and i.

$$\begin{array}{ccc} D_j\Sigma^{di}E \wedge D_k\Sigma^{di}E & \xrightarrow{\alpha_{j,k}} & D_{j+k}\Sigma^{di}E \\ \downarrow{\scriptstyle \xi_{j,i}\wedge\xi_{k,i}} & & \downarrow{\scriptstyle \xi_{j+k,i}} \\ \Sigma^{dji}E \wedge E^{dki}E & \xrightarrow{\phi} & \Sigma^{d(j+k)i}E \end{array} \qquad \begin{array}{ccc} D_jD_k\Sigma^{di}E & \xrightarrow{\beta_{j,k}} & D_{jk}\Sigma^{di}E \\ \downarrow{\scriptstyle D_j\xi_{k,i}} & & \downarrow{\scriptstyle \xi_{jk,i}} \\ D_j\Sigma^{dki}E & \xrightarrow{\xi_{j,ki}} & \Sigma^{djki}E \end{array}$$

and

$$\begin{array}{ccccc} D_j(\Sigma^{dh}E\wedge\Sigma^{di}E) & & \xrightarrow{\delta_j} & & D_j\Sigma^{dh}E \wedge D_j\Sigma^{di}E \\ \downarrow{\scriptstyle D_j\phi} & & & & \downarrow{\scriptstyle \xi_{j,h}\wedge\xi_{j,i}} \\ D_j\Sigma^{d(h+i)}E & \xrightarrow{\xi_{j,h+i}} & \Sigma^{dj(h+i)}E & \xleftarrow{\phi} & \Sigma^{djh}E\wedge\Sigma^{dji}E \end{array}$$

Here the maps ϕ are obtained by suspension from the product $\xi_{2,0}\iota_2$ on E. A map $f:E \to F$ between H_∞^d ring spectra is an H_∞^d ring map if $\xi_{j,i} \circ D_j\Sigma^{di}f = \Sigma^{dji}f \circ \xi_{j,i}$ for all j and i.

Remarks 4.4. (i) Taking $i = 0$, we see that E is an H_∞ ring spectrum. The last diagram is a consequence of the first two when $i = 0$ but is independent otherwise.
(ii) Since $D_0E = S$ for all spectra E, there is only one map $\xi_{i,0}$, namely the unit $e:S^0 \to E$.
(iii) As in Proposition 3.4(iii), the following diagram commutes.

$$\begin{array}{ccc} (\Sigma^{di}E)^{(j)} & \xrightarrow{\iota_j} & D_j\Sigma^{di}E \\ & {\scriptstyle \phi}\searrow \quad \swarrow{\scriptstyle \xi_{j,i}} & \\ & \Sigma^{dji}E & \end{array}$$

(iv) As in Proposition 3.4(ii), the smash product of an H_∞^d ring spectrum E and an H_∞ ring spectrum F is an H_∞^d ring spectrum with structural maps the composites

$$D_j(\Sigma^{di}E\wedge F) \xrightarrow{\delta_j} D_j\Sigma^{di}E \wedge D_jF \xrightarrow{\xi_{j,i}\wedge\xi_j} \Sigma^{dji}E \wedge F.$$

(v) The last diagram in the definition involves a permutation of suspension coordinates, hence one would expect a sign to appear. However, as McClure will explain in VII.6.1, $\pi_0 E$ necessarily has characteristic two when d is odd.

Given this last fact, precisely the same proof as that of Proposition 3.6 yields the following result.

Proposition 4.5. Let R be a commutative ring. If R has characteristic two, then HR admits a unique and functorial H_∞^1 ring structure. In general, HR admits a unique and functorial H_∞^2 ring structure. If E is a connective H_∞^d ring spectrum and $i:E \to H(\pi_0 E)$ is the unique map which induces the identity homomorphism on π_0, then i is an H_∞^d ring map.

At this point, most of the main definitions are on hand, but only rather simple examples. We survey the examples to be obtained later in the rest of this section.

We have three main techniques for the generation of examples. The first, and most down to earth where it applies, is due to McClure and will be explained in chapter VII. The idea is this. In nature, one does not encounter spectra E with E_i homeomorphic to ΩE_{i+1} but only prespectra T consisting of spaces T_i and maps $\sigma_i:\Sigma T_i \to T_{i+1}$. There is a standard way of associating a spectrum to a prespectrum, and McClure will specify concrete homotopical conditions on the spaces T_{di} and composites $\Sigma^d T_{di} \to T_{d(i+1)}$ which ensure that the associated spectrum is an H_∞^d ring spectrum. Curiously, the presence of d is essential. We know of no such concrete way of recognizing H_∞ ring spectra which are not H_∞^d ring spectra for some $d > 0$.

McClure will use this technique to show that the most familiar Thom spectra and K-theory spectra are H_∞^d ring spectra for the appropriate d. While this technique is very satisfactory where it applies, it is limited to the recognition of H_∞^d ring spectra and demands that one have reasonably good calculational control over the spaces T_{di}. The first limitation is significant since, as McClure will explain, the sphere spectrum, for example, is not an H_∞^d ring spectrum for any d. The second limitation makes the method unusable for generic classes of examples.

Our second method is at the opposite extreme, and depends on the black box of infinite loop space machinery. In [71], Nigel Ray, Frank Quinn, and I defined the notion of an E_∞ ring spectra. Intuitively, this is a very precise point-set level notion, of which the notion of an H_∞ ring spectrum is a cruder and less structured up to homotopy analog. Of course, E_∞ ring spectra determine H_∞ ring spectra by neglect of structure. There are also notions of E_∞ space and H_∞ ring space which bear the same relationship of one to the other. Just as the zeroth space of an H_∞ ring spectrum is an H_∞ ring space, so the zeroth space of an E_∞ ring spectrum is an E_∞ ring space. In general, given an H_∞ ring space, there is not the slightest reason to believe that it is equivalent, or nicely related, to the zeroth space of an H_∞ ring spectrum. However, the machinery of [71,73] shows that E_∞ ring spaces functorially determine E_∞ ring spectra the zeroth spaces of which are, in a suitable sense, ring completions of the original semiring spaces. Precise definitions and proofs of the relationship between E_∞ ring theory and H_∞ ring theory will be given in the sequel.

As explained in detail in [73], which corrects [71], the classifying spaces of categories with suitable internal structure, namely bipermutative categories, are E_∞ ring spaces. Among other examples, there result E_∞ ring structures and therefore H_∞ ring structures on the connective spectra of the algebraic K-theory of commutative rings.

The E_∞ and H_∞ ring theories summarized above are limiting cases of E_n and H_n theories for $n \geq 1$, to which the entire discussion applies verbatim. The full theory of extended powers and structured ring spaces and spectra entails the use of operads, namely sequences $\mathcal{C}$ of suitably related Σ_j-spaces $\mathcal{C}_j$. An action of $\mathcal{C}$ on a spectrum E consists of maps $\xi_j: \mathcal{C}_j \ltimes_{\Sigma_j} E^{(j)} \to E$ such that appropriate diagrams commute. For an action up to homotopy, the same diagrams are only required to homotopy commute. If each $\mathcal{C}_j$ has the Σ_j-equivariant homotopy type of the configuration space of j-tuples of distinct points in R^n, then $\mathcal{C}$ is said to be an E_n operad. E_n or H_n ring spectra are spectra with actions or actions up to homotopy by an E_n operad. The notions of E_n and H_n ring space require use of a second operad, assumed to be an E_∞ operad, to encode the additive structure which is subsumed in the iterated loop structure on the spectrum level. E_n ring spaces naturally give rise to E_n and thus H_n ring spectra, and interesting examples of E_n ring spaces have been discovered by Cohen, Taylor, and myself [29] in connection with our study of generalized James maps.

Our last technique for recognizing E_n and H_n ring spectra lies halfway between the first two, and may be described as the brute force method. It consists of direct appeal to the precise definition of extended powers of spectra to be given in the sequel. One class of examples will be given by Steinberger's construction of free $\mathcal{C}$-spectra. Another class of examples will be given in Lewis' study of generalized Thom spectra.

CHAPTER II

MISCELLANEOUS APPLICATIONS IN STABLE HOMOTOPY THEORY

by J. P. May
with contributions by R. R. Bruner, J. E. McClure, and M. Steinberger

A number of important results in stable homotopy theory are very easy consequences of quite superficial properties of extended powers of spectra. We give several such applications here.

The preservation properties of equivariant half-smash products (e.g. in I.1.2) do not directly imply such properties for extended powers since the jth power functor from spectra to Σ_j-spectra tends not to enjoy such properties. We illustrate the point in section 1 by analyzing the structure of extended powers of wedges and deriving useful consequences about extended powers of sums of maps. These results are largely spectrum level analogs of results of Nishida [90] about extended powers of spaces, but the connection with transfer was suggested by ideas of Segal [96].

Reinterpreting Nishida's proof [90], we show in section 2 that the nilpotency of the ring π_*S of stable homotopy groups of spheres (or "stable stems") is an immediate consequence of the Kahn-Priddy theorem and our analysis of extended powers of wedges. The implication depends only on the fact that the sphere spectrum is an H_∞ ring spectrum. This proof gives a very poor estimate of the order of nilpotency. Nishida also gave a different proof [90] which applies only to elements of order p but gives a much better estimate of the order of nilpotency. In section 6, we show that this too results by specialization to S of a result valid for general H_∞ ring spectra. Here the key step is an application of a splitting theorem that Steinberger will prove by use of homology operations in the next chapter. His theorem will make clear to what extent this method of proof applies to elements of order p^i with $i > 1$.

The material discussed so far dates to 1976-77 (and was described in [72]). The material of sections 3-5 is much more recent, dating from 1982-83. The ideas here are entirely due to Miller, Jones, and Wegmann, who saw applications of extended powers that we had not envisaged. (However, all of the information about extended powers needed to carry out their ideas was already explicit or implicit in [72] and the 1977 theses [23, 101] of Bruner and Steinberger.) Jones and Wegmann [44] constructed new homology and cohomology theories from old ones by use of systems of extended powers and showed that theorems of Lin [53] and Gunawardena [38] imply that these theories specialize to give exotic descriptions of stable homotopy and stable cohomotopy. Jones [43] later gave a remarkably ingenious proof of the Kahn-Priddy theorem in terms of these theories. The papers [43, 44] only treated the case $p = 2$, and we give the details for all primes in sections 3 and 4. (In fact, much of the work goes through for non-prime integers.) The idea for the Jones-Wegmann theories grew out of Haynes Miller's unpublished observation that systems of extended powers can be used to realize cohomologically a basic algebraic construction introduced by Singer [52, 98]. We explain this fact and its relationship to the cited theorems of Lin and Gunawardena in section 5.

§1. Extended powers of wedges and transfer maps

Fix positive integers j and k and spectra Y_i for $1 \leq i \leq k$. Let $Y = Y_1 \vee \cdots \vee Y_k$ and let $\nu_i : Y_i \to Y$ be the inclusion. For a partition $J = (j_1, \ldots, j_k)$ of j, $j_i \geq 0$ and $j_1 + \cdots j_k = j$, write $\alpha_J = \alpha_{j_1,\ldots,j_k}$ and let f_J denote the composite

$$D_{j_1}Y_1 \wedge \cdots \wedge D_{j_k}Y_k \xrightarrow{D_{j_1}\nu_1 \wedge \cdots \wedge D_{j_k}\nu_k} D_{j_1}Y \wedge \cdots \wedge D_{j_k}Y \xrightarrow{\alpha_J} D_jY .$$

For later use, note that permutations $\sigma \in \Sigma_k$ act on partitions and that I.2.8 implies the equivariance formula $f_J = f_{\sigma J} \circ \sigma$. Note too that, for maps $h_i : Y_i \to E$ with wedge sum $h : Y \to E$, the following diagram commutes by the naturality of α_J.

$$\begin{array}{ccc} D_{j_1}Y_1 \wedge \cdots \wedge D_{j_k}Y_k & \xrightarrow{f_J} & D_jY \\ {\scriptstyle D_{j_1}h_1 \wedge \cdots \wedge D_{j_k}h_k}\downarrow & & \downarrow{\scriptstyle D_jh} \\ D_{j_1}E \wedge \cdots \wedge D_{j_k}E & \xrightarrow{\alpha_J} & D_jE \end{array}$$

Theorem 1.1. Let $Y = Y_1 \vee \cdots \vee Y_k$. Then the wedge sum

$$f_j : \bigvee_J D_{j_1}Y_1 \wedge \cdots \wedge D_{j_k}Y_k \longrightarrow D_jY$$

of the maps f_J is an equivalence of spectra.

Proof. By the distributivity of smash products over wedges,

$$Y^{(j)} \cong \bigvee_I Y_{i_1} \wedge \cdots \wedge Y_{i_j} ,$$

where I runs over all sequences $(i_1,\ldots,i_j)$ such that $1 \le i_r \le k$. Say that $I \in J$ if there are exactly j_s entries i_r equal to s for each s from 1 to k. For each partition J of j, let $\Sigma_J = \Sigma_{j_1} \times \cdots \times \Sigma_{j_k}$ and define

$$Y_J = \bigvee_{I\in J} Y_{i_1} \wedge \cdots \wedge Y_{i_j} \cong \Sigma_j \ltimes_{\Sigma_J} (Y_1^{(j_1)} \wedge \cdots \wedge Y_k^{(j_k)}).$$

(Here the isomorphism would be obvious on the space level and holds on the spectrum level by direct inspection of the definitions in [Equiv. II §§3-4].) Then Y_J is a Σ_j-subspectrum of $Y^{(j)}$ and $Y^{(j)} = \bigvee_J Y_J$. Now

$$D_j Y = \bigvee_J E\Sigma_j \ltimes_{\Sigma_j} Y_J \quad \text{and} \quad E\Sigma_j \ltimes_{\Sigma_j} Y_J = E\Sigma_j \ltimes_{\Sigma_J} (Y^{(j_1)} \wedge \cdots \wedge Y^{(j_k)})$$

by I.1.2(i) and I.1.4. Clearly f_J has image in $E\Sigma_j \ltimes_{\Sigma_j} Y_J$ and factors as the composite

$$\begin{array}{c}
(E\Sigma_{j_1} \ltimes_{\Sigma_{j_1}} Y_1^{(j_1)}) \wedge \cdots \wedge (E\Sigma_{j_k} \ltimes_{\Sigma_{j_k}} Y_k^{(j_k)}) \\
\big\downarrow \alpha \\
(E\Sigma_{j_1} \times \cdots \times E\Sigma_{j_k}) \ltimes_{\Sigma_J} (Y^{(j_1)} \wedge \cdots \wedge Y^{(j_k)}) \\
\big\downarrow i \ltimes 1 \\
E\Sigma_j \ltimes_{\Sigma_J} (Y^{(j_1)} \wedge \cdots \wedge Y^{(j_k)}) .
\end{array}$$

Here α is an isomorphism. (Technically, the smash product in its domain is "internal" while that in its range is "external"; see [Equiv, II§3].) The map $i: E\Sigma_{j_1} \times \cdots \times E\Sigma_{j_k} \longrightarrow E\Sigma_j$ is given by the commutation with products and naturality of the functor E and is a Σ_J-equivalence. Therefore $i \ltimes 1$ is an equivalence (by [Equiv, VI.1.15]). The conclusion follows.

Our interest is mainly in finite wedges, but precisely the same argument applies to give an analog for infinite wedges.

Theorem 1.2. Let $\{Y_i\}$ be a set of spectra indexed on a totally ordered set of indices and let $Y = \bigvee_i Y_i$. For a strictly increasing sequence $I = \{i_1,\ldots,i_k\}$ of indices and a partition $J = (j_1,\ldots,j_k)$ of j with each $j_i > 0$ (hence $k \le j$), let

$$f_{J,I}: D_{j_1}Y_{i_1} \wedge \cdots \wedge D_{j_k}Y_{i_k} \to D_j Y$$

be the composite of f_J and the evident inclusion. Then the wedge sum

$$f_J : \bigvee_{J,I} D_{j_1}Y_{i_1} \wedge \cdots \wedge D_{j_k}Y_{i_k} \longrightarrow D_jY$$

of the maps $f_{J,I}$ is an equivalence of spectra.

Parenthetically, this leads to an attractive alternative version of the definition, I.4.3, of an H_∞^d ring spectrum.

Proposition 1.3. An H_∞^d ring structure on E determines and is determined by an H_∞ ring structure on the wedge $\bigvee_i \Sigma^{di}E$.

Proof. If $\bigvee_i \Sigma^{di}E$ is an H_∞ ring spectrum with structural maps ξ_j, then the evident composites

$$\xi_{j,i} : D_j\Sigma^{di}E \longrightarrow D_j(\bigvee_h \Sigma^{dh}E) \xrightarrow{\xi_j} \bigvee_h \Sigma^{dh}E \longrightarrow \Sigma^{di}E$$

give E an H_∞^d ring structure. If E is an H_∞^d ring spectrum with structural maps $\xi_{j,i}$, then the maps

$$\xi_j : D_j(\bigvee_i \Sigma^{di}E) \xrightarrow{f_j^{-1}} \bigvee_{J,I} D_{j_1}\Sigma^{di_1}E \wedge \cdots \wedge D_{j_k}\Sigma^{di_k}E \longrightarrow \bigvee_i \Sigma^{di}E$$

determined by the composites

$$D_{j_1}\Sigma^{di_1}E \wedge \cdots \wedge D_{j_k}\Sigma^{di_k}E \xrightarrow{\xi_{j_1,i_1} \wedge \cdots \wedge \xi_{j_k,i_k}} \Sigma^{dj_1i_1}E \wedge \cdots \wedge \Sigma^{dj_ki_k}E \xrightarrow{\phi} \Sigma^{dr}E,$$

$r = \sum_{a=1}^{k} j_a i_a$, give $\bigvee_i \Sigma^{di}E$ on H_∞ ring structure. These correspondences are inverse to one another.

Returning to the context of Theorem 1.1, let

$$g_J : D_j(Y_1 \vee \cdots \vee Y_k) \longrightarrow D_{j_1}Y_1 \wedge \cdots \wedge D_{j_k}Y_k$$

denote the Jth component of f_j^{-1}. Thus g_J is the composite of the projection to $E\Sigma_j \ltimes_{\Sigma_J} Y_J$ and the inverse of the equivalence $(i \ltimes 1)\alpha$ in the proof of the theorem.

The theorem is of particular interest when $Y_1 = \cdots = Y_k$, hence we change notations and consider a spectrum Y and its k-fold wedge sum, which we denote by $^{(k)}Y$. Recall that finite wedges are finite products in the stable category and let

$$\Delta : Y \longrightarrow {}^{(k)}Y \quad \text{and} \quad \nabla : {}^{(k)}Y \longrightarrow Y$$

denote the diagonal and folding maps.

Definition 1.4. Define $\tau_J : D_jY \to D_{j_1}Y \wedge \cdots \wedge D_{j_k}Y$ to be the composite

$$D_jY \xrightarrow{D_j\Delta} D_j({}^{(k)}Y) \xrightarrow{g_J} D_{j_1}Y \wedge \cdots \wedge D_{j_k}Y .$$

Explicitly, let $\pi_J : ({}^{(k)}Y)^{(j)} \to \bigvee_{I \in J} Y^{(j)}$ be the projection and let τ_J also denote the map

$$E\Sigma_j \ltimes_{\Sigma_j} (\pi_J \Delta^{(j)}) : E\Sigma_j \ltimes_{\Sigma_j} Y^{(j)} \longrightarrow E\Sigma_j \ltimes_{\Sigma_j} (\bigvee_{I \in J} Y^{(j)}) = E\Sigma_j \ltimes_{\Sigma_J} Y^{(j)}.$$

Our original map τ_J is the composite of this map and the equivalence $[(i \ltimes 1)\alpha]^{-1}$. We write τ_j for $\tau_J : D_jY \to Y^{(j)}$ when $k = j$ and each $j_s = 1$.

We think of τ_J as a kind of spectrum level transfer map. When $Y = \Sigma^\infty X^+$ for a space X and $\pi \subset \Sigma_j$, we have

$$E\Sigma_j \ltimes_\pi Y^{(j)} \cong \Sigma^\infty(E\Sigma_j \ltimes_\pi (X^+)^{(j)}) = \Sigma^\infty(E\Sigma_j \times_\pi X^j)^+$$

by I.1.1. We shall prove the following result in the sequel.

Theorem 1.5. When $Y = \Sigma^\infty X^+$, the map

$$\tau_J : E\Sigma_j \ltimes_{\Sigma_j} Y^{(j)} \to E\Sigma_j \ltimes_{\Sigma_J} Y^{(j)}$$

is the transfer associated to the natural cover

$$E\Sigma_j \times_{\Sigma_J} X^j \to E\Sigma_j \times_{\Sigma_j} X^j .$$

We do not wish to overemphasize this result. As we shall see, the spectrum level maps τ_J, for general Y, are quite easily studied directly.

The importance of these maps is that they measure the deviation from additivity of the functor D_jY.

For maps $h_i : Y \to E$, $h_1 + \cdots + h_k$ is defined to be $\nabla(h_1 \vee \cdots \vee h_k)\Delta$. Thinking now in cohomological terms, consider the h_i as elements of the Abelian group $E^0Y = [Y,E]$ of maps $Y \to E$ in $\overline{h\mathcal{S}}$.

Corollary 1.6. $D_j(h_1 + \cdots + h_k) = \sum_J \tau_J^*(\alpha_J(D_{j_1}h_1 \wedge \cdots \wedge D_{j_k}h_k))$. Moreover, the following equivariance formula holds for $\sigma \in \Sigma_k$.

$$\tau_J^*(\alpha_J(D_{j_1}h_1 \wedge \cdots \wedge D_{j_k}h_k)) = \tau_{\sigma J}^*(\alpha_{\sigma J}(D_{j_{\sigma^{-1}(1)}}h_{\sigma^{-1}(1)} \wedge \cdots \wedge D_{j_{\sigma^{-1}(k)}}h_{\sigma^{-1}(k)})).$$

Proof. By Theorem 1.1 and the naturality diagram preceding it, the following diagram commutes.

$$\begin{array}{ccccccc}
D_jY & \xrightarrow{D_j\Delta} & D_j({}^{(k)}Y) & \xrightarrow{D_j(h_1\vee\cdots\vee h_k)} & D_j({}^{(k)}E) & \xrightarrow{D_j\nabla} & D_jE \\
\downarrow{\scriptstyle \Delta} & & \downarrow{\scriptstyle (g_J)} & & \downarrow{\scriptstyle (g_J)} & & \uparrow{\scriptstyle \nabla} \\
\bigvee_J D_jY & \xrightarrow{\vee\tau_J} & \bigvee_J D_{j_1}Y\wedge\cdots\wedge D_{j_k}Y & \xrightarrow{\vee D_{j_1}h_1\wedge\cdots\wedge D_{j_k}h_k} & \bigvee_J D_{j_1}E\wedge\cdots\wedge D_{j_k}E & \xrightarrow{\vee\alpha_J} & \bigvee_J D_jE
\end{array}$$

The equivalence follows from I.2.8, the formula $f_J = f_{\sigma J}\circ\sigma$, and the fact that $\sigma\Delta = \Delta$.

Taking each h_i to be the identity map, we obtain the following special case.

Corollary 1.7. $D_j(k) = \sum_J \tau_J^*(\alpha_J)$, and $\tau_J^*(\alpha_J)$ depends only on the conjugacy class of J under the action of Σ_j.

When j is a prime number p and $k = p^iq$ with $i \geq 1$ and q prime to p, a simple combinatorial argument demonstrates that every conjugacy class of partitions has p^is elements for some $s \geq 1$ except for the conjugacy class of the partition $J(k) = (1,\ldots,1,0,\ldots,0)$, p values 1, which has $(p,k-p)$ elements. Of course, p^{i-1} but not p^i divides this binomial coefficient. A trivial diagram chase based on use of the projection ${}^{(k)}Y \to {}^{(p)}Y$ shows that $\tau_{J(k)}$ coincides with $\tau_{J(p)} = \tau_p : D_pY \to Y^{(p)}$. Also, by I.2.7 and I.2.11, $\alpha_{J(p)} = \iota_p : E^{(p)} \to D_pE$. Putting these observations together, we obtain the following result.

Corollary 1.8. If $k = p^iq$ with p prime, $i \geq 1$, and q prime to p, then $D_pk : D_pY \to D_pY$ can be expressed in the form $p^i\lambda + (p,k-p)\iota_p\tau_p$ for some map λ.

In favorable cases, the following three lemmas will lead to a more precise calculation of D_p on general sums.

Lemma 1.9. The following diagram commutes for all Y, j, and k and all partitions J of j.

$$\begin{array}{ccc}
D_jY & \xrightarrow{\tau_J} & D_{j_1}Y\wedge\cdots\wedge D_{j_k}Y \\
\downarrow{\scriptstyle \tau_j} & & \downarrow{\scriptstyle \tau_{j_1}\wedge\cdots\wedge\tau_{j_k}} \\
Y^{(j)} & = & Y^{(j_1)}\wedge\cdots\wedge Y^{(j_k)}
\end{array}$$

Proof. This follows from a straightforward diagram chase which boils down to the factorization of $\Delta:Y \to {}^{(j)}Y$ as the composite

$$Y \xrightarrow{\Delta} {}^{(k)}Y \xrightarrow{\Delta \vee \cdots \vee \Delta} {}^{(j_1)}Y \vee \cdots \vee {}^{(j_k)}Y$$

(where $\Delta:Y \to {}^{(0)}Y = S$ is interpreted as the zero map if any $j_r = 0$).

Lemma 1.10. The composite $\tau_j \iota_j : Y^{(j)} \to Y^{(j)}$ is the sum over $\sigma \in \Sigma_j$ of the permutation maps $\sigma : Y^{(j)} \to Y^{(j)}$.

Proof. This is an easy direct inspection of definitions and may be viewed as a particularly trivial case of the double coset formula.

Lemma 1.11 For any ordinary homology theory H_*, the composite

$$H_*D_jY \xrightarrow{\tau_{J*}} H_*(D_{j_1}Y \wedge \cdots \wedge D_{j_k}Y) \xrightarrow{\alpha_{J*}} H_*D_jY$$

is multiplication by the multinomial coefficient $(j_1,\ldots,j_k)$. In particular, $\iota_{j*}\tau_{j*}$ is multiplication by $j!$.

Proof. We may assume that Y is a CW-spectrum and exploit I.2.1. Since $\pi_i\Delta \simeq 1:Y \to Y$, where $\pi_i:{}^{(k)}Y \to Y$ is the ith projection, $\Delta_*:C_*Y \to C_*({}^{(k)}Y) = C_*Y \oplus \cdots \oplus C_*Y$ is chain homotopy equivalent to the algebraic diagonal. With $Y_1 = \cdots = Y_k = Y$, the composite $(i \ltimes 1)\alpha$ in the proof of Theorem 1.1 induces α_J upon passage to orbits over Σ_j (rather than over $\Sigma_{j_1} \times \cdots \times \Sigma_{j_k}$). Therefore $\alpha_J \circ \tau_J$ is just the composite

$$W_j \ltimes_{\Sigma_j} Y^{(j)} \xrightarrow{1 \ltimes \Delta^{(j)}} W_j \ltimes_{\Sigma_j} ({}^{(k)}Y)^{(j)} \xrightarrow{1 \ltimes \pi_J} W_j \ltimes_{\Sigma_j} (\bigvee_{I \in J} Y^{(j)}) \xrightarrow{1 \ltimes \nabla} W_j \ltimes_{\Sigma_j} Y^{(j)}.$$

Since there are $(j_1,\ldots,j_k)$ sequences $I \in J$ and thus $(j_1,\ldots,j_k)$ wedge summands here, the conclusion clearly holds on the level of cellular chains.

§2. Power operations and Nishida's nilpotency theorem

Let E be an H_∞ ring spectrum and Y be any spectrum. Recall from I.4.1 that we have power operations $\mathcal{P}_j:E^0Y \to E^0D_jY$ specified by $\mathcal{P}_j(h) = \xi_jD_j(h)$. We use the results of section 1 to derive additivity formulas for these operations and apply these formulas to derive the nilpotency of π_*S.

<u>Lemma 2.1.</u> For $h_i \in E^0Y$, $\mathcal{P}_j(h_1 + \cdots + h_k) = \sum_J \tau_J^*(\mathcal{P}_{j_1}(h_1) \wedge \cdots \wedge \mathcal{P}_{j_k}(h_k))$, where the product $\wedge$ is the external product in E-cohomology and the sum extends over all partitions $J = (j_1,\ldots,j_k)$ of j.

<u>Proof.</u> This is immediate from Corollary 1.6 and the commutative diagram

$$\begin{array}{ccc} D_{j_1}E \wedge \cdots \wedge D_{j_k}E & \xrightarrow{\alpha_J} & D_jE \\ {\scriptstyle \xi_{j_1} \wedge \cdots \wedge \xi_{j_k}} \downarrow & & \downarrow {\scriptstyle \xi} \\ E \wedge \cdots \wedge E & \xrightarrow{\phi} & E \end{array}$$

Here the terms with one $j_i = j$ and the rest zero give the sum of the $\mathcal{P}_j(h_i)$. When j is a prime number p, the remaining error term simplifies. The full generality of the following result is due to McClure.

<u>Proposition 2.2.</u> Let $h_i \in E^0Y$. If $p = 2$, then

$$\mathcal{P}_2(h_1 + \cdots + h_k) = \mathcal{P}_2(h_1) + \cdots + \mathcal{P}_2(h_k) + \sum_{1 \leq i < j \leq k} \tau_2^*(h_i \wedge h_j).$$

If p is an odd prime and Y and E are p-local, then

$$\mathcal{P}_p(h_1 + \cdots + h_k) = \mathcal{P}_p(h_1) + \cdots + \mathcal{P}_p(h_k) + \tau_p^*(\frac{1}{p!}[(h_1 + \cdots + h_k)^p - (h_1^p + \cdots + h_k^p)]).$$

In particular, $\mathcal{P}_p(kh) = k\mathcal{P}_p(h) + \frac{1}{p!}(k^p - k)\tau_p^*(h^p)$ in both cases.

<u>Proof.</u> We must show that

$$j_1! \cdots j_k!\, \tau_J^*(\mathcal{P}_{j_1}(h_1) \wedge \cdots \wedge \mathcal{P}_{j_k}(h_k)) = \tau_p^*(h_1^{j_1} \wedge \cdots \wedge h_k^{j_k})$$

for a partition $J = (j_1,\ldots,j_k)$ of p with no $j_i = p$. By Lemma 1.9, $\tau_p^* = \tau_J^*(\tau_{j_1} \wedge \cdots \wedge \tau_{j_k})^*$. Thus it suffices to show that

$$j!\, \mathcal{P}_j(h) = \tau_j^*(h^j)$$

for any $j \geq 0$ and $h \in E^0(Y)$. If $j = 0$, $h^{(0)}$ and $D_0(h)$ are to be interpreted as the identity map of S and the conclusion is trivial. If $j = 1$, the conclusion is also trivial. There are no more cases if $p = 2$, so assume that $p > 2$ and $1 < j < p$. By Lemma 1.11, the composite

$$D_jY \xrightarrow{\tau_j} Y^{(j)} \xrightarrow{\iota_j} D_jY$$

induces multiplication by $j!$ in ordinary homology. It is thus an equivalence since Y and hence also D_jY is p-local. Therefore $\iota_j^*:E^*(D_jY) \to E^*(Y^{(j)})$ is a monomorphism and we need only check that

$$j!\iota_j^*\mathcal{P}_j(h) = \iota_j^*\tau_j^*(h^j).$$

The left side is $j!h^j$. By Lemma 1.10, the right side is the sum over $\sigma \in \Sigma_p$ of $\sigma_*(h^j)$. The commutativity of E implies that $\sigma_*(h^j) = h^j$ for all σ, j, and h, and the conclusion follows.

Now recall from I.4.2 that elements $\alpha \in E_r(D_pS^q)$ determine homotopy operations $\tilde{\alpha}:\pi_qE \to \pi_rE$ via the formula $\tilde{\alpha}(h) = \alpha/\mathcal{P}_p(h)$.

<u>Corollary 2.3</u>. Let $\alpha \in E_r(D_pS^q)$ and $h \in \pi_qE$, where q is even and E is p-local if p is odd. Then

$$\tilde{\alpha}(kh) = k\tilde{\alpha}(h) + \frac{1}{p!}(k^p - k)(\Sigma^{-pq}\tau_{p*}(\alpha))h^p,$$

where the product is the multiplication in π_*E.

<u>Proof</u>. The following diagram is easily seen to commute.

$$\begin{array}{ccccccc}
S^r & \xrightarrow{\alpha} & D_pS^q \wedge E & \xrightarrow{\tau_p \wedge 1} & S^{pq} \wedge E & \xrightarrow{h^p \wedge 1} & E \wedge E \\
\| \wr & & & & & & \searrow \phi \\
\Sigma^{-pq}S^r \wedge S^{pq} & & \xrightarrow{\Sigma^{-pq}((\tau_p \wedge 1)\alpha) \wedge h^p} & & \Sigma^{-pq}(S^{pq} \wedge E) \wedge E \cong E \wedge E & \xrightarrow{\phi} & E
\end{array}$$

Thus $\alpha/\tau_p^*(h^p) = (\Sigma^{-pq}\tau_{p*}(\alpha))h^p$. The conclusion follows from the last statement of the previous proposition.

Assuming that E is p-local (when $p = 2$ as well as when p is odd), we obtain the following immediate corollaries.

<u>Corollary 2.4</u>. If $p^ih = 0$, then $p^{i-1}(\Sigma^{-pq}\tau_{p*}(\alpha))h^{p+1} = 0$ for all α.

Here we have multiplied by h to kill $p^i\tilde{\alpha}(h)$. Of course, this may not be necessary.

<u>Corollary 2.5</u>. If both $p^ih = 0$ and $p^i\alpha = 0$, then $p^{i-1}(\Sigma^{-pq}\tau_{p*}(\alpha))h^p = 0$.

One can also arrive at the last two corollaries by direct diagram chases from Corollary 1.8 and the definition of an H_∞ ring spectrum, without bothering with additivity formulae. (That approach was taken in [72], following Nishida [90, §8]).

These relations specialize to give nilpotency assertions, the sharpest estimate being as follows.

Corollary 2.6. Let $x \in \pi_q E$ satisfy $p^i x = 0$, where $i > 0$ and q is even if $p > 2$. Suppose that $x = \Sigma^{-pq}\tau_{p*}(\alpha)$ for some $\alpha \in E_{pq+q}(D_p S^q)$. Then $p^{i-1}x^{p+2} = 0$. Moreover, if $p^i\alpha = 0$, then $p^{i-1}x^{p+1} = 0$.

The problem, of course, is to study $E_*(D_p S^q)$ and τ_{p*}. Everything above applies to an arbitrary H_∞ ring spectrum E, but to compute τ_{p*} we must specialize. If $E = MO$, for example, then every element of π_*E has order 2 and no element is nilpotent, hence $\tau_{2*}:MO_*(D_2S^q) \to MO_*(S^{2q})$ must be the zero homomorphism for all q. This does not contradict the following assertion.

Conjecture 2.7. Any element of finite order in the kernel of the (integral) Hurewicz homomorphism $\pi_*E \to H_*E$ is nilpotent.

We shall prove the conjecture for elements of order exactly p in section 6, but the methods there fail for general elements of order p^i with $i > 1$.

When we specialize to $E = S$, we find that the Kahn-Priddy theorem gives appropriate input for application of the results above.

Theorem 2.8. If $p = 2$, let $\phi(k)$ be the number of integers j such that $0 < j \leq k$ and $j \equiv 0,1,2,$ or $4 \bmod 8$. If $p > 2$, let $\phi(k) = [k/2(p-1)]$. Let q be an integer such that $q \equiv 0 \bmod p^{\phi(k)}$, where q is even if $p > 2$. Then $\tau_{p*}:\pi_r D_p S^q \to \pi_r S^{pq}$ is a (split) epimorphism for $pq < r < pq+k(p-1)$.

We shall prove this in section 4. Actually, the purely stable methods we use will give surjectivity without giving a splitting. For this reason, we are really only entitled to use Corollary 2.4, rather than Corollary 2.5. This doesn't change the heuristic picture, but to give the correct estimate of the order of nilpotency, we assume the splitting (from [46, 95, or 27]) in the discussion to follow.

Theorem 2.9. Let $x \in \pi_n S$ satisfy $p^i x = 0$, where $i > 0$ and n is even if $p > 2$. Let m be minimal such that $mn \equiv 0 \bmod p^{\phi([n/p-1]+1)}$. Then $p^{i-1}x^{mp+1} = 0$. Inductively, some power of x is zero.

Proof. Let $q = mn$. Since $n < ([n/p-1]+1)(p-1)$, there exists $\alpha \in \pi_{pq+n}D_pS^q$ such that $\Sigma^{-pq}\tau_{p*}(\alpha) = x$. With $h = x^m$, Corollary 2.4 gives $p^{i-1}x^{mp+2} = 0$. Using $p^i\alpha = 0$, Corollary 2.5 gives $p^{i-1}x^{mp+1} = 0$.

Unfortunately, m increases rapidly with n (although our estimate for $p > 2$ is sharper than Nishida's since he only knew Theorem 2.8 for $r < pq+k$). For example, the first stem in which an interesting element x of order 2 occurs is the 14-stem ("interesting" meaning that x is neither in π_*J nor a product of Hopf maps). Here m = 64 and we can only conclude that $x^{129} = 0$, a truly stratospheric estimate. So far, and granting that our stemwise calculations still extend through only a very small range, we have no reason to disbelieve that $x^4 = 0$ if $2x = 0$. Corollary 2.6 seems to suggest that this answer might be correct. However, as pointed out to me by Bruner, $\tau_{2*}:\pi_*D_2S^q \to \pi_*S^{2q}$ is not always an epimorphism and thus Corollary 2.6 cannot be used to prove this answer.

§3. The Jones-Wegmann homology and cohomology theories

The next three sections will all make heavy use of certain twisted diagonal maps implicit in the general properties of extended powers.

Definition 3.1. Let π be a subgroup of Σ_j and let W be a free π-CW complex. For a based CW complex X and a CW spectrum Y, define a map of spectra

$$\Delta:(W \ltimes_\pi Y^{(j)}) \wedge X \to W \ltimes_\pi (Y \wedge X)^{(j)}$$

by passage to orbits over π from the π-map

$$(W \ltimes Y^{(j)}) \wedge X \xrightarrow{1 \wedge \Delta} (W \ltimes Y^{(j)}) \wedge X^{(j)} \cong W \ltimes (Y \wedge X)^{(j)}.$$

Here the isomorphism is given by I.1.2(ii) and the shuffle π-isomorphism $Y^{(j)} \wedge X^{(j)} \cong (Y \wedge X)^{(j)}$. Note that Δ is the identity map when $X = S^0$ and that the following transitivity and commutativity diagram commutes, where X' is another based CW complex.

$$\begin{array}{ccccc}
(W \ltimes_\pi Y^{(j)}) \wedge X \wedge X' & & \xrightarrow{1 \wedge \tau} & & (W \ltimes_\pi Y^{(j)}) \wedge X' \wedge X \\
\swarrow{\scriptstyle \Delta \wedge 1} & & & & \\
W \ltimes_\pi (Y \wedge X)^{(j)} \wedge X' & \downarrow{\scriptstyle \Delta} & & & \downarrow{\scriptstyle \Delta} \\
\searrow{\scriptstyle \Delta} & & & & \\
W \ltimes_\pi (Y \wedge X \wedge X')^{(j)} & & \xrightarrow{1 \ltimes (1 \wedge \tau)^{(j)}} & & W \ltimes_\pi (Y \wedge X' \wedge X)^{(j)}
\end{array}$$

With $\pi = \Sigma_j$ and $W = E\Sigma_j$, we obtain

$$\Delta:(D_jY) \wedge X \to D_j(Y \wedge X).$$

Although not strictly relevant to the business at hand, we record the relationship between these maps and the maps ι_j, $\alpha_{j,k}$, $\beta_{j,k}$, and δ_j of I§2 and use them to construct new examples of H_∞ ring spectra.

Lemma 3.2. The following diagrams commute for spectra Y and Z and spaces X. The unlabeled arrows are obvious composites of shuffle maps and the diagonal on X.

$$\begin{array}{ccc} Y^{(j)}\wedge X & \xrightarrow{\iota_j\wedge 1} & D_jY\wedge X \\ \downarrow & & \downarrow \Delta \\ (Y\wedge X)^{(j)} & \xrightarrow{\iota_j} & D_j(Y\wedge X) \end{array} \qquad \begin{array}{ccc} D_j(Y\wedge Z)\wedge X & \xrightarrow{\delta_j\wedge 1} & D_jY\wedge D_jZ\wedge X \\ \Delta\downarrow & & \downarrow \\ D_j(Y\wedge Z\wedge X) & & D_jY\wedge X\wedge D_jZ\wedge X \\ \downarrow & & \downarrow \Delta\wedge\Delta \\ D_j(Y\wedge X\wedge Z\wedge X) & \xrightarrow{\delta_j} & D_j(Y\wedge X)\wedge D_j(Z\wedge X) \end{array}$$

$$\begin{array}{ccc} D_jY\wedge D_kY\wedge X & \xrightarrow{\alpha_{j,k}\wedge 1} & D_{j+k}Y\wedge X \\ \downarrow & & \\ D_jY\wedge X\wedge D_kY\wedge X & & \downarrow \Delta \\ \downarrow \Delta\wedge\Delta & & \\ D_j(Y\wedge X)\wedge D_k(Y\wedge X) & \xrightarrow{\alpha_{j,k}} & D_{j+k}(Y\wedge X) \end{array} \qquad \begin{array}{ccc} D_jD_kY\wedge X & \xrightarrow{\beta_{j,k}\wedge 1} & D_{jk}Y\wedge X \\ \downarrow \Delta & & \\ D_j(D_kY\wedge X) & & \downarrow \Delta \\ \downarrow D_j\Delta & & \\ D_jD_k(Y\wedge X) & \xrightarrow{\beta_{j,k}} & D_{jk}(Y\wedge X) \end{array}$$

I learned the following lemma from Miller and McClure.

Lemma 3.3. Let X be an unbased space and E be an H_∞ ring spectrum. Then the function spectrum $F(X^+,E)$ is an H_∞ ring spectrum with structural maps the adjoints of the composites

$$D_jF(X^+,E)\wedge X \xrightarrow{\Delta} D_j(F(X^+,E)\wedge X^+) \xrightarrow{D_j\varepsilon} D_jE \xrightarrow{\xi_j} E,$$

where ε is the evaluation map. In particular, the dual $F(X^+,S)$ of $\Sigma^\infty X^+$ is an H_∞ ring spectrum.

Proof. If $j = 0$, $\Delta: S\wedge X^+ \simeq \Sigma^\infty X^+ \longrightarrow \Sigma^\infty S^0 = S$ is to be interpreted as $\Sigma^\infty\delta$, where $\delta: X^+ \to S^0$ is the discretization map sending X to the non-basepoint. The diagrams of I.3.1 are easily checked to commute by use of the diagrams of the previous lemma.

Returning to the business at hand, observe that, with $X = S^1$, we obtain a natural map $\Delta:\Sigma D_jY \to D_j\Sigma Y$. Thus, for any integer n (positive or negative), we have the map

$$\Sigma^n\Delta:\Sigma^{n+1}D_j\Sigma^{-n-1}Y = \Sigma^n\Sigma D_j\Sigma^{-1}\Sigma^{-n}Y \longrightarrow \Sigma^n D_j\Sigma^{-n}Y.$$

We shall be interested in the resulting inverse system

$$\cdots \longrightarrow \Sigma^n D_j \Sigma^{-n} Y \longrightarrow \cdots \longrightarrow \Sigma^1 D_j \Sigma^{-1} Y \longrightarrow D_j Y \longrightarrow \Sigma^{-1} D_j \Sigma Y \longrightarrow \cdots \longrightarrow \Sigma^{-n} D_j \Sigma^n Y \longrightarrow \cdots$$

(where $n \geq 0$). By the diagram in Definition 3.1, the maps

$$\Sigma^n \Delta : (\Sigma^n D_j S^{-n}) \wedge X \cong \Sigma^n (D_j S^{-n} \wedge X) \longrightarrow \Sigma^n D_j (S^{-n} \wedge X) \cong \Sigma^n D_j \Sigma^{-n} \Sigma^\infty X$$

specify a morphism of systems, again denoted Δ,

$$\{(\Sigma^n D_j S^{-n}) \wedge X\} \longrightarrow \{\Sigma^n D_j \Sigma^{-n} \Sigma^\infty X\}.$$

We shall study the homological and homotopical properties of these systems. In this section, we consider any $j \geq 2$. We shall obtain calculational results when j is a prime in the following two sections.

Let E_* and E^* denote the homology and cohomology theories represented by a spectrum E. For spectra Y, define

$$E_*^{(j)} Y = \lim E_*(\Sigma^n D_j \Sigma^{-n} Y) \quad \text{and} \quad E_{(j)}^* Y = \operatorname{colim} E^*(\Sigma^n D_j \Sigma^{-n} Y)$$

$$F_*^{(j)} Y = \lim E_*(\Sigma^n D_j S^{-n} \wedge Y) \quad \text{and} \quad F_{(j)}^* Y = \operatorname{colim} E^*(\Sigma^n D_j S^{-n} \wedge Y).$$

Upon restriction to spaces (that is, to $Y = \Sigma^\infty X$), we obtain induced natural transformations

$$\Delta_* : F_*^{(j)} X \longrightarrow E_*^{(j)} X \quad \text{and} \quad \Delta^* : E_{(j)}^* X \longrightarrow F_{(j)}^* X,$$

and these reduce to identity homomorphisms when $X = S^0$. It is clear that $F_*^{(j)}$ is a homology theory and $F_{(j)}^*$ is a cohomology theory on finite CW spectra. Passage to colimits from the homomorphisms

$$(\Sigma^{n-1}\Delta)^* : E^{i+1}(\Sigma^n D_j \Sigma^{-n} \Sigma Y) \cong E^i(\Sigma^{n-1} D_j \Sigma^{-n+1} Y) \longrightarrow E^i(\Sigma^n D_j \Sigma^{-n} Y)$$

yields suspension isomorphisms

$$E_{(j)}^{i+1} \Sigma Y \longrightarrow E_{(j)}^i Y,$$

and Δ^* is easily seen to commute with suspension. The analogous assertions hold for $E_*^{(j)}$. With these notations, the main theorems of Jones and Wegmann [44] read as follows (although they only consider primes j and only provide proofs when $j = 2$).

<u>Theorem 3.4.</u> The functor $E_{(j)}^*$ is a cohomology theory on finite CW spectra, hence $\Delta^* : E_{(j)}^* X \to F_{(j)}^* X$ is an isomorphism for all finite CW complexes X.

Theorem 3.5. Let E be connective and j-adically complete, with π_*E of finite type over the j-adic integers $\hat{Z}_j = \underset{p|j}{\times} \hat{Z}_p$. Then $E_*^{(j)}$ is a homology theory on finite CW spectra, hence $\Delta_*: F_*^{(j)}X \to E_*^{(j)}X$ is an isomorphism for all finite CW complexes X.

We defer the proofs for a moment. As Jones and Wegmann point out, these results are no longer valid for infinite CW complexes.

Recall that $D_jS^0 = \Sigma^\infty B\Sigma_j^+$ and the discretization map $B\Sigma_j^+ \to S^0$ induces $\xi_j: D_jS^0 \to S^0$. Upon smashing with Y, the composites

$$\Sigma^n D_j S^{-n} \xrightarrow{\Delta} D_j S^0 \xrightarrow{\xi_j} S^0$$

give a morphism from the system $\{\Sigma^n D_j S^{-n} \wedge Y\}$ to the constant system at Y. We call this map of systems ξ_j and obtain a map of cohomology theories

$$\xi_j^*: E^*Y \longrightarrow F_{(j)}^*Y,$$

commutation with the suspension isomorphisms being easily checked. We shall shortly prove a complement to this observation.

Proposition 3.6. Let E be an H_∞ ring spectrum. Then the composites of the functions

$$\mathcal{P}_j: E^nY = [\Sigma^{-n}Y, E] \longrightarrow [D_j\Sigma^{-n}Y, E] = E^n(\Sigma^n D_j \Sigma^{-n}Y)$$

and the natural homomorphisms $E^n(\Sigma^n D_j\Sigma^{-n}Y) \to E^n_{(j)}Y$ specify a map of cohomology theories

$$\mathcal{P}_j: E^*Y \longrightarrow E^*_{(j)}Y.$$

We thus have the triangle of cohomology theories

$$\begin{array}{ccc} & E^*X & \\ {}^{\mathcal{P}_j}\swarrow & & \searrow^{\xi_j^*} \\ E^*_{(j)}X & \xrightarrow{\Delta^*} & F^*_{(j)}X \end{array}$$

on finite CW complexes X. Since $\mathcal{P}_j(x) = \xi_j \circ D_j(x)$, we see immediately that $\Delta^*\mathcal{P}_j(1) = \xi_j^*(1)$, where $1 \in E^0(S^0)$ is the identity element. It does not follow that $\Delta^*\mathcal{P}_j = \xi_j^*$ in general. As we shall see in the next section, this fails, for example when E = MO. However, as observed by Jones and Wegmann [44], this implication does hold for E = S.

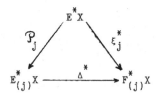

Proposition 3.7. The following diagram commutes for any finite CW complex X.

$$\begin{array}{ccc} & \pi^*X & \\ {\scriptstyle \mathcal{P}_j}\swarrow & & \searrow{\scriptstyle \xi_j^*} \\ \operatorname{colim} \pi^*(\Sigma^n D_j \Sigma^{-n} X) & \xrightarrow{\Delta^*} & \operatorname{colim} \pi^*(\Sigma^n D_j S^{-n} \wedge X) \end{array}$$

Proof. Since $\Delta^* \mathcal{P}_j$ and ξ_j^* are morphisms of cohomology theories, they are equal for all X if they are equal for $X = S^0$. Any morphism $\phi: E^*X \to F^*X$ of cohomology theories is given by morphisms of π^*S^0-modules. When $E^* = \pi^*$ and $X = S^0$, $\phi(x) = \phi(1\cdot x) = \phi(1)\cdot x$, so that ϕ is determined by its behavior on the unit $1 \in \pi^0(S^0)$.

For general E and $X = S^0$, it is obvious that $\xi_j^*(x) = \xi_j^*(1)x$. It is not at all obvious that $(\Delta^*\mathcal{P}_j)(x) = \Delta^*\mathcal{P}_j(1)\cdot x$ We now have this relation for E = S, and we shall use it to prove the Kahn-Priddy theorem in the next section. As we shall explain in section 5, theorems of Lin when p = 2 and of Gunawardena when p > 2 imply that ξ_p^* and thus $\mathcal{P}_p$ in Proposition 3.7 are actually isomorphisms. We complete this section by giving the deferred proofs, starting with that of Proposition 3.6. We need two lemmas.

Lemma 3.8. The following diagram commutes for any partition $J = (j_1,\dots,j_k)$ of j.

$$\begin{array}{ccccc} D_jY\wedge X & \xrightarrow{\tau_J\wedge 1} & D_{j_1}Y\wedge\cdots\wedge D_{j_k}Y\wedge X & \xrightarrow{(\text{shuffle})(1\wedge\Delta)} & D_{j_1}Y\wedge X\wedge\cdots\wedge D_{j_k}Y\wedge X \\ \downarrow{\scriptstyle\Delta} & & & & \downarrow{\scriptstyle\Delta\wedge\cdots\wedge\Delta} \\ D_j(Y\wedge X) & & \xrightarrow{\tau_J} & & D_{j_1}(Y\wedge X)\wedge\cdots\wedge D_{j_k}(Y\wedge X) \end{array}$$

Proof. The "transfer" τ_J is specified in Definition 1.4, and the proof is an easy naturality argument.

Lemma 3.9. For an H_∞ ring spectrum E, the composite

$$[Y,E] \xrightarrow{\mathcal{P}_j} [D_jY,E] \xrightarrow{\Delta^*} [\Sigma D_j\Sigma^{-1}Y,E]$$

is a homomorphism.

Proof. By Lemma 2.1, we have the formula

$$\mathcal{P}_j(x+y) = \mathcal{P}_j(x) + \mathcal{P}_j(y) + \sum_{i=1}^{p-1} \tau^*_{i,p-i}(\mathcal{P}_i(x)\wedge\mathcal{P}_{j-i}(x)).$$

With $X = S^1$, Lemma 3.8 and the fact that $\Delta: S^1 \to S^1 \wedge S^1$ is null homotopic imply that $\tau_{i,j-i}\Delta$ is null homotopic.

Thus $\mathcal{P}_j$ in Proposition 3.6 is a natural homomorphism. It is easily checked that $\mathcal{P}_j$ commutes with suspension and this proves the proposition.

Finally, we turn to the proofs of Theorems 3.4 and 3.5. Clearly it only remains to show that $E^*_{(j)}$ and $E^{(j)}_*$ satisfy the exactness axiom on finite CW pairs (Y,B). Although not strictly necessary, we insert a general observation which helps explain the idea and will be used later.

<u>Lemma 3.10</u>. Let $f: B \to Y$ be a map of CW spectra with cofibre Cf. There is a map $\psi: CD_jf \to D_jCf$, natural in f, such that the diagram

$$\begin{array}{ccccc} D_jY & \xrightarrow{i} & CD_jf & \xrightarrow{\partial} & \Sigma D_jB \\ \| & & \downarrow{\psi} & & \downarrow{\Delta} \\ D_jY & \xrightarrow{D_ji} & D_jCf & \xrightarrow{D_j\partial} & D_j\Sigma B \end{array}$$

commutes, where $i: Y \to Cf$ and $\partial: Cf \to \Sigma B$ are the canonical maps. If f is the inclusion of a subcomplex in a CW spectrum, then the diagram

$$\begin{array}{ccc} CD_jf & \xrightarrow{\psi} & D_jCf \\ \pi\downarrow & & \downarrow{D_j\pi} \\ D_jY/D_jB & \xrightarrow{\psi} & D_j(Y/B) \end{array}$$

also commutes, where the maps π are the canonical (quotient) equivalences and the bottom map ψ is induced by the quotient map $Y \to Y/B$.

<u>Proof</u>. $CD_jf = D_jY \cup_{D_jf} CD_jB$ and $D_jCf = D_j(Y \cup_f CB)$; ψ is induced by the inclusion $D_jY \to D_jCf$ and the composite of $\Delta: CD_jB \to D_jCB$ and the inclusion $D_jCB \to D_jCf$. The diagrams are easily checked.

Of course, the bottom row in the first diagram is not a cofibre sequence and ψ is not an equivalence. Now let (Y,B) be a finite CW pair. For notational simplicity, set

$$D_j(Y,B) = D_jY/D_jB \quad \text{and} \quad Z = Y/B.$$

As n varies, the maps

$$\Sigma^n\psi: \Sigma^n D_j(\Sigma^{-n}Y, \Sigma^{-n}B) \longrightarrow \Sigma^n D_j \Sigma^{-n} Z$$

specify a map of inverse systems, again denoted ψ, and we shall prove the following result.

Proposition 3.11. For any pair (Y,B) of finite CW spectra,

$$\psi^*: E^*_{(j)}Z \longrightarrow \operatorname{Colim} E^*\Sigma^n D_j(\Sigma^{-n}Y, \Sigma^{-n}B)$$

and, under the hypotheses of Theorem 3.5,

$$\psi_*: \lim E_*\Sigma^n D_j(\Sigma^{-n}Y, \Sigma^{-n}B) \longrightarrow E_*^{(j)}Z$$

are isomorphisms.

Note that the assumptions on E in Theorem 3.5 imply that all groups in sight are finitely generated $\hat{Z}_j$-modules and thus that all inverse limits in sight preserve exact sequences. Given the proposition, the required $E^*_{(j)}$ and $E_*^{(j)}$ exact sequences of the pair (Y,B) are obtained by passage to colimits and limits from the E^* and E_* exact sequences of the pairs $(\Sigma^n D_j \Sigma^{-n}Y, \Sigma^n D_j \Sigma^{-n}B)$.

Following ideas of Bruner (which he uses in a much deeper way in chapters V and VI), we prove Proposition 3.11 by filtering $Y^{(j)}$. For $0 \leq s \leq j$, define

$$\Gamma_s = \Gamma_s(Y,B) = \bigcup Y_1 \wedge \cdots \wedge Y_j,$$

where $Y_r = Y$ or $Y_r = B$ and s of the Y_r are equal to B. We have

$$B^{(j)} = \Gamma_j \subset \Gamma_{j-1} \subset \cdots \subset \Gamma_0 = Y^{(j)}.$$

Each inclusion is a Σ_j-equivariant cofibration, and we define

$$\Pi_s = \Pi_s(Y,B) = \Gamma_s(Y,B)/\Gamma_{s+1}(Y,B).$$

Then $\Pi_0 = Z^{(j)}$ and, for $0 < s < j$, Π_s breaks up as the wedge of its (s,j-s) distinct subspectra of the form $Z_1 \wedge \cdots \wedge Z_j$, where $Z_r = Z$ or $Z_r = B$ and s of the Z_r are equal to B. It follows that Π_s is the free Σ_j-spectrum generated by the $(\Sigma_s \times \Sigma_{j-s})$-spectrum $B^{(s)} \wedge Z^{(j-s)}$. That is,

$$\Pi_s \cong \Sigma_j \ltimes_{\Sigma_s \times \Sigma_{j-s}} B^{(s)} \wedge Z^{(j-s)}.$$

The functor $E\Sigma_j \ltimes_{\Sigma_j} (?)$ converts Σ_j-cofibrations to cofibrations and commutes with quotients, hence we have cofibre sequences

$$(*) \qquad E\Sigma_j \ltimes_{\Sigma_j} \Gamma_s/\Gamma_t \longrightarrow E\Sigma_j \ltimes_{\Sigma_j} \Gamma_r/\Gamma_t \longrightarrow E\Sigma_j \ltimes_{\Sigma_j} \Gamma_r/\Gamma_s$$

for $0 \leq r < s < t \leq j$. For a based space X, the map $\Delta: D_jY \wedge X \to D_j(Y \wedge X)$ induces compatible maps

$$\Delta:[E\Sigma_j \ltimes_{\Sigma_j} \Gamma_s(Y,B)]\wedge X \longrightarrow E\Sigma_j \ltimes_{\Sigma_j} \Gamma_s(Y\wedge X, B\wedge X)$$

and similarly for Π_s on passage to quotients. The following simple observation is the crux of the matter.

Lemma 3.12. For $0 < s < j$, there is a natural equivalence

$$\alpha: D_sB \wedge D_{j-s}Z \to E\Sigma_j \ltimes_{\Sigma_j} \Pi_s(Y,B)$$

such that the following diagram commutes for any X.

$$\begin{array}{ccccc}
(D_sB\wedge D_{j-s}Z)\wedge X & \xrightarrow{(1\wedge\tau\wedge 1)\Delta} & D_sB\wedge X\wedge D_{j-s}Z\wedge X & \xrightarrow{\Delta\wedge\Delta} & D_s(B\wedge X)\wedge D_{j-s}(Z\wedge X)\\
\downarrow{\scriptstyle \alpha\wedge 1} & & & & \downarrow{\scriptstyle\alpha}\\
[E\Sigma_j \ltimes_{\Sigma_j}\Pi_s(Y,B)]\wedge X & & \xrightarrow{\Delta} & & E\Sigma_j\ltimes_{\Sigma_j}\Pi_s(Y\wedge X,B\wedge X)
\end{array}$$

In particular, the bottom map Δ is null homotopic when $X = S^1$.

Proof. By I.1.4 and the description of $\Pi_s(Y,B)$ above, we have

$$E\Sigma_j \ltimes_{\Sigma_j} \Pi_s(Y,B) \cong E\Sigma_j \ltimes_{\Sigma_s \times \Sigma_{j-s}} B^{(s)} \wedge Z^{(j-s)}.$$

As in the proof of Theorem 1.1, we may replace $E\Sigma_j$ by $E\Sigma_s \times E\Sigma_{j-s}$ on the right side, and it then becomes isomorphic to $D_sB\wedge D_{j-s}Z$. The diagram is easily checked.

Now apply Σ^n to the cofibre sequence (*) for the pair $(\Sigma^{-n}Y, \Sigma^{-n}B)$ with quotient $\Sigma^{-n}Z$. We obtain an inverse system of cofibre sequences for $0 \leq r < s < t \leq j$. On passage to E^* and then to colimits (or to E_* and then to limits), there results a long exact sequence. For $0 < s < j$, the maps between terms of the system

$$\{\Sigma^n E\Sigma_j \ltimes_{\Sigma_j} \Pi_s(\Sigma^{-n}Y, \Sigma^{-n}B)\}$$

are null homotopic, hence its colimit of cohomologies is zero. Inductively, we conclude from the long exact sequences that the colimits of cohomologies associated to the quotients Γ_s/Γ_t with $s > 0$ are all zero and that the maps of colimits of cohomologies associated to the quotient maps $\Gamma_0/\Gamma_t \to \Gamma_0/\Gamma_s$ are all isomorphisms. With $s = 1$ and $t = p$, this proves Proposition 3.11.

§4. Jones' proof of the Kahn-Priddy theorem

We prove Theorem 2.8 here. The proof for p = 2 is due to Jones [43] and we have adapted his idea to the case p > 2. We begin more generally than necessary by relating the cofibre sequences (*) above Lemma 3.12 to the maps $\tau_j: D_jY \to Y^{(j)}$ of Definiton 1.4. The idea here is again due to Bruner. Thus let (Y,B) be a pair of finite CW spectra with quotient Z = Y/B. The map τ_j is obtained by applying the functor $E\Sigma_j \ltimes_{\Sigma_j} (?)$ to the composite

$$Y^{(j)} \xrightarrow{\Delta^{(j)}} ({}^{(j)}Y)^{(j)} \xrightarrow{\pi_J} \Sigma_j \ltimes Y^{(j)},$$

J = (1,...,1), and using the equivalence $E\Sigma_j \ltimes Y^{(j)} \simeq Y^{(j)}$ of nonequivariant spectra (where, technically, the smash product is external on the left and internal on the right; see [Equiv. II §3]). The spectrum $\Sigma_j \ltimes Y^{(j)}$ is a wedge of isomorphic copies of $Y^{(j)}$ indexed on the elements of Σ_j, and $\pi_J\Delta^{(j)}$ is just the sum of the j! permutation maps. It follows that $\pi_J\Delta^{(j)}$ restricts to a Σ_j-equivariant map $\Gamma_s \to \Sigma_j \times \Gamma_s$ for $0 \leq s \leq j$. Upon passage to subquotients and application of the functor $E\Sigma_j \ltimes_{\Sigma_j}(?)$, we obtain maps of cofibre sequences

$$\begin{array}{ccccc}
E\Sigma_j \ltimes_{\Sigma_j} \Gamma_s/\Gamma_t & \longrightarrow & E\Sigma_j \ltimes_{\Sigma_j} \Gamma_r/\Gamma_t & \longrightarrow & E\Sigma_j \ltimes_{\Sigma_j} \Gamma_r/\Gamma_s \\
\downarrow \tau_j & & \downarrow \tau_j & & \downarrow \tau_j \\
\Gamma_s/\Gamma_t & \longrightarrow & \Gamma_r/\Gamma_t & \longrightarrow & \Gamma_r/\Gamma_s
\end{array}$$

for $0 \leq r < s < t \leq j$. With t = s+1, the left map τ_j is nicely related to the equivalence α of Lemma 3.12, as can easily be checked by inspection of definitions.

<u>Lemma 4.1</u>. The following diagram commutes for 0 < s < j, where ρ is the projection onto the unpermuted wedge summand.

$$\begin{array}{ccc}
D_sB \wedge D_{j-s}Z & \xrightarrow{\quad\alpha\quad} & E\Sigma_j \ltimes_{\Sigma_j} \Pi_s(Y,B) \\
\downarrow \tau_s \wedge \tau_{j-s} & & \downarrow \tau_j \\
B^{(s)} \wedge Z^{(j-s)} & \xleftarrow{\ \rho\ } \Sigma_j \ltimes_{\Sigma_s \times \Sigma_{j-s}} B^{(s)} \wedge Z^{(j-s)} \cong & \Pi_s(Y,B)
\end{array}$$

When j = 2, there is only one map of cofibre sequences above, and we obtain the following conclusion.

<u>Proposition 4.2</u>. For CW pair (Y,B) with quotient Z = Y/B,

$$B \wedge Z \xrightarrow{\iota_2'} D_2Y/D_2B \xrightarrow{\psi} D_2Z \xrightarrow{\tau_2'} \Sigma B \wedge Z$$

is a cofibre sequence, where ψ is induced by the quotient map $Y \to Z$, ι_2' is the composite

$$B \wedge Z = (B \wedge Y)/(B \wedge B) \xrightarrow{i \wedge 1} (Y \wedge Y)/(B \wedge B) \xrightarrow{\iota_2} D_2Y/D_2B,$$

and τ_2' is the composite

$$D_2Z \xrightarrow{\tau_2} Z \wedge Z \simeq (Y \cup CB) \wedge Z \xrightarrow{\partial \wedge 1} \Sigma B \wedge Z.$$

Proof. Combine the cofibre sequence

$$E\Sigma_2 \ltimes_{\Sigma_2} \Pi_1(Y,B) \longrightarrow D_2Y/D_2B \longrightarrow D_2Z \longrightarrow \Sigma E\Sigma_2 \ltimes_{\Sigma_2} \Pi_1(Y,B)$$

with the equivalence $\alpha: B \wedge Z \to E\Sigma_2 \ltimes_{\Sigma_2} \Pi_1(Y,B)$ and check that the resulting maps are those specified.

Our main interest is in the pair (CY,Y).

Corollary 4.3. The following is a cofibre sequence.

$$\Sigma(Y \wedge Y) \xrightarrow{\Sigma\iota_2} \Sigma D_2Y \xrightarrow{\Delta} D_2\Sigma Y \xrightarrow{\tau_2} \Sigma Y \wedge \Sigma Y.$$

Proof. Use the evident equivalence $D_2CY/D_2Y \simeq \Sigma D_2Y$ and check the maps, using Lemma 3.10 for the middle one.

For $j > 2$, we have too many cofibre sequences in sight. Henceforward, let p be a prime and localize all spaces and spectra at p without change of notation. We shall show that, for odd primes p and pairs (CS^q,S^q), our system of cofibre sequences collapses to a single one like that in the previous corollary. Recall from Lemma 1.10 that $\tau_r\iota_r: Y^{(r)} \to Y^{(r)}$ is the sum of permutations map and $\iota_r\tau_r: D_rY \to D_rY$ induces multiplication by r! on ordinary homology. In particular, for $1 < r < p$, D_rY is a wedge summand of $Y^{(r)}$.

Lemma 4.4. For $1 < r < p$, D_rS^{2q+1} is equivalent to the trivial spectrum and $\iota_r: S^{2qr} \to D_rS^{2q}$ is an equivalence with inverse $\frac{1}{r!}\tau_r$.

Proof. When $Y = S^{2q}$, $\tau_r\iota_r$ induces multiplication by r! on homology; when $Y = S^{2q+1}$, it induces zero. The conclusions follow.

Thus, when Y is a sphere spectrum, most of the spectra

$$E\Sigma_p \ltimes_{\Sigma_p} \Pi_s(CY,Y) \simeq D_sY \wedge D_{p-s}\Sigma Y$$

are trivial.

Corollary 4.5. Let $p > 2$ and let q be an even integer. Then there are cofibre sequences

$$S^{pq-1} \longrightarrow \Sigma D_p S^{q-1} \xrightarrow{\Delta} D_p S^q \xrightarrow{\tau_p} S^{pq}$$

and

$$S^{pq+1} \longrightarrow \Sigma D_p S^q \xrightarrow{\Delta} D_p S^{q+1} \longrightarrow S^{pq+2} .$$

Proof. Let $\Gamma_s = \Gamma_s(CY,Y)$ and $\Pi_s = \Gamma_s/\Gamma_{s+1}$. If $Y = S^{q-1}$, then $E\Sigma_p \ltimes_{\Sigma_p} \Pi_s$ is trivial for $2 \le s < p$, hence $E\Sigma_p \ltimes_{\Sigma_p} \Gamma_r/\Gamma_s$ is trivial for $2 \le r < s \le p$. Thus $\Gamma_1/\Gamma_p \to \Pi_1$ and $\Gamma_0/\Gamma_p \to \Gamma_0/\Gamma_2$ induce equivalences upon application of $E\Sigma_p \ltimes_{\Sigma_p} (?)$ and there results a cofibre sequence

$$E\Sigma_p \ltimes_{\Sigma_p} \Pi_1 \longrightarrow E\Sigma_p \ltimes_{\Sigma_p} \Gamma_0/\Gamma_p \longrightarrow E\Sigma_p \ltimes_{\Sigma_p} \Pi_0 \longrightarrow \Sigma E\Sigma_p \ltimes_{\Sigma_p} \Pi_1 .$$

This gives the first sequence upon interpreting the terms and maps (by use of Lemmas 3.10, 3.12, 4.1, and 4.4). Similarly, if $Y = S^q$, then $E\Sigma_p \ltimes_{\Sigma_p} \Pi_s$ is trivial for $1 \le s < p-1$, hence $E\Sigma_p \ltimes_{\Sigma_p} \Gamma_r/\Gamma_s$ is trivial for $1 \le r < s \le p-1$. Thus $\Gamma_0/\Gamma_{p-1} \to \Pi_0$ and $\Pi_{p-1} \to \Gamma_1/\Gamma_p$ induces equivalences upon application of $E\Sigma_p \ltimes_{\Sigma_p} (?)$ and there results a cofibre sequence

$$E\Sigma_p \ltimes_{\Sigma_p} \Pi_{p-1} \longrightarrow E\Sigma_p \ltimes_{\Sigma_p} \Gamma_0/\Gamma_p \longrightarrow E\Sigma_p \ltimes_{\Sigma_p} \Pi_0 \longrightarrow \Sigma E\Sigma_p \ltimes_{\Sigma_p} \Pi_{p-1} .$$

This gives the second sequence.

One can also check these cofibre sequences by direct homological calculation; compare Lemma 5.6 below. We need some further information about the spectra $\Sigma^n D_p S^{-n}$ in order to use these sequences to prove Theorem 2.8. Proofs of the claims to follow will be given by Bruner in V§2.

If $p = 2$, let $L = \Sigma^\infty RP^\infty$ with its standard cell structure. (We write L rather than the usual P for uniformity with the case $p > 2$.) If $p > 2$, let L be a CW spectrum of the p-local homotopy type of $\Sigma^\infty B\Sigma_p$ such that L has one cell in each positive dimension $q \equiv 0$ or $-1 \mod 2(p-1)$. The existence and essential uniqueness of such an L was pointed out by Adams [7,2.2]. Let L^k be the k-skeleton of L and let $L_n = L/L^{n-1}$ and $L_n^{n+k} = L^{n+k}/L^{n-1}$ for $k \ge 0$. Let $\phi(k)$ be as in Theorem 2.8 (and recall that it depends on p). If $p = 2$, then

$$L_n^{n+k} \simeq \Sigma^{n-m} L_m^{m+k} \quad \text{for } m \equiv n \mod 2^{\phi(k)}.$$

If $p > 2$, $\varepsilon = 0$ or 1, and $k \ge \varepsilon$, then

$$L_{2n+\varepsilon}^{2n+k} \simeq \Sigma^{2(n-m)} L_{2m+\varepsilon}^{2m+k} \quad \text{for } m \equiv n \mod p^{\phi(k)}.$$

We use this periodicity to define spectra L_n^{n+k} for non-positive n, so that these equivalences hold for all integers m and n. We then have that

$$L_n^{n+k} \text{ is } (-1)\text{-dual to } L_{-n-1-k}^{-n-1} .$$

Our interest in these spectra comes from the following result (proven by Bruner in V§2).

Theorem 4.6. For any integer n, $\Sigma^{-n}D_pS^n$ is p-locally equivalent to $L_{n(p-1)}$.

We define $D_p^kS^n = \Sigma^nL_{n(p-1)}^{n(p-1)+k}$. If p = 2, we may view $D_p^kS^n$ as $S^k \ltimes_{\Sigma_2} S^{2n}$. If p > 2, no model for $E\Sigma_p$ has few enough cells to give as convenient a filtration of D_pS^n. We shall shortly prove the following result.

Proposition 4.7. If $\rho:L_{-k}^0 \to S^0$ is the projection onto the top cell, then

$$\rho^*:\pi^{-q}(S^0) \to \pi^{-q}(L_{-k}^0)$$

is zero for $0 < q < k(p-1)$.

Since ρ is (-1)-dual to the inclusion $\iota:S^{-1} \to L_{-1}^{k-1}$ of the bottom cell, $\iota_*:\pi_q(S^{-1}) \to \pi_q(L_{-1}^{k-1})$ is zero for $0 \le q < k(p-1)-1$. The cofibre sequences of Corollaries 4.3 and 4.5 restrict to give cofibre sequences

$$S^{-1} \xrightarrow{\iota} L_{-1}^{k-1} \xrightarrow{\Delta} L_0^{k-1} \xrightarrow{\tau_p} S^0 .$$

Thus, $\tau_{p*}:\pi_q(L_0^{k-1}) \to \pi_q(S^0)$ is an epimorphism for $0 < q < k(p-1)$. Now let k go to infinity. Of course, $L_0 \simeq \Sigma^\infty B\Sigma_p^+$ splits as the wedge $\Sigma^\infty B\Sigma_p \vee S^0$. Since $\tau_p\iota_p:S^0 \to S^0$ has degree p!, the finiteness of π_*S^0 allows us to deduce the following version of the Kahn-Priddy Theorem.

Theorem 4.8. The restriction $\tau_p:\Sigma^\infty B\Sigma_p \to S^0$ induces an epimorphism $\pi_q(\Sigma^\infty B\Sigma_p) \to \pi_q(S^0) \otimes Z_{(p)}$ for q > 0.

To prove Theorem 2.8, consider the following diagram, where $q \equiv 0 \bmod p^{\phi(k)}$ and q is even if p > 2.

$$\begin{array}{ccccccc}
S^{pq-1} & \xrightarrow{\Sigma^{pq}\iota} & \Sigma^{pq}L_{-1}^{k-1} & \xrightarrow{\Sigma^{pq}\Delta} & \Sigma^{pq}L_0^{k-1} & \xrightarrow{\Sigma^{pq}\tau_p} & S^{pq} \\
\| & & \downarrow{\scriptstyle\nu} & & \downarrow{\scriptstyle\omega} & & \| \\
S^{pq-1} & \longrightarrow & \Sigma^qL_{(p-1)q-1}^{(p-1)q+k-1} & \xrightarrow{\Delta} & \Sigma^qL_{(p-1)q}^{(p-1)q+k-1} & \xrightarrow{\tau_p} & S^{pq}
\end{array}$$

The bottom cofibre sequence is obtained by restriction from sequences in Corollaries 4.3 and 4.5. Periodicity gives an equivalence ν such that the left square commutes. Standard cofibration sequence arguments then give an equivalence ω such that the remaining squares commute. The bottom map τ_p factors through $\tau_p : D_p S^q \to S^{pq}$ and is an epimorphism in the range stated in Theorem 2.8.

It remains to prove Proposition 4.7. For amusement, we proceed a bit more generally. Recall the not necessarily commutative diagram

below Proposition 3.6, where E is an H_∞ ring spectrum. With $E = S$ and $X = S^0$, the following result is Proposition 4.7.

Proposition 4.9. Let X be a finite CW complex of dimension less than $k(p-1)-q$, where $0 < q < k(p-1)$. Then

$$(\rho \wedge 1)^* : E^{-q}X = E^{-q}(S^0 \wedge X) \longrightarrow E^{-q}(L^0_{-k} \wedge X)$$

is zero if E is a connective H_∞ ring spectrum such that $\Delta^* \mathcal{P}_p = \xi_p^*$.

Proof. For $n \geq k$, the cofibre of $\Delta : \Sigma^{n+1} D_p S^{-n-1} \to \Sigma^n D_p S^{-n}$ has dimension at most $-k(p-1)$, and it follows that the colimit $F^{-q}_{(p)}X$ is attained as $E^{-q}(\Sigma^k D_p S^{-k} \wedge X)$. Let $\iota : L^0_{-k} \to L_{-k} \simeq \Sigma^k D_p S^{-k}$ be the inclusion and consider the following diagram, where x is any map $X \to \Sigma^{-q}E$.

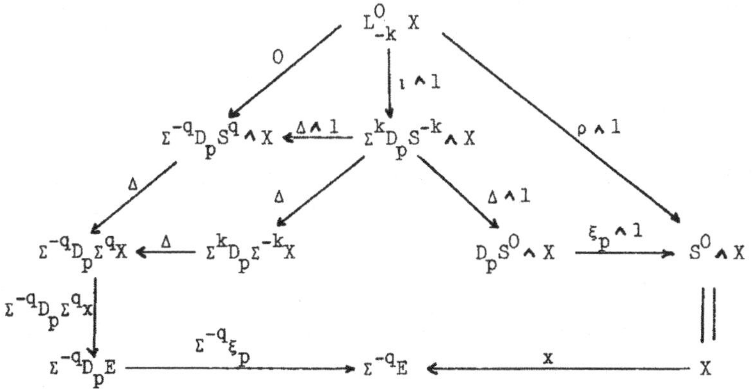

Since $\Delta^*\mathcal{P}_p = \xi_p^*$, the bottom part commutes. We have

$$\xi_p \Delta\iota \simeq \rho : L_{-k}^0 \to S^0$$

since the composite is obviously null homotopic on L_{-k}^{-1} and of degree one on the top cell. We have

$$\Delta\iota \simeq 0 : L_{-k}^0 \to \Sigma^{-q} D_p S^q$$

since $\Sigma^{-q} D_p S^q$ is 0-connected. The conclusion follows.

Replacing S by E in the deductions from Proposition 4.7 and using the results of section 2, we conclude that, for $q > 0$, all p-torsion elements of $\pi_p E$ are nilpotent if $\Delta^*\mathcal{P}_p = \xi_p^*$. This implies our earlier claim that $\Delta^*\mathcal{P}_2 \neq \xi_2^*$ when $E = MO$.

§5. The Singer construction and theorems of Lin and Gunawardena

Singer introduced a remarkable algebraic functor R_+ from A-modules to A-modules, where A is the mod p Steenrod algebra, and Miller began the study of the cohomology theories in section 3 by making the following basic observation. All homology and cohomology is to be taken with mod p coefficients.

Theorem 5.1. Let Y be a spectrum such that H_*Y is bounded below and of finite type. Then $\operatorname{colim} H^*(\Sigma^n D_p \Sigma^{-n} Y)$ is isomorphic to $\Sigma^{-1} R_+ H^* Y$.

We shall prove this and some related observations after explaining its relationship to the following theorems of Lin [53, 54] and Gunawardena [38, 39]. Let $\hat{\pi}^*$ and $\hat{\pi}_*$ denote the p-adic completions of stable cohomotopy and stable homotopy.

Theorem 5.2. The map $\xi_p^* : \hat{\pi}^* Y \to \operatorname{colim} \hat{\pi}^*(\Sigma^n D_p S^{-n} \wedge Y)$ is an isomorphism for all finite CW spectra Y.

As we shall explain shortly, $\lim \hat{\pi}_{-1}(\Sigma^n D_p S^{-n}) = \hat{Z}_p$. Realizing the unit by a compatible system of maps $\xi^p : S^{-1} \to \Sigma^n D_p S^{-n}$ and smashing with Y, we obtain a compatible system of maps

$$\xi^p : \Sigma^{-1} Y \cong S^{-1} \wedge Y \to \Sigma^n D_p S^{-n} \wedge Y.$$

Theorem 5.3. The map $\xi_*^p : \hat{\pi}_* \Sigma^{-1} Y \to \lim \hat{\pi}_*(\Sigma^n D_p S^{-n} \wedge Y)$ is an isomorphism for all finite CW spectra Y.

Since ξ_p^* is a map of cohomology theories and ξ_*^p is a map of homology theories, it suffices to prove these isomorphisms for $Y = S^0$. Since

$$\Sigma^n D_p^{k(p-1)-1} S^{-n} \text{ is } (-1)\text{-dual to } \Sigma^{n+k} D_p^{k(p-1)-1} S^{-n-k},$$

the theorems are esentially dual to one another. Indeed, using the $\lim^1$ exact sequence and waving one's hands at certain compatibility questions, one finds the following chain of isomorphisms, where $m(p-1) > q$.

$$\begin{aligned}
\operatorname*{colim}_n \hat{\pi}^q(\Sigma^n D_p S^{-n}) &= \hat{\pi}^q(\Sigma^m D_p S^{-m}) \\
&= \lim_k \hat{\pi}^q(\Sigma^m D_p^{k(p-1)-1} S^{-m}) \\
&= \lim_k \hat{\pi}_{-q-1}(\Sigma^{m+k} D_p^{k(p-1)-1} S^{-m-k}) \\
&= \lim_n \hat{\pi}_{-q-1}(\Sigma^n D_p S^{-n})
\end{aligned}$$

There is a map of A-modules $\varepsilon: R_+ Z_p \to Z_p$, and the main point of the work of Lin and Gunawardena can be reformulated as follows; see Adams, Gunawardena, and Miller [9].

<u>Theorem 5.4.</u> $\varepsilon^*: \mathrm{Ext}_A(Z_p, Z_p) \to \mathrm{Ext}_A(R_+ Z_p, Z_p)$ is an isomorphism.

An inverse system $\{Y_n\}$ of bounded below spectra Y_n of finite type gives rise to an inverse limit

$$\{E_r\} = \lim \{E_r Y_n\}$$

of Adams spectral sequences, where $\{E_r Y\}$ denotes the classical Adams spectral sequence for the computation of $\hat{\pi}_* Y$. Clearly

$$E_2 \cong \mathrm{Ext}_A(\operatorname{colim} H^* Y_n, Z_p).$$

As pointed out in [74], $\{E_r\}$ converges strongly to $\lim \hat{\pi}_* Y_n$. We apply this with $Y_n = \Sigma^n D_p S^{-n}$. Here Theorems 5.1 and 5.4 give

$$E_2 \cong \mathrm{Ext}_A(\Sigma^{-1} Z_p, Z_p).$$

From this and convergence, it is easy to check that $\lim \hat{\pi}_{-1}(\Sigma^n D_p S^{-n}) = \hat{Z}_p$. The compatible system of maps $\xi^p: S^{-1} \to \Sigma^n D_p S^{-n}$ then induces a map of spectral sequences

$$\{E_r \xi^p\}: \{E_r S^{-1}\} \to \{E_r\}.$$

By Theorem 5.4 again, $E_2 \xi^p$ is an isomorphism, and Theorem 5.3 follows by convergence. Theorem 5.2 can be obtained by a similar Adams spectral sequence argument (as in Lin [53] and Gunawardena [38]) or by dualization.

The crux of the proof of Theorem 5.1 is the following result of Steinberger, which is proven in VIII.3.2 of the sequel. For spaces, it is due to Nishida [89]; see also [68, 9.4]. Let π be the cyclic group of order p. We assume familiarity with the mod p homology $H_*D_\pi Y$, its determination being a standard exercise in the homology of groups in view of I.2.3 (see e.g. [68, §1]). Suffice it to say that $H_*D_\pi Y$ has a basis consisting of elements of the form $e_0 \otimes x_1 \otimes \cdots \otimes x_p$ and $e_i \otimes x^p$, $i \geq 0$. Here the x_i and x run through basis elements of H_*Y, the x_i are not all equal, and the $x_1 \otimes \cdots \otimes x_p$ and x^p together run through a set of π-generators for $(H_*Y)^p$. Restricting to those i of the form $(2s-q)(p-1)-\varepsilon$, where $q = \deg(x)$ and $\varepsilon = 0$ or 1, and to a set of Σ_p-generators for $(H_*Y)^p$, we obtain a basis for H_*D_pY. At least if H_*Y is bounded below and of finite type, we have analogous dual bases for $H^*D_\pi Y$ and H^*D_pY with typical elements denoted $w_0 \otimes y_1 \otimes \cdots \otimes y_p$ and $w_i \otimes y^p$.

<u>Theorem 5.5</u>. Assume that H_*Y is bounded below and of finite type. The subspace of $H^*D_\pi Y$ spanned by $\{w_0 \otimes y_1 \otimes \cdots \otimes y_p\}$ is closed under Steenrod operations and, modulo this subspace, the following relations hold for $y \in H^qY$.

(i) For p = 2,

$$Sq^s(w_j \otimes y^2) = \sum_i \binom{j+q-i}{s-2i} w_{j+s-2i} \otimes (Sq^i y)^2.$$

(ii) For p > 2, let $\delta(2n+\varepsilon) = \varepsilon$, $m = \frac{1}{2}(p-1)$, and $\alpha(q) = -(-1)^{mq}m!$; then

$$P^s(w_j \otimes y^p) = \sum_i \binom{[j/2]+qm-(p-1)i}{s-pi} w_{j+2(s-pi)(p-1)} \otimes (P^i y)^p$$
$$+ \delta(j-1)\alpha(q) \sum_i \binom{[j/2]+qm-(p-1)i-1}{s-pi-1} w_{j-p+2(s-pi)(p-1)} \otimes (\beta P^i y)^p.$$

(iii) For p > 2, $\Delta(w_{2j-1} \otimes y^p) = w_{2j} \otimes y^p$.

We also need to know $\Delta^*: H^*D_\pi Y \to H^*(\Sigma D_\pi \Sigma^{-1}Y)$. Let $\Sigma^n: H^q(Y) \to H^{q+n}(\Sigma^n Y)$ denote the iterated suspension isomorphism for any integer n.

<u>Lemma 5.6</u>. For $y \in H^qY$,

$$\Delta^*(w_j \otimes y^p) = (-1)^{j+1}\alpha(q)\Sigma(w_{j+p-1} \otimes (\Sigma^{-1}y)^p).$$

<u>Proof</u>. We first compute $\Delta_*: H_*(\Sigma D_\pi Y) \to H_*(D_\pi \Sigma Y)$. Take f to be the identity map of Y and replace D_p by D_π in Lemma 3.10. We find that the composite of Δ_* and the homology suspension Σ_* is the suspension associated to the zero sequence

$$C_*(D_\pi Y) \longrightarrow C_*(D_\pi CY) \longrightarrow C_*(D_\pi \Sigma X).$$

By I.2.3 and [68,§1], we may instead use the zero sequence

$$W \otimes_\pi C_*(Y)^p \longrightarrow W \otimes_\pi C_*(CY)^p \longrightarrow W \otimes_\pi C_*(\Sigma Y)^p,$$

where W is the standard π-free resolution of Z_p. A direct chain level computation, details of which are in [68,p. 166-167], gives the formula

$$\Delta_* \Sigma_* e_{j+p-1} \otimes x^p = (-1)^{j+1}\alpha(q)e_j \otimes (\Sigma_* x)^p$$

for $x \in H_{q-1}(Y)$. Clearly $\Delta_*\Sigma_*(e_0 \otimes x_1 \otimes \cdots \otimes x_p) = 0$ for all x_i. The conclusion follows upon dualization (and a careful check of signs).

The results above determine colim $H^*(\Sigma^n D_\pi \Sigma^{-n} Y)$ as an A-module, and similarly with D_π replaced by D_p. To compare the answer to the Singer construction, we must first recall the definition of the latter [98,52]. When $p = 2$, $\Sigma^{-1}R_+M$ is additively isomorphic to $\Lambda \otimes M$, where Λ is the Laurent series ring $Z_2[v,v^{-1}]$, deg $v = 1$. Its Steenrod operations are specified by

$$Sq^s(v^r \otimes x) = \sum_i \binom{r-i}{s-2i} v^{r+s-i} \otimes Sq^i x.$$

When $p > 2$, $\Sigma^{-1}R_+M$ is additively isomorphic to $\Lambda \otimes M$, where $\Lambda = E\{u\} \times Z_p[v,v^{-1}]$, deg $u = 2p-3$ and deg $v = 2p-2$. Its Steenrod operations are specified by

$$P^s(u^\varepsilon v^{r-\varepsilon} \otimes x) = \sum_i (-1)^{s+i}\binom{(p-1)(r-i)-\varepsilon}{s-pi} u^\varepsilon v^{r+s-i-\varepsilon} \otimes P^i x$$

$$+ (1-\varepsilon)\sum_i (-1)^{s+i}\binom{(p-1)(r-i)-1}{s-pi-1} uv^{r+s-i-1} \otimes \beta P^i x$$

and

$$\beta(u^\varepsilon v^{r-\varepsilon} \otimes x) = \varepsilon(v^r \otimes x).$$

We can now prove Theorem 5.1. We define an isomorphism

$$\omega : \text{colim } H^*(\Sigma^n D_p \Sigma^{-n} Y) \to \Sigma^{-1} R_+ H^* Y$$

as follows. For $p = 2$ and $y \in H^q(Y)$, let

$$\omega(\Sigma^n(w_{r-q+n} \otimes (\Sigma^{-n}y)^2 = v^r \otimes y.$$

For $p > 2$ and $y \in H^q(Y)$, let

$$\omega(\Sigma^n(w_{(2r+n-q)(p-1)-\varepsilon} \otimes (\Sigma^{-n}y)^p) = (-1)^{r+q+(\varepsilon+1)n}\nu(q-n)^{-1}u^\varepsilon v^{r-\varepsilon} \otimes y,$$

where $\nu(2j+\varepsilon)=(-1)^j(m!)^\varepsilon$. Note that

$$\alpha(q)\nu(q-1)^{-1} = \nu(q)^{-1} \quad\text{and}\quad (-1)^q\nu(q)^{-1} = (-1)^{mq}\nu(q).$$

By Lemma 5.6, these ω induce a well-defined isomorphism on passage to colimits. by Theorem 5.5, we see that our constants have been so chosen that ω is an isomorphism of A-modules.

Remark 5.7. When $p > 2$, there are two variants of the Singer construction. We are using the smaller one appropriate to D_p. This is a summand of the larger variant, for which Theorem 5.1 is true with D_p replaced by D_π. See Gunawardena [39,9] for details (but note that his signs don't quite agree with ours).

With $Y = S^0$, Theorem 5.1 specializes to an isomorphism

$$\Lambda = \Sigma^{-1}R_+Z_p \cong \operatorname{colim} H^*(\Sigma^n D_p S^{-n}).$$

Since Λ is an A-module, $\Lambda \otimes M$ admits the diagonal A action, which is evidently quite different from that originally specified on $\Sigma^{-1}R_+M$. For finite CW complexes X, we have the isomorphism

$$\Delta^*:\operatorname{colim} H^*(\Sigma^n D_p \Sigma^{-n}X) \longrightarrow \operatorname{colim} H^*(\Sigma^n D_p S^{-n}\wedge X)$$

of Theorem 3.2. We next obtain an explicit description of the resulting isomorphism

$$\Delta^*:\Sigma^{-1}R_+\tilde{H}^*X \to \Lambda \otimes \tilde{H}^*X.$$

Thus consider $\Delta:D_\pi Y\wedge X \to D_\pi(Y\wedge X)$. When $X = S^1$, we computed Δ_* in the proof of Lemma 5.6. When $Y = S$, $D_\pi Y = \Sigma^\infty B\pi^+$ and the effect of Δ_* is implicit in the definition of the Steenrod operations; see Steenrod and Eptein [100] (or, for correct signs, [68, 9.1]). The following result is a common generalization of these calculations.

Propsition 5.8. Let $x \in \tilde{H}_k(X)$ and $y \in H_q(Y)$. If $p = 2$,

$$\Delta_*(e_r \otimes y^2 \otimes x) = \sum_i e_{r+2i-k} \otimes (y \otimes Sq^i_* x)^2.$$

if $p > 2$, let $\nu(2j+1) = (-1)^j(m!)^\varepsilon$ and $\varepsilon(2j+\varepsilon) = \varepsilon$; then

$$\Delta_*(e_r \otimes y^p \otimes x) = (-1)^{mkq}\nu(k) \sum_i (-1)^i e_{r+(2pi-k)(p-1)} \otimes (y \otimes P^i_* x)^p$$
$$-(-1)^{q+m(k-1)q}\delta(r)\nu(k-1) \sum_i (-1)^i e_{r+p+(2pi-k)(p-1)} \otimes (y \otimes P^i_*\beta x)^p.$$

Proof. Modulo shuffling in $C_*(Y)^p$, which introduces the signs depending on q when $p > 2$, Δ_* is computable from the map obtained by quotienting out the action of π from the π-map

$$\Phi \otimes 1 : C_*(W) \otimes \tilde{C}_*(X) \otimes C_*(Y)^p \longrightarrow C_*(W) \otimes \tilde{C}_*(X)^p \otimes C_*(Y)^p$$

induced by a π-equivariant approximation Φ of $1 \otimes \Delta'_*$, where Δ' is a cellular approximation of the diagonal $X \to X^p$; see e.g. [100, V§3] or [68,7.1]. The essential point is that Y acts like a dummy variable, so that the standard calculation for $Y = S^0$ of [68, 9.1] implies the general result.

Dualizing, and paying careful attention to signs, we obtain the following version in cohomology.

Proposition 5.9. Assume that H_*X and H_*Y are of finite type and that H_*Y is bounded below. Let $x \in \tilde{H}^k(X)$ and $y \in H^q(Y)$. If $p = 2$,

$$\Delta^*(w_j \otimes (y \otimes x)^2) = \sum_i w_{j+k-i} \otimes y^2 \otimes Sq^i x.$$

If $p > 2$,

$$\Delta^*(w_j \otimes (y \otimes x)^p) = (-1)^{mk(q+1)}\nu(k) \sum_i (-1)^i w_{j+(k-2i)(p-1)} \otimes y^p \otimes P^i x$$

$$-(-1)^{q+mk(q+1)}\delta(j-1)\nu(k) \sum_i (-1)^i w_{j+(k-2i)(p-1)-1} \otimes y^p \otimes \beta P^i x.$$

A check of constants gives the following consequence.

Corollary 5.10. For $M = \tilde{H}^*X$, the formula

$$\Delta^*(v^r \otimes x) = \sum_i v^{r-i} \otimes Sq^i x$$

if $p = 2$ and

$$\Delta^*(u^\varepsilon v^{r-\varepsilon} \otimes x) = \sum_i u^\varepsilon v^{r-i-\varepsilon} \otimes P^i x - (1-\varepsilon) \sum_i uv^{r-i-1} \otimes \beta P^i x$$

if $p > 2$ specifies a morphism of A-modules $\Delta^* : \Sigma^{-1}R_+M \to \Lambda \otimes M$.

The same formulae give a morphism of A-modules for all A-modules M which are either unstable or bounded above, either assumption ensuring that the relevant sums are finite. In the bounded above case, but not in general in the unstable case, this morphism is an isomorphism. See [98, 52, 82].

Define $\varepsilon: R_+M \to M$ by the formulas

$$\varepsilon\Sigma(v^{r-1} \otimes x) = Sq^r x$$

if $p = 2$ (where $Sq^r(x) = 0$ if $r < 0$) and

$$\varepsilon\Sigma(uv^{r-1} \otimes x) = P^r x \quad \text{and} \quad \varepsilon\Sigma(v^r \otimes x) = -\beta P^r x$$

if $p > 2$. By [98,3.4] and [52,3.5], ε is a well-defined morphism of A-modules. When Δ^* is defined, ε is the composite

$$R_+M \xrightarrow{\Sigma\Delta^*} \Sigma(\Sigma^{-1}R_+Z_p \otimes M) \xrightarrow{\Sigma(\varepsilon \otimes 1)} \Sigma(\Sigma^{-1}Z_p \otimes M) = M.$$

Generalizing Theorem 5.4, Adams, Gunawardena, and Miller [9] proved that ε is an Ext-isomorphism for any M. This leads to a generalization of Theorem 5.3 to a version appropriate to $(Z_p)^k$ for any $k \geq 1$, and this generalization is the heart of the proof of the Segal conjecture for elementary Abelian p-groups. See [9,74].

§6. Nishida's second nilpotency theorem.

If $x \in \pi_n E$ has order p, then x extends over the Moore spectrum $M^n = S^n \cup_p CS^n$. The idea of Nishida's second nilpotency theorem is to exploit this extension by showing that $D_j M^n$ splits as a wedge of Eilenberg-MacLane specta in a range of dimensions. The relevant splitting is a special case of the following result which, as we shall explain shortly, is in turn a special case of the general splitting theorem to be proven by Steinberger in the next chapter.

<u>Theorem 6.1</u>. Let Y be a spectrum obtained from S^n by attaching cells of dimension greater than n. Assume that $\pi_n Y$ is Z or Z_{p^i} and let $\nu \in H^n(Y;Z_p)$ be a generator. Assume one of the following further hypotheses.

(a) $p = 2$ and either n is odd or $\beta(\nu) \neq 0$.

(b) $p > 2$, n is even, and $\beta(\nu) \neq 0$.

(c) $p = 2$ and $Sq^3(\nu) \neq 0$.

(d) $p > 2$, n is even, and $\beta P^1(\nu) \neq 0$.

Then $D_j Y$ splits p-locally as a wedge of suspensions of Eilenberg-MacLane spectra through dimensions $r < nj + \frac{1}{p}(2p-3)(j+1)-1$. In cases (a) and (b), only suspensions of HZ_p are needed.

Before discussing the proof, we explain how to use these splittings to obtain relations in the homotopy groups of H_∞ ring spectra. Let Y and ν be as in the theorem above and localize all spectra at p.

Theorem 6.2. Let E be an H_∞ ring spectrum, let F be a connective spectrum, and let $\phi:E\ \ F \to E$ be any map (for example, the product when F = E or the identity when F = S). Let $x \in \pi_n E$ and assume one of the following hypotheses.

(a) $p = 2$ and n is odd; here let $Y = S^n$.

(b) $p \geq 2$, n is even, and x has order 2; here let $Y = M^n$.

(c) $p = 2$, n is even, and x extends over some Y with $Sq^3(\nu) \neq 0$.

(d) $p > 2$, n is even, and x extends over some Y with $\beta P^1(\nu) \neq 0$.

Let $R = Z_p$ in cases (a) and (b) and $R = \pi_n Y$ in cases (c) and (d) and let $y \in \pi_n F$ be in the kernel of the Hurewicz homomorphism $\pi_q F \to H_q(F;R)$. Then $x^j y = 0$ if $q < \frac{1}{p}(2p-3)(j+1)-1$.

Proof. Our hypotheses ensure that $H^{nj}(D_j Y;R) \cong R$. We can choose a generator μ such that the composite

$$S^{nj} \xrightarrow{\iota_j} D_j S^n \xrightarrow{D_j f} D_j Y \xrightarrow{\mu} \Sigma^{nj} HR$$

is $\Sigma^{nj} e$, where $f:S^n \to Y$ is the inclusion of the bottom cell and $e:S \to HR$ is the unit. Choose $\tilde{x}:Y \to E$ such that $\tilde{x}f = x$. Then the solid arrow part of the following diagram commutes and the top composite is $x^j y$.

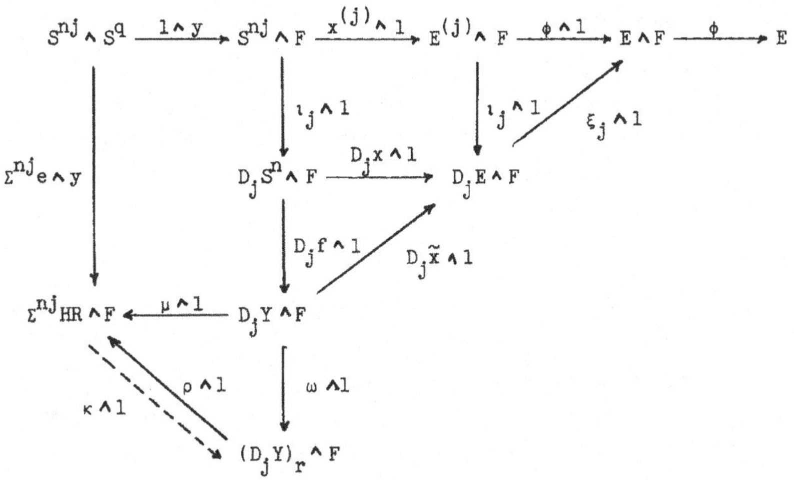

Here $r = nj+q$, $\omega: D_j Y \to (D_j Y)_r$ is the rth stage of a Postnikov decomposition of $D_j Y$, and $\rho:(D_j Y)_r \to \Sigma^{nj} HR$ is the unique cohomology class such that $\rho\omega = \mu$. The previous theorem gives $\kappa:\Sigma^{nj} HR \to (D_j Y)_r$ such that $\rho\kappa = 1$. The complementary wedge summand of $\Sigma^{nj} HR$ in $(D_j Y)_r$ is (nj)-connected, and it follows that $\kappa \cdot \Sigma^{nj} e = \omega \cdot D_j f \cdot \iota_j$. Since F is connective, $\omega \wedge 1$ induces an isomorphism on π_{nj+q}. Since y is in the kernel of the Hurewicz homomorphism and the latter is induced by $e\wedge 1: F = S\wedge F \to HR\wedge F$, $\Sigma^{nj}e\wedge y = 0$. Chasing the diagram, we conclude that $x^j y = 0$.

In particular, with $F = E$, $q = n$, and $y = x$, we obtain $x^{j+1} = 0$. With $E = S$ and $n > 0$, case (b) applies to any even degree element of order p. As observed by Steinberger, when $p = 2$ case (a) applies to any odd degree element and gives a better estimate of the order of nilpotency than that obtained by applying case (b) to x^2. While this result gives a much better estimate of the order of nilpotency of elements of order p in $\pi_* S$ than does Theorem 2.9, the estimate is presumably still far from best possible. For example, if $p = 2$ and $n = 14$, the estimate is now $x^{30} = 0$. Cases (c) and (d) apply to some elements of order p^i with $i > 1$. The idea is to add further cells to S^n, or to $S^n \cup_{p^i} CS^n$, so as to obtain a spectrum Y for which the relevant Steenrod operation is non-zero. However, a given element x need not extend over any such Y. (Conceivably some power of x must so extend.) This explains why Nishida's second method fails to give the full nilpotency theorem and why we cannot yet prove Conjecture 2.7.

We must still explain how to prove Theorem 6.1. The idea is to approximate D_j through the specified range by a spectrum with additional structure and then use homology operations to split the latter. The approximation is based on the following observation about mod p homology.

<u>Proposition 6.3</u>. Let Y be an (n-1)-connected spectrum with $H_n Y = Z_p$, where n is even if $p > 2$. Let $f: S^n \to Y$ induce an isomorphism on H_n. Then the homomorphism $H_i \Sigma^n D_q Y \to H_i D_{q+1} Y$ induced by the composite

$$D_q Y \wedge S^n \xrightarrow{1\wedge f} D_q Y \wedge Y \xrightarrow{\alpha_{q,1}} D_{q+1} Y$$

is a monomorphism for all i and is an isomorphism if $i < n(q+1) + \frac{1}{p}(2p-3)(q+1)$.

For spaces X, a self-contained calculation of $H_* D_q X$ for all q is given in [28,I§4-5]. The generalization to spectra is given by McClure in Chapter IX, and the conclusion is easily read off from these calculations.

With the proposition as a hint, we construct the approximating spectra as follows.

<u>Definition 6.4</u>. Let (Y,f) be a spectrum together with a map $f: S^n \to Y$ for some integer n and define $D(Y,f) = \text{tel}\, \Sigma^{-nq} D_q Y$, where the nth map of the system is obtained by applying $\Sigma^{-n(q+1)}$ to the composite

$$D_q Y \wedge S^n \xrightarrow{1\wedge f} D_q Y \wedge Y \xrightarrow{\alpha_{q,1}} D_{q+1} Y.$$

Now the previous proposition has the following consequence.

Corollary 6.5. With Y and f as in the proposition, assume further that Y is p-local of finite type. Then the natural map $D_jY \to \Sigma^{nj}D(Y,f)$ is an equivalence through dimensions less than $nj + \frac{1}{p}(2p-3)(j+1) - 1$.

Proof. By the proposition, the maps $\Sigma^{-n(q+1)}(\alpha_{q,1} \circ 1 \wedge f)$ used to construct D(Y,f) induce isomorphisms in mod p homology and thus in p-local homology in degrees less than $\frac{1}{p}(2p-3)(q+1)$. This fact for $q \geq j$ implies the conclusion (with the usual loss of a dimension as one passes from homology to homotopy).

Thus, to prove Theorem 6.1, we need only split D(Y,f).

The following ad hoc definition, which generalizes Nishida's notion of a Γ-spectrum [90,1.5], allows us to describe the structure present on the spectra D(Y,f). In the rest of this section we shall refer to weak maps and weakly commutative diagrams when the domain is a telescope and phantom maps are to be ignored.

Definition 6.6. A spectrum E is a pseudo H_∞ ring spectrum if

(i) E is the telescope of a sequence of connective spectra E_q, $q \geq 0$;

(ii) E is a weak ring spectrum with unit induced from a map $S \to E_0$ and product induced from a unital, associative, and commutative system of compatible maps $E_q \wedge E_r \to E_{q+r}$; and

(iii) For each $j \geq 0$ and $q \geq 0$, there exists an integer $d = d(j,q)$ and a map $\xi_j : D_j\Sigma^{dq}E_q \to \Sigma^{djq}E_{jq}$ whose composite with $\iota_j : \Sigma^{djq}E_q^{(j)} \cong (\Sigma^{dq}E_q)^{(j)} \to D_j\Sigma^{dq}E_q$ is the $(djq)^{th}$ suspension of the interated product $E_q^{(j)} \to E_{jq}$.

Examples 6.7. (i) With each $E_q = E$ and each $d(j,q) = 0$, a connective H_∞ ring spectrum may be viewed as a pseudo H_∞ ring spectrum.

(ii) With each $E_q = E$ and each $d(j,q) = d$, a connective H_∞^{d*} ring spectrum may be viewed as a pseudo H_∞ ring spectrum; since E has structural maps ξ_j for all q, negative as well as positive, we could obtain a different pseudo structure with each $d(j,q) = -d$.

(iii) For an (n-1)-connected spectrum Y and map $f:S^n \to Y$ such that either $2 = 0:Y \to Y$ or n is even, D(Y,f) is a pseudo H_∞ ring spectrum with q^{th} term $\Sigma^{-nq}D_qY$. Its product is induced by the maps

$$\Sigma^{-nq}D_qY \wedge \Sigma^{-nr}D_rY \cong \Sigma^{-n(q+r)}(D_qY \wedge D_rY) \xrightarrow{\Sigma^{-n(q+r)}\alpha_{q,r}} \Sigma^{-n(q+r)}D_{q+r}Y \ ,$$

these forming a unital, associative, commutative, and compatible system by I.2.6 and I.2.8 and our added hypothesis, which serves to eliminate signs coming from permutations of spheres. With all $d(j,q) = n$, its structural maps are

$$\xi_j = \beta_{j,q}: D_j\Sigma^{nq}(\Sigma^{-nq}D_qY) = D_jD_qY \to D_{jq}Y = \Sigma^{njq}(\Sigma^{-njq}D_{jq}Y).$$

The following analog of I.3.6 and I.4.5 admits precisely the same simple cohomological proof.

Proposition 6.8. Let E be a pseudo H_∞ ring spectrum with char $\pi_0E = 2$ or all $d(j,q)$ even. Assume that $\pi_0E = \pi_0E_q$ for all $q \geq q_0$ and, for such q, let $i:E_q \to H(\pi_0E)$ be the unique map which induces the identity homomorphism on π_0. Then the following diagrams commute, where $d = d(j,q)$:

$$\begin{array}{ccc} D_j\Sigma^{dq}E_q & \xrightarrow{D_j\Sigma^{dq}i} & D_j\Sigma^{dq}H(\pi_0E) \\ \downarrow{\scriptstyle \xi_j} & & \downarrow{\scriptstyle \xi_j} \\ \Sigma^{djq}E_{jq} & \xrightarrow{\Sigma^{djq}i} & \Sigma^{djq}H(\pi_0E) \end{array}$$

In the next chapter, Steinberger will use a computation of the homology operations of the H_∞ ring spectrum $\bigvee_q \Sigma^{dq}HZ_p$ to prove the following generalization of Nishida's result [90,3.2].

Theorem 6.9. Let E be a p-local pseudo H_∞ ring spectrum. If $\pi_0E = Z_p$, then E splits as a wedge of suspensions of HZ_p. If $\pi_0E = Z_{p^r}$, $r > 1$, or $\pi_0E = Z_{(p)}$ and if $p = 2$ and $Sq^3i \neq 0$ or $p > 2$ and $\beta P^1 i \neq 0$, where i generates $H^0(E;Z_p)$, then E splits as a wedge of suspensions of HZ_{p^s}, $s \geq 1$, and $HZ_{(p)}$.

Considering the natural map $\Sigma^{-n}Y \to D(Y,f)$, and using the formula $\beta(w_0 \otimes \nu^2) = nw_1 \otimes \nu^2$ of Theorem 5.5 for case (a), we easily check that the theorem applies to split $D(Y,f)$ for Y as in Theorem 6.1.

We complete this section with some remarks about the role played by Definition 6.4 in the general theory of H_∞ ring spectra.

Remarks 6.10. Let (E,e) be a spectrum with unit $e:S \to E$. Let $DE = D(E,e)$ and let $\eta:E = D_1E \to DE$ be the natural inclusion. By I.2.7, I.2.9, and I.2.13, the maps $\beta_{j,k}:D_jD_kE \to D_{jk}E$ induce a natural weak map $\mu_k:DD_kE \to DE$ such that the following diagrams (weakly) commute:

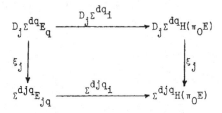

$$\begin{array}{ccc} D_kE & \xrightarrow{\eta} & DD_kE \\ & \searrow & \downarrow{\mu_k} \\ & & DE \end{array} \qquad \text{and} \qquad \begin{array}{ccc} DD_jD_kE & \xrightarrow{D\beta_{j,k}} & DD_{jk}E \\ \downarrow{\mu_j} & & \downarrow{\mu_{jk}} \\ DD_kE & \xrightarrow{\mu_k} & DE \end{array}$$

If E is an H_∞ ring spectrum, then, by Proposition 1.3, the maps $\xi_j:D_jE \rightarrow E$ determine a weak map $\xi:DE \rightarrow E$ such that the following diagrams (weakly) commute.

$$\begin{array}{ccc} E & \xrightarrow{\eta} & DE \\ & \diagdown\!\!\diagdown & \downarrow{\xi} \\ & & E \end{array} \qquad \text{and} \qquad \begin{array}{ccc} DD_kE & \xrightarrow{D\xi_k} & DE \\ \downarrow{\mu_k} & & \downarrow{\xi} \\ DE & \xrightarrow{\xi} & E \end{array}$$

Conversely, by the same result, if $\psi:DE \rightarrow E$ makes these diagrams weakly commute, then its restrictions $\xi_j:D_jE \rightarrow E$ give E a structure of H_∞ ring spectrum. These assertions are analogous to, but weaker than, the assertions that D is a monad and that an H_∞ ring spectrum is an algebra over this monad (compare [69, §2]). The point is that the μ_k fail to satisfy the requisite compatibility to determine a weak map $\mu:DDE \rightarrow DE$. By I.2.11 and I.2.15, the compatibility they do have is described by the weakly commutative diagram

$$\begin{array}{ccccc} DD_kE \wedge S & \longrightarrow & DD_kE \wedge DS & \longrightarrow & DD_kE \\ \| & & \downarrow{\nu_k} & & \downarrow{D(\alpha_{k,1} \circ 1 \wedge e)} \\ DD_kE & \longrightarrow & DE & \longleftarrow & DD_{k+1}E \end{array}$$

where ν_k is induced by the composites

$$D_jD_kE \wedge D_jS \xrightarrow{\beta_{j,k} \wedge D_je} D_{jk}E \wedge D_jE \xrightarrow{\alpha_{jk,j}} D_{jk+j}E$$

and $\delta:DF \rightarrow DF \wedge DS$ is induced by the maps $\delta_j:D_j(F \wedge S) \rightarrow D_jF \wedge D_jS$.

CHAPTER III.

HOMOLOGY OPERATIONS FOR H_∞ AND H_n RING SPECTRA

by Mark Steinberger

Since H_∞ ring spectra are analogs of H_∞ spaces and H_n ring spectra are analogs up to homotopy of n-fold loop spaces, it is to be expected that their homologies admit operations analogous to those introduced by Araki and Kudo [12], Browder [22], Dyer and Lashof [33] and Cohen [28]. We define such operations in section 1 for H_∞ ring spectra and in section 3 for H_n ring spectra.

As an amusing example, we end section 1 with the observation, due independently to Haynes Miller and Jim McClure, that our homology operations in $H_*F(X^+,S) = H^*X$ coincide with the Steenrod operations when X is a finite complex.

For connective H_∞ ring spectra, we show that the resulting ring of operations is precisely the Dyer-Lashof algebra. Moreover, if X is an H_∞ space with zero (as in II.1.7), then the new operations for the H_∞ ring spectrum $\Sigma^\infty X$ coincide with the space level operations of $\tilde{H}_*X$.

As will be shown by Lewis in the sequel, the Thom spectrum Mf of an n-fold or infinite loop map $f:X \to BF$ is an H_n or H_∞ ring spectrum and the Thom isomorphism carries the space level operations to the new operations in H_*Mf. This applies in particular to the Thom spectra of the classical groups (although a simpler argument could be used here).

In section 2 we present calculations of the new operations in less obvious cases (with the proofs deferred until sections 5 and 6). Our central calculations concern Eilenberg-MacLane spectra, where , in contrast to the additive homology operations for Eilenberg-MacLane spaces, these operations are highly nontrivial. In fact, they provide a conceptual framework for the splittings of various cobordism spectra into wedges of Eilenberg-MacLane spectra or Brown-Peterson spectra. The proofs of these splittings in the literature are based on computations of the Steenrod operations on the Thom class. We show in section 4 that the presence of an H_n ring structure, $n \geq 2$ ($n \geq 3$ for the BP splittings), reduces these computations to a check of at most one low dimensional operation, depending on the type of splitting. In addition, we have placed these splitting theorems in a more general context which, as explained in the previous chapter, leads to a reproof of Nishida's bound on the order of nilpotency of an element of order p in the stable stems. All of our splittings are deduced directly from our computation of the new operations in the homology of Eilenberg-MacLane spectra.

I wish to thank Peter May for his help and encouragement and to thank Arunas Liulevicius for helpful conversations, and for sharing the result listed as Proposition 5.1.

§1. Construction and properties of the operations

Just as the space level operations of Araki and Kudo, Browder, and Dyer and Lashof are based on maps

$$E\Sigma_j \times_{\Sigma_j} X^j \to X,$$

so our new spectrum level operations are based on the structural maps

$$\xi_j : D_j E \to E$$

of H_∞ ring spectra (see I.3.1). We consider homology with mod p coefficients for a prime p. The following omnibus theorem describes our operations. Properties of the operations at the prime 2 which are distinct from the properties at odd primes are indicated in square brackets. As usual, β denotes the homology Bockstein operation, and P^r_* denotes the dual of the Steenrod operations P^r, with $P^r = Sq^r$ if p = 2.

Theorem 1.1. For integers s there exist operations Q^s in the homology of H_∞ ring spectra E. They enjoy the following properties.

(1) The Q^s are natural homomorphisms.

(2) Q^s raises degree by 2s(p-1) [by s].

(3) $Q^s x = 0$ if 2s < degree(x) [if s < degree(x)].

(4) $Q^s x = x^p$ if 2s = degree(x) [if s = degree(x)].

(5) $Q^s 1 = 0$ for $s \neq 0$, where $1 \in H_0 X$ is the algebraic unit element of H_*X.

(6) The external and internal Cartan formulas hold:

$$Q^s(x \times y) = \sum_{i+j=s} Q^i x \times Q^j y \quad \text{for } x \times y \in H_*(E \wedge F);$$

$$Q^s(xy) = \sum_{i+j=s} (Q^i x)(Q^j y) \quad \text{for } x,y \in H_*E.$$

(7) The Adem relations hold: if $p \geq 2$ and $r > ps$, then

$$Q^r Q^s = \sum_i (-1)^{r+i}(pi - r,\ r - (p-1)s - i - 1) Q^{r+s-i} Q^i;$$

if $p > 2$ and $r \geq ps$, then

$$Q^r\beta Q^s = \sum_i (-1)^{r+i}(pi - r, r - (p - 1)s - i)\beta Q^{r+s-i}Q^i$$

$$- \sum_i (-1)^{r+i}(pi - r - 1), r - (p - 1)s - i)Q^{r+s-i}\beta Q^i.$$

(8) The Nishida relations hold: For $p \geq 2$ and n sufficiently large,

$$P_*^r Q^s = \sum_i (-1)^{r+i}(r - pi, p^n + s(p - 1) - pr + pi)Q^{s-r+i}P_*^i .$$

In particular, for $p = 2$, $\beta Q^s = (s - 1)Q^{s-1}$. For $p > 2$ and n sufficiently large,

$$P_*^r \beta Q^s = \sum_i (-1)^{r+i}(r - pi, p^n + s(p - 1) - pr + pi - 1)\beta Q^{s-r+i}P_*^i$$

$$- \sum_i (-1)^{r+i}(r - pi-1, p^n + s(p - 1) - pr + pi)Q^{s-r+i}P_*^i\beta.$$

(9) The homology suspension $\sigma:\tilde{H}_*E_0 \to H_*E$ carries the operations given by the multiplicative H_∞ space structure of E_0 to the operations in the homology of E.

(10) If $E = \Sigma^\infty X$ for an $H_{\infty 0}$-space X, then the operations in H_*E agree with the space level operations in $\tilde{H}_*X$.

The statement here is identical to that for the space level operations except that operations of negative degree can act on homology classes of negative degree and that a high power of p is added to the right entry in the binomial coefficients appearing in the Nishida relations. For spaces, the same answer is obtained with or without the power of p because of the restrictions on the degrees of dual Steenrod operations acting nontrivially on a given homology class. Our conventions are that (a,b) is zero if either $a < 0$ or $b < 0$ and is the binomial coefficient $(a + b)!/a!b!$ otherwise. The Nishida relations become cleaner when written in terms of classical binomial coefficients since

$$(a, p^n + b) = \binom{p^n+a+b}{a} = \binom{a+b}{a} \quad \text{for } a < p^n \text{ and } b \geq 0.$$

The Q^s and βQ^s generate an algebra of operations. If we restrict attention to the operations on connective H_∞ ring spectra, then the resulting algebra is precisely the Dyer-Lashof algebra in view of relations (3) and (8) and application of (10) to the $H_{\infty 0}$ space obtained by adjoining a disjoint basepoint to the additive H_∞ space structure on QS^0.

We sketch the proof of the theorem in the rest of this section. With the exception of the proof of the Nishida relations, the argument is precisely parallel to the treatment of the space level homology operations in [28] and is based on the general algebraic approach to Steenrod type operations developed in [68] and summarized by Bruner in IV§2.

Let π be the cyclic group of order p embedded as usual in Σ_p and let W be the standard π-free resolution of Z_p (see IV.2.2). Let $C_*(E\Sigma_p)$ be the cellular chains of the standard Σ_p-free contractible space $E\Sigma_p$ and choose a morphism $j: W \to C_*(E\Sigma_p)$ of π-complexes over Z_p. We may assume that our H_∞ ring spectrum E is a CW-spectrum with cellular structure maps $\xi_j : D_j E \to E$. By I.2.1, $D_j E$ is a CW-spectrum with cellular chains isomorphic to $C_*(E\Sigma_j) \otimes_{\Sigma_j} (C_*E)^j$. Thus we have a composite chain map

$$W \otimes_\pi (C_*E)^p \xrightarrow{j \otimes 1} C_*(E\Sigma_p) \otimes_{\Sigma_p} (C_*E)^p \cong C_*(D_pE) \xrightarrow{\xi_*} C_*E.$$

The homology of the domain has typical elements $e_i \otimes x^p$ (and $e_0 \otimes x_1 \otimes \cdots \otimes x_p$), where $x \in H_*E$, and we let $Q_i(x) \in H_*E$ be the image of $e_i \otimes x^p$. Let x have degree q. If p = 2 define

$$Q^s(x) = 0 \text{ if } s < q \quad \text{and} \quad Q^s(x) = Q_{s-q}(x) \text{ if } s \geq q.$$

for p > 2, define

$$Q^s(x) = 0 \text{ if } 2s < q \quad \text{and} \quad Q^s(x) = (-1)^s \nu(q) Q_{(2s-q)(p-1)}(x) \text{ if } 2s \geq q$$

where $\nu(q) = (-1)^{q(q-1)m/2}(m!)^q$, with $m = \frac{1}{2}(p-1)$. By [68] the Q^s and βQ^s account for all non-trivial Q_i when p > 2. Since ξ_p restricts on $E^{(p)}$ to the p-fold product of E and since the unit $e: S \to E$ is an H_∞-map, parts (1)-(5) of the theorem are immediate from [68].

It is proven in the sequel [Equiv, VIII.2.9] that the maps ι_j, $\alpha_{j,k}$, $\beta_{j,k}$, and δ_j discussed in I§2 have the expected effect on cellular chains. For example, δ_{j*} can be identified with the homomorphism

$$C_*(E\Sigma_j) \otimes (C_*E \otimes C_*E)^j \xrightarrow{(1 \otimes t \otimes 1)(\Delta'_* \otimes u)} C_*(E\Sigma_j) \otimes (C_*E)^j \otimes C_*(E\Sigma_j) \otimes (C_*E)^j$$

where Δ' is a cellular approximation to the diagonal of $E\Sigma_j$ and u and t are shuffle and twist isomorphisms (with the usual signs). The Cartan formula and Adem relations follow. For the former, the smash product of H_∞ ring spectra E and F is an H_∞ ring spectrum with structural maps the composites

$$D_j(E \wedge F) \xrightarrow{\delta_j} D_jE \wedge D_jF \xrightarrow{\xi_j \wedge \xi_j} E \wedge F,$$

and the product $E \wedge E \to E$ of an H_∞ ring spectrum is an H_∞ map; see I.3.4. For the latter, we use the case j = k = p of the second diagram in the definition, I.3.1, of an H_∞ ring spectrum. The requisite algebra is done once and for all in [68].

The Steenrod operations in $H_*(D_\pi E)$ are computed in [Equiv. VIII §3], and the Nishida relations follow by naturality. (See also II.5.5 and VIII §3 here.)

Since $\sigma_*:\tilde{H}_*(E_0) \to H_*E$ is the composite of the identification $\tilde{H}_*(E_0) \cong H_*(\Sigma^\infty E_0)$ and the natural map $\varepsilon_*:H_*(\Sigma^\infty E_0) \to H_*E$ and since $\varepsilon:\Sigma^\infty E_0 \to E$ is an H_∞ map when E is an H_∞ ring spectrum, by I.3.10, part (9) of the theorem is a consequence of part (10). In turn, part (10) is an immediate comparison of definitions in view of I.2.2 and I.3.8. The essential point is that the isomorphism $D_\pi\Sigma^\infty X \cong \Sigma^\infty D_\pi X$ induces the obvious identification on passage to cellular chains, by [Equiv. VIII.2.9].

As promised, we have the following observation of Miller and McClure.

Remark 1.2. Let X be a finite CW complex. By II.3.2, the dual $F(X^+,S)$ of $\Sigma^\infty X^+$ is an H_∞ ring spectrum with p^{th} structural map the adjoint of the composite

$$D_pF(X^+,S)\wedge X^+ \xrightarrow{\Delta} D_p(F(X^+,S)\wedge X^+) \xrightarrow{D_p\varepsilon} D_pS \xrightarrow{\xi_p} S.$$

Here Δ_* is computed in II.5.8, ε_* is the Kronecker product $H^*X \otimes H_*X \to Z_p$, and ξ_{p*} is the identity in degree zero and is zero in positive degrees. For $y \in H_{-q}F(X^+,S) = H^qX$, we find by a simple direct calculation that $Q^{-s}y = P^sy$ for all $s \geq 0$. A more conceptual proof by direct comparison of McClure's abstract definitions of homology and cohomology operations is also possible; see VIII §3.

§2. Some calculations of the homology operations

For R a commutative ring, let HR be the spectrum representing ordinary cohomology with coefficients in R. We wish to compute the operations on the homology of HZ_p and some related spectra. We shall state our results here, but shall present proofs of the computations for HZ_p in sections 5 and 6. Recall that the mod p homology of HZ_p is A_*, the dual of the Steenrod algebra.

Notations 2.1. We shall adopt the notations of Milnor in our analysis of A_* [86]. Thus, at the prime 2, A_* has algebra generators ξ_i of degree 2^i-1 for $i \geq 1$. At odd primes, A_* has generators ξ_i of degree $2p^i-2$ for $i \geq 1$ and generators τ_i of degree $2p^i-1$ for $i \geq 0$. We shall denote the conjugation in A_* by χ.

We have the following theorems.

Theorem 2.2. For $p = 2$, A_* is generated by ξ_1 as an algebra over the Dyer-Lashof algebra. In fact, for $i > 1$,

$$Q^{2^i-2}\xi_1 = \chi\xi_i \ .$$

Moreover, $Q^s\xi_1$ is nonzero for each $s > 0$ and, for $i > 1$,

$$Q^s\chi\xi_i = \begin{cases} Q^{s+2^i-2}\xi_1 & \text{if } s \equiv 0 \text{ or } -1 \bmod 2^i \\ 0 & \text{otherwise.} \end{cases}$$

In particular, $Q^{2^i}\chi\xi_i = \chi\xi_{i+1}$ for $i > 0$.

Theorem 2.3. For $p > 2$, A_* is generated by τ_0 as an algebra over the Dyer-Lashof algebra. In fact, for $i > 0$

$$Q^{\rho(i)}\tau_0 = (-1)^i\chi\tau_i \quad \text{and}$$

$$\beta Q^{\rho(i)}\tau_0 = (-1)^i\chi\xi_i \ ,$$

where $\rho(i) = (p^i-1)/(p-1)$. Moreover, $\beta Q^s\tau_0$ is nonzero for each $s > 0$ and, for $i > 0$,

$$Q^s\chi\xi_i = \begin{cases} (-1)^i\beta Q^{s+\rho(i)}\tau_0 & \text{if } s \equiv -1 \bmod p^i \\ (-1)^{i+1}\beta Q^{s+\rho(i)}\tau_0 & \text{if } s \equiv 0 \bmod p^i \\ 0 & \text{otherwise,} \end{cases}$$

while

$$Q^s\chi\tau_i = \begin{cases} (-1)^{i+1}Q^{s+\rho(i)}\tau_0 & \text{if } s \equiv 0 \bmod p^i \\ 0 & \text{otherwise.} \end{cases}$$

In particular, $Q^{p^i}\chi\xi_i = \chi\xi_{i+1}$ for $i > 0$ and $Q^{p^i}\chi\tau_i = \chi\tau_{i+1}$ for $i \geq 0$.

Thus, for $p \geq 2$, the operations on the higher degree generators are determined by the operations on the generator of degree one. A complete determination of the operations on this degree one generator does not seem feasible. However, we do have a conceptual determination of these classes. For $p \geq 2$, let ξ be the total ξ class

$$\xi = 1 + \xi_1 + \xi_2 + \cdots$$

For $p > 2$, let τ be the total τ class

$$\tau = 1 + \tau_0 + \tau_1 + \cdots$$

Since the component of these classes in degree zero is one, we may take arbitrary powers of these classes.

Theorem 2.4. For $p = 2$ and $s > 0$,

$$Q^s\xi_1 = (\xi^{-1})_{s+1};$$

that is, $Q^s\xi_1$ is the $(s+1)$-st coordinate of the inverse of the total ξ class. For $p > 2$ and $s > 0$,

$$Q^s\tau_0 = (-1)^s(\xi^{-1}\tau)_{2s(p-1)+1}, \quad \text{and}$$

$$\beta Q^s\tau_0 = (-1)^s(\xi^{-1})_{2s(p-1)},$$

that is, $Q^s\tau_0$ is $(-1)^s$ times the $(2s(p-1)+1)$-st coordinate of the product of the total τ class and the inverse of the total ξ class, and $\beta Q^s\tau_0$ is $(-1)^s$ times the $(2s(p-1))$th coordinate of the inverse of the total ξ class.

Here we are using the H_∞ ring structure on HZ_p derived in I.3.6. In the following corollaries, we consider connective ring spectra E together with morphisms of ring spectra $i:E \to HZ_p$ which induce monomorphisms on mod p homology. When E is an H_∞ ring spectrum, i is an H_∞ ring map by I.3.6.

For $p > 2$, the homology of HZ or $HZ_{(p)}$ embeds as the subalgebra of A_* generated by $\chi\xi_i$ and $\chi\tau_i$ for $i \geq 1$. For $p = 2$, the homology of HZ or $HZ_{(2)}$ embeds as the subalgebra of A_* generated by ξ_1^2 and $\chi\xi_i$ for $i > 1$.

Corollary 2.5. For $p > 2$, the homology of HZ or $HZ_{(p)}$ is generated by $\chi\xi_1$ and $\chi\tau_1$ as an algebra over the Dyer-Lashof algebra. For $p = 2$, the homology of HZ or $HZ_{(2)}$ is generated by ξ_1^2 and $\chi\xi_2$ as an algebra over the Dyer-Lashof algebra.

Similarly, at the prime 2, the homology of kO, the spectrum representing real connective K-theory, embeds as the subalgebra of A_* generated by ξ_1^4, $\chi\xi_2^2$ and $\chi\xi_i$ for $i > 2$. The homology of kU embeds as the subalgebra of A_* generated by ξ_1^2, $\chi\xi_2^2$ and $\chi\xi_i$ for $i > 2$.

Corollary 2.6. At the prime 2, the homology of kO is generated by ξ_1^4, $\chi\xi_2^2$ and $\chi\xi_3$ as an algebra over the Dyer-Lashof algebra, while the homology of kU is generated by ξ_1^2 and $\chi\xi_3$ as an algebra over the Dyer-Lashof algebra.

Proof. By the Cartan formula,

$$Q^4\xi_1^2 = (Q^2\xi_1)^2 = \chi\xi_2^2 .$$

We have analogous results for the p-local Brown-Peterson spectrum BP. Let $i:BP \to HZ_p$ be the unique map of ring spectra. By the Cartan formula, if $p = 2$, or by Theorem 2.4, if $p > 2$, i_* embeds H_*BP as a subalgebra of A_* which is closed under the action of the Dyer-Lashof algebra.

Corollary 2.7. For $p > 2$, H_*BP is generated by $\chi\xi_1$ as an algebra over the Dyer-Lashof algebra. For $p = 2$, H_*BP is generated by ξ_1^2 as an algebra over the Dyer-Lashof algebra.

It is not known whether or not BP is an H_∞ ring spectrum. However, suppose that E is a connective H_∞ ring spectrum and that $f:E \to BP$ has the property that $if:H \to HZ_p$ induces a ring homomorphism on π_0. Then if is an H_∞ ring map, so that $(if)_*$ commutes with the operations. Since i_* is a monomorphism, so does f_*.

We shall also examine the operations on the homology of HZ_{p^n} for $n > 1$. Let B_* be the homology of HZ and let $x \in H_1HZ_{p^n}$ be the element dual to the n-th Bockstein operation on the fundamental cohomology class (so that $\beta_n x = -1$). Then $H_*HZ_{p^n}$ is the truncated polynomial algebra

$$H_*HZ_{p^n} = B_*[x]/(x^2),$$

as an algebra over the dual Steenrod operations. Here the inclusion of B_* in $H_*HZ_{p^n}$ is induced by the natural map $HZ \to HZ_{p^n}$, x maps to zero in the homology of HZ_p, and x is annihilated by the dual Steenrod operations.

Corollary 2.8. For $p > 2$, $H_*HZ_{p^n}$ is generated by x and the elements $\chi\xi_1$ and $\chi\tau_1$ of B_* as an algebra over the Dyer-Lashof algebra. For $p = 2$, $H_*HZ_{p^n}$ is generated by x and the elements ξ_1^2 and $\chi\xi_2$ of B_* as an algebara over the Dyer-Lashof algebra. For $p \geq 2$, the element x is annihilated by all of the operations Q^s.

Proof. For the last assertion, note that $Q^s x$ is an element of B_*x for all s since $Q^s x$ maps to zero in A_*. Since x is annihilated by the dual Steenrod operations, the Nishida relations reduce to

$$P_*^r Q^s x = (-1)^r (r, p^m + s(p-1) - pr) Q^{s-r} x,$$

and

$$P_*^r \beta Q^s x = (-1)^r (r, p^m + s(p-1) - pr - 1) \beta Q^{s-r} x$$

for $p > 2$. Since B_*x is isomorphic to B_* as a module over the dual Steenrod operations, and since no nontrivial element of B_* is annihilated by P_*^r for $r > 0$, and β if $p > 2$, $Q^s x = 0$ by induction.

§3. Homology operations for H_n ring spectra, $n < \infty$

Cohen, [28], by computing the equivariant homology of the space $\mathcal{C}_n(j)$ of j little n-cubes, completed the theory of homology operations for n-fold loop spaces begun by Araki and Kudo, Browder and Dyer and Lashof. Since an H_n ring spectrum (cf. [I,§4]) E is defined by structure maps $\mathcal{C}_n(j) \ltimes_{\Sigma_j} E^{(j)} \to E$, we can use Cohen's calculations to obtain analogous theorems for H_n ring spectra.

Theorem 3.1. For integers s there are operations Q^s in the homology of H_n ring spectra. $Q^s x$ is defined when $2s - \text{degree}(x) < n-1$ [$s - \text{degree}(x) < n-1$] and the operations satisfy properties (1)-(8) of Theorem 1.1 and the analogues of (9) and (10) for $n < \infty$. Moreover, these operations are compatible as n increases.

The Browder operation, λ_{n-1}, is also defined for H_n ring spectra.

Theorem 3.2. There is a natural homomorphism $\lambda_{n-1}: H_qE \otimes H_rE \to H_{q+r+n-1}E$, which satisfies the following properties.

(1) If E is an H_{n+1} ring spectrum, λ_{n-1} is the zero homomorphism,

(2) $\lambda_0(x,y) = xy - (-1)^{qr}yx$,

(3) $\lambda_{n-1}(x,y) = (-1)^{qr+1+(n-1)(q+r+1)}\lambda_{n-1}(y,x)$; $\lambda_{n-1}(x,x) = 0$ if $p = 2$,

(4) $\lambda_{n-1}(1,x) = 0 = \lambda_{n-1}(x,1)$, where $1 \in H_*E$ is the algebraic unit,

(5) The analog of the external and internal Cartan formulas hold:

$$\lambda_{n-1}(x \otimes y, x' \otimes y') = (-1)^{|x'|(|y|+n-1)}\, xx' \otimes \lambda_{n-1}(y,y') + (-1)^{|y|(|x'|+|y'|+n-1)}\lambda_{n-1}(x,x') \otimes yy',$$

where $|z|$ denotes the degree of z,

$$\begin{aligned}\lambda_{n-1}(xy,x'y') &= x\lambda_{n-1}(y,x')y' \\ &+ (-1)^{|y|(n-1+|x'|)}\lambda_{n-1}(x,x')yy' \\ &+ (-1)^{|x'|(n-1+|x|+|y|)}x'x\lambda_{n-1}(y,y') \\ &+ (-1)^{|y|(n-1+|y'|)+|x'||y'|}\lambda_{n-1}(x,y')yx'\end{aligned}$$

(6) The Jacobi identity holds:

$$(-1)^{(q+n-1)(s+n-1)}\lambda_{n-1}(x,\lambda_{n-1}(y,z)) + (-1)^{(r+n-1)(q+n-1)}\lambda_{n-1}(y,\lambda_{n-1}(z,x)) + (-1)^{(s+n-1)(r+n-1)}\lambda_{n-1}(z,\lambda_{n-1}(x,y)) = 0$$

for $x \in H_qE$, $y \in H_rE$, $z \in H_sE$; $\lambda_{n-1}(x,\lambda_{n-1}(x,x)) = 0$ for all x if $p = 3$.

(7) $P^s_*\lambda_{n-1}(x,y) = \sum_{i+j=s} \lambda_{n-1}(P^i_*x \otimes P^j_*y)$,

and

$$\beta\lambda_{n-1}(x,y) = \lambda_{n-1}(\beta x,y) + (-1)^{|x|+n-1}\lambda_{n-1}(x,\beta y)$$

(8) $\lambda_{n-1}(x,Q^s y) = 0$.

There is also a "top" operation, ξ_{n-1}.

Theorem 3.3. There is a function $\xi_{n-1}: H_qE \to H_{q+(n-1+q)(p-1)}E$ $[H_qE \to H_{2q+n-1}]$ defined when $q+n-1$ is even [for all q], which is natural with respect to maps of H_n ring spectra and satisfies the following properties. Here $ad(x)(y) = \lambda_{n-1}(y,x)$, $ad^i(x)(y) = ad(x)(ad^{i-1}(x)(y))$, and $\zeta_{n-1}x$ is defined, for $p > 2$, by the formula $\zeta_{n-1}x = \beta\xi_{n-1}x - ad^{p-1}(x)(\beta x)$.

(1) If E is an H_{n+1} ring spectrum, $\xi_{n-1}x = Q^{(n-1+q)/2}x$ $[\xi_{n-1}x = Q^{n-1+q}x]$, hence $\zeta_n x = \beta Q^{(n-1+q)/2}x$ for $x \in H_qE$.

(2) If we let $Q^{(n-1+q)/2}x$ $[Q^{n-1+q}x]$ denote $\xi_{n-1}x$, then $\xi_{n-1}x$ satisfies formulas (3)-(5) of Theorem 1.1, the external Cartan formula, the Adem relations, and the following analogue of the internal Cartan formula:

$$\xi_{n-1}(xy) = \sum_{i+j=s} Q^i x Q^j y + \sum_{\substack{0<i+j<p \\ 0<i,j}} x^i y^j \Gamma_{ij} \qquad \text{for } n > 1,$$

where $s = \frac{n-1+q}{2}$ $[n-1+q]$, $q = \text{degree}(xy)$, and Γ_{ij} is a function of x and y specified in [28, III.1.3(2)]. In particular, if $p = 2$,

$$\xi_{n-1}(xy) = \sum_{i+j=s} Q^i x Q^j y + x\lambda_{n-1}(x,y)y.$$

Moreover, the Nishida relations for ξ_{n-1} are the usual ones plus an unstable error term given by sums of Pontrjagin products which contain nontrivial iterated Browder operations.

(3) $\lambda_{n-1}(x,\xi_{n-1}y) = ad^p(y)(x)$ and $\lambda_{n-1}(x,\zeta_{n-1}y) = 0$.

(4) $\xi_{n-1}(x + y) = \xi_{n-1}x + \xi_{n-1}y$ + a sum of iterated Browder operations specified in [28, III.1.3(5)].

In the remainder of this section we sketch the proofs of these theorems.

After replacing E by a CW spectrum and replacing $\mathcal{C}_n(j)$ by the geometric realization of its total singular complex, we have that $\mathcal{C}_n(j) \ltimes_\pi E^{(j)}$, is a CW spectrum, for any $\pi \subset \Sigma_j$, with cellular chains naturally isomorphic to

$C_* \mathcal{C}_n(j) \otimes_\pi (C_*E)^j$ (cf. [Equiv., VIII. 2.9]). With field coefficients, $(C_*E)^j$ is equivariantly chain homotopy equivalent to $(H_*E)^j$, so we can apply Cohen's calculations. We define $Q_i x$ to be the image under the structure map of $e_i \otimes x^p$, where $e_i \in H_i\ \mathcal{C}_n(p)/\pi_p$ is Cohen's class, $\pi_p \subset \Sigma_p$ the cyclic group of order p. Define $Q^s x$ and $\xi_{n-1}x$ by the formula in §1. Since $\mathcal{C}_n(2)$ is homotopy equivalent to S^{n-1}, we can define $\lambda_{n-1}(x,y)$ to be the image under the structure map of $(-1)^{(n-1)q+1}\iota \otimes x \otimes y$, where $\iota \in H_{n-1}\ \mathcal{C}_n(2)$ is the fundamental class and $x \in H_qE$.

As noted by Cohen, Theorem 3.1 is a consequence of Theorem 3.3, with 3.3(1) immediate from the definition. With the exception of those statements involving Steenrod operations, all of the statements in Theorems 3.2 and 3.3 follow from equalities between the images under the structure map γ of the operad $\mathcal{C}_n$ of the classes in the equivariant homology of the $\mathcal{C}_n(j)$ which induce the stipulated operations. These equalities follow from Cohen's work. This leaves Theorem 3.2(7), the Nishida relations, and the verification that $\zeta_{n-1}x$ is the image under the structure map of the appropriate multiple of $e_{(n-1)(p-1)} \otimes x^p$, this last giving the definition of $\zeta_{n-1}x$ which Cohen uses in deriving his formulas.

Since the Browder operation is defined nonequivariantly, Theorem 3.2(7) follows from the Cartan formula for Steenrod operations. The Nishida relations follow from the computation of the Steenrod operations in $H_*D_{\pi_p}E$ [Equiv, VIII §3], together with the fact that the kernel of $H_*(\mathcal{C}_n(p) \ltimes_{\pi_p} E) \to H_*D_{\pi_p}E$ consists of classes which are carried to sums of Pontrjagin products of the type stated [28, III §5 and 12.3].

For the last statement, we calculate $\beta(e_{(n-1)(p-1)} \otimes x^p)$. Let ε be a chain in $C_* \mathcal{C}_n(p)$ which projects to a cycle in $C_* \mathcal{C}_n(p)/\pi_p$ representing $e_{(n-1)(p-1)}$ and let a be a chain in the integral cellular chains of E, representing x mod p. Let $da = pb$. Let $N = 1 + \alpha + \cdots + \alpha^{p-1}$ in $Z[\pi_p]$, where α is a generator of π_p. Then

$$d(a^p) = pNba^{p-1},$$

so that

$$d(\varepsilon \otimes a^p) = p\varepsilon N \otimes ba^{p-1} + (d\varepsilon) \otimes a^p.$$

Since ε projects to a cycle mod p in $C_* \mathcal{C}_n(p)/\pi_p$, the transfer homomorphism shows that εN is a cycle mod p in $C_* \mathcal{C}_n(p)$. Thus, $\varepsilon N \otimes ba^{p-1}$ gives rise to a sum of Pontrjagin products of Browder operations in βx and x [28, III. 12.3], which, by the space level calculation, must be the appropriate multiple of $ad^{p-1}(x)(\beta x)$. Since $d\varepsilon$ projects to zero in the mod p chains of $\mathcal{C}_n(p)/\pi_p$, and since a^p is fixed under the action of π_p, we can find a chain δ such that

$$(d\varepsilon) \otimes a^p = \delta N \otimes a^p = \delta \otimes Na^p = p\delta \otimes a^p$$

for all a. By naturality and the space level result, δ must project to a cycle representing $e_{(n-1)(p-1)-1}$ in $H_*(\mathcal{C}_n(p)/\pi_p)$, so that $\delta \otimes a^p$ reduces mod p to a representative of $e_{(n-1)(p-1)} \otimes x^p$.

§4. The Splitting Theorems

We present simple necessary and sufficient conditions for a more general class of spectra than previously mentioned to split as wedges of p-local Eilenberg-MacLane spectra or as wedges of suspensions of BP. The spectra we consider are pseudo H_n ring spectra, defined as in Definition II.6.6, but with $D_j\Sigma^{dq}E_q$ replaced by $\mathcal{C}_n(j) \ltimes_{\Sigma_j} (\Sigma^{dq}E_q)^{(j)}$, with $n \geq 2$.

Fix a pseudo H_n ring spectrum $E = \mathrm{Tel}\, E_q$, and assume that π_*E is of finite type over π_0E and that $\pi_0E = \pi_0E_q$ for q sufficiently large. Let $i: E \to HZ_p$ be such that $ie: S^0 \to HZ_p$ is the unit of HZ_p and regard i as an element of $H^0(E;Z_p)$; under our hypotheses i will be unique. Let $Z_{(p)}$ be the integers localized at p.

Theorem 4.1. If $\pi_0E = Z_p$, then E splits as a wedge of suspensions of HZ_p.

Theorem 4.2. If $\pi_0E = Z_{p^r}$, $r > 1$, or $\pi_0E = Z_{(p)}$ and if $p = 2$ and $Sq^3i \neq 0$ or $p > 2$ and $\beta P^1 i \neq 0$, then E splits as a wedge of suspensions of HZ_{p^s}, $s \geq 1$, and $HZ_{(p)}$.

Theorem 4.3. Let $n \geq 3$. If $\pi_0E = Z_{(p)}$ and $H_*(E;Z_{(p)})$ is torsion free and if $p = 2$ and $Sq^2i \neq 0$ or $p > 2$ and $P^1i \neq 0$, then E splits as a wedge of suspensions of the p-local Brown-Peterson spectrum BP.

Remarks 4.4. The various known splittings of Thom spectra are direct consequences of these theorems. Obviously the splitting of MO and the other Thom spectra of unoriented cobordism theories follow from Theorem 4.1. When $\pi_0MG = Z_{(p)}$, the mod p Thom isomorphism commutes with the Bockstein. At 2, the splittings of MSO and of the Thom spectra into which MSO maps follow from Theorem 4.2 and the facts that Sq^2i is the image of w_2 under the Thom isomorphism and that $Sq^1w_2 = w_3$ in H^*BSO. The BP splittings of MU at all primes and of MSO and MSU at odd primes follow from Theorem 4.3 and similar trivial calculations. Most strikingly perhaps, the splitting of MSF at odd primes follows trivially from Theorem 4.2. Indeed, P^1i is nonzero by consideration of the first Wu class in MSO. Since the p-component of $\pi_q^s = \pi_qSF = \pi_{q+1}BSF$ is Z_p for $q = 2p-3$ and zero for $0 < q < 2p-3$,

$$H_q(BSF;Z_{(p)}) = \begin{cases} Z_p & \text{for } q = 2p-2 \\ 0 & \text{for } 0 < q < 2p-2. \end{cases}$$

Thus, $H_{2p-2}(BSF;Z_p) = Z_p$, and the Bockstein

$$\beta : H_{2p-1}(BSF;Z_p) \to H_{2p-2}(BSF;Z_p)$$

is an epimorphism. Thus, the dual cohomology Bockstein is a monomorphism.

We turn to the proof of the splitting theorems. Define

$$HZ_p[x,x^{-1}] = \bigvee_{q \in Z} \Sigma^{dq} HZ_p ,$$

where $d = 1$ if $p = 2$ and $d = 2$ if $p > 2$. As pointed out in I.4.5 and II.1.3, $HZ_p[x,x^{-1}]$ is an H_∞ ring spectrum. We think of it as the Laurent series spectrum on HZ_p.

Let $A_* \subset H_*(HZ_p[x,x^{-1}])$ be the homology of the zero-th wedge summand HZ_p. Since HZ_p is a sub-H_∞ ring spectrum of $HZ_p[x,x^{-1}]$, we know the operations on A_*. Moreover, if $x \in H_d HZ_p[x,x^{-1}]$ comes from the canonical generator of $H_d \Sigma^d HZ_p$, then the homology of $HZ_p[x,x^{-1}]$ is isomorphic as an algebra over the dual Steenrod operations to $A_*[x,x^{-1}]$, the ring of Laurent polynomials in x over A_*. We could easily calculate the operations on the powers, x^n, of x by use of the techniques of the next section. However, remarkably, we shall only need the p-th power operation on x. We should remark that multiplication by x,

$$H_*\Sigma^{dq}HZ_p \to H_*\Sigma^{d(q+1)}HZ_p,$$

is the homology suspension.

Lemma 4.7. In $A_*[x,x^{-1}]$, for $p \geq 2$, $i > 0$ and q an integer

$$Q^{pq+p^i}(\chi\xi_i \cdot x^{pq}) = \chi\xi_{i+1} \cdot x^{p^2 q} ,$$

hence

$$Q^{p^2q+p^{i+1}}(\chi\xi_i^{\,p} \cdot x^{p^2q}) = \chi\xi_{i+1}^{p} \cdot x^{p^3q} .$$

For $p > 2$, $i \geq 0$ and q an integer,

$$Q^{pq+p^i}(\chi\tau_i \cdot x^{pq}) = \chi\tau_{i+1} \cdot x^{p^2q}.$$

Proof. The internal Cartan formula, together with the degree of $\chi\xi_i$ and of x^{pq} gives

$$Q^{pq+p^i}(\chi\xi_i \cdot x^{pq}) = (Q^{p^i}\chi\xi_i)(Q^{pq}x^{pq}) + Q^{p^i-1}\chi\xi_i)(Q^{pq+1}x^{pq}) .$$

By the Cartan formula, $Q^{pq+1}x^{pq} = 0$. Of course, $Q^{pq}x^{pq} = x^{p^2q}$ (Theorem 1.2.(4)). The first statement follows from Theorem 2.2 or Theorem 2.3 and the fact $A_* \subset A_*[x,x^{-1}]$ is a subalgebra over the Dyer-Lashof algebra. Since $\chi\xi_i^{\,p} \cdot x^{p^2q} =$

$(\chi\xi_i \cdot x^{pq})^p$, the second statement now follows by the Cartan formula. The proof of the third statement is almost identical to the proof of the first.

It should be noted that the full strength of Theorems 2.2 and 2.3 is quite unnecessary for the computations above. They could be derived quite simply and directly. We shall apply these computations to the proofs of the splitting theorems by means of the following commutative diagram, analogous to that of II.6.8.

$$\begin{array}{ccc} \mathcal{C}_n(j) \ltimes_{\Sigma_j} (\Sigma^{dq}E_q)^{(j)} & \xrightarrow{\;1 \ltimes \Sigma^{dq} i_q\;} & \mathcal{C}_n(j) \ltimes_{\Sigma_j} (\Sigma^{dq}HZ_p)^{(j)} \\ \downarrow{\scriptstyle \xi_j} & & \downarrow{\scriptstyle \xi_j} \\ \Sigma^{djq}E_{jq} & \xrightarrow{\;\Sigma^{djq} i_{jq}\;} & \Sigma^{djq}HZ_p \end{array}$$

Here, i_s is the restriction of $i:E \to HZ_p$ to E_s, the right-hand map ξ_j is the induced H_n ring structure of $HZ_p[x,x^{-1}]$ restricted to the (dq)-th wedge summand. The commutativity of the diagram is an easy cohomology calculation provided tht $E_q \to E_s$ induces an isomorphism of π_0 for $s > q$.

The key step in the proofs of Theorems 4.1, 4.2 and 4.3 is the following result.

<u>Proposition 4.8</u>. Let $E = \mathrm{Tel}\, E_q$ satisfy the hypotheses of Theorem 4.1, 4.2 or 4.3. For the first two cases, let $j:E \to H\pi_0E$ be such that $je:S \to H\pi_0E$ is the unit. In the third case, let $j:E \to BP$ be a lift of j above to BP. Then j induces a monomorphism of p-primary cohomology.

<u>Proof</u>. We shall show that j induces an epimorphism of p-primary homology. Recall that i is the projection of j above into HZ_p. In the second case, if $\pi_0E = Z_{p^r}$ for $r > 1$, the nontriviality of the r-th Bockstein operation on i shows that the generator $x \in H_*HZ_{p^r} = B_*[x]/(x^2)$ is in the image of j_*. (Here $B_* = H_*HZ_{(p)}$.) Thus, for the second case as a whole, it suffices to show that $B_* \subset A_*$ is in the image of i_*. Similarly, for the third case, it suffices to show that $H_*BP \subset A_*$ is in the image of i_*. The hypotheses of the theorems give us the following conclusions. In Theorem 4.1, the nontriviality of the Bockstein operation on i_q, for q sufficiently large, shows that τ_0, if $p > 2$, or ξ_1, if $p = 2$, is in the image of i_{q*}. In Theorem 4.2, the nontriviality of P^1i and $\beta P^1 i$, for $p > 2$, or of $Sq^2 i$ and $Sq^3 i$, for $p = 2$, shows that for q sufficiently large, $\chi\xi_1$ and $\chi\tau_1$, for $p > 2$, or ξ_1^2 and $\chi\xi_2$ for $p = 2$, are in the image of i_{q*}. In Theorem 4.3, the nontriviality of $P^1 i$, for $p > 2$, or of $Sq^2 i$, for $p = 2$, shows that for q sufficiently large, $\chi\xi_1$,

for $p > 2$ or ξ_1^2, for $p = 2$, is in the image of i_{q*}. Thus, the following consequences of Lemma 4.7 and the diagram preceding the statement will suffice.

(1) If $p = 2$ or if $p > 2$ and $n \geq 3$ and if $\chi\xi_i$ is in the image of i_{dpq*}, then $\chi\xi_{i+1}$ is in the image of i_{dp^2q*}.

(2) If $p > 2$ and $\chi\tau_i$ is in the image of i_{dpq*}, then $\chi\tau_{i+1}$ is in the image of i_{dp^2q*}.

(3) If $p = 2$, $n \geq 3$, and $\chi\xi_i^2$ is in the imge of i_{4q*}, then $\chi\xi_{i+1}^2$ is in the imge of i_{8q*}.

The conditions on n are just enough to ensure that $H_*(\mathcal{C}_n(p) \ltimes_{\Sigma_p} \Sigma^{dq}E_q)$ contains preimages of the operations needed to carry out the argument.

The passage from the proposition above to the splitting theorems is well known and has been exploited in the literature to prove the splittings of the cobordism theories. Theorems 4.1 and 4.3 follow from the algebraic splitting theorem of Milnor and Moore [87] together with standard properties of HZ_p and BP. For Theorem 4.2, H^*E splits as a direct sum of suspensions of $A/A\beta$ and of A as a module over the Steenrod algebra A. However, the E_2 term of the Bockstein spectral sequence of H^*E is spanned by the A-module generators of the summands isomorphic to $A/A\beta$. By pairing up these generators with respect to their higher order Bocksteins, we may construct a map of E into a wedge of p-local cyclic Eilenberg-MacLane spectra which induces an isomorphism on mod p cohomology. In all cases, the hypothesis on π_0E ensures that E is p-local, and the cohomology isomorphisms yield equivalences.

§5. Proof of Theorem 2.4; Some low-dimensional calculations

We shall exploit the following observation of Liulevicius.

Proposition 5.1. Let $C = Z_2[x,x^{-1}]$ be the algebra over the Steenrod algebra A which is obtained by inverting the polynomial generator of H^*RP^∞. Let C_* be the dual of C, with a generator e_t in degree t. Let $f_t:C_* \to A_*$ be the unique nontrivial morphism of A_* comodules of degree -t (i.e., $f_te_t = 1$). Then f_te_n is the component of the t-th power of the total ξ class in degree n-t:

$$f_te_n = (\xi^t)_{n-t}.$$

Proof. Let $\lambda:C \to C \hat{\otimes} A_*$ be the dual of the module structure of C_* over the dual operations. Recall that for $c \in C$ and $a \in A$, if $\lambda c = \sum c_i \otimes \alpha_i$, then $ac = \sum \langle a,\alpha_i\rangle c_i$. Here $\langle\ ,\ \rangle:A \otimes A_* \to Z_2$ is the Kronecker product. In particular, if $\lambda x^t = \sum x^i \otimes \alpha_i$, then $f_t e_n = \alpha_n$: for $a \in A$,

$$\begin{aligned}\langle a, f_t e_n\rangle &= \langle f_t^* a, e_n\rangle \\ &= \langle a x^t, e_n\rangle \\ &= \langle\langle a,\alpha_n\rangle x^n, e_n\rangle \\ &= \langle a,\alpha_n\rangle,\end{aligned}$$

since $\langle x^n, e_n\rangle = 1$. However, λ is an algebra map, and Milnor has shown that

$$\lambda x = \sum_{i \geq 0} x^{2^i} \otimes \xi_i = \sum_{i \geq 1} x^i \otimes (\xi)_{i-1}.$$

Thus

$$\lambda x^t = \sum_{i \geq t} x^i \otimes (\xi^t)_{i-t}.$$

We also have an odd primary analogue.

Proposition 5.2. For $p > 2$, let C be the A-algebra obtained by inverting the polynomial generator in the cohomology of the lens space L^∞. Thus, C is the tensor product of an exterior algebra on a generator x of degree one and an inverted polynomial algebra on $y = \beta x$. Let C_* be the dual of C and let $e_{2n} \in C_*$ be dual to y^n and let $e_{2n+1} \in C_*$ be dual to xy^n. Let $f_t : C_* \to A_*$ be the A_* comodule map such that $f_t e_t = 1$.

(1) If $t = 2s$, then $f_t e_n$ is $(-1)^n$ times the (n-t)-th component of the s-th power of the total ξ class:

$$f_t e_n = (-1)^n (\xi^s)_{n-t}.$$

(2) If $t = 2s+1$, then $f_t e_n$ is the (n-t)-th component of the product of the total τ class with the s-th power of the total ξ class:

$$f_t e_n = (\xi^s \tau)_{n-t}.$$

Proof. Let $z_i \in C$ be the dual of e_i. Suppose that $\lambda z_t = \sum z_i \otimes \alpha_i$. The sign convention here is that for $a \in A$,

$$a z_t = \sum (-1)^{i(i-t)} \langle a, \alpha_i\rangle z_i .$$

A similar argument to that when $p = 2$ shows that $f_t e_n = (1)^{n(n-t)} \alpha_n$. Here, Milnor's calculations are that

$$\lambda x = x \otimes 1 + \sum_{i \geq 1} y^i \otimes (\tau)_{2i-1} \quad \text{and}$$

$$\lambda y = \sum_{i \geq 1} y^i \otimes (\xi)_{2i-2} .$$

Thus

$$\lambda y^s = \sum_{i \geq s} y^i \otimes (\xi^s)_{2i-2s} \quad \text{and}$$

$$\lambda (xy^s) = \sum_{i \geq 2s+1} z_i \otimes (\xi^s \tau)_{i-2s-1} .$$

In the remainder of this section and in the next, we shall need to evaluate binomial coefficients mod p. The standard technique is the following.

Lemma 5.3. Let $a = \sum a_i p^i$ and $b = \sum b_i p^i$ be the p-adic expansions of a and b. Then $(a,b) \equiv 0 \bmod p$ unless $a_i + b_i < p$ for all i, when

$$(a,b) \equiv \prod_i (a_i, b_i) \bmod p.$$

Moreover, for $a \leq p^n - 1$,

$$(a, p^n - 1 - a) \equiv (-1)^a \bmod p.$$

We shall not bother to quote the first statment, but shall use it implicitly.

The following proposition is the key step in proving Theorem 2.4.

Proposition 5.4. For $p = 2$, the map $f: C_* \to A_*$ given by

$$fe_n = \begin{cases} Q^n \xi_1 & \text{for } n > 0 \\ \xi_1 & \text{for } n = 0 \\ 1 & \text{for } n = -1 \\ 0 & \text{otherwise} \end{cases}$$

is a map of A_* coalgebras. For $p > 2$, the map $f: C_* \to A_*$ given by

$$fe_n = \begin{cases} (-1)^s Q^s \tau_0 & \text{if } n = 2s(p-1) \\ (-1)^s \beta Q^s \tau_0 & \text{if } n = 2s(p-1)-1 \\ -\tau_0 & \text{for } n = 0 \\ 1 & \text{for } n = -1 \\ 0 & \text{otherwise} \end{cases}$$

is a map of A_* coalgebras. Thus, in either case, the map f coincides with the map f_{-1} described above.

Proof. Of course $f:C_* \to A_*$ is a map of A_* comodules if and only if $f^*:A \to C$ is a map of A-modules. But this latter condition is equivalent to the statement that f_* commutes with the action of the dual Steenrod operations $P_*^{p^k}$ for $k \geq 0$ and also commutes with the Bockstein β when $p > 2$

For $p > 2$, $\beta e_{2s} = e_{2s-1}$ and $\beta\tau_0 = -1$. (We have adopted the covention that for $y \in H^qX$ and $x \in H_{q+1}X$, $\langle x, \beta x\rangle = (-1)^{q+1}\langle\beta y, x\rangle$.) Moreover, the subspace of C_* spanned by $e_{2s(p-1)}$ and $e_{2s(p-1)-1}$ for s an integer is a direct summand of C_* as a module over the dual Steenrod operations. We have specified that $f = 0$ on the complementary summand. Thus, for $p \geq 2$, it will suffice to show that the dual Steenrod operations in C_* agree under f with the Nishida relations on the pertinent homology operations on ξ_1 or τ_0.

For symmetry, we shall write y for the polynomial generator of C when $p = 2$. For $p \geq 2$, the computation is divided into three cases. First, those e_i which are carried by $P_*^{p^k}$ to an element of positive degree, second, those which have image in degree zero, and third, those which have image in degree -1.

In the first case, we show that for $p = 2$ and $2^k < s$,

$$P_*^{2^k} e_s = (2^k, s-2^{k+1})e_{s-2^k},$$

and that for $p > 2$ and $p^k < s$,

$$P_*^{p^k} e_{2s(p-1)} = (p^k, s(p-1) - p^{k+1})e_{2(s-p^k)(p-1)}.$$

Let $d = 1$ when $p = 2$ and let $d = 2$ when $p > 2$. Then the statements above reduce to

$$P_*^{p^k} e_{ds(p-1)} = (p^k, s(p-1) - p^{k+1})e_{d(s-p^k)(p-1)}$$

for $p \geq 2$. However, since C was obtained from the cohomology of RP^∞ or L^∞,

$$P^r y = \begin{cases} y & \text{for } r = 0 \\ y^p & \text{for } r = 1 \\ 0 & \text{otherwise} \end{cases}$$

Thus, for $n > 0$, $P^r y^n = (r, n-r)y^{n+r(p-1)}$ by the Cartan formula. Our claim follows from the calculation

$$\langle y^{d(s-p^k)(p-1)}, P_*^{p^k} e_{ds(p-1)}\rangle = \langle P^{p^k} y^{d(s-p^k)(p-1)}, e_{ds(p-1)}\rangle$$

$$= (p^k, s(p-1) - p^{k+1}).$$

For $p > 2$ and $s > p^k$, we have similarly that

$$P_*^{p^k} e_{2s(p-1)-1} = (p^k, s(p-1) - p^{k+1} - 1) e_{2(s-p^k)(p-1) - 1}.$$

Here, $P^r x = 0$ for $r > 0$, so that

$$\langle xy^{s(p-1)-p^k(p-1)-1}, P_*^{p^k} e_{2s(p-1)-1}\rangle = (p^k, s(p-1) - p^{k+1} - 1).$$

On the other hand, the Nishida relations give us, for $s > p^k$,

$$P_*^{2^k} Q^s \xi_1 = (2^k, 2^m + s - 2^{k+1}) Q^{s-2^k} \xi_1$$

for $p = 2$, and, for $p > 2$,

$$P_*^{p^k} Q^s \tau_0 = -(p^k, p^m + s(p-1) - p^{k+1}) Q^{s-p^k} \tau_0 ,$$

and

$$P_*^{p^k} \beta Q^s \tau_0 = -(p^k, p^m + s(p - 1) - p^{k+1} - 1) \beta Q^{s-p^k} \tau_0 .$$

Here, the initial -1 is cancelled by the conventions in the definition of f, and the additional high power of p in the right-hand side does not alter the binomial coefficients unless the right-hand side would otherwise be negative. Thus, we must check that for $s > p^k$, if $s(p-1) < p^{k+1}$, then $(p^k, p^m + s(p-1) - p^{k+1})$ and $(p^k, p^m + s(p-1) - p^{k+1} - 1)$ are zero. Since $s(p-1) \leq p^{k+1} - 1$, we have $s \leq \rho(k+1) = 1 + p + \cdots + p^k$. But since $p^k < s$, we have $s = p^k + t$ with $0 < t \leq \rho(k)$. Thus, $s(p-1) = p^k(p-1) + t_1$, with $0 < t_1 < p^k$. Thus, the specified coefficients are zero.

It remains to check those operations $P_*^{p^k}$ whose images have degree 0 or -1 in C_*. However, e_0 may not be in the image of any $P_*^{p^k}$, as $P^r 1 = 0$ for $r > 0$. $P_*^r Q^r \xi_1$ and $P_*^r Q^r \tau_0$ are zero by the Nishida relations. (Q_0 kills ξ_1 or τ_1.) For the remaining case, we shall show that for $p = 2$,

$$P_*^{2^k} e_{2^k-1} = e_{-1} ,$$

and for $p > 2$,

$$P_*^{p^k} e_{2p^k(p-1)-1} = -e_{-1}.$$

To do this, we must compute the Steenrod operations on y^{-1} when $p = 2$ and on xy^{-1}

when $p > 2$. For $p \geq 2$ and $r > 0$,

$$0 = P^r(yy^{-1}) = (P^0 y)(P^r y^{-1}) + (P^1 y)(P^{r-1} y^{-1})$$

$$= yP^r y^{-1} + y^p P^{r-1} y^{-1}$$

by the Cartan formula. Thus, $P^r y^{-1} = -y^{p-1} P^{r-1} y^{-1}$, so that

$$P^r y^{-1} = (-1)^r y^{r(p-1)-1} ,$$

by induction. For $p > 2$, since $P^r x = 0$ for $r > 0$,

$$P^r(xy^{-1}) = (-1)^r x y^{r(p-1)-1}.$$

Thus, for $p = 2$,

$$\langle y^{-1}, P_*^{2^k} e_{2^k-1} \rangle = \langle y^{2^k-1}, e_{2^k-1} \rangle = 1$$

and for $p > 2$

$$\langle xy^{-1}, P_*^{p^k} e_{2p^k(p-1)-1} \rangle = (-1)^{p^k} \langle xy^{p^k(p-1)-1}, e_{2p^k(p-1)-1} \rangle = -1.$$

The following lemma will complete the proof.

Lemma 5.5. For $p = 2$,

$$P_*^{s+1} Q^s \xi_1 = 1.$$

For $p > 2$,

$$P_*^s \beta Q^s \tau_0 = (-1)^{s-1}.$$

Proof. For $p = 2$, the Nishida relations reduce to

$$P_*^{s+1} Q^s \xi_1 = (s-1, 2^n - s) Q^0 P_*^1 \xi_1 = 1,$$

by Lemma 4.3. For $p > 2$, the Nishida relations reduce to

$$P_*^s \beta Q^s \tau_0 = -(s-1, p^n - s) Q^0 P_*^0 \beta \tau_0$$

$$= (-1)^{s-1}$$

by Lemma 4.3, since $\beta\tau_0 = -1$.

<u>Proof of Theorem 2.4</u>. For $p = 2$ and $s > 0$, the fact that

$$Q^s\xi_1 = (\xi^{-1})_{s+1}$$

follows immediately from Propositions 5.1 and 5.4. For $p > 2$ and $s > 0$, the fact that

$$Q^s\tau_0 = (-1)^s(\xi^{-1}\tau)_{2s(p-1)+1} \quad \text{and}$$
$$\beta Q^s\tau_0 = (-1)^s(\xi^{-1}\tau)_{2s(p-1)}$$

follows immediately from Proposition 5.2 and 5.4. However, all of the even degree coordinates of $\xi^{-1}\tau$ come from ξ^{-1}. Thus,

$$\beta Q^s\tau_0 = (-1)^s(\xi^{-1})_{2s(p-1)} \cdot$$

One can identify certain algorithms such as the following curiosity when $p = 2$:

$$Q^{2^i}\xi_1 = \sum_{j=1}^{2^{i-1}-1} (Q^j\xi_1)(Q^{2^i-j-1}\xi_1)$$

Thus, the actual computations can get quite ugly. We have the following low-dimensional computations of $Q^s\xi_1$ for $p = 2$. In the next section we shall show that $Q^{2t-1}\xi_1 = (Q^{t-1}\xi_1)^2$. Thus, we shall only list $Q^{2t}\xi_1$. We shall write $\chi\xi_i = \beta_i$ for $i \geq 1$.

$Q^{2t}\xi_1$ for $0 < t \leq 15$, where $p = 2$:

s	$Q^s\xi_1$
2	β_2
4	$\beta_1^2\beta_2$
6	β_3
8	$\beta_1^6\beta_2 + \beta_1^2\beta_3 + \beta_2^3$
10	$\beta_1^4\beta_3$
12	$\beta_2^2\beta_3$
14	β_4
16	$\beta_1^2\beta_4 + \beta_2\beta_3^2 + \beta_1^4\ \beta_2^2\beta_3 + \beta_1^2\beta_2^5 + \beta_1^8(\beta_1^6\beta_2 + \beta_1^2\beta_3 + \beta_2^3)$
18	$\beta_1^4\beta_4 + \beta_1^{12}\beta_3 + \beta_2^4\beta_3$
20	$\beta_1^8\beta_2^2\beta_3 + \beta_2^2\beta_4 + \beta_3^3$
22	$\beta_1^8\beta_4$
24	$\beta_1^4\ \beta_2^2\ \beta_4 + \beta_1^4\ \beta_3^3 + \beta_2^6\ \beta_3$
26	$\beta_2^4\beta_4$
28	$\beta_3^2\beta_4$
30	β_5

§6. Proofs of Theorems 2.2 and 2.3

We shall compute the operations on $H_*HZ_p = A_*$. The elements of A_* are completely determined by the effect of the dual Steenrod operations $P_*^{p^k}$ for $k \geq 0$, along with the Bockstein operation if $p > 2$. Thus, our computations will be based on induction arguments using the Nishida relations.

Theorems 2.2 is the composite of Lemma 5.5 and Propsitions 6.4 and 6.7. Theorem 2.3 is the composite of Lemma 5.5, Propositions 6.4, 6.7 and 6.9, and Corollary 6.5.

We begin by recalling some basic facts about the dual Steenrod operations in A_*.

Lemma 6.1. The following equalities hold in A_*. For $p \geq 2$ and $i > 0$,

$$P_*^r \chi\xi_1 = \begin{cases} -\chi\xi_{i-k}^{p^k} & \text{if } r = \rho(k) \\ 0 & \text{otherwise} \end{cases}$$

(Recall that $\rho(k) = \frac{p^k - 1}{p - 1}$.) For $p > 2$ and $i \geq 0$,

$$P_*^r \chi\tau_i = 0 \quad \text{for } r > 0,$$

and

$$\beta\chi\tau_i = \chi\xi_i \,.$$

Here, ξ_0 is identified with the unit, 1, of A_*.

Remarks 6.2. Notice that the added high power of p in the right-hand side of the binomial coefficients in the Nishida relations allows us to make the following simplification. For $p \geq 2$,

$$P_*^{p^k} Q^s = \sum_i (-1)^{i+1}(p^k - pi, s(p-1) - pi) Q^{s-p^k+i} P_*^i \,.$$

For $p > 2$,

$$P_*^{p^k} \beta Q^s = \sum_i (-1)^{i+1}(p^k - pi, s(p-1) + pi - 1)\beta Q^{s-p^k+i} P_*^i$$
$$+ \sum_i (-1)^{i+1}(p^k - pi - 1, s(p-1) + pi) Q^{s-p^k+i} P_*^i \beta \,.$$

One of the key observations in our calculations is the following.

Lemma 6.3. (The p-th power lemma). For $p = 2$ and $s > 1$,

$$Q^{2s-1}\xi_1 = (Q^{s-1}\xi_1)^2.$$

For $p > 2$ and $s > 0$,

$$\beta Q^{ps}\tau_0 = (\beta Q^s\tau_0)^p.$$

Proof. We argue by induction on s. We shall show that both sides of the proposed equalities agree under $P_*^{p^k}$ for $k \geq 0$ and under β when $p > 2$. Of course, β is no problem, and both sides of both equations vanish under P_*^1. For the right hand side, this follows from the Cartan formula. For the left-hand side, the Nishida relations give

$$P_*^1 Q^s = (s-1)Q^{s-1}\text{, and for } p > 2$$

$$P_*^1\beta Q^s = s\beta Q^{s-1} - Q^{s-1}\beta\ .$$

Thus, we may restrict attention to $P_*^{p^k}$ for $k > 0$. If $s = p^{k-1}$, Lemma 5.5 and the Cartan formula show that both sides of the equations are carried to 1 by $P_*^{p^k}$. Thus, the lemma is true for $p = 2$ and $s = 2$, and for $p > 2$ and $s = 1$. In the remaining cases, $k > 0$ and $s > p^{k-1}$. Here for $p = 2$,

$$P_*^{2^k} Q^{2s-1}\xi_1 = (2^k, 2s-1)Q^{2s-2^k-1}\xi_1\ ,$$

while

$$\begin{aligned} P_*^{2^k}(Q^{s-1}\xi_1)^2 &= (P_*^{2^{k-1}} Q^{s-1}\xi_1)^2 \\ &= (2^{k-1}, s-1)(Q^{s-2^{k-1}-1}\xi_1)^2 \\ &= (2^{k-1}, s-1)Q^{2s-2^k-1}\xi_1\ , \end{aligned}$$

by the Cartan formula, the Nishida relations and induction. For $p > 2$,

$$\begin{aligned} P_*^{p^k}(\beta Q^s\tau_0)^p &= (P_*^{p^{k-1}}\beta Q^s\tau_0)^p \\ &= -(p^{k-1}, s(p-1) - 1)(\beta Q^{s-p^{k-1}}\tau_0)^p \\ &= -(p^{k-1}, s(p-1) - 1)\beta Q^{ps-p^k}\tau_0\ , \end{aligned}$$

by the Cartan formula, the Nishida relations and induction. The conclusion follows easily from Lemma 5.3.

We can now evaluate certain of the operations.

Proposition 6.4. For $p = 2$ and $i > 1$,

$$Q^{2^i-2}\xi_1 = \chi\xi_i .$$

For $p > 2$ and $i > 0$,

$$\beta Q^{\rho(i)}\tau_0 = (-1)^i\chi\xi_i .$$

(Again $\rho(i) = \frac{p^i-1}{p-1}$.)

Proof. We argue by induction on i. Again it will be sufficient to show that both sides of the equations agree under $P_*^{p^k}$ for $k \geq 0$. For $p = 2$,

$$P_*^{2^k}Q^{2^i-2}\xi_1 = (2^k, 2^i-2)Q^{2^i-2-2^k}\xi_1 .$$

For $0 < k < i$, the binomial coefficient is zero, while for $k \geq i$, $Q^{2^i-2-2^k}\xi_1 = 0$ for dimensional reasons. Thus, the only nontrivial operation is

$$P_*^1 Q^{2^i-2}\xi_1 = Q^{2^i-3}\xi_1 .$$

For $i = 2$, $Q^{2^i-3}\xi_1 = Q^1\xi_1 = \xi_1^2$. Since $\xi_1 = \chi\xi_1$, the proposition is true for $i = 2$ by Lemma 6.1. For $i > 2$,

$$Q^{2^i-3}\xi_1 = (Q^{2^{i-1}-2}\xi_1)^2$$

$$= (\chi\xi_{i-1})^2 ,$$

by the p-th power lemma and induction. Lemma 6.1 is again sufficient. For $p > 2$, let $i = 1$. Then

$$P_*^1\beta Q^{\rho(1)}\tau_0 = P_*^1\beta Q^1\tau_0 = 1$$

by Lemma 5.5. Thus, $\beta Q^1\tau_0 = -\chi\xi_1$. For $i > 1$,

$$P_*^{p^k}\beta Q^{\rho(i)}\tau_0 = -(p^k, \rho(i)(p-1) - 1)\beta Q^{\rho(i)-p^k}\tau_0$$

$$= -(p^k, p^i-2)\beta Q^{\rho(i)-p^k}\tau_0,$$

by the p-th power lemma and induction. The result follows from Lemma 6.1.

Corollary 6.5. For $p > 2$ and $i > 0$,

$$Q^{\rho(i)}\tau_0 = (-1)^i \chi\tau_1 .$$

Proof. We have just shown that $Q^{\rho(i)}\tau_0$ and $(-1)^i\chi\tau_i$ have the same Bockstein. However,

$$\begin{aligned} P_*^{p^k} Q^{\rho(i)}\tau_0 &= -(p^k,\rho(i)(p-1))Q^{\rho(i)-p^k}\tau_0 \\ &= -(p^k,p^i-1)Q^{\rho(i)-p^k}\tau_0 . \end{aligned}$$

For $k < i$, $(p^k,p^i-1) = 0$, while for $k \geq i$, $Q^{\rho(i)-p^k}\tau_0 = 0$ for dimensional reasons. The result follows from Lemma 6.1.

We wish now to compute the operations on the higher degree generators. By the Nishida relations and Lemma 6.1,

$$\begin{aligned} P_*^{p^k} Q^s \chi\xi_i &= -(p^k,s(p-1))Q^{s-p^k}\chi\xi_i \\ &\quad + \sum_{j\geq 1} (-1)^{j+1}(p^k - p\rho(j),s(p-1) + p\rho(j))\cdot Q^{s-p^k+\rho(j)}(-\chi\xi_{i-j}^{p^j}) , \end{aligned}$$

and for $p > 2$,

$$\begin{aligned} P_*^{p^k} \beta Q^s\chi\tau_i &= -(p^k,s(p-1) - 1)\beta Q^{s-p^k}\chi\tau_i - (p^k-1,s(p-1))Q^{s-p^k}\chi\xi_i \\ &\quad + \sum_{j\geq 1} (-1)^{j+1}(p^k - p\rho(j) - 1,s(p-1) + p\rho(j))Q^{s-p^k+\rho(j)}(-\chi\xi_{i-j}^{p^j}). \end{aligned}$$

However, we may simplify this expression considerably.

Lemma 6.6. For $p \geq 2$ and $i > 0$,

$$P_*^{p^k} Q^s\chi\xi_i = -(p^k,s(p-1))Q^{s-p^k}\chi\xi_i - (p^k - p,s(p-1) + p)Q^{s-p^k+1}\chi\xi_{i-1}^p .$$

For $p > 2$ and $i \geq 0$,

$$P_*^{p^k} \beta Q^s\chi\tau_i = -(p^k,s(p-1) - 1)\beta Q^{s-p^k}\chi\tau_i - (p^k-1,s(p-1))Q^{s-p^k}\chi\xi_i .$$

Moreover, the following additional simplifications hold for particular values of s. For $p > 2$, $s \not\equiv 0 \bmod p$ and $k > 0$,

$$P_*^{p^k} \beta Q^s\chi\tau_i = -(p^k,s(p-1) - 1)\beta Q^{s-p^k}\chi\tau_i .$$

For $p \geq 2$, $s \not\equiv -1 \bmod p^2$ and $k > 1$,

$$P_*^{p^k} Q^s \chi\xi_i = -(p^k, s(p-1)) Q^{s-p^k} \chi\xi_i .$$

Proof. The assertion is true for $k = 0$ or $k = 1$ because of the left-hand term of the binomial coefficients. We shall assume $k > 1$. If $s \not\equiv -1 \bmod p$ and $j > 0$, then $s - p^k + \rho(j) \not\equiv -1 \bmod p$. By the Cartan formula (or Theorem 1.2(5) if $i = j$), $Q^{s-p^k+\rho(j)} \chi\xi_{i-j}^{p^k} = 0$. If $s \not\equiv -1 \bmod p$, $p > 2$, $k > 0$ and $j \geq 0$, $p^k - p\rho(j) - 1 \equiv -1 \bmod p$, while $s(p-1) + p\rho(j) \not\equiv 0 \bmod p$. Thus,

$$(p^k - p\rho(j) - 1,\ s(p-1) + p\rho(j)) = 0.$$

For $s \equiv -1 \bmod p$, but $s \not\equiv -1 \bmod p^2$ (here $p \geq 2$), $s \equiv tp-1 \bmod p^2$ for $0 < t < p$. Thus

$$s(p-1) + p\rho(j) \equiv (p-t)p+1 \bmod p^2,$$

while

$$p^k - p\rho(j) \equiv (p-1)p \bmod p^2.$$

Thus,

$$(p^k - p\rho(j), s(p-1) + p\rho(j)) = 0.$$

It suffices to assume $s \equiv -1 \bmod p^2$. Here, for $j > 1$ (and $k > 1$),

$$s - p^k + \rho(j) \equiv p \bmod p^2 .$$

By the Cartan formula (or Theorem 1.2(5) if $i = j$),

$$Q^{s-p^k+\rho(j)} \chi\xi_{i-j}^{p^j} = 0.$$

Proposition 6.7. For $p = 2$, $i > 0$ and $s > 0$,

$$Q^s \chi\xi_i = \begin{cases} Q^{s+2^i-2}\xi_1 & \text{if } s \equiv 0 \text{ or } -1 \bmod 2^i \\ 0 & \text{otherwise} \end{cases} .$$

For $p > 2$, $i > 0$ and $s > 0$,

$$Q^s \chi\xi_i = \begin{cases} (-1)^i \beta Q^{s+\rho(i)} \tau_0 & \text{if } s \equiv -1 \bmod p^i \\ (-1)^{i+1} \beta Q^{s+\rho(i)} \tau_0 & \text{if } j \equiv 0 \bmod p^i \\ 0 & \text{otherwise} \end{cases} .$$

Proof. We argue by induction on s and i. For p = 2, the assertion is trivial for i = 1. For $p \geq 2$, and $0 < s \leq p^i-1$ the assertion holds by dimensional reasons and the p-th power lemma. Of course, we shall show that both sides of the equations agree under $P_*^{p^k}$ for $k \geq 0$ and under β when p > 2. Clearly both sides agree under P_*^1, and when p > 2, Lemma 6.1 implies that $\beta Q^s\chi\xi_i = 0$ for all i and s by induction and the Nishida relations. Thus, it suffices to check $P_*^{p^k}$ for k > 0.

Case 1. $s \equiv 0 \bmod p$, but $s \not\equiv 0 \bmod p^i$.

By the preceding lemma,

$$P_*^{p^k} Q^s\chi\xi_i = -(p^k, s(p-1))Q^{s-p^k}\chi\xi_i .$$

By induction $Q^{s-p^k}\chi\xi_i = 0$ unless $s - p^k \equiv 0 \bmod p^i$. Since $s \not\equiv 0 \bmod p^i$, this means $k < i$ and $s \equiv p^k \bmod p^i$. Here $(p^k,s(p-1)) = (p^k,p^k(p-1)) = 0$. Thus $Q^s\chi\xi_i = 0$.

Case 2. $s \equiv 0 \bmod p^i$.

Again

$$P_*^{p^k} Q^s\chi\xi_i = -(p^k,s(p-1))Q^{s-p^k}\chi\xi_i$$

$$= \begin{cases} 0 & \text{if } k < i \text{ or } p^k \geq s \\ (-1)^i(p^k,s(p-1))\beta Q^{s-p+\rho(i)}\tau_0 & \text{if } s > p^k \geq p^i,\ p > 2 \\ (2^k,s)Q^{s+2^i-2-2^k}\xi_1 & \text{if } s > 2^k \geq 2^i,\ p = 2 \end{cases}$$

by induction. On the other hand,

$$P_*^{p^k}\beta Q^{s+\rho(i)}\tau_0 = -(p^k,s(p-1) + p^i - 2)\beta Q^{s+\rho(i)-p^k}\tau_0 \quad \text{if } p > 2,$$

and

$$P_*^{2^k} Q^{s+2^i-2}\xi_1 = (2^k,s+2^i - 2)Q^{s+2^i-2-2^k}\xi_1 \quad \text{if } p = 2.$$

Since $s \equiv 0 \bmod p^i$,

$$(p^k,s(p-1) + p^i - 2) = \begin{cases} 0 & \text{for } 1 \leq k < i \\ (p^k,s(p-1)) & \text{for } k \geq i \end{cases}$$

It suffices to show that $P_*^{p^k}\beta Q^{s+\rho(i)}\tau_0 = 0$ for $s \leq p^k < s + \rho(i)$, when p > 2, and that $P_*^{2^k} Q^{s+2^i-2}\xi_1 = 0$ for $s \leq 2^k < s+2^i-2$. These inequalities imply that $s = p^k$, so that $(p^k,s(p-1)) = 0$.

Case 3. $s \not\equiv 0$ or $-1 \bmod p$.

Again,

$$P_*^{p^k} Q^s \chi\xi_i = -(p^k, s(p-1)) Q^{s-p^k} \chi\xi_i = 0$$

by induction.

Case 4. $s \equiv -1 \bmod p^i$

Here,

$$P_*^{p^k} Q^s \chi\xi_i = -(p^k, s(p-1)) Q^{s-p^k} \chi\xi_i - (p^k - p, s(p-1) + p)(Q^{((s+1)/p)-p^{k-1}} \chi\xi_{i-1})^p$$

by Lemma 6.6 and the Cartan formula.

For $1 \leq k < i$, $Q^{s-p^k}\chi\xi_i = 0$ by induction. Since $\frac{s+1}{p} - p^{k-1} \equiv -p^{k-1} \bmod p^{i-1}$, $Q^{((s+1)/p)-p^{k-1}} \chi\xi_{i-1} = 0$ for $1 < k < i$. For $k = 1 < i$,

$$P_*^p Q^s \chi\xi_i = \begin{cases} (-1)^i (\beta Q^{((s+1)/p)-1+\rho(i-1)} \tau_0)^p = (-1)^i \beta Q^{s-p+\rho(i)} \tau_0 & \text{for } p > 2 \\ (Q^{((s+1)/2)-1+2^{i-1}-2} \xi_1)^2 = Q^{s+2^i-4} \xi_1 & \text{for } p = 2 \end{cases}$$

by induction and the p-th power lemma. On the other hand, for $p^k < s + \rho(i)$ and $p > 2$,

$$P_*^{p^k} \beta Q^{s+\rho(i)} \tau_0 = -(p^k, s(p-1) + p^i - 2) \beta Q^{s+\rho(i)-p^k} \tau_0$$

and for $p = 2$ and $2^k < s + 2^i - 2$,

$$P_*^{2^k} Q^{s+2^i-2} \xi_1 = (2^k, s+2^i-2) Q^{s+2^i-2-2^k} \xi_1 .$$

Since $s \equiv -1 \bmod p^i$, the right-hand side of the binomial coefficient is congruent to $p^i - p - 1 \bmod p^i$. Thus, if $1 < k < i$, the coefficient is zero and if $k = 1$, the coefficient is -1.

For $s > p^k \geq p^i$ and $i > 1$,

$$P_*^{p^k} Q^s \chi\xi_i = \begin{cases} -[(p^k, s(p-1)) + (p^k - p, s(p-1)+p](-1)^i \beta Q^{s+\rho(i)-p^k} \tau_0 & \text{for } p > 2 \\ [(2^k, s) + (2^k - 2, s+2)] Q^{s-2^k+2^i-2} \xi_1 & \text{for } p = 2, \end{cases}$$

by induction and the p-th power lemma. Thus, for these values of k, it suffices to check that

$$(p^k,s(p-1)) + (p^k - p,s(p-1) + p) = (p^k,s(p-1) + p^i - 2),$$

which the reader may verify (or c.f. [101, p.54]).

For $p > 2$, $i = 1$ and $s > p^k$,

$$P_*^{p^k} Q^s \chi\xi_1 = -(p^k,s(p-1))(-\beta Q^{s+1-p^k}\tau_0)$$

by induction, while

$$P_*^{p^k} \beta Q^{s+1}\tau_0 = -(p^k,(s+1)(p-1))\beta Q^{s+1-p^k}\tau_0 ,$$

and the binomial coefficients here are equal.

For $s < p^k \leq s+\rho(i)$, when $p > 2$, or for $s < 2^k \leq s + 2^i-2$, when $p = 2$, a simple calculation shows that $s = p^k-1$. Here

$$P_*^{p^k} \beta Q^{p^k-1+\rho(i)}\tau_0 = -(p^k,p^k(p-1) + p(p^{i-1}-1))\beta Q^{\rho(i)-1}\tau_0 \qquad \text{for } p > 2$$

$$P_*^{2^k} Q^{2^k-1+2^i-2}\xi_1 = (2^k,2^k+2^i-3)Q^{2^i-3}\xi_1 \qquad \text{for } p = 2$$

Since $k \geq i > 1$, the binomial coefficient is zero.

<u>Case 5</u>. $s \equiv -1 \bmod p$, but $s \not\equiv -1 \bmod p^2$, $i > 1$ and $k > 1$.

Here,

$$P_*^{p^k} Q^s \chi\xi_i = -(p^k,s(p-1))Q^{s-p^k}\chi\xi_i$$

by Lemma 6.6. But $s-p^k \not\equiv -1 \bmod p^2$, so that $Q^{s-p^k}\chi\xi_i = 0$.

<u>Case 6</u>. $s \equiv -1 \bmod p^2$, but $s \not\equiv -1 \bmod p^i$; or $s \equiv -1 \bmod p$ but $s \not\equiv -1 \bmod p^2$, $k = 1$ and $i > 1$.

Here,

$$P_*^{p^k} Q^s \chi\xi_i = -(p^k,s(p-1))Q^{s-p^k}\chi\xi_i - (p^k-p,s(p-1)+p)(Q^{((s+1)/p)-p^{k-1}}\chi\xi_{i-1})^p.$$

Now $s - p^k \equiv -1 \bmod p^i$ if and only if $\frac{s+1}{p} - p^{k-1} \equiv 0 \bmod p^{i-1}$. Since $\frac{s+1}{p} \not\equiv 0 \bmod p^{i-1}$ either $Q^{s-p^k}\chi\xi_i$ and $(Q^{((s+1)/p)-p^{k-1}}\chi\xi_{i-1})^p$ are both zero or they are both equal to the appropriate operation on τ_0 if $p > 2$ or ξ_1 if $p = 2$. In the latter case, the coefficients cancel as $k < i$ and $s \equiv p^k-1 \bmod p^i$.

Lemma 6.8. For $p > 2$, $i \geq 0$ and $s > 0$,

$$\beta Q^s \chi\tau_i = \begin{cases} Q^s\chi\xi_i = (-1)^{i+1}\beta Q^{s+\rho(i)}\tau_0 & \text{if } s \equiv 0 \bmod p^i \\ 0 & \text{otherwise .} \end{cases}$$

Proof. We argue by induction on s and i. The lemma is trivial for $i = 1$ or for $0 < s < p^i$. Again, both sides agree under β and P^1_*. We shall show that both sides agree under $P^{p^k}_*$ for $k > 0$.

Case 1: $s \equiv 0 \bmod p$.

Here $\beta Q^{s-p^k}\chi\tau_i = Q^{s-p^k}\chi\xi_i$ by induction. By Lemma 6.6,

$$\begin{aligned} P^{p^k}_* \beta Q^s\chi\tau_i &= -[(p^k,s(p-1) - 1) + (p^k - 1,s(p-1))]Q^{s-p^k}\chi\xi_i \\ &= -(p^k,s(p-1))Q^{s-p^k}\chi\xi_i \\ &= P^{p^k}_* Q^s\chi\xi_i \ . \end{aligned}$$

Therefore, $\beta Q^s\chi\tau_i = Q^s\chi\xi_i$.

Case 2. $s \not\equiv 0 \bmod p$.

Here, by Lemma 6.6,

$$P^{p^k}_* \beta Q^s\chi\tau_i = -(p^k,s(p-1) - 1)\beta Q^{s-p^k}\chi\tau_i \ ,$$

but $\beta Q^{s-p^k}\chi\tau_i = 0$ by induction.

Proposition 6.9. For $p > 2$, $s > 0$ and $i \geq 0$,

$$Q^s\chi\tau_i = \begin{cases} (-1)^{i+1}Q^{s+\rho(i)}\tau_0 & \text{if } s \equiv 0 \bmod p^i \\ 0 & \text{otherwise.} \end{cases}$$

Proof. We have shown that both sides of the prospective equation agree under the Bockstein. By Lemma 6.1,

$$P^{p^k}_* Q^s\chi\tau_i = -(p^k,s(p-1))Q^{s-p^k}\chi\tau_i \ .$$

For fixed i, we argue by induction on s that $P_*^{p^k}$ agree on both sides of the prospective equation. Again the assertion is a triviality for $i = 0$, for $k = 0$, or for $0 < s < p^i$.

Case 1: $s \not\equiv 0 \mod p$.

Here, $Q^{s-p^k}\chi\tau_i = 0$ by induction.

Case 2: $s \equiv 0 \mod p$ but $s \not\equiv 0 \mod p^i$.

By induction, $Q^{s-p^k}\tau_i = 0$ unless $k < i$ and $s \equiv p^k \mod p^i$. Here

$$(p^k, s(p-1)) = (p^k, p^k(p-1)) = 0.$$

Case 3: $s \equiv 0 \mod p^i$.

Here $Q^{s-p^k}\tau_i = 0$ by induction for $k < i$. Again by induction,

$$P_*^{p^k} Q^s \chi\tau_i = -(p^k, s(p-1))(-1)^{i+1} Q^{s-p^k+\rho(i)}\tau_0 ,$$

for $i \leq k < s$. We have

$$P_*^{p^k} Q^{s+\rho(i)}\tau_0 = -(p^k, s(p-1) + p^i - 1) Q^{s-p^k+\rho(i)}\tau_0 .$$

Since $s \equiv 0 \mod p^i$,

$$(p^k, s(p-1) + p^i - 1) = \begin{cases} 0 & \text{for } 0 < k < i \\ (p^k, s(p-1)) & \text{for } k \geq i. \end{cases}$$

For $s \leq p^k < s+\rho(i)$, $s = p^k$ and

$$P_*^{p^k} Q^{s+\rho(i)}\tau_0 = -(p^k, p^k(p-1)) Q^{\rho(i)}\tau_0$$

$$= 0 .$$

CHAPTER IV

THE HOMOTOPY THEORY OF H_∞ RING SPECTRA

by Robert R. Bruner

Around 1960, Liulevicius [55] and Novikov [91] introduced Steenrod operations into the cohomology of cocommutative Hopf algebras, in particular the E_2 term of the Adams spectral sequence converging to the p-component of π_*S^0. During the 1960's, Barratt and Mahowald (unpublished) studied the quadratic construction, using it to construct homotopy operations and to derive relations in homotopy. Toda [106] studied the mod p analog, the extended p^{th} power construction, and used it to derive relations in the odd primary components of π_*S^0. Early in the study of the quadratic construction, it was conjectured that the quadratic construction could be used to provide maps representing Steenrod operations. This was proved by D. S. Kahn [45]. He also showed that this determined some differentials in the Adams spectral sequence and related the homotopy operations to Steenrod operations. Milgram [81] reformulated Kahn's work in a form which generalizes to the mod p case, this formulation being exactly analogous to the reformulation necessary to define mod p Steenrod operations. He also showed how to derive many more differentials from the geometric construction of the Steenrod operations in the Adams spectral sequence. In particular, he showed that the Hopf invariant one differentials follow in this way. Milgram's work was confined to a range in which it is possible to act as if one were operating on a permanent cycle. At about the same time, Mäkinen [62], working at the prime 2, showed how to account for the fact that one may not be operating on a permanent cycle.

In order to construct the Steenrod operations geometrically, a map from an extended power of a sphere to the pth power of that sphere is needed. Kahn, Milgram and Mäkinen obtained such maps by using coreductions of the extended powers of spheres. As usual when studying stable phenomena on the space level, such coreductions exist only in a range of dimensions, but, by suspending everything an appropriate number of times, that range can be made arbitrarily large. This makes it appear that we should be working with spectra. To do this, however, extended powers of spectra are required. With this motivation and others, May [72] showed how to construct them. In place of a coreduction, this allows us to use the structure map $D_pY \to Y$ of an H_∞ ring spectrum Y. This permits us to construct homotopy operations which are related to Steenrod operations in the Adams spectral sequence for Y. In addition, we get differentials in the Adams spectral sequence and relations in the homotopy groups of any such spectrum.

We can now indicate that part of the present work which is new. First, everything we do applies to all H_∞ ring spectra, not only the sphere spectrum. Second, we have done the homological algebra necessary to produce Steenrod operations in the generalized Adams spectral sequence and have shown that they come from the H_∞ ring structure just as in the ordinary mod p Adams spectral sequence. Third, we have included a reasonably thorough account of the homotopy operations and the relations between them. Undoubtedly, some of these results, especially in the mod 2 case, are known, although difficult to find in the literature. Passing references to Barratt and Mahowald are found in [45] and some related results exist in [106], [104], [80] and [79]. Fourth, we have generalized the results of Mäkinen to the odd primary case, producing new formulas for differentials in the Adams spectral sequence. This involves a detailed study of the homotopy of extended powers of cells. Finally, it is our hope that the present account has benefitted sufficiently from the process of refinement that occurs with each extension or generalization of previous work, that it is simpler and clearer than previous accounts and that this will make the results more accessible. In this spirit, we have attempted to include all nontrivial details.

We have tried to maximize the extent to which all of this carries over to arbitrary homotopy functors $[X,-]_*$ besides the traditional $\pi_* = [S^0,-]_*$. Of particular interest is the case in which X is a Moore space. Much of the work in [92] can be interpreted as calculations of the homotopy operations which apply when X is a Moore space. The generalization to arbitrary X is only partially carried out. The difficulty in extending it lies in our ignorance about the extended powers of spaces other than spheres. Note, however, that VI §2 contains results which facilitate the anlysis of extended powers of other spaces. Finally, we should point out the remarkable fact that the key differentials needed for the computation of the stable homotopy groups of spheres from the cohomology of the Steenrod algebra are direct consequences of the H_∞ ring structure of the sphere spectrum. It is appealing to think of the H_∞ ring structure as a machine which encodes the destruction of Steenrod operations, which exist uniformly in E_2, converting them into more complicated relations in homotopy. In this vein, we point out in section VI §1 that our analysis of the differentials can be used to compute extensions which are hidden in E_∞. In summary, we feel that the results contained here should be a part of everyone's Adams spectral sequence toolkit, and we hope that the present exposition will make this possible.

We have organized this paper so that the general theory is in Chapter IV, explicit computations and relations in homotopy are in Chapter V, and formulas for differentials are in Chapter VI.

Chapter IV is organized as follows. In §1 we introduce $\mathrm{Ext}_A(N,M)$ for comodules N and M over a commutative Hopf algebroid A. In §2 we define and study products and

Steenrod operations in $\mathrm{Ext}_A(N,M)$ when N is a coalgebra and M is an algebra in the category of A-comodules. In §3 we set up the Adams spectral sequence. In §4 we set up an external smash product pairing in the Adams spectral sequence and use it to define an internal product in the Adams spectral sequence converging to $[X,Y]_*$ when X is a suspension spectrum and Y is a ring spectrum. In §5 we derive the main conceptual result of the chapter: the H_∞ ring structure map $D_pY \to Y$ naturally induces the (algebraically defined) Steenrod operations in $\mathrm{Ext}_{E_*E}(E_*X,E_*Y)$, the E_2 term of the Adams spectral sequence converging to $[X,Y]^E_*$. Thus, for H_∞ ring spectra Y, the Steenrod operations in E_2 reflect structure which exists in $[X,Y]_*$. In §7 we define the homotopy operations in π_*Y derived from $D_pY \to Y$ and use a spectral sequence originally due to Milgram to identify operations in $\mathrm{Ext}(\pi_*E,E_*Y)$ which correspond to homotopy operations and relations between them. In §6 the spectral sequence is defined and its relevant properties are derived.

I have benefitted from conversations with many people in the preparation of this material. Of special importance are Peter May, Arunas Liulevicius, Daniel Kahn, Mark Mahowald, Jim Milgram, Jim McClure, Jim Stasheff, Mark Steinberger, and Bob Wellington.

§1. Cohomology of Hopf Algebroids

Let k be a commutative ring with unit. A Hopf algebroid (R,A) is a cogroupoid in the category of graded commutative k-algebras. Thus R and A are graded commutative k-algebras and there are k-algebra homomorphisms $\eta_L,\eta_R:R \to A$, $\varepsilon:A \to R$, $\psi:A \to A \otimes_R A$, and $\chi:A \to A$. The simplest way to recall the diagrams these satisfy is to dualize the diagrams satisfied by a groupoid with "objects" R and "morphisms" A. The left and right units η_L and η_R are dual to the source and target, the augmentation ε is dual to the morphism which assigns each "object" its identity "morphism", the conjugation χ is dual to the inverse, the coproduct ψ is dual to composition, and the product $\phi:A \otimes_k A \to A$ is dual to the diagonal.

The two units, η_L and η_R give A two R-module structures: a left R-action $r\cdot a = \eta_L(r)a$ and a right R-action $a\cdot r = a\eta_R(r)$. Therefore we shall find the category of R-R-bimodules more appropriate than the category of R-modules. The commutativity of R enables us to embed the category of (either left or right) R-modules as the full subcategory whose objects are those R-R-bimodules which satisfy $r\cdot x = (-1)^{|x||r|}x\cdot r$ for all elements x. There are two forgetful functors from R-R-bimodules to left or right R-modules which simply forget the R-action on one side or the other. We let M_R be the right R-module whose R-action equals the right R-action on M. Of course, the above embedding gives M_R a left R-action which agrees up to sign with the right R-action. For example, here is the left R-action on A_R:

$$
\begin{aligned}
r \cdot a &= (-1)^{|r||a|} a \cdot r \\
&= (-1)^{|r||a|} a\eta_R(r) \\
&= \eta_R(r)a \ .
\end{aligned}
$$

Similarly, M_L will denote M with its right action forgotten.

We let $\otimes_R$ denote any of the tensor products

$$
\begin{array}{lll}
\text{R-R-bimodules} & \times \text{ R-R-bimodules} \longrightarrow & \text{R-R-bimodules} \\
\text{Right R-modules} & \times \text{ R-R-bimodules} \longrightarrow & \text{Right R-modules} \\
\text{R-R-bimodules} & \times \text{ Left R-modules} \longrightarrow & \text{Left R-modules.}
\end{array}
$$

Thus, $M \otimes_R N$ gets a left action from M if it has one, gets a right action from N if it has one, and amalgamates the right action on M with the left action on N. It is necessary to distinguish these three tensor products and to avoid automatically embedding one sided R-modules in R-R-bimodules because the embeddings do not commute with tensor products. The next paragraph contains a telling example of this. We let $\times = \times_R$ in the rest of this section.

A <u>right A-comodule</u> is a right R-module M with an R-linear map $\psi_M: M \to M \times A$ making

$$
\begin{array}{ccc}
M & \xrightarrow{\psi_M} & M \otimes A \\
 & \diagdown\!\!\diagdown & \downarrow{\scriptstyle 1 \otimes \varepsilon} \\
 & & M
\end{array}
\qquad\qquad
\begin{array}{ccc}
M & \xrightarrow{\psi_M} & M \otimes A \\
{\scriptstyle \psi_M}\downarrow & & \downarrow{\scriptstyle \psi_M \otimes 1} \\
M \otimes A & \xrightarrow{1 \otimes \psi} & M \otimes A \otimes A
\end{array}
$$

commute. The algebra R is a right A-comodule with $\psi_R = \eta_R$ and a left A-comodule with $\psi_R = \eta_L$. The coproduct $\psi: A \to A \otimes A$ makes A_R a right A-comodule and A_L a left A-comodule. The module $M \otimes A$ exemplifies the lack of commutativity between $\otimes$ and the embeddings of R-modules into R-R-bimodules. If we tensor with A, then embed we get a bimodule whose left and right actions agree, whereas, if we convert M to a bimodule then tensor with A we get a bimodule with different left and right actions. This prevents us from viewing ψ_M as a bimodule homomorphism unless we replace the codomain by $(M \otimes A)_R$. It is simpler to think of $\psi_M: M \to A \otimes M$ as existing in the category of right R-modules. There is one situation in which we will automatically view a one sided module as a bimodule. If N is a right R-module and we write $M \otimes N$, we mean to imply that N is first converted to a bimodule so that the tensor product is one of the three discussed above.

We assume henceforth that <u>A is R-flat</u> (on either side; the two conditions are equivalent). Then the category A-Comod of right A-comodules has kernels (which may be computed in R-Mod) and is therefore abelian.

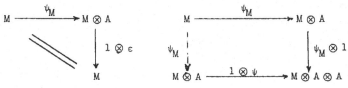

If P and Q are right R-modules then $\mathrm{Hom}_R(P,Q)$ is the graded R-module whose degree t component consists of homomorphisms which raise degrees by t. If M and N are right A-comodules then $\mathrm{Hom}_A(M,N)$ is the k-submodule of $\mathrm{Hom}_R(M,N)$ consisting of comodule homomorphisms. It is an R submodule for all M and N if and only if $\eta_L = \eta_R$.

The forgetful functor A-Comod → R-Mod (which we denote by $m \mapsto \hat{m}$ and $f \mapsto \hat{f}$) has a right adjoint

$$(?) \otimes A : \text{R-Mod} \to \text{A-Comod}$$

which sends a right R-module to the right A-comodule $P \otimes A$ with coproduct $1 \otimes \psi$. We call such comodules extended. The adjunction

$$\mathrm{Hom}_R(\hat{M},P) \cong \mathrm{Hom}_A(M,P \otimes A)$$

sends $f:\hat{M} \to P$ to $(f \otimes 1)\psi_M$ and sends $f:M \to P \otimes A$ to $(1 \otimes \varepsilon)f$.

Retracts of the extended comodules form an injective class relative to the R-split exact sequences, and we have the usual

Comparison Theorem: If $0 \to M \to X_0 \to X_1 \to \cdots$ is an R-split exact sequence of right A-comodules and $0 \to N \to Y_0 \to Y_1 \to \cdots$ is a complex of injective right A-comodules then for each A-homomorphism $f:M \to N$ there is a unique chain homotopy class of A-homomorphisms $F:X \to Y$ extending f.

We note for future reference that we may choose the splitting homomorphisms

$$M \xleftarrow{\varepsilon} X_0 \xleftarrow{\sigma_0} X_1 \xleftarrow{\sigma_1} \cdots$$

so that $\varepsilon\sigma_0 = 0$ and $\sigma_i\sigma_{i+1} = 0$.

We define Ext_A^i to be the ith right derived functor of Hom_A relative to injective comodules and R-split exact sequences.

The tensor product $M \otimes N$ of right A-comodules can be made a right A-comodule by the diagonal coproduct

$$M \otimes N \xrightarrow{\psi \otimes \psi} M \otimes A \otimes N \otimes A \xrightarrow{1 \otimes \tau \otimes 1} M \otimes N \otimes A \otimes A \xrightarrow{1 \otimes 1 \otimes \phi} M \otimes N \otimes A.$$

[The alert reader will notice that the separate maps here are well defined only if $\otimes = \otimes_k$ rather than $\otimes_R$. The composite, however, is well defined with $\otimes = \otimes_R$.] When $N = A_R$ we have the right A-comodule $M \otimes A_R$ with diagonal coproduct, in contrast to the extended coproduct on $M \otimes A$. Nevertheless, $M \otimes A_R$ is isomorphic to $M \otimes A$ as a right A-comodule. The isomorphism $\theta : M \otimes A_R \to M \otimes A$ is the adjoint of the R-homomorphism $1 \otimes \varepsilon : M \otimes A_R \to M$. Explicitly, $\theta(m \otimes a) = \sum m' \otimes a''a$ and

$\theta^{-1}(m \otimes a) = \sum m' \otimes \chi(a'')a$ if $\psi(m) = \sum m' \otimes a''$. The isomorphism θ makes the diagram

$$\begin{array}{ccc} & \xrightarrow{1 \otimes \eta_R} & M \otimes A_R \\ M & & \downarrow \theta \\ & \xrightarrow{\psi_M} & M \otimes A \end{array}$$

commute. Both $1 \otimes \eta_R$ and ψ_M are R-split by $1 \otimes \varepsilon$. Thus we may take either as our canonical R-split monomorphisms into an injective comodule. We choose $1 \otimes \eta_R$ because it will relate well to the Kunneth homomorphism later. It also allows the following convenient description of the canonical injective resolution. Let $p: A_R \to \overline{A}$ be $\mathrm{Cok}(\eta_R)$, and write $\overline{a}$ for $p(a)$. Define $t: \overline{A} \to A_R$ by $tp = 1 - \eta_R\varepsilon$. Then for any right A-comodule M, there is a short exact sequence

$$0 \longrightarrow M \underset{1 \otimes \varepsilon}{\overset{1 \otimes \eta_R}{\rightleftarrows}} M \otimes A_R \underset{1 \otimes t}{\overset{1 \otimes p}{\rightleftarrows}} M \otimes \overline{A} \longrightarrow 0$$

of right A-comodules (solid arrows), which is R-split (dotted arrows).

Definition 1.1: Let M be a right A-comodule. The normalized canonical resolution $C(A,M)$ of M is the R-split differential graded right A-comodule

$$0 \longrightarrow C_0 \underset{\sigma_0}{\overset{d_0}{\rightleftarrows}} C_1 \underset{\sigma_1}{\overset{d_1}{\rightleftarrows}} \cdots$$

where $C_s = M \otimes \overline{A}^s \otimes A_R$, $d_s = (1 \otimes \eta_R)(1 \otimes p)$ and $\sigma_s = (1 \otimes t)(1 \otimes \varepsilon)$. We write $m[a_1| \cdots |a_s]a$ for $m \otimes \overline{a}_1 \otimes \cdots \otimes \overline{a}_s \otimes a \in C_s$, and assign it homological degree s, internal degree $t = |m| + \sum |a_i| + |a|$, bidegree (s,t) and total degree $t-s$. If N is also a right A-comodule, the canonical complex $C(N,A,M)$ has

$$C_{s,t}(N,A,M) = \mathrm{Hom}_A^t(N, C_s(A,M)).$$

Proposition 1.2. $\mathrm{Ext}_A(N,M) = H(C(N,A,M))$.

Proof. If we let $\eta: M \to C(A,M)$ and $\varepsilon: C(A,M) \to M$ be

$$1 \otimes \eta_R: M \to M \otimes A = C_0(A,M) \quad \text{and}$$

$$1 \otimes \varepsilon: C_0(A,M) = M \otimes A \to M,$$

then it is easy to check that $d^2 = d\eta = 0$, $\sigma^2 = \varepsilon\sigma = 0$ and $d\sigma + \sigma d = 1 - \eta\varepsilon$. Thus

$$0 \longrightarrow M \xrightarrow{\eta} C_0 \xrightarrow{d_0} C_1 \xrightarrow{d_1} C_2 \xrightarrow{d_2} \cdots$$

is an injective resolution R-split by σ and ε, which implies the proposition. //

Note that we use t-s as our total degree rather than t+s. This (t-s) is the topologically significant degree in the Adams spectral sequence

$$\mathrm{Ext}^{s,t}_{E_*E}(E_*X,E_*Y) \Longrightarrow [X,Y]_{t-s}.$$

If we regrade C(A,M) by nonpositive superscripts, it really is the total degree in the sense of being the sum of the internal and homological degrees.

§2. Products and Steenrod Operations in Ext

We begin this section with a quick description of the product in the Ext module we have just defined. The rest of the section is devoted to the development of the Steenrod operations in this context. The main point is to show how the development of Steenrod operations in [68] is carried over to the cobar complex C(N,A,M) in the setting appropriate to generalized homology theories.

The indexing we have chosen for Steenrod operations disagrees with that of [55],[68] and [81]. Our reason is this: as noted in section 1, the appropriate total degree for $\mathrm{Ext}^{s,t}$ is t-s rather than t+s. This change converts the grading of [55] and [68] to the grading we have chosen. With our grading, the operation P^i raises the geometrically significant total degree t-s by 2i(p-1) if p > 2 and by i if p = 2. This conforms to the pattern established by the Steenrod operations in cohomology and the Dyer-Lashof operations in homology. This is not merely an analogy. We shall see that the Adams spectral sequence connects the Steenrod operations in Ext with homotopy operations. Under the Hurewicz homomorphism these homotopy operations correspond to Dyer-Lashof operations and our choice of indexing leads to precise compatibility with these Dyer-Lashof operations.

In this section we let $\otimes = \otimes_k$.

In order to introduce products and Steenrod operations into $\mathrm{Ext}_A(N,M)$ we require more structure on N and M. The necessary definition follows.

Definition 2.1. Let $\mathcal{C}$ be the category whose objects are triples (N,A,M) such that

1) (R,A) is a Hopf algebroid over k,
2) M is a commutative unital A-algebra (that is, an algebra with unit η_M: R → M in the category of A-comodules), and
3) N is a cocommutative unital A-coalgebra (that is, a coalgebra with counit ε_N:N → R in the category of A-comodules)

and whose morphisms (N,A,M) → (N',A',M') are triples (f,λ,g) such that

1) λ:(R,A) → (R',A') is a morphism of Hopf algebroids,
2) f:M → M' is an algebra homomorphism preserving units and a λ-equivariant comodule homomorphism (f(mr) = f(m)λ(r) and $\psi_{M'}f = (f \otimes \lambda)\psi_M$), and

3) $g:N' \to N$ (note reverse direction) is a coalgebra homomorphism preserving counits and a λ-equivariant comodule homomorphism ($g(n'\lambda(r)) = g(n')r$ and $(1 \otimes \lambda)\psi_N g = (g \otimes 1)\psi_{N'}$).

If (N,A,M) is in $\mathcal{C}$, we write ϕ: $M^r \to M$ and Δ: $N \to N^r$ for the iterated product and coproduct.

Note that (R,A,R) is in $\mathcal{C}$ and that the unit η_M and counit ε_N induce a homomorphism

$$(\varepsilon_N, 1, \eta_M):(R,A,R) \to (N,A,M)$$

in $\mathcal{C}$. In turn, this induces a unit

$$(*) \qquad \mathrm{Ext}_A(R,R) \to \mathrm{Ext}_A(N,M).$$

If (R,A) and (R',A') are Hopf algebroids over k then the obvious structure maps make $(R \otimes R', A \otimes A')$ a Hopf algebroid which we will usually call $A \otimes A'$. The functor $\otimes$ defines a functor

$$A\text{-Comod} \times A'\text{-Comod} \to A \otimes A'\text{-Comod}.$$

Thus, if M is an A-comodule and M' is an A'-comodule, then $C(A,M) \otimes C(A',M')$ is a differential graded $A \otimes A'$-comodule with differential $d \otimes 1 + 1 \otimes d$, unit $\eta \otimes \eta$, augmentation $\varepsilon \otimes \varepsilon$, and contracting homotopy $\sigma \otimes 1 + \eta\varepsilon \otimes \sigma$. By the comparison theorem, there is a unique chain homotopy class of $A \otimes A'$-homomorphisms

$$C(A,M) \otimes C(A',M') \to C(A \otimes A', M \otimes M')$$

extending the identity of $M \otimes M'$. If C is an A-comodule, let $C^n = C \otimes \cdots \otimes C$ with n factors C. Regard C^n as an A-comodule by means of the iterated product ϕ: $A^n \to A$ in the usual way. For each integer n there is a unique chain homotopy class of A-homomorphisms

$$\phi:C(A,M)^n \to C(A,M)$$

extending the product $\phi:M^n \to M$. This implies that $C(A,M)$ is a homotopy associative and commutative differential graded A-comodule algebra (DGA in A-Comod). Finally, if $(N,A,M) \in \mathcal{C}$, the homomorphism

$$\begin{array}{ccc}
\mathrm{Hom}_A(N,C(A,M))^n & = & C(N,A,M)^n \\
\downarrow \otimes & & \\
\mathrm{Hom}_A(N^n,C(A,M)^n) & & \downarrow \\
\downarrow \mathrm{Hom}(\Delta,\phi) & & \\
\mathrm{Hom}_A(N,C(A,M)) & = & C(N,A,M)
\end{array}$$

makes C(N,A,M) into a homotopy associative and commutative differential graded k-algebra. (There is an Alexander-Whitney map which makes C(A,M) and C(N,A,M) strictly associative.) This product on C(N,A,M) makes $\mathrm{Ext}_A(N,M)$ into a bigraded commutative associative algebra over $\mathrm{Ext}_A(R,R)$ with unit (*) induced by $\mathrm{Hom}(\varepsilon,\eta):\mathrm{Hom}_A(R,R) \to \mathrm{Hom}_A(N,M)$.

We can now summarize the development of Steenrod operations given in [68]. Let $k = Z_p$ and let $\pi \subset \Sigma_p$ be the cyclic p-Sylow subgroup generated by the permutation $\alpha = (1\ 2 \cdots p)$. Recall the usual $k\pi$ free resolution of k.

<u>Definition 2.2.</u> Let $\mathscr{W}_i$ be free over $k\pi$ on one generator e_i, let

$$d(e_{2i}) = (1 + \alpha + \alpha^2 + \cdots + \alpha^{p-1})e_{2i-1} \text{ and } d(e_{2i+1}) = (\alpha - 1)e_{2i},$$

and let $\mathscr{W}_0 \to k$ send $\alpha^i e_0$ to 1.

Let $\mathscr{V}$ be any $k\Sigma_p$ free resolution of k and let $j:\mathscr{W} \to \mathscr{V}$ be a $k\pi$ chain map covering the identity map of k. Let π and Σ_p act trivially on a chain complex K, by permuting factors on K^p, and diagonally on $\mathscr{W} \otimes K^p$ and $\mathscr{V} \otimes K^p$ respectively. We let $\mathscr{W}_i \otimes (K^p)_n$ have degree n-i, n being the total degree if K is bigraded. Then we can define Steenrod operations in H(K) if K is a homotopy associative differential k-algebra with a $k\pi$ morphism $\theta: \mathscr{W} \otimes K^p \to K$ such that

(i) $\theta | e_0 \otimes K^p$ is the iterated product $K^p \to K$ associated in some fixed order, and

(ii) ϕ is $k\pi$-homotopic to $\mathscr{W} \otimes K^p \xrightarrow{j \otimes 1} \mathscr{V} \otimes K^p \xrightarrow{\phi} K$ for some $k\Sigma_p$-homomorphism ϕ.

A morphism $(K,\theta) \to (K',\theta')$ is a morphism $f:K \to K'$ of differential k-modules such that $f\theta$ is $k\pi$-homotopic to $\theta'(1 \otimes f^p)$. The tensor product $(K,\theta) \otimes (K',\theta')$ is defined in an evident way and the Steenrod operations satisfy the (internal) Cartan formula if the product $K \otimes K \to K$ defines a morphism $(K,\theta) \otimes (K,\theta) \to (K,\theta)$. Let $\mathscr{U}$ be a $k\Sigma_{p^2}$ free resolution of k and let $\tau = Z_p \wr Z_p \subset \Sigma_{p^2}$ be the p-Sylow subgroup. Let $\omega: \mathscr{W} \otimes \mathscr{W}^p \to \mathscr{U}$ be a $k\tau$-homomorphism extending the identity $k \to k$ where $\mathscr{W} \otimes \mathscr{W}^p$ is given the evident τ action. Then the Steenrod operations in H(K) satisfy the Adem relations if there is a $k\Sigma_{p^2}$-homomorphism $\xi: \mathscr{U} \otimes K^{p^2} \to K$ such that

$$\begin{array}{ccccc} (\mathscr{W} \otimes \mathscr{W}^p) \otimes K^{p^2} & \xrightarrow{\omega \otimes 1} & \mathscr{U} \otimes K^{p^2} & & \\ \Big\downarrow{\scriptstyle \text{shuffle}} & & & \searrow{\scriptstyle \xi} & \\ & & & & K \\ & & & \nearrow{\scriptstyle \theta} & \\ \mathscr{W} \otimes (\mathscr{W} \otimes K^p)^p & \xrightarrow{1 \otimes \theta^p} & \mathscr{W} \otimes K^p & & \end{array}$$

is $k\tau$ homotopy commutative.

The following lemma will imply that θ, ϕ, and ξ exist and make the appropriate diagrams homotopy commute when $K = C(N,A,M)$, giving us Steenrod operations in $\mathrm{Ext}_A(N,M)$.

<u>Lemma 2.3</u>: Let ρ be a subgroup of Σ_r. Let $\mathcal{V}$ be any $k\rho$ free resolution of k such that $\mathcal{V}_0 = k\rho$ with generator e_0. Let M and N be A-comodules. Let

$$0 \longrightarrow M \underset{\varepsilon}{\overset{\eta}{\rightleftarrows}} K_0 \underset{\sigma}{\overset{d}{\rightleftarrows}} K_1 \underset{\sigma}{\overset{d}{\rightleftarrows}} \cdots$$

be an R-split exact sequence of A-comodules and let

$$0 \longrightarrow N \xrightarrow{\eta} L_0 \xrightarrow{d} L_1 \xrightarrow{d} \cdots$$

be a complex of extended A-comodules. Let $f: M^r \to N$ be a ρ-equivariant A-comodule homomorphism, where ρ acts trivially on N and by permuting factors on M^r. Let ρ also act on K^r by permuting factors, and on $\mathcal{V} \otimes K^r$ by the diagonal action. Give $\mathcal{V} \otimes K^r$ the A-comodule structure induced by that of K^r and let $\mathcal{V}_i \otimes (K^r)_{j,t}$ have bidegree $(j-i,t)$. Then there is a unique ρ-equivariant chain homotopy class of ρ-equivariant A-comodule chain homomorphisms $\Phi: \mathcal{V} \otimes K^r \to L$ which extend f:

$$\begin{array}{ccc} M^r & \xrightarrow{\quad f \quad} & N \\ {\scriptstyle \eta^r}\downarrow & & \downarrow{\scriptstyle \eta} \\ K_0^r \cong \langle e_0 \rangle \otimes K_0^r & \xrightarrow{\quad \Phi \quad} & L_0 . \end{array}$$

Some such Φ satisfies $\Phi(\mathcal{V}_i \otimes (K^r)_j) = 0$ if $ri > (r-1)j$.

<u>Proof</u>. We will define ρ-equivariant A-comodule homomorphisms from $\mathcal{V}_i \otimes (K^r)_j$ to extended comodules by specifying their adjoint R-maps on elements $v \otimes k$ with v in a chosen ρ basis of $\mathcal{V}_i$. It is easy to check that we get the same homomorphism by extending by equivariance and then taking adjoints as we get by first taking adjoints and then extending by equivariance.

Write $\Phi_{i,j}$ for $\Phi | \mathcal{V}_i \otimes (K^r)_j$. We define $\Phi_{i,j}$ by induction on i and subsidiary induction on j. The existence of $\Phi_{0,*} | \langle e_0 \rangle \otimes K^r \to L$ follows from the comparison theorem, so we may assume $\Phi_{i',j}$ constructed for all $i' < i$. If $j < i$ then $\Phi_{i,j} = 0$ since L is a nonnegative complex, so we may assume $\Phi_{i,j'}$ constructed for $j' < j$. If $\tilde{\Phi}$ is the adjoint of Φ, we let

$$\text{(a)} \qquad \tilde{\Phi}_{i,j} = (\widetilde{d\Phi_{i,j-1}} - \widetilde{\Phi_{i-1,j-1}(d \otimes 1)})(1 \otimes S)$$

on elements $v \otimes k$ with v in a chosen ρ-basis of $\mathcal{V}_i$, where

$$S = \sum_i (\eta\varepsilon)^i \otimes \sigma \otimes 1^{r-i-1}$$

is the contracting homotopy of K^r (so that $dS + Sd = 1 - (\eta\varepsilon)^r$). To show that this makes Φ a chain homomorphism we must show that

(b) $$d\Phi_{i,j-1} = \Phi_{i-1,j-1}(d \otimes 1) + \Phi_{i,j}(1 \otimes d).$$

It suffices to show that the adjoint of (b) is true on our chosen ρ-basis, and we may assume (b) holds for smaller i and j. Thus, letting the adjunctions be understood and using (a), we have

$$\Phi_{i-1,j-1}(d \otimes 1) + \Phi_{i,j}(1 \otimes d)$$

$$= \Phi_{i-1,j-1}(d \otimes 1) + d\Phi_{i,j-1}(1 \otimes Sd) - \Phi_{i-1,j-1}(d \otimes Sd)$$

$$= d\Phi_{i,j-1} + (\Phi_{i-1,j-1}(d \otimes 1) - d\Phi_{i,j-1})(1 \otimes dS + 1 \otimes (\eta\varepsilon)^r).$$

Applying (b) inductively twice shows that

$$(\Phi_{i-1,j-1}(d \otimes 1) - d\Phi_{i,j-1})(1 \otimes d) = -dd\Phi_{i,j-2} = 0.$$

If we let $\overline{f}: \mathcal{V}_0 \otimes M^r \to N$ be ρ-equivariant and satisfy $f = \overline{f}|\langle e_0\rangle \otimes M^r$ then $\Phi_{0,0}(1 \otimes \eta^r) = \eta\overline{f}$. Then $(\Phi_{i-1,j-1}(d \otimes 1) - d\Phi_{i,j-1})(1 \otimes (\eta\varepsilon)^r) = 0$ because $\overline{f}(d \otimes 1) = 0$ by ρ-equivariance of f and because $d\eta = 0$. This completes the inductive construction of Φ. Now let us show that the Φ we have constructed satisfies $\Phi_{i,j} = 0$ if $ri > (r-1)j$. This is trivial if $i = 0$ or $j < i$ so we use induction on i and a subsidiary induction on j. When $ri > (r-1)j$ the induction hypothesis implies that (a) reduces to $\Phi_{i,j} = -\Phi_{i-1,j-1}(d \otimes S)$. This implies the result, again by the induction hypothesis, except when $j = rn+1$ and $i = j-n$. In this case we iterate (a) to obtain

(c) $$\tilde{\Phi}_{i,j} = (-1)^r\tilde{\Phi}_{i-r,j-r}(d \otimes S)p_1(d \otimes S)p_2 \cdots p_{r-1}(d \otimes S),$$

where each p_i is a sum of permutations of the factors of K^r coming from the equivariance of Φ. The number of factors c_i of $c_1 \otimes \cdots \otimes c_r \in K^r$ which are annihilated by $\sigma: K \to K$ increases by at least one each time we apply S; this is where we require $\sigma^2 = \varepsilon\sigma = 0$. Since permutations preserve this property and since $d \otimes S$ occurs r times in (c), it follows that $\Phi_{i,j} = 0$ in this case also, completing the induction.

Finally, we show that Φ is unique up to ρ-equivariant chain homotopy. Suppose $\Phi, \Theta: \mathcal{V} \otimes K^r \to L$ both extend f. We define $H_{i,j}: \mathcal{V}_i \otimes (K^r)_j \to L_{j-i-1}$ by letting its adjoint be

$$\begin{cases} 0 & \text{if } j < i+1 \text{ or } i < 0 \\ (\tilde{\Phi}_{i,j-1} - \tilde{\Theta}_{i,j-1} - dH_{i,j-1} - H_{i-1,j-1}(d \otimes 1))(1 \otimes S) & \text{otherwise} \end{cases}$$

on elements $v \otimes c$ with v in a chosen ρ-basis of $\mathscr{V}_i$. We must show

$$dH_{i,j-1} + H_{i-1,j-1}(d \otimes 1) + H_{i,j}(1 \otimes d) = \Phi_{i,j-1} - \Theta_{i,j-1} \,.$$

The definition of $H_{i,j}$ implies that, on the ρ-basis, the adjoint of $H_{i,j}(1 \otimes d)$ is the desired expression minus

$$(\Phi_{i,j-1} - \Theta_{i,j-1} - dH_{i,j-1} - H_{i-1,j-1}(d \otimes 1))(1 \otimes dS + 1 \otimes (\eta\varepsilon)^r).$$

Now, $1 \otimes dS = 0$ unless $j \geq 2$ and $1 \otimes (\eta\varepsilon)^r = 0$ unless $j = 1$. If $j = 1$ then everything is zero unless $i = 0$, when we get $(\Phi_{0,0} - \Theta_{0,0})(1 \otimes \eta^r)(1 \otimes \varepsilon^r)$. Since $\Phi_{0,0}(1 \otimes \eta^r) = \eta\bar{f} = \Theta_{0,0}(1 \otimes \eta^r)$ the result follows when $j \leq 1$. When $j \geq 2$ we find by induction that

$$(\Phi_{i,j-1} - \Theta_{i,j-1} - dH_{i,j-1} - H_{i-1,j-1}(d \otimes 1))(1 \otimes d) = 0.$$

Hence H is a ρ-equivariant A-comodule chain homotopy $\Phi \simeq \Theta$. //

<u>Remark</u>: Since Φ is determined up to chain homotopy by $f\colon M^r \to N$ it is easy to see that Φ is natural in M and N up to chain homotopy.

Suppose (N,A,M) is a triple in $\mathscr{C}$ defined over $k = Z_p$. The product $M^p \to M$ is commutative, hence Lemma 2.3 with $\rho = \pi$ and $r = p$ implies that there is a unique π-equivariant chain map $\Phi\colon \mathscr{W} \otimes C^p \to C$, where $C = C(A,M)$. Since Φ is an A-homomorphism we also have a homomorphism

$$\begin{array}{ccc} \mathscr{W} \otimes \mathrm{Hom}_A(N,C)^p & = & \mathscr{W} \otimes C(N,A,M)^p \\ \downarrow & & \\ \mathrm{Hom}_A(N^p, \mathscr{W} \otimes C^p) & & \Big\downarrow \Phi \\ \Big\downarrow \mathrm{Hom}(\Delta,\Phi) & & \\ \mathrm{Hom}_A(N,C) & = & C(N,A,M) \end{array}$$

and since $\Delta\colon N \to N^p$ is cocommutative, this Φ is also π-equivariant.

<u>Definition 2.4</u>: With the notation of the preceding paragraph, let $x \in \mathrm{Ext}_A^{s,t}(N,M)$. If $p = 2$ define

(i) $$Sq^i(x) = P^i(x) = \Phi_*(e_{i-t+s} \otimes x^2) \quad \text{if } i \geq t-s.$$

If $p > 2$, define

(ii) $$\begin{cases} P^i(x) = (-1)^i \nu(t-s)\Phi_*(e_{(2i-t+s)(p-1)} \otimes x^p) & \text{if } 2i \geq t-s \\ \beta P^i(x) = (-1)^i \nu(t-s)\Phi_*(e_{(2i-t+s)(p-1)-1} \otimes x^p) & \text{if } 2i > t-s \,, \end{cases}$$

where $m = (p-1)/2$ and $\nu(n) = (-1)^j (m!)^\varepsilon$ if $n = 2j+\varepsilon$.

Note that βP^i is a single symbol, <u>a priori</u> unrelated to P^i (however, see Theorem 2.5 vii). By [68], the P^i and βP^i account for all the nonzero operations of the form $x \longmapsto \Phi_*(e_i \otimes x^p)$.

If (N,A,M) is an object of $\mathcal{C}$ defined over $k = Z$ such that N,A and M are all torsion free, let $\overline{N} = N \otimes Z_p$, $\overline{A} = A \otimes Z_p$ and $\overline{M} = M \otimes Z_p$. Then $(\overline{N},\overline{A},\overline{M}) \in \mathcal{C}$ and, as usual, the sequence $Z_p \to Z_{p^2} \to Z_p$ induces a Bockstein homomorphism

$$\beta : \mathrm{Ext}_{\overline{A}}^{s,t}(\overline{N},\overline{M}) \to \mathrm{Ext}_{\overline{A}}^{s+1,t}(\overline{N},\overline{M})$$

which we will use in Theorem 2.5 (vii).

We are now ready to apply Lemma 2.3 and [68] to produce the main result of this section.

<u>Theorem 2.5</u>: The P^i and βP^i are natural homomorphisms with the following properties.

(i) $\beta^\varepsilon P^i : \mathrm{Ext}_A^{s,t} \to \mathrm{Ext}_A^{s+(t-2i)(p-1)+\varepsilon,pt}$ $\quad(\varepsilon = 0$ if $p = 2)$

(ii) When $p = 2$, $P^i = 0$ unless $t-s \leq i \leq t$. When $p > 2$, $P^i = 0$ unless $t-s \leq 2i \leq t$ and $\beta P^i = 0$ unless $t-s+1 \leq 2i \leq t$.

(iii) $P^i(x) = x^p$ if $p = 2$ and $i = t-s$ or if $p > 2$ and $2i = t-s$.

(iv) The internal and external Cartan formulas hold:

$$P^n(x \otimes y) = \sum_i P^i(x) \otimes P^{n-i}(y) \quad \text{and}$$

$$\beta P^n(x \otimes y) = \sum_i \beta P^i(x) \otimes P^{n-i}(y) + \sum_i (-1)^{|x|} P^i(x) \otimes \beta P^{n-i}(y) \,.$$

(v) The Adem relations hold: if $a > pb$ and $\varepsilon = 0$ or 1 ($\varepsilon = 0$ if $p = 2$) then

$$\beta^{\varepsilon}P^{a}P^{b} = \sum_{i} (-1)^{a+i}(pi-a, a-(p-1)b - i-1)\beta^{\varepsilon}P^{a+b-i}P^{i} \ ;$$

if $p > 2$, $a \geq pb$ and $\varepsilon = 0$ or 1 then

$$\beta^{\varepsilon}P^{a}\beta P^{b} = (1-\varepsilon) \sum_{i} (-1)^{a+i}(pi-a, a-(p-1)b - i)\beta P^{a+b-i}P^{i}$$

$$- \sum_{i} (-1)^{a+i}(pi-a-1, a-(p-1)b-i)\beta^{\varepsilon}P^{a+b-i}\beta P^{i}$$

(vi) Suppose $f:(N,A,M) \to (N'',A'',M'')$ and $g:(N',A',M') \to (N,A,M)$ are morphisms in $\mathcal{C}$ such that $C(fg):C(N',A',M') \to C(N'',A'',M'')$ is zero on the cokernels of the units. Then $\sigma P^{i} = P^{i}\sigma$ and $\sigma\beta P^{i} = -\beta P^{i}\sigma$ where σ is the suspension

$$\sigma: \mathrm{Ext}_{A'}^{s,t}(N',M') \to \mathrm{Ext}_{A''}^{s-1,t}(N'',M'')$$

defined as $C(f)d^{-1}C(g)$ on representative cycles.

(vii) If (N,A,M) is the mod p reduction of a torsion free triple defined over Z then $\beta P^{i+1} = iP^{i}$ if $p = 2$ while βP^{i} is the composite of β and P^{i} if $p > 2$.

Proof. Let $C = C(N,A,M)$. Lemma 2.3 produces the necessary chain homomorphism $\Phi: \mathcal{W} \otimes C^{p} \to C$ and, if $\mathcal{V}$ is a $k\Sigma_{p}$ free resolution of k, $\Theta: \mathcal{V} \otimes C^{p} \to C$. The uniqueness of Φ implies that Φ factors through Θ up to chain homotopy. Hence the Steenrod operations are defined and satisfy (i), (iii), and (vii). Naturality follows from the uniqueness of Φ. Lemma 2.3 also shows $\Phi = 0$ in the cases relevant to (ii). Commutativity of $\phi: M \otimes M \to M$ and the uniqueness clause of Lemma 2.3 imply that $C(N,A,M)$ is a Cartan object and an Adem object. Hence (iv) and (v) hold. To prove (vi) we must construct Φ, Φ' and Φ'' such that equality holds in $f\Phi = \Phi''(1 \otimes f^{p})$ and $g\Phi' = \Phi(1 \otimes g^{p})$ rather than just chain homotopy. It is easy to check that this will be true if we construct Φ, Φ' and Φ'' as in Lemma 2.3, because $C(N,A,M)$ is functorial. //

§3. The Adams Spectral Sequence

This section begins with some technical lemmas about homotopy exact couples and the associated spectral sequences for use in VI. We end the section by setting up the Adams spectral sequence.

We will work in the graded stable category $\bar{h}_{*}\mathcal{S}$. This is obtained from the stable category $\bar{h}\mathcal{S}$ specified in I§1 by introducing maps of nonzero degrees. The category $\bar{h}_{*}\mathcal{S}$ has the same objects as $\bar{h}\mathcal{S}$, and its morphisms from X to Y are the elements of the graded abelian group $[X,Y]_{*}$ with $[X,Y]_{n} = [\Sigma^{n}X,Y]$.

Definition 3.1: Consider inverse sequences

$$Y_0 \xleftarrow{i_0} Y_1 \xleftarrow{i_1} Y_2 \xleftarrow{i_2} \cdots$$

such that each Y_s is a CW spectrum and each i_s is the inclusion of a subcomplex. (This restriction is imposed purely for technical convenience. It represents no real restriction since any inverse sequence can be replaced by an equivalent one of this form by means of CW approximation and mapping telescopes.) Define $i_{s,r} = i_s i_{s+1} \cdots i_{s+r-1} : Y_{s+r} \to Y_s$ and $Y_{s,r} = Y_s/Y_{s+r} \simeq Ci_{s,r}$ and let

$$\begin{array}{ccc} Y_{s+r} & \xrightarrow{i_{s,r}} & Y_s \\ {\scriptstyle \partial_{s,r}}\nwarrow & & \swarrow{\scriptstyle p_{s,r}} \\ & Y_{s,r} & \end{array}$$

be a cofiber sequence with $\partial_{s,r}$ of degree -1.

Given a spectrum X we obtain an exact couple

$$\begin{array}{ccc} \bigoplus_{s,t} [X,Y_s]_{t-s} & \xrightarrow{i_*} & \bigoplus_{s,t} [X,Y_s]_{t-s} \\ {\scriptstyle \partial_*}\nwarrow & & \swarrow{\scriptstyle p_*} \\ & \bigoplus_{s,t} [X,Y_{s,1}]_{t-s} & \end{array}$$

and hence a spectral sequence. The term $E_r^{s,t}$ has many descriptions, of which we will find

$$E_r^{s,t} = \frac{\operatorname{im}([X,Y_{s,r}]_{t-s} \to [X,Y_{s,1}]_{t-s})}{\ker([X,Y_{s,1}]_{t-s} \to [X,Y_{s-r+1,r}]_{t-s})}$$

particularly convenient.

If $x \in [X,Y_{s,r}]_n$, we let $\tilde{x}$ denote its image in $E_r^{s,n+s}$. The following lemma gives minimal hypotheses needed to recognize differentials in the spectral sequence.

Lemma 3.2: Let $f \in [X,Y_{s,1}]_{n+1}$ and $g \in [X,Y_{s+r,1}]_n$ satisfy $p\partial f = i'g \in [X,Y_{s+1,r}]_n$, where i' is induced by $i_{s+1,r-1}$. Then

$$d_k\tilde{f} = 0 \text{ if } k < r$$

$$d_r\tilde{f} = \tilde{g}$$

and $$d_k\tilde{g} = 0 \text{ for all } k.$$

The next two technical lemmas will be used repeatedly.

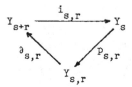

<u>Lemma 3.3</u>: If $f \in [X,Y_{s,r+p}]_n$ and $g \in [X,Y_{s+r+p+q}]_{n-1}$ are such that $i\partial f = ig \in [X,Y_{s+r}]_{n-1}$, then there exist

$$H \in [X,Y_{s+r,p}]_n \quad \text{and} \quad f' \in [X,Y_{s,r+p+q}]_n$$

such that

$$if = if' \in [X,Y_{s,r}]_n ,$$

$$\tilde{f} = \tilde{f}' \in E_k^{s,n+s} \quad \text{for } k \leq r+p+q ,$$

$$d_r\tilde{H} = \widetilde{\partial(f)} , \quad \text{and}$$

$$d_{r+p+q}\tilde{f} = d_{r+p+q}\tilde{f}' = \tilde{g} .$$

<u>Lemma 3.4</u>: Assume $p \leq q < r$ and suppose given $f \in [X,Y_{s+p,r-p}]_n$, $g \in [X,Y_{s+q,r-q}]_n$ and $h \in [X,Y_{s,p}]_{n+1}$ such that $\partial f = \partial g \in [X,Y_{s+r}]_{n-1}$ and $p\partial h = f - ig \in [X,Y_{s+p,r-p}]_n$. Then

$$\text{if } p < q , \ d_p\tilde{h} = \tilde{f} \ \text{ and } \ d_{r-q}\tilde{g} = \partial f = \partial g ,$$

while if $\quad p = q, \ d_p\tilde{h} = \tilde{f} - \tilde{g} \ \text{ and } \ d_{r-p}\tilde{f} = \partial f = \partial g = d_{r-q}\tilde{g}$.

Now we turn our attention to the Adams spectral sequence based on a commutative ring spectrum E with unit. We shall use $\otimes$ to denote $\otimes_{\pi_*E}$ <u>We assume that E_*E is flat as a (right or left) module over</u> π_*E. This ensures that (π_*E,E_*E) is a Hopf algebroid over π_0E and that E_*X is an E_*E comodule for any spectrum X. Here $E_*X = \pi_*(X \wedge E)$. The structural homomorphisms are defined as follows. Let $\eta:S \to E$ and $\mu:E \wedge E \to E$ be the unit and product of E, and let $\tau:A \wedge B \to B \wedge A$ be the twist map. Then $\eta_R = E_*(\eta) = (\eta \wedge 1)_*$, $\eta_L = (1 \wedge \eta)_*$, $\varepsilon = \mu_*$ and $\chi = \tau_*$ while $\phi:\pi_*E \otimes_{\pi_0E} \pi_*E \to \pi_*E$ is given by $\phi(\alpha \otimes \beta) = \mu(\alpha \wedge \beta)$, and $\phi:E_*E \otimes_{\pi_0E} E_*E \to E_*E$ is given by $\phi(\alpha \otimes \beta) = (\mu \wedge \mu)(1 \wedge \tau \wedge 1)(\alpha \wedge \beta)$. The coproducts ψ_X and ψ_E are defined as $\theta_2^{-1}E_*(1 \wedge \eta)$ in the following diagram. In it, the homomorphisms θ_1 and θ_2 are defined by

$$\theta_1(\alpha \otimes \beta) = (1 \wedge 1 \wedge \mu)(1 \wedge \tau \wedge 1)(\alpha \wedge \beta)$$

$$\theta_2(\alpha \otimes \beta) = (1 \wedge \mu \wedge 1)(\alpha \wedge \beta)$$

while θ is the algebraic isomorphism defined in §1. (Recall that $(E_*E)_R$ means E_*E with only its right π_*E action.) Adams [6, Lemma 12.5] shows θ_2 is an isomorphism since E_*E is flat over π_*E.

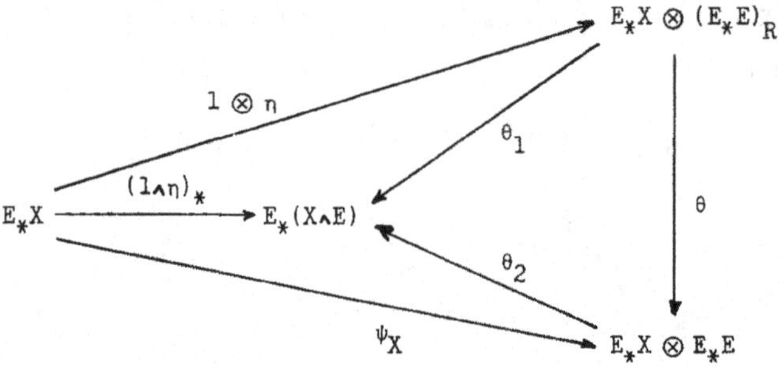

We have seen in §2 that θ is an isomorphism. It follows that θ_1 is also an isomorphism. Note that θ_1 is the Kunneth homomorphism for X and E.

Definition 3.5. An Adams resolution of a spectrum Y is an inverse sequence

$$Y \simeq Y_0 \xleftarrow{i_0} Y_1 \xleftarrow{i_1} Y_2 \longleftarrow \cdots$$

as in Definition 3.1 such that, for each s

(i) $Y_{s,1}$ is a retract of $X_s \wedge E$ for some spectrum X_s, and

(ii) $E_*Y_s \to E_*Y_{s,1}$ is a π_*E-split monomorphism.

A map of Adams resolutions is a map of inverse systems. The canonical Adams resolution is defined inductively by letting $Y_0 = Y$, $Y_{s+1} = Y_s \wedge \overline{E}$ and

$$i_s = 1 \wedge i\colon Y_s \wedge \overline{E} \to Y_s \wedge S \simeq Y_s$$

where the unit $S \to E$ is the cofiber of $i\colon \overline{E} \to S$. The Adams spectral sequence for $[X,Y]_*$ is the spectral sequence of the homotopy exact couple obtained by applying $[X,-]_*$ to an Adams resolution of Y. It is denoted by $E_r^{*,*}(X,Y)$.

Condition (i) ensures that $E_*Y_{s,1}$ is a direct summand of an extended comodule and condition (ii) ensures that the sequences $E_*Y_s \to E_*Y_{s,1} \to E_*\Sigma Y_{s+1}$ are π_*E split short exact sequences. Splicing them, we obtain an injective resolution (*) of E_*Y:

(*)

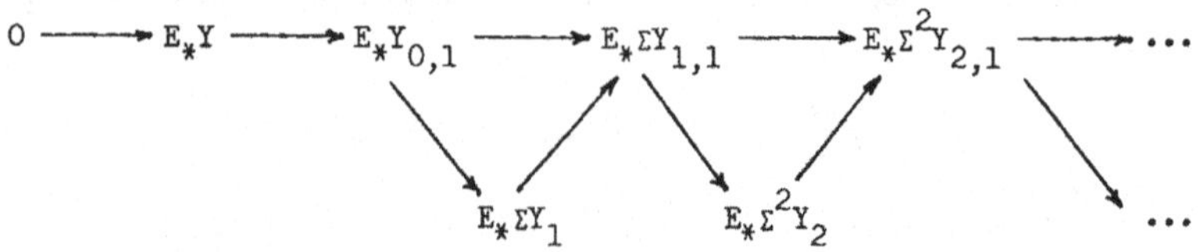

To proceed, we need another assumption on E.

Condition 3.6.

$$[X, Y \wedge E]_* \cong \mathrm{Hom}_{E_*E}(E_*X, E_*(Y \wedge E)) \cong \mathrm{Hom}_{\pi_*E}(E_*X, E_*Y)$$

for any Y when E_*X is π_*E projective. By [6, Prop. 13.4 and Thm. 13.6] this holds for $E = S$, HZ_p, MO, MU, MSp, K, KO and BP. Note that Condition 3.6 will be satisfied if we have a universal coefficient spectral sequence.

$$\mathrm{Ext}_{\pi_*E}(E_*X, \pi_*F) \Longrightarrow F^*X$$

for the module spectra $F = Y \wedge E$ over E. Also note that Condition 3.6 will be satisfied for all Y if it is satisfied for $Y = S$, using the argument of [6, Lemma 12.5]. Thus we have the following equivalent form of Condition 3.6:

if E_*X is π_*E projective then $E^*X \cong \mathrm{Hom}_{\pi_*E}(E_*X, \pi_*E)$.

Finally, if Condition 3.6 holds then the isomorphism in 3.6 will also hold with $Y \wedge E$ replaced by any retract (wedge summand) of $Y \wedge E$.

Given Condition 3.6, Definition 3.5(i) implies that if E_*X is π_*E projective then $[X, Y_{s,1}] \cong \mathrm{Hom}_{E_*E}(E_*X, E_*Y_{s,1})$. Hence E_2 of the Adams spectral sequence is $\mathrm{Ext}_{E_*E}(E_*X, E_*Y)$ in this case. By [6, Thm. 15.1], under appropriate hypotheses the spectral sequence converges to $[X,Y]_*^E$, where $[\ ,\ ]_*^E$ denotes homotopy classes of maps in the category obtained from the stable category $\overline{h}_*\mathcal{S}$ by inverting E equivalences.

For future references we note the following lemma.

Lemma 3.7. The resolution (*) obtained from the canonical Adams resolution is isomorphic to the cobar resolution $C(E_*E, E_*Y)$ of Definition 1.1. If E_*X is π_*E projective then the E_1 term of the resulting spectral sequence is isomorphic to $C(E_*X, E_*E, E_*Y)$.

Proof. The isomorphism θ_1 converts the cobar resolution into (*). If E_*X is π_*E projective we use the natural isomorphism 3.6. //

In the next section we will need the following result on maps of Adams resolutions.

Proposition 3.8: Suppose E_*X is π_*E projective, $\{X_i\}$ and $\{Y_i\}$ are Adams resolutions of X and Y, and each E_*X_s is π_*E projective. Let $f: X \to Y$ and let $\tilde{f}$ be a chain homomorphism extending f_*:

$$\begin{array}{ccccccccc}
0 & \longrightarrow & E_*X & \longrightarrow & E_*X_{0,1} & \longrightarrow & E_*X_{1,1} & \longrightarrow & \cdots \\
 & & \downarrow{\scriptstyle f_*} & & \downarrow{\scriptstyle \tilde{f}_0} & & \downarrow{\scriptstyle \tilde{f}_1} & & \\
0 & \longrightarrow & E_*Y & \longrightarrow & E_*Y_{0,1} & \longrightarrow & E_*Y_{1,1} & \longrightarrow & \cdots
\end{array} .$$

Then there is a map of Adams resolutions extending f and inducing $\tilde{f}$.

Proof. Since all E_*X_i are π_*E projective, so are all $E_*X_{i,1}$ ($\cong E_*X_i \oplus E_*X_{i+1}$). Hence

$$[X_{i,1},Y_{i,1}]_* \cong \mathrm{Hom}(E_*X_{i,1},E_*Y_{i,1})$$

and the $\tilde{f}_i$ correspond to unique maps $\overline{f}_i:X_{i,1} \to Y_{i,1}$ such that $(\overline{f}_i)_* = \tilde{f}_i$. We construct $f_i:X_i \to Y_i$ commuting with f_{i-1} and $\overline{f}_i$ by induction. When $i = 0$ we let $f_0 = f$. This commutes with $\overline{f}_0$ since it commutes after applying E_* and

$$[X,Y_{0,1}]_* \cong \mathrm{Hom}(E_*X,E_*Y_{0,1}).$$

Assume $f_0,f_1,\ldots,f_{i-1}$ have been constructed. Let f_i be a map which makes the following diagram commute.

$$\begin{array}{ccccccc}
X_{i-1,1} & \longrightarrow & X_i & \longrightarrow & X_{i-1} & \longrightarrow & X_{i-1,1} \\
\downarrow{\scriptstyle \overline{f}_{i-1}} & & \vdots{\scriptstyle f_i} & & \downarrow{\scriptstyle f_{i-1}} & & \downarrow{\scriptstyle \overline{f}_{i-1}} \\
Y_{i-1,1} & \longrightarrow & Y_i & \longrightarrow & Y_{i-1} & \longrightarrow & Y_{i-1,1}
\end{array}$$

To see that f_i commutes with $\overline{f}_i$ we need only check that it commutes after applying E_*, and this holds because it holds after composing with the epimorphism $E_*X_{i-1,1} \to E_*X_i$. This completes the induction. //

§4. Smash Products in the Adams Spectral Sequence

We are now ready to introduce smash products into the Adams spectral sequence. Our main result is

Theorem 4.4: There is a pairing of Adams spectral sequences

$$E_r^{**}(X,Y) \otimes E_r^{**}(X',Y') \to E_r^{**}(X\wedge X',Y\wedge Y')$$

converging to the smash product

$$[X,Y]_* \otimes [X',Y']_* \to [X\wedge X',Y\wedge Y']_* .$$

If E_*X and E_*X' are π_*E projective then the pairing on E_2 is the external product

$$\mathrm{Ext}(E_*X,E_*Y) \otimes \mathrm{Ext}(E_*X',E_*Y') \to \mathrm{Ext}(E_*X \otimes E_*X',E_*Y \otimes E_*Y')$$

composed with the homomorphisms induced by

$$E_*(X\wedge X') \xrightarrow{\cong} E_*X \otimes E_*X'$$

and

$$E_*Y \otimes E_*Y' \longrightarrow E_*(Y\wedge Y').$$

(Note that the preceding isomorphism is the inverse of the external product $E_*X \otimes E_*X' \to E_*(X\wedge X')$, and is an isomorphism because E_*X and E_*X' are π_*E projective.)

As a corollary we have

Corollary 4.5: (i) $\{E_r(S,S)\}$ is a spectral sequence of bigraded commutative algebras.

(ii) $E_r(X,Y)$ is a differential $E_r(S,S)$ module.

(iii) If $X = \Sigma^\infty Z$ for some space Z, and if Y is a commutative ring spectrum then $\{E_r(X,Y)\}$ is a spectral sequence of bigraded commutative $\{E_r(S,S)\}$ algebras whose product converges to the smash product internalized by means of the diagonal $\Delta: X \to X\wedge X$ and the product $\mu: Y\wedge Y \to Y$. If Z has a disjoint basepoint, then the $E_r(X,Y)$ are unital.

In the ordinary Adams spectral sequence ($E = HZ_p$, p prime) these results are quite easy. If $\{Y_i\}$ and $\{Y_i'\}$ are Adams resolutions of Y and Y', then their smash product $\{Y_i\}\wedge\{Y_i'\}$ (to be defined shortly) is an Adams resolution of $Y\wedge Y'$. The pairing in Theorem 4.4 is then obtained by simply taking the smash product of representative maps. To get the internal product of Corollary 4.5 we need only note that the product $Y\wedge Y \to Y$ is covered by a map of Adams resolutions $\{Y_i\}\wedge\{Y_i\} \to \{Y_i\}$. In the general case, this plan of proof encounters two obstacles. First, the smash product of Adams resolutions may or may not be an Adams resolution. Second, a map $X \to Y$ may or may not be covered by a map from a given Adams resolution of X to a given Adams resolution of Y. There are two facts which enable us to avoid these difficulties. First, for spectra which have π_*E projective E-homology, everything works as in the ordinary case. Second, all the Adams resolutions we need have the following form: spectrum to be resolved smashed with an Adams resolution of a sphere. This enables us to reduce to the case of the sphere spectrum, for which everything works as in the ordinary case, since E_*S is π_*S projective. The details follow.

Lemma 4.1. Let (X,A,U) and (Y,B,V) be CW triples. The geometric boundary ∂ makes the following diagram commute.

$$\begin{array}{ccc}
\dfrac{X \wedge Y}{A \wedge Y \cup X \wedge B} & \cong & \dfrac{X}{A} \wedge \dfrac{Y}{B} \\
\Big\downarrow \partial & & \Big\downarrow (\partial \quad 1) \quad (1 \quad \partial) \\
\dfrac{A \wedge Y \cup X \wedge B}{U \wedge Y \cup A \wedge B \cup X \wedge V} & \cong & \left(\dfrac{A}{U} \wedge \dfrac{Y}{B}\right) \vee \left(\dfrac{X}{A} \wedge \dfrac{B}{V}\right)
\end{array}$$

Definition 4.2. Let $X_0 \xleftarrow{i_0} X_1 \xleftarrow{i_1} X_2 \xleftarrow{i_2} \cdots$

and $Y_0 \xleftarrow{j_0} Y_1 \xleftarrow{j_1} Y_2 \xleftarrow{j_2} \cdots$

be inverse systems in which each map is the inclusion of a subcomplex. The product $\{X_i\} \wedge \{Y_i\}$ is the inverse system

$$Z_0 \xleftarrow{k_0} Z_1 \xleftarrow{k_1} \cdots$$

where $Z_n = \bigcup_{i+j=n} X_i \wedge Y_j$.

Proposition 4.3: Let $\{X_i\}$ and $\{Y_i\}$ be Adams resolutions of X and Y. Then $\{X_i\} \wedge \{Y_i\}$ is an Adams resolution of $X \wedge Y$ if either

(a) E_*X and E_*X_i for each i are π_*E projective

or (b) $\{X_i\}$ and $\{Y_i\}$ are the canonical Adams resolutions.

The resolution of $E_*(X \wedge Y)$ associated to $\{X_i\} \wedge \{Y_i\}$ is, respectively,

(a) the tensor product of the resolutions associated to $\{X_i\}$ and $\{Y_i\}$,

or (b) $E_*(X \wedge Y) \otimes C(E_*E, \pi_*E) \otimes C(E_*E, \pi_*E) \cong C(E_*E, E_*(X \wedge Y)) \otimes C(E_*E, \pi_*E)$

$$\cong C(E_*E \otimes E_*E, E_*(X \wedge Y)).$$

(Recall that $C(A,M) = M \otimes C(A,R)$. Also, in case (a) note that the split exact sequences

$$0 \longrightarrow E_*X_n \longrightarrow E_*X_{n,1} \longrightarrow E_*X_{n+1} \longrightarrow 0$$

show that if two of $X_n, X_{n,1}$ and X_{n+1} have π_*E projective E-homology, so does the third. Hence, if E_*X is π_*E projective, then X has Adams resolutions $\{X_i\}$ in which each E_*X_i is π_*E projective. The canonical Adams resolution is one such.)

Proof: Use the notation of Definition 4.2. The equivalence

$$Z_{n,1} \simeq \bigvee_{p+q=n} X_{p,1} \wedge Y_{q,1}$$

implies that Definition 3.5.(i) is satisfied in either case.

Suppose E_*X_n is π_*E projective for each n. Then $E_*X_{n,1}$ is also π_*E projective for each n. Hence $E_*(X_{p,1} \wedge Y_{q,1}) \cong E_*X_{p,1} \otimes E_*X_{q,1}$. This and Lemma 4.1 imply that

$$0 \longrightarrow E_*(X \wedge Y) \longrightarrow E_*Z_{0,1} \longrightarrow E_*Z_{1,1} \longrightarrow \cdots$$

is the tensor product of the resolutions associated to $\{X_i\}$ and $\{Y_i\}$, and is therefore π_*E split since each of the factors is. This implies that $\{X_i\} \wedge \{Y_i\}$ satisfies Definition 3.5(ii) and is therefore an Adams resolution of $X \wedge Y$. This completes case (a).

Let $\{E_i\}$ be the canonical Adams resolution of S, and let $\{F_i\} = \{E_i\} \wedge \{E_i\}$. By (a), this is also an Adams resolution of S and its associated resolution of E_*S is $C(E_*E,\pi_*E) \otimes C(E_*E,\pi_*E)$ (by Lemma 3.7). The canonical Adams resolutions of X and Y are $X \wedge \{E_i\} = \{X \wedge E_i\}$ and $Y \wedge \{E_i\}$, and their smash product is $X \wedge Y \wedge \{F_i\}$. Since each E_*F_i is π_*E projective, (b) follows immediately. //

Proof of Theorem 4.4. Let $\{E_n\}$ be the canonical Adams resolution of S and let $\{F_n\} = \{E_n\} \wedge \{E_n\}$. Let $\gamma = \{\gamma_i\}:\{F_n\} \to \{E_n\}$ be a map of Adams resolutions which extends the equivalence $S \wedge S \to S$. Define a pairing of spectral sequences

$$E_r(X,Y) \otimes E_r(X',Y') \to E_r(X \wedge X', Y \wedge Y')$$

by composing the smash product

$$[X,Y \wedge E_{s,r}]_n \otimes [X',Y' \wedge E_{s',r}]_{n'} \longrightarrow [X \wedge X', Y \wedge E_{s,r} \wedge Y' \wedge E_{s',r}]_{n+n'}$$

with the homomorphism induced by

$$\begin{array}{ccccc} Y \wedge E_{s,r} \wedge Y' \wedge E_{s',r} & \xrightarrow{1 \wedge \tau \wedge 1} & Y \wedge Y' \wedge E_{s,r} \wedge E_{s',r} & \subset & Y \wedge Y' \wedge F_{s+s',r} \\ & & & & \downarrow{\scriptstyle 1 \wedge \bar{\gamma}} \\ & & & & Y \wedge Y' \wedge E_{s+s',r} \end{array}$$

where $\bar{\gamma}$ is a map of cofibers induced by γ. According to [64], this induces a pairing of spectral sequences if

(1) the pairing on E_r induces that on E_{r+1}

and, (2) d_r acts as a derivation with respect to it.

Condition (1) is obviously satisfied, and condition (2) is an immediate consequence of Lemma 4.1 and the fact that $(1\wedge\partial)(f\wedge g) = (-1)^{|f|}f\wedge\partial g$.

It is clear that this pairing converges to the smash product.

That the pairing on E_2 is as stated when E_*X and E_*X' are π_*E projective follows from the commutativity and naturality of the following diagram

$$\begin{array}{ccc}
[X,Y]_n \otimes [X',Y']_n & \xrightarrow{\ \wedge\ } & [X\wedge X', Y\wedge Y']_{n+n'} \\
\Big\downarrow{\scriptstyle E_*\otimes E_*} & & \Big\downarrow{\scriptstyle E_*} \\
\mathrm{Hom}^n(E_*X,E_*Y)\otimes \mathrm{Hom}^{n'}(E_*X',E_*Y') & & \mathrm{Hom}^{n+n'}(E_*(X\wedge X'),E(Y\wedge Y')) \\
\Big\downarrow{\scriptstyle \otimes} & & \Big\downarrow{\scriptstyle \kappa^*} \\
\mathrm{Hom}^{n+n'}(E_*X\otimes E_*X',E_*Y\otimes E_*Y') & \xrightarrow{\ \kappa_*\ } & \mathrm{Hom}^{n+n'}(E_*X\otimes E_*X',E_*(Y\wedge Y'))
\end{array}$$

(Here $\kappa: E_*X\otimes E_*Y \to E_*(X\wedge Y)$ is the Kunneth homomorphism.) //

§5. Extended Powers in the Adams Spectral Sequence

We are now prepared to show that if Y is a commutative ring spectrum whose r^{th} power map $Y^{(r)} \to Y$ extends to a map

$$\xi: D_rY \to Y,$$

then ξ can be used to construct a homomorphism of the type used in §2 to define Steenrod operations in $\mathrm{Ext}_{E_*E}(M,E_*Y)$. Assume given such a spectrum Y and map ξ throughout this section. As a consequence, we obtain in Corollary 5.4 an explicit representative map for $\beta^\varepsilon P^j x$ given a representative map for x. In chapter VI this will enable us to compute some differentials on $\beta^\varepsilon P^j x$.

Let $\pi \subset \Sigma_r$ and let W_n be the n-skeleton of a contractible π free CW complex W. Assume that $W_0 = \pi$. The skeletal filtration of W induces a filtration $D^i_\pi Y = W_i \ltimes_\pi Y^{(r)}$ of $D_\pi Y = W \ltimes_\pi Y^{(r)}$ and, more generally, a filtration $W_i \ltimes_\pi X$ of $W\ltimes_\pi X$, where X is any π spectrum.

Let E be a ring spectrum which satisfies Condition 3.6 and for which E_*E is π_*E flat. Let

$$Y \simeq Y_0 \leftarrow Y_1 \leftarrow \cdots$$

be an Adams resolution with respect to E. Let $\{F_s\}$ be the r-fold smash product $\{Y_s\}^{(r)}$. The π action on $F_0 = Y_0^{(r)}$ is cellular and F_{s+1} is a π subcomplex of F_s for each s. Thus we may define

$$Z = D_\pi Y_0 = W \ltimes_\pi Y_0^{(r)} = W \ltimes_\pi F_0$$

and

$$Z_{i,s} = W_i \ltimes_\pi F_s .$$

<u>Lemma 5.1</u>: Let $B_i = W_i/\pi$.

(i) $Z_{i-1,s}$ and $Z_{i,s+1}$ are subcomplexes of $Z_{i,s}$

(ii) $$\frac{Z_{i,s}}{Z_{i-1,s}} \simeq \frac{B_i}{B_{i-1}} \wedge F_s$$

(iii) $$\frac{Z_{i,s}}{Z_{i-1,s} \cup Z_{i,s+1}} \simeq \frac{B_i}{B_{i-1}} \wedge \frac{F_s}{F_{s+1}}$$

(iv) The following diagram commutes.

$$\begin{array}{ccc}
\dfrac{Z_{i,s}}{Z_{i-1,s} \cup Z_{i,s+1}} & \simeq & \dfrac{B_i}{B_{i-1}} \wedge \dfrac{F_s}{F_{s+1}} \\
\Big\downarrow \partial & & \Big\downarrow \partial \wedge 1 \vee 1 \wedge \partial \\
\dfrac{Z_{i-1,s}}{Z_{i-2,s} \cup Z_{i-1,s+1}} \vee \dfrac{Z_{i,s+1}}{Z_{i-1,s+1} \cup Z_{i,s+2}} & \simeq & \dfrac{B_{i-1}}{B_{i-2}} \wedge \dfrac{F_s}{F_{s+1}} \vee \dfrac{B_i}{B_{i-1}} \wedge \dfrac{F_{s+1}}{F_{s+2}}
\end{array}$$

<u>Proof</u>. Parts (i), (ii) and (iii) are in Theorem I.1.3. Part (iv) is much more delicate and is proved in [Equiv, VI. 4.9 and VIII. 2.7]. //

<u>Theorem 5.2</u>: If E_*Y_s is π_*E projective for each s then there exist maps $\xi_{i,s}\colon Z_{i,s} \to Y_{s-i}$ which make the following diagrams commute.

$$\begin{array}{ccc}
D_\pi Y & \longleftarrow & Z_{i,s} \\
\xi \downarrow & & \downarrow \xi_{i,s} \\
Y & \longleftarrow & Y_{s-i}
\end{array}
\qquad
\begin{array}{ccccc}
Z_{i,s-1} & \longleftarrow & Z_{i,s} & \longleftarrow & Z_{i-1,s} \\
\downarrow \xi_{i,s-1} & & \downarrow \xi_{i,s} & & \downarrow \xi_{i-1,s} \\
Y_{s-i-1} & \longleftarrow & Y_{s-i} & \longleftarrow & Y_{s-i+1}
\end{array}$$

<u>Proof</u>. Since $W_0 = \pi$, $Z_{0,s} = F_s$. Thus we may let $\xi_{0,s}$ be the map of Adams resolutions which Proposition 4.4 ensures us is induced by $Y^{(r)} \to Y$. For induction we may suppose $\xi_{i,s}$ constructed satisfying the theorem for $i < k$. The maps $\xi_{k,s}$ for $s < k$ are defined to be

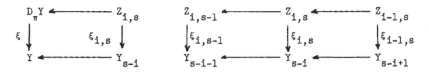

$$Z_{k,s} \longrightarrow W \ltimes_{\pi} F_0 \xrightarrow{\ \xi\ } Y.$$

Hence we may also assume that $\xi_{k,s'}$ has been constructed satisfying the theorem for $s' < s$. To construct $\xi_{k,s}$ compatible with $\xi_{k,s-1}$ and $\xi_{k-1,s}$, we need $\xi_{k,s}$ to make the following diagram commute.

$$\begin{array}{ccccc}
Z_{k-1,s} & \longrightarrow & Z_{k-1,s-1} & \xrightarrow{\ \xi_{k-1,s-1}\ } & Y_{s-k} \\
\big\downarrow & & \big\downarrow & \overset{\xi_{k,s}}{\dashrightarrow} \ (\text{from } Z_{k,s}) & \big\downarrow i_{s-k-1} \\
Z_{k,s} & \longrightarrow & Z_{k,s-1} & \xrightarrow{\ \xi_{k,s-1}\ } & Y_{s-k-1}
\end{array}$$

The obstruction to the existence of such a $\xi_{k,s}$ lies in $[Z_{k,s}/Z_{k-1,s}, Y_{s-k-1,1}]$ and by naturality lies in the image of $[Z_{k,s-1}/Z_{k-1,s-1}, Y_{s-k-1,1}]$. By Lemma 5.1.(ii), $E_*(Z_{i,s}/Z_{i-1,s})$ is π_*E projective for each i, and hence

$$[Z_{k,s}/Z_{k-1,s}, Y_{s-k-1,1}] \cong \mathrm{Hom}_{E_*E}(E_*(Z_{k,s}/Z_{k-1,s}), E_*Y_{s-k-1,1}).$$

The equivalence 5.1.(ii) converts the inclusion $Z_{k,s}/Z_{k-1,s} \to Z_{k,s-1}/Z_{k-1,s-1}$ into $1 \wedge j_{s-1}$ where j_{s-1} is the inclusion $F_s \to F_{s-1}$. Since $E_*j_{s-1} = 0$, the obstruction to the existence of $\xi_{k,s}$ is 0. //

If we define $\mathcal{W}_k$ to be $\pi_k(W_k/W_{k-1})$ and $d\colon \mathcal{W}_k \to \mathcal{W}_{k-1}$ to be ∂_*, we obtain a $Z[\pi]$-free resolution of Z with $\mathcal{W}_0 = Z[\pi]$. Let $C_{s,t} = E_{t-s}Y_{s,1}$. Then $0 \to C_0 \to C_1 \to C_2 \to \cdots$ is the resolution of E_*Y associated to $\{Y_s\}$. If each E_*Y_s is π_*E projective then the Kunneth homomorphism is an isomorphism from C^r to the resolution associated to $\{F_s\}$. Let $h_E\colon \pi_* \to E_*$ be the Hurewicz homomorphism, κ the Kunneth homomorphism, and assume $\pi_0 E = Z_p$.

Corollary 5.3. If Φ is defined to make the diagram

$$\begin{array}{ccc}
\mathcal{W}_k \otimes \bigotimes_i C_{s_i,t_i} & \xrightarrow{\ h_E \otimes 1\ } & E_k(W_k/W_{k-1}) \otimes \bigotimes_i E_{t_i-s_i}Y_{s_i,1} \\
 & & \big\downarrow \kappa \\
\Phi \,\big\downarrow\, (\text{dashed}) & & E_{t-s+k}(W_k/W_{k-1} \wedge F_{s,1}) \\
 & & \big\downarrow \xi_{k,s\,*} \\
C_{s-k,t} & = & E_{t-s+k}Y_{s-k,1}
\end{array}$$

commute (where $t = t_1 + \cdots + t_r$ and $s = s_1 + \cdots + s_r$), then Φ is in the chain homotopy class described in Lemma 2.3.

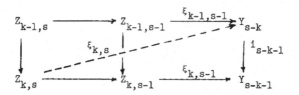

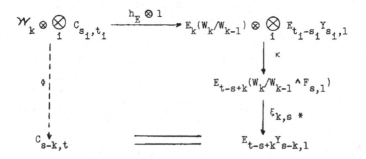

Proof. The E_*E comodule structure, the π action, and the differential on $\mathcal{W} \otimes C^r$ are specified in Lemma 2.3. By 5.1.(iv), Φ respects the differential. Since ξ restricts to the product $\{Y_s\}^r \to \{Y_s\}$, Φ restricts to the product $C^r \to C$. Both $\xi_{k,s*}$ and κ are comodule maps, while $h_E \otimes 1$ is a comodule map because the image of h_E is primitive. Φ is π-equivariant because $\xi_{k,s}$ is defiend on the orbit spectrum $W^k \ltimes_\pi F_s$. //

Now assume X is a spectrum with a coproduct $\Delta: X \to X \wedge X$. For example, X could be a suspension spectrum with its natural diagonal. Assume also that E_*X is π_*E projective so that $\kappa:(E_*X)^r \to E_*(X^{(r)})$ is an isomorphism.

Corollary 5.4: If $e \in \mathcal{W}_k$ and $f_j \in [X, Y_{s_j,1}]_{t_j-s_j}$ then $\Phi_*(e \otimes f_{1*} \otimes \cdots \otimes f_{r*})$ is represented by the composite

$$\begin{array}{ccccc}
\Sigma^{t-s+k}X & - - - - - - - - - - - - - - \dashrightarrow & & & Y_{s-k,1} \\
\Big\downarrow \Sigma^{t-s+k}\Delta & & & & \Big\uparrow \xi_{k,s} \\
\Sigma^{t-s+k}X^{(r)} & \longrightarrow & \Sigma^k S \wedge \bigwedge_j \Sigma^{t_j-s_j}X & \xrightarrow[e \wedge \bigwedge_j f_j]{} & W_k/W_{k-1} \wedge \bigwedge_j Y_{s_j,1}
\end{array}$$

Proof. Consider the following diagram

$$\begin{array}{ccc}
\mathcal{W}_k \otimes \bigotimes_j \mathrm{Hom}^{t_j}(E_*X, C_{s_j}) & \cong & \pi_k(W_k/W_{k-1}) \otimes \bigotimes_j [X, Y_{s_j,1}]_{t_j-s_j} \\
\Big\downarrow \otimes & & \Big\downarrow \wedge \\
\mathcal{W}_k \otimes \mathrm{Hom}^t((E_*X)^r, \bigotimes_j C_{s_j}) & & [X^{(r)}, (W_k/W_{k-1}) \wedge \bigwedge_j Y_{s_j,1}]_{t-s+k} \\
\wr\| & & \\
\mathrm{Hom}^t((E_*X)^r, \mathcal{W}_k \otimes \bigotimes_j C_{s_j}) & & \Big\downarrow [\Delta, \xi_{k,s}] \\
\Big\downarrow \mathrm{Hom}(\Delta, \Phi) & & \\
\mathrm{Hom}^t(E_*X, C_{s-k}) & \cong & [X, Y_{s-k,1}]_{t-s+k}
\end{array}$$

The left column is the homomorphism Φ used in §2 to define Steenrod operations in $\mathrm{Ext}_{E_*E}(E_*X, E_*Y)$. The right column sends $e \otimes f_1 \otimes \cdots \otimes f_r$ to the composite which the corollary asserts represents $\Phi_*(e \otimes f_1 \otimes \cdots \otimes f_{r*})$. Thus we need only show that this diagram commutes. This is an easy diagram chase from the following two facts. First, there is the relation between $\otimes$ and $\wedge$ expressed by the diagram at the end of §4. Second, the homomorphism

$$\alpha: \mathcal{W}_k \otimes \mathrm{Hom}(M,N) \to \mathrm{Hom}(M, \mathcal{W}_k \otimes N)$$

given by $\alpha(e \otimes f)(m) = e \otimes f(m)$, when composed with $\mathrm{Hom}(1, h_E \otimes 1)$, sends $e \otimes f$ to

$$M \cong E_*S \otimes M \xrightarrow{e_* \otimes f} E_*(W_k/W_{k-1}) \otimes N. \qquad //$$

Remark 5.5: When $Y = S$ we are in the situation studied by Kahn [45], Milgram [81] and Mäkinen [62]. They worked unstably, and in place of the H_∞ structure map $\xi: D_pS \to S$, used coreductions $\theta_{i,n}: D_p^iS^n \to S^{np}$. (A coreduction is a map which, together with the inclusion $\iota: S^{np} \to D_pS^n$, splits off the bottom cell.) Such coreductions exist for n even and congruent to 0 modulo a power of p increasing with i (Theorems V.2.9 and V.2.14). They can be obtained by "destabilizing" ξ as follows. In V §2 we will show that $D_2^iS^n \simeq \Sigma^nP_n^{n+i}$ and that $P_n^{n+i} \simeq \Sigma^nP_0^i$ if $n \equiv 0 \ (2^{\phi(i)})$ (and similarly for odd primes). Thus, the following composite is a coreduction.

$$\begin{array}{ccccc}
D_2^iS^n & & \dashrightarrow & & S^{2n} \\
\| \wr & & & & \| \\
\Sigma^nP_n^{n+i} & & & & \\
\| \wr & & & & \\
\Sigma^{2n}P_0 & & & & \\
\| \wr & & & & \\
\Sigma^{2n}D_2^iS & \longrightarrow & \Sigma^{2n}D_2S & \xrightarrow{\Sigma^{2n}\xi} & \Sigma^{2n}S.
\end{array}$$

This implies that we are looking at the same structure they were considering.

§6. Milgram's Generalization of the Adams Spectral Sequence

In [81] and [80], Milgram introduced a generalization of the Adams spectral sequence and used it to study differentials of the form $d_r\beta^\epsilon P^jX$ in the mod p Adams spectral sequence for π_*S^0. The essential idea behind the spectral sequence is this. The Adams spectral sequence for maps into Z arises from a geometric construction of a resolution of H_*Z. Suppose that we have a filtration of H_*Z of the form

$$H_*Z \supset H_*Z_1 \supset H_*Z_2 \supset \cdots$$

for some sequence (usually finite) of maps $Z \leftarrow Z_1 \leftarrow Z_2 \leftarrow \cdots$. Milgram's idea was to construct a geometric resolution of Z in which we delay the resolution of H_*Z_i so that it begins in filtration i. The Adams spectral sequence is then the special case defined by $Z \leftarrow * \leftarrow * \leftarrow \cdots$. When Z_i is the N-i skeleton of an N dimensional complex Z, the differentials are determined by and provide a clear picture of the attaching maps.

Continue to assume that E is a ring spectrum such that E_*E is π_*E flat and which satisfies Condition 3.6.

Theorem 6.1: Let

$$\mathfrak{z} = \{Z = Z_0 \xleftarrow{f_0} Z_1 \xleftarrow{f_1} Z_2 \xleftarrow{f_2} \cdots\}$$

be a sequence in which E_*Z_i is π_*E projective and E_*f_i is a π_*E split monomorphism for each i. Then

(i) there exists a spectral sequence $E_r^{**}(X,\mathfrak{z})$, natural with respect to maps of such sequences, such that

$$E_2^{s,t}(X,\mathfrak{z}) = \bigoplus_i E_2^{s-i,t-i}(X,Cf_i) ,$$

where $E_r^{**}(X,Cf_i)$ is the Adams spectral sequence converging to $[X,Cf_i]_*^E$;

(ii) if E_*Y' is π_*E projective and

$$\mathfrak{z} \wedge Y' = \{Z_0 \wedge Y' \xleftarrow{f_0 \wedge 1} Z_1 \wedge Y' \xleftarrow{f_1 \wedge 1} \cdots\}$$

there is a pairing

$$\begin{array}{ccc} E_r^{**}(X,\mathfrak{z}) \otimes E_r^{**}(X',Y') & \longrightarrow & E_r^{**}(X \wedge X', \mathfrak{z} \wedge Y') \\ \Downarrow & & \Downarrow \\ [X,Z]_*^E \otimes [X',Y']_*^E & \xrightarrow{\wedge} & [X \wedge X', Z \wedge Y']_*^E \end{array}$$

which is the direct sum of the smash product pairings on E_2

$$E_2^{**}(X,Cf_i) \otimes E_2^{**}(X',Y') \longrightarrow E_2^{**}(X \wedge X', Cf_i \wedge Y');$$

(iii) if

$$\begin{array}{ccccc} & Z_0 & \xleftarrow{f_0} & Z_1 & \xleftarrow{f_1} \cdots \\ & {\scriptstyle c_0}\downarrow & & \downarrow{\scriptstyle c_1} & \\ Y = & Y_0 & \xleftarrow{i_0} & Y_1 & \xleftarrow{i_1} \cdots \end{array}$$

is a map from $\mathfrak{z}$ into an Adams resolution of Y, then there is a homomorphism c of spectral sequences

$$\begin{array}{ccc} E_r^{**}(X,\mathfrak{z}) & \Longrightarrow & [X,Z]_*^E \\ {\scriptstyle c}\downarrow & & \downarrow{\scriptstyle c_{0*}} \\ E_r^{**}(X,Y) & \Longrightarrow & [X,Y]_*^E \end{array}$$

which maps the pairing in (ii) to the smash product pairing

$$\begin{array}{ccc} E_r^{**}(X,\mathfrak{Z}) \otimes E_r^{**}(X',Y') & \longrightarrow & E_r^{**}(X\wedge X',\mathfrak{Z}\wedge Y') \\ \downarrow{\scriptstyle c\,\otimes\,1} & & \downarrow{\scriptstyle c} \\ E_r^{**}(X,Y) \otimes E_r^{**}(X',Y') & \longrightarrow & E_r^{**}(X\wedge X',Y\wedge Y') \end{array} ;$$

(iv) the spectral sequence $E_r^{**}(X,\mathfrak{Z})$ converges to $[X,Z]_*^E$ if

(1) E and Z satisfy Adams' condition for convergence of the Adams spectral sequence $E_r^{**}(X,Z) \Longrightarrow [X,Z]_*^E$ (stated below) and

(2) $E_*(\text{Mic } Z_i) = 0$, where Mic Z_i is the microscope, or homotopy inverse limit of the Z_i.

<u>Remarks</u>: Adams' conditions for the convergence of $E_r^{**}(X,Z) \Longrightarrow [X,Z]_*^E$ are

(a) Z is bounded below,

(b) E is connective and $\mu_*: \pi_0 E \underset{Z}{\otimes} \pi_0 E \to \pi_0 E$ is an isomorphism,

(c) if $R \subset Q$ is maximal such that the natural ring homomorphism $Z \to \pi_0 E$ extends to $R \to \pi_0 E$ then $H_r E$ is finitely generated as an R-module for all r;

see [6].

The proof of the convergence will show that $E_*\text{Mic } Z_i = 0$ is equivalent to $\lim E_* Z_i = 0 = \lim^1 E_* Z_i$.

<u>Proof</u>. First we will construct a new inverse system into which Z maps and from which $E_r^{**}(X,\mathfrak{Z})$ will be obtained by applying $[X,?]_*^E$. Then we will show that E_2^{**} splits as stated. Next we will prove a statement which will imply naturality of the spectral sequence and the first part of (iii) simultaneously. The next step is to construct the smash product pairing and prove (ii) and the last statement in (iii). Finally we prove convergence.

To construct the inverse system from which the spectral sequence will be obtained, we begin by choosing Adams resolutions

$$Cf_i = Z_{i,0} \xleftarrow{f_{i,0}} Z_{i,1} \xleftarrow{f_{i,1}} Z_{i,2} \xleftarrow{f_{i,2}} \cdots .$$

Let $\pi_{i,j}: Z_{i,j} \to Cf_{i,j}$ be the natural map. Since E_*Cf_i is a direct summand of E_*Y_i (as π_*E modules), E_*Cf_i is π_*E projective. Thus we may assume that $E_*Z_{i,j}$ is π_*E projective for all i and j. We will inductively construct spectra $\overline{Z}_i$ and maps $\overline{f}_i: \overline{Z}_{i+1} \to \overline{Z}_i$ and $e_i: Z_i \to \overline{Z}_i$ such that we have a map of inverse systems

$$\begin{array}{ccccccc}
Z_0 & \xleftarrow{f_0} & Z_1 & \xleftarrow{f_1} & Z_2 & \xleftarrow{f_2} & \cdots \\
\downarrow{\scriptstyle e_0} & & \downarrow{\scriptstyle e_1} & & \downarrow{\scriptstyle e_2} & & \\
\overline{Z}_0 & \xleftarrow{\overline{f}_0} & \overline{Z}_1 & \xleftarrow{\overline{f}_1} & \overline{Z}_2 & \xleftarrow{\overline{f}_2} & \cdots
\end{array}$$

To start the induction, let $\overline{Z}_0 = Z_0$ and $e_0 = 1$. Assume for induction that we have constructed $\overline{Z}_0 \leftarrow \cdots \leftarrow \overline{Z}_k$ and $e_0,\ldots,e_k$ such that for each i, $0 \le i \le k$, there is an E_*E comodule isomorphism

$$E_*\overline{Z}_i \cong E_*Z_i \oplus E_*Z_{i-1,1} \oplus \cdots \oplus E_*Z_{0,i}$$

under which e_i is inclusion of the first summand. This implies that $E_*\overline{Z}_k$ is π_*E projective. Thus we may define

$$\pi_k : \overline{Z}_k \longrightarrow Cf_{k,0} \vee \cdots \vee Cf_{0,k}$$

by requiring that $\pi_{k*} \cong (\pi_{k,0}\phi_k)_* \oplus \pi_{k-1,1\,*} \oplus \cdots \oplus \pi_{0,k\,*}$ under the isomorphism, where $\phi_k : Z_k \to Cf_k = Z_{k,0}$ is the natural map. Define $\overline{f}_k : \overline{Z}_{k+1} \to \overline{Z}_k$ to be the fiber of π_k. The definition of π_k implies that the following diagram commutes, thereby inducing e_{k+1}.

$$\begin{array}{ccccc}
Z_{k+1} & \xrightarrow{f_k} & Z_k & \xrightarrow{\phi_k} & Cf_k \\
\downarrow{\scriptstyle e_{k+1}} & & \downarrow{\scriptstyle e_k} & & \downarrow{\scriptstyle (\pi_{k,0},0,\ldots,0)} \\
\overline{Z}_{k+1} & \xrightarrow{\overline{f}_k} & \overline{Z}_k & \xrightarrow{\pi_k} & Cf_{k,0} \vee \cdots \vee Cf_{0,k}
\end{array}$$

Let $C = C(e_kf_k)$ and $D = Ce_{k+1}$ be the indicated cofibers, and consider the following braid of cofibrations.

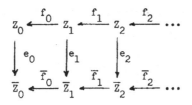

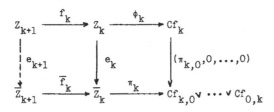

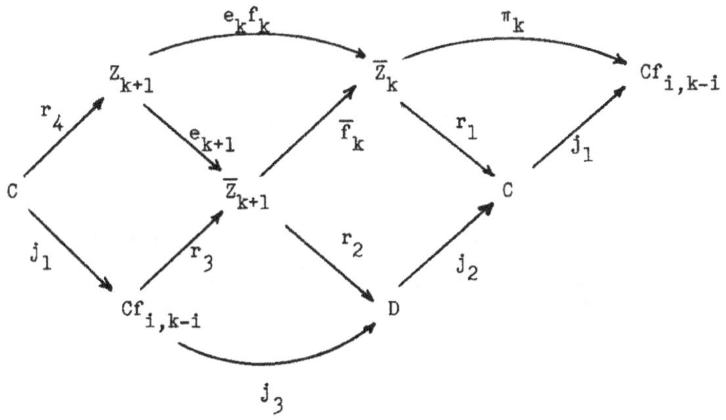

Since $(e_kf_k)_*$ is a monomorphism, r_{1*} is an epimorphism and hence

$$E_*C \cong E_*Z_{k,0} \oplus E_*Z_{k-1,1} \oplus \cdots E_kZ_{0,k}.$$

Since $j_1r_1 = \pi_k$ we must have $j_{1*} = \oplus \pi_{i,k-i\,*}$. This is a monomorphism and, hence, $j_{3\,*}$ is an epimorphism. It follows that

$$E_*D \cong E_*Z_{k,1} \oplus \cdots \oplus E_*Z_{0,k+1}\ .$$

Now $r_{4\,*} = 0$ because $(e_kf_k)_*$ is a monomorphism. It follows that $(r_3j_1)_* = 0$. This implies that there is a unique homomorphism $r: E_*D \to E_*\overline{Z}_{k+1}$ such that $rj_{3\,*} = r_{3\,*}$. Thus $r_{2*}r = 1$, from which it follows that $E_*\overline{Z}_{k+1} \cong E_*Z_{k+1} \oplus E_*D$ and that $e_{k+1\,*}$ is inclusion of the first summand. This completes the induction. We define $E_r^{**}(X,\mathfrak{Z})$ to be the spectral sequence obtained by applying $[X,?]_*^E$ to the inverse system $\{\overline{f}_i\}$. It is clear that

$$E_1^{s,t}(X,\mathfrak{Z}) = \bigoplus_i E_1^{s-i,t-i}(X,Cf_i).$$

To show that the same splitting applies to E_2^{**} we need only show that d_1 is the direct sum $\bigoplus_i d_1$. For each k, $C\overline{f}_k = Cf_{i,k-i}$ is a retract of $C_k \wedge E$ for some C_k and $E_*C\overline{f}_k$ is π_*E projective. Therefore, the map $C\overline{f}_k \to \overline{Z}_{k+1} \to C\overline{f}_{k+1}$ is completely determined by its induced homomorphism $E_*C\overline{f}_k \to E_*C\overline{f}_{k+1}$, which splits as desired by construction. In other words, the sequence

$$0 \to E_*C\overline{f}_0 \to E_*C\overline{f}_1 \to \cdots$$

is the direct sum of the sequences

$$0 \to E_*Cf_{i,0} \to E_*Cf_{i,1} \to \cdots$$

with the ith sequence delayed until homological degree i before it begins:

$$\begin{array}{ccccccccc}
0 & \longrightarrow & E_*Cf_{0,0} & \longrightarrow & E_*Cf_{0,1} & \longrightarrow & E_*Cf_{0,2} & \longrightarrow & \cdots \\
 & & \oplus & & \oplus & & \oplus & & \\
 & & 0 & \longrightarrow & E_*Cf_{1,0} & \longrightarrow & E_*Cf_{1,1} & \longrightarrow & \cdots \\
 & & & & \oplus & & \oplus & & \\
 & & & & 0 & \longrightarrow & E_*Cf_{2,0} & \longrightarrow & \cdots
\end{array}$$

To prove naturality and the first part of (iii), we suppose given a map of inverse systems

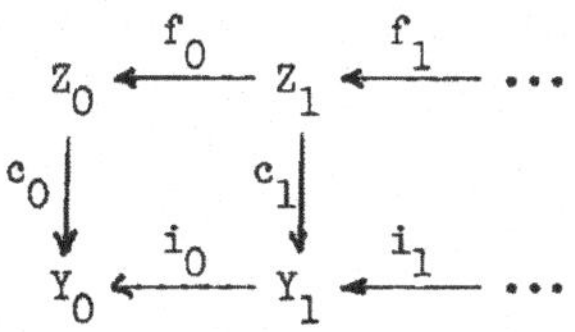

where, for each k, Ci_k is a retract of $C_k \wedge E$ for some C_k. We shall factor this map through the inverse system $\{\overline{f}_k\}$. That is, we shall construct maps $c'_k : \overline{Z}_k \to Y_k$ such that the following diagram commutes.

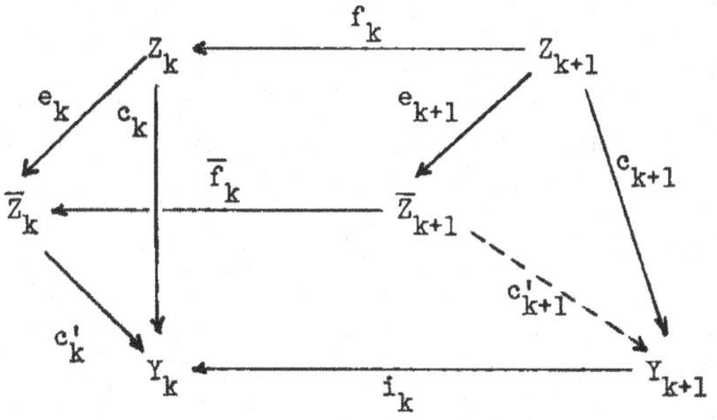

We proceed by induction. Let $c'_0 = c_0$ and assume inductively that $c'_0, \ldots, c'_k$ have been constructed. We seek c'_{k+1} such that

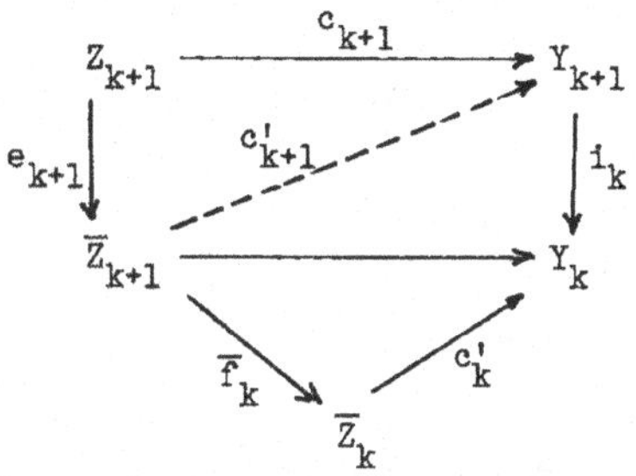

commutes. With the notations used in the braid of cofibrations, the obstruction to the existence of c'_{k+1} lies in the image of $j_2^* : [C, Ci_k] \to [D, Ci_k]$. This image is zero because $0 = j_{2*} : E_*C \to E_*D$ while E_*D is π_*E projective and Ci_k is a retract of $C_k \wedge E$.

This completes the inductive construction of the c'_k. Now (iii) follows by assuming $Y_0 \leftarrow Y_1 \leftarrow \cdots$ is an Adams resolution. For naturality, suppose given $\mathfrak{Z} \to \mathfrak{Z}' = \{Z'_0 \leftarrow Z'_1 \leftarrow \cdots\}$. Let $Y_i = \overline{Z}'_i$ and let $c_i : Z_i \to \overline{Z}'_i$ be the composite $Z_i \to Z'_i \to \overline{Z}'_i$; then apply the preceding paragraph.

Our next step is to construct the smash product pairing. First note that $\mathfrak{Z} \wedge Y'$ satisfies the hypotheses of the theorem, so gives rise to a spectral sequence $E_2^{**}(X, \mathfrak{Z} \wedge Y') = \bigoplus_i E_2^{**}(X, Cf_i \wedge Y') \Longrightarrow [X, Z \wedge Y']_*^E$ for any X. Choose an Adams resolution of Y' with each $E_*Y'_i$ projective over π_*E

$$Y' = Y'_0 \xleftarrow{g_0} Y'_1 \xleftarrow{g_1} \cdots$$

and let

$$F_n = \bigcup_{i+j=n} \overline{Z}_i \wedge Y'_j .$$

As in the derivation of smash products in the Adams spectral sequence (Theorem 4.5), we have a pairing from

$$E_r^{**}(X,\mathfrak{z}) \otimes E_r^{**}(X',Y')$$

to the spectral sequence obtained from $F_0 \leftarrow F_1 \leftarrow \cdots$ by applying $[X \wedge X', ?]_*^E$, and this pairing converges to the smash product. Thus, to show the existence of the pairing

$$E_r^{**}(X,\mathfrak{z}) \otimes E_r^{**}(X',Y') \to E_r^{**}(X \wedge X', \mathfrak{z} \wedge Y'),$$

we need only show that the sequence $F_0 \leftarrow F_1 \leftarrow \cdots$ is equivalent to the sequence $\overline{Z_0 \wedge Y'} \leftarrow \overline{Z_1 \wedge Y'} \leftarrow \cdots$ derived from $\mathfrak{z} \wedge Y'$. To construct the latter sequence, we need Adams resolutions of $C(f_i \wedge 1) \simeq Cf_i \wedge Y'$. If we use the smash product of our chosen Adams resolutions of Cf_i and of Y' then both $F_{n+1} \to F_n$ and $\overline{Z_{n+1} \wedge Y'} \to \overline{Z_n \wedge Y'}$ have cofiber

$$\bigvee_{i+j+k=n} Cf_{i,j} \wedge Cg_k .$$

Starting with $F_0 = Z_0 \wedge Y' = \overline{Z_0 \wedge Y'}$ we obtain an equivalence

$$\begin{array}{ccccc} F_0 & \longleftarrow & F_1 & \longleftarrow & \cdots \\ \wr\| & & \wr\| & & \\ \overline{Z_0 \wedge Y'} & \longleftarrow & \overline{Z_1 \wedge Y'} & \longleftarrow & \cdots \end{array}$$

by induction. This proves the existence of the pairing. It is immediate that it operates componentwise on E_2^{**} because the pairing is defined by taking the smash product of representative maps. This completes the proof of (ii). The second half of (iii) is also immediate because the maps $c'_k \wedge 1$ induce a map from $F_0 \leftarrow F_1 \leftarrow \cdots$ to the smash product of the Adams resolutions of Y and Y'.

To prove convergence we refer the reader to [6, Theorem 15.1] for the body of the proof and indicate only the changes needed to adapt Adams' proof to our situation. The essential step is to show that $(\prod_i \overline{Z}_i) \wedge E \simeq \prod_i (\overline{Z}_i \wedge E)$ so that we will have a short exact sequence

$$(*) \qquad 0 \longrightarrow \lim{}^1 E_*\overline{Z}_i \longrightarrow E_* \operatorname{Mic} \overline{Z}_i \longrightarrow \lim E_*\overline{Z}_i \longrightarrow 0.$$

We will also want this result with $\overline{Z}_i$ replaced by Z_i throughout. By Adams'

Theorem 15.2 it suffices to show $\pi_r\overline{Z}_i = 0$ for $r < n_1$ where n_1 is independent of i, and similarly for $\pi_r Z_i$.

Since $Z = Z_0$ is bounded below we may assume $\pi_r Z_0 = 0$ for $r < n_1$. Then the Hurewicz theorem and the Kunneth theorem imply that $E_r Z_0 = \pi_r(Z_0 \wedge E) = 0$ for $r < n_1$. Since $E_* Z_i \to E_* Z_0$ is a monomorphism, $E_r Z_i = 0$ for $r < n_1$. The Hurewicz and Kunneth theorems now imply that $H_r Z_i \otimes \pi_0 E = 0$ for $r < n_1$, but Adams shows $H_r Z_i \otimes \pi_0 E \cong H_r Z_i$. We conclude that $\pi_r Z_i = 0$ for $r < n_1$. We therefore have a short exact sequence

$$0 \to \lim{}^1 E_* Z_i \to E_* \mathrm{Mic}\ Z_i \to \lim E_* Z_i \to 0.$$

By hypothesis (6.1.iv.2), $E_* \mathrm{Mic}\ Z_i = 0$ and hence $\lim^1 E_* Z_i = 0 = \lim E_* Z_i$.

By construction of $\overline{f}_i\colon \overline{Z}_{i+1} \to \overline{Z}_i$ we see that the inverse system $E_*\overline{Z}_0 \leftarrow E_*\overline{Z}_1 \leftarrow \cdots$ is the direct sum of $E_* Z_0 \leftarrow E_* Z_1 \leftarrow \cdots$ and an inverse system all of whose maps are 0. It follows that $\lim^1 E_*\overline{Z}_i = 0 = \lim E_*\overline{Z}_i$. Thus, once we have the exact sequence (*) we will know $E_*\ \mathrm{Mic}\ \overline{Z}_i = 0$ from which convergence follows as an Adams' Theorem 15.1.

It remains only to show $\pi_r\overline{Z}_i = 0$ for $r < n_1 - 1$. Since $\pi_r Z_i = 0$ for $r < n_1$ and all i, the exact sequence $\pi_r Z_i \to \pi_{r-1} Z_{i+1}$ implies $\pi_r Cf_i = 0$ for $n < n_1$ and all i. This easily implies that $\pi_r Cf_{ij} = 0$ for $r < n_1$. Suppose, for induction, that $\pi_r\overline{Z}_i = 0$ for $r < n_1 - 1$. The exact sequence

$$\pi_{r+1} \bigvee_j Cf_{j,i-1} \longrightarrow \pi_r\overline{Z}_{i+1} \longrightarrow \pi_r\overline{Z}_i$$

implies that $\pi_r\overline{Z}_{i+1} = 0$ for $r < n_1 - 1$, completing the inductive step and, hence, the proof of convergence. //

7. Homotopy Operations for H_∞ Ring Spectra

In this section we define the homotopy operations which can be obtained from H_∞ ring structures and derive their purely formal properties. Calculations of extended powers of spheres will enable us to give concrete results about thiese operations in Chapter V. Most of our applications will deal with the case k = 1 of the following definition.

Definition 7.1. If $\alpha \in Y_m(D_{j_1}S^{n_1} \cdots D_{j_k}S^{n_k})$, define $\alpha^*: \pi_{n_1} \times \cdots \times \pi_{n_k} \to \pi_m$ by letting $\alpha^*(f_1,\ldots,f_k)$ be the composite

$$S^m \xrightarrow{\alpha} D_{j_1}S^{n_1} \wedge \cdots \wedge D_{j_k}S^{n_k} \wedge Y \xrightarrow{D_{j_1}f_1 \wedge \cdots \wedge D_{j_k}f_k \wedge 1} D_{j_1}Y \wedge \cdots \wedge D_{j_k}Y \wedge Y \xrightarrow{\xi} Y.$$

Remarks 7.2. (i) We write ξ for any composite of the maps $\xi: D_jY \to Y$, $\alpha_J: D_{j_1}Y \wedge \cdots \wedge D_{j_n}Y \to D_jY$ where $j = j_1 + \cdots j_n$, and $\mu: Y^{(n)} \to Y$, since they are all homotopic.

(ii) We can obtain similar operations

$$[X,Y]_{n_1} \times \cdots \times [X,Y]_{n_k} \xrightarrow{\alpha^*} [X,Y]_m$$

parameterized by $\alpha \in Y_m(D_{j_1}S^{n_1} \wedge \cdots \wedge D_{j_k}S^{n_k})$ for any space X by defining $\alpha^*(f_1,\ldots,f_k)$ to be

$$\begin{array}{ccc} X \wedge S^m \xrightarrow{1 \wedge \alpha} X \wedge \bigwedge_{i=1}^{k} D_{j_i}S^{n_i} \wedge Y & \xrightarrow{\Delta \wedge 1} & (\bigwedge_{i=1}^{k} D_{j_i}\Sigma^{n_i}X) \wedge Y \\ \searrow^{\alpha(f_1^*,\ldots,f_k)} & & \downarrow \wedge D_{j_i}f_i \wedge 1 \\ & Y & \xleftarrow{\xi} \bigwedge_{i=1}^{k} D_{j_i}Y \wedge Y \end{array}$$

where Δ is the composite of the diagonal $X \to X^{(k)}$, a shuffle map, and the natural transformation $X \wedge D_jS^n \to D_j\Sigma^nX$. This is a direct generalization of the classical derivation of Steenrod operations from the map $X \wedge D_2S \to D_2X$.

The next proposition records fairly obvious properties of these operations. Recall from I.§1 the natural transformation

$$\delta: D_\pi(X \wedge Y) \to D_\pi X \wedge D_\pi Y.$$

The product $\mu: Y \wedge Y \to Y$ induces products

$$Y_iA \otimes Y_jB \longrightarrow Y_{i+1}(A \wedge B)$$

and

$$\pi_iY \otimes \pi_jY \longrightarrow \pi_{i+j}Y,$$

both of which we denote by juxtaposition.

<u>Proposition 7.3</u>. (i) α^* is natural with respect to H_∞ ring maps

(ii) $(\alpha + \beta)^* = \alpha^* + \beta^*$

(iii) The natural inclusion $\iota: S^{nj} \to D_j S^n$ induces $\iota^*(x) = x^j$.

(iv) If $\alpha_1 \in Y_{m_1}(D_{j_1} S^{n_1} \wedge \cdots \wedge D_{j_k} S^{n_k})$ and $\alpha_2 \in Y_{m_2}(D_{j_{k+1}} S^{n_{k+1}} \wedge \cdots \wedge D_{j_\ell} S^{n_\ell})$

then $(\alpha_1\alpha_2)^*(f_1,\ldots,f_\ell) = \alpha_1^*(f_1,\ldots,f_k)\alpha_2^*(f_{k+1},\ldots,f_\ell)$.

(v) $\alpha^*(xy) = (\delta_*(\alpha))^*(x,y)$. Thus if $\delta_*(\alpha) = \sum \alpha_i\beta_i$ then

$\alpha^*(xy) = \sum \alpha_i^*(x)\beta_i^*(y)$.

<u>Proof</u>. (i) and (ii) are obvious. (iii) follows from the fact that

$$Y^j \xrightarrow{\ \iota\ } D_j Y \xrightarrow{\ \xi\ } Y$$

is the j-fold product. (iv) is also obvious from the definitions. (v) follows from the commutativity of the following diagram

$$\begin{array}{ccccccccc}
S^n & \xrightarrow{\alpha} & D_j(S^n \wedge S^m)\wedge Y & \xrightarrow{D_j(x\wedge y)\wedge 1} & D_j(Y\wedge Y)\wedge Y & \xrightarrow{D_j(\mu)\wedge 1} & D_j Y \wedge Y & \xrightarrow{\xi} & Y \\
 & \searrow^{\delta_*(\alpha)} & \downarrow{\scriptstyle \delta\wedge 1} & & \downarrow{\scriptstyle \delta\wedge 1} & & & & \downarrow{\scriptstyle \mu} \\
 & & D_j S^n \wedge D_j S^m \wedge Y & \xrightarrow{D_j x\wedge D_j Y\wedge 1} & D_j Y \wedge D_j Y & & \xrightarrow{\xi\wedge\xi\wedge 1} & & Y\wedge Y\wedge Y
\end{array}$$

Commutativity of the rectangle at the right follows from the definition of H_∞ ring spectrum and commutativity of the diagram

$$\begin{array}{ccc}
D_j(Y\wedge Y) & \xrightarrow{D_j(\iota)} & D_j D_2 Y \\
\downarrow{\scriptstyle \delta} & & \downarrow{\scriptstyle \beta_{j,2}} \\
D_j Y \wedge D_j Y & \xrightarrow[\alpha_{j,j}]{} & D_{2j} Y
\end{array}$$

by Lemma I.2.12. //

As should be expected from their essentially multiplicative origin, the operations α^* are far from being additive. In fact, their behaviour on sums is determined by the transfer maps

$$\tau_J : D_j Y \to D_{j_1} Y \wedge \cdots \wedge D_{j_k} Y$$

defined in II.1.4 for each partition $J = (j_1,\dots,j_k)$ of j. When $J = (1,1,\dots,1)$ we write $\tau_j = \tau_J : D_jY \to Y^{(j)}$.

Theorem 7.4. If $\alpha \in Y_*D_jS_n$ then

$$\alpha^*(x_1 + \cdots + x_k) = \sum_J [\tau_{J*}(\alpha)]^*(x_1,\dots,x_k)$$

where the sum is taken over all length k partitions J of j. If j is a prime p and we localize at p, then for $\alpha \in Y_*D_pS^n$

$$\alpha^*(\sum x_i) = \sum \alpha^*(x_i) + \begin{cases} 0 & p > 2 \text{ and } n \text{ odd} \\ \frac{1}{p!}\tau_{p\,*}(\alpha)((\sum x_i)^p - \sum x_i{}^p) & p > 2 \text{ and } n \text{ even} \\ \tau_{2\,*}(\alpha)(\sum_{i<j} x_ix_j) & p = 2 \end{cases}$$

(all unindexed sums are over $i = 1,\dots,k$).

Proof. This is an immediate consequence of Proposition II.2.2. //

In the rest of this section we shall use the spectral sequence of §6 together with the filtered maps obtained from §5 to describe the behaviour of homotopy operations in the Adams spectral sequence. Let us adopt the following notations. Let $x \in \pi_nY$ be detected by $\overline{x} \in E_2^{s,n+s}(S,Y)$, the Adams spectral sequence based on a ring theory E. Let $\mathcal{D}$ be the sequence

$$D_\pi^{ps}S^n \to D_\pi^{ps-1}S^n \to \cdots \to D_\pi^1S^n \to S^{np}$$

where π is cyclic of order p and $D_\pi^iS^n = W_i \ltimes_\pi S^{n\,(p)}$ is the extended power of S^n based on the i-skeleton W_i of the standard free π CW complex ($W_{2i-1} = S^{2i-1}$). By Theorem 5.2, $\xi D_\pi(x)$ induces compatible maps

$$D_\pi^iS^n \xrightarrow{D_\pi(x)} D_\pi^iY_s \xrightarrow{\xi_{i,ps}} Y_{ps-i}$$

(if E_*Y_j is π_*E projective for each j), and hence, by 6.1(iii), a homomorphism

$$\mathcal{P}(x) : E_r^{**}(S,\mathcal{D}) \to E_r^{**}(S,Y)$$

of spectral sequences provided the domain spectral sequence exists. Similarly, smashing with Y and multiplying, we have compatible maps

$$(D_\pi^iS^n) \wedge Y \longrightarrow Y_{ps-i}$$

and, hence, a homomorphism

$$\mathcal{P}(x) : E_r^{**}(S, \mathcal{D} \wedge Y) \longrightarrow E_r^{**}(S,Y).$$

Proposition 7.5. If $E_*D_\pi^{i-1}S^n \to E_*D_\pi^iS^n$ is a π_*E split monomorphism for each $i \le ps$ then the spectral sequence $E_r^{**}(S,\mathcal{D})$ exists and $E_2(S,\mathcal{D})$ is free over $E_2(S,S)$ on generators $e_i \in E_2^{ps-i,ps+pn}(S,\mathcal{D})$. Similarly, $E_r^{**}(S,\mathcal{D}\wedge Y)$ exists and $E_2(S,\mathcal{D}\wedge Y)$ is free over $E_2(S,Y)$ on the images of the e_i under the map induced by the unit $S \to Y$.

Proof. The cofiber of $D_\pi^{i-1}S^n \to D_\pi^iS^n$ is $W_i/W_{i-1}\wedge S^{n(p)} \simeq S^{np+i}$, so $E_*D_\pi^iS^n$ is a free π_*E module. Thus, Theorem 6.1(i) implies that the spectral sequence exists and

$$E_2^{s',t}(S,\mathcal{D}) \cong \bigoplus_j E_2^{s'-j,t-j}(S,S^{np+sp-j})$$

$$\cong \bigoplus_j E_2^{s'-j,t-np-sp}(S,S).$$

We let e_i be the generator in $E_2^{0,0}(S,S)$ for the $j = ps-i$ summand. //

We think of e_i as the $np+i$ cell of $D_\pi S^n$, or alternatively, as $e_i \otimes \iota_n \otimes \cdots \otimes \iota_n$ (this is its name in the cellular chains of $D_\pi S^n$).

Note that $\mathcal{D}$ satisfies the hypotheses of the proposition when $E = HZ_p$. Recall the function ν from 2.4 $(\nu(2j+\varepsilon) = (-2)^j(m!)^\varepsilon)$.

Theorem 7.6. Assume in addition to the hypotheses of 7.5, that E_*Y is π_*E projective. Then $\mathcal{P}(x)$ sends e_i to $\Phi_*(e_i \otimes \bar{x}^p)$. Thus, when $p = 2$, $\mathcal{P}(x)$ sends e_i to $P^{i+n}(\bar{x})$ and when $p > 2$, $\mathcal{P}(x)$ sends $(-1)^j\nu(n)e_i$ to $\beta^\varepsilon P^j\bar{x}$ if $i = (2j-n)(p-1)-\varepsilon$ and to 0 if i does not have this form.

Proof. The definition of $\mathcal{P}(x)$ implies that $\mathcal{P}(x)(e_i)$ is the composite

$$S^{np+i} \xrightarrow{e_i} \frac{W_i}{W_{i-1}} \wedge S^{n(p)} \xrightarrow{1\wedge \bar{x}^{(p)}} \frac{W_i}{W_{i-1}} \wedge Y_{s,1}^{(p)} \xrightarrow{\xi_{i,ps}} Y_{ps-i,1}.$$

We choose as generator e_i the map

$$S^{np+i} \simeq S^i \wedge S^{n\,(p)} \xrightarrow{e_i \wedge 1} \frac{W_i}{W_{i-1}} \wedge S^{n\,(p)}$$

in which $e_i \in \pi_i(W_i/W_{i-1}) = \mathcal{H}_i$ is the usual generator. Thus $\mathcal{P}(x)(e_i)$ is exactly the map which Corollary 5.4 asserts represents $\Phi_*(e_i \otimes \bar{x}^p)$. //

Since $\mathcal{P}(x)$ annihilates elements e_i with i not of the form $(2j-n)(p-1)-\varepsilon$, we will ignore them too. In V.§2 we will see that this amounts to restricting attention to a wedge summand of $D_\pi S^n$ which is p-equivalent to D_pS^n.

Convergence of the spectral sequence $E_r(S,\mathcal{B})$ to $\pi_* D_\pi^{ps} S^n$ implies that any $\alpha \in \pi_* D_\pi^{ps} S^n$ is detected by an element $\sum a_k e_k \in E_2(S,\mathcal{B})$, $a_k \in E_2(S,S)$. Applying $\mathcal{P}(x)$, we find that $\alpha^*(x)$ is detected by $\sum a_k \Phi_*(e_k \otimes \overline{x}^p)$. Similarly, for $\alpha \in Y_* D_\pi^{ps} S^n$ detected by $\sum a_k e_k \in E_2(S,\mathcal{B}\wedge Y)$, $a_k \in E_2(S,Y)$, except that if Y is not bounded below we have no guarantee that $E_r(S,\mathcal{B}\wedge Y)$ will converge to $Y_* D_\pi^{ps} S^n$.

<u>Corollary 7.7</u>. If $\alpha \in Y_* D_\pi{}^{ps} S^n$ is detected by $\sum a_k e_k$ in $E_2(S,\mathcal{B}\wedge Y)$ then $\alpha^*(x)$ is detected by $\sum_k a_k P^{k+n}\,\overline{x}$ if $p = 2$ or by $\sum (-1)^j \nu(n)^{-1} a_k \beta^\varepsilon P^j\,\overline{x}$ if $p > 2$ and $k = (2j-n)(p-1)-\varepsilon$.

The map $\mathcal{P}(x):\{E_r^{**}(S,\mathcal{B}\wedge Y)\} \to \{E_r^{**}(S,Y)\}$ also enables us to translate differentials in $\{E_r(S,\mathcal{B}\wedge Y)\}$ into differentials on Steenrod operations.

<u>Corollary 7.8</u>. If $d_r(ae_k) = \sum a_i e_{k_i}$ in $E_r(S,\mathcal{B}\wedge Y)$ then

$$d_r(aP^{k+n}\,\overline{x}) = \sum a_i P^{k_i+n}\,\overline{x} \qquad \text{if } p = 2 \quad \text{and}$$

$$d_r(a\beta^\varepsilon P^j\,\overline{x}) = \sum (-1)^{j+j_i} a_i \beta^{\varepsilon_i} P^{j_i}\,\overline{x} \qquad \text{if } p > 2\ ,$$

where $k = (2j-n)(p-1)-\varepsilon$ and $k_i = (2j_i-n)(p-1)-\varepsilon_i$. In particular, if ae_k is a permanent cycle, then so is $aP^{k+n}\,\overline{x}$ (if $p = 2$) or $a\beta^\varepsilon P^j\overline{x}$ (if $p > 2$).

Note that Corollary 7.8 only applies to permanent cycles $\overline{x}$. Much more general results will be obtained in chapter VI.

The next result says that in the ordinary mod p Adams spectral sequence ($E = HZ_p$), a homotopy operation cannot lower filtration.

<u>Proposition 7.9</u>. Let $E = HZ_p$. If $x_i \in \pi_* Y$ has filtration s_i and $\alpha \in Y_*(D_{j_1} S^{n_1} \wedge \cdots \wedge D_{j_k} S^{n_k})$ then $\alpha^*(x_1,\ldots,x_k)$ has filtration at least $s_1 + \cdots + s_k$.

<u>Proof</u>. First, it suffices to show that

$$D_j Y_s \longrightarrow D_j Y \xrightarrow{\xi} Y$$

lifts to Y_s, for then $\alpha^*(x_1,\ldots,x_k)$ will factor through

$$Y_{s_1} \wedge \cdots \wedge Y_{s_k} \wedge Y \longrightarrow Y_{s_1+\cdots+s_k} \wedge Y \longrightarrow Y$$

To obtain the lifting we need to factor $D_j Y_s \to D_j Y$ as the composite of s maps which are zero in homology. But this is easy. The factorization

$$D_jY_s \to D_jY_{s-1} \to \cdots \to D_jY_1 \to D_jY$$

suffices since the natural isomorphism

$$H_*(D_jX) \cong H_*(\Sigma_j;H_*X^{(j)})$$

and the fact that $H_*Y_{i+1} \to H_*Y_i$ is zero imply that $H_*D_jY_{i+1} \to H_*D_jY_i$ is zero for each i. //

Note that the proposition will hold in the E Adams spectral sequence whenever E is such that if $E_*X \to E_*Y$ is zero then $E_*D_jX \to E_*D_jY$ is also zero. The spectral sequence

$$H_*(\Sigma_j;E_*X^{(j)}) \Longrightarrow E_*D_jX$$

only gives us this on an associated graded to E_*D_jX and E_*D_jY. I have no reason to believe or disbelieve the result for general E.

Remark 7.10. There are two variants of $\mathcal{B}$ which are also useful. First, taking into account the fact that all of $D_\pi S^n$ will be mapped into $Y = Y_0$ by the composite

$$D_\pi S^n \xrightarrow{D_\pi x} D_\pi Y_s \longrightarrow D_\pi Y \xrightarrow{\xi} Y$$

we can replace $D_p^{ps}S^n$ in $\mathcal{B}$ by all of D_pS^n, giving $\mathcal{B}'$:

$$D_\pi S^n \leftarrow D_\pi^{ps-1}S^n \leftarrow D_\pi^{ps-2}S^n \leftarrow \cdots D_\pi^1 S^n \leftarrow S^{np} .$$

We still get $\mathcal{P}(x):E_r(S\,\mathcal{B}') \to E_r(S,Y)$ for any $x \in \pi_*Y$. To get $E_2(S,\mathcal{B}')$ from $E_2(S,\mathcal{B})$ simply replace the summand $E_2(S,S^{(n+s)p})$ by $E_2(S,\Sigma^n L^\infty_{n(p-1)+ps})$, which can be obtained (through a range of dimensions) from Mahowald's tables [59] when p = 2. Mahowald's tables have the virtue that they are derived from the cellular filtration of the stunted projective space, so that elements are named by giving an element of $E_2(S,S)$ and the cell on which it occurs. Thus Theorem 7.6 and Corollaries 7.7 and 7.8 can be used with $E_r(S,\mathcal{B}')$ as easily as with $E_r(S,\mathcal{B})$.

The other variant of $\mathcal{B}$ requires that $E = HZ_p$. It takes account of Proposition 7.9 by putting everything into filtrations between s and ps, rather than 0 and ps as $\mathcal{B}'$ does. That is, $\mathcal{B}''$ is the sequence

$$D_\pi S^n = \cdots = D_\pi S^n \leftarrow D_\pi^{(p-1)s-1}S^n \leftarrow \cdots \leftarrow D_\pi^1 S^n \leftarrow S^{np}$$

with $D_\pi S^n$ in filtrations 0 through s. Its E_2 term is similar to $E_2(S,\mathcal{B}')$. It has a copy of $E_2(S,S)$ for each cell from np to np + (p-1) - 1 together with an copy of

$E_2(S,\Sigma^n L^\infty_{(n+s)(p-1)})$. The spectral sequence $E_r(S,\mathcal{B}'')$ is optimal in the sense that it has all homotopy operations (unlike $E_r(S,\mathcal{B})$ which only uses the bottom ps cells of $D_\pi S^n$) and puts them into as high a filtration as they will go universally.

CHAPTER V

THE HOMOTOPY GROUPS OF H_∞ RING SPECTRA

By Robert R. Bruner

§1. Explicit homotopy operations and relations

This section contains statements of our results on homotopy operations as well as some applications of these results. The proofs depend on material in §2 and will be given in §3.

Note that, aside from the computations in π_*S at the end of this section, all the results here apply to the homotopy of any H_∞ ring spectrum Y. Let $\xi: D_pY \to Y$ denote the structure map.

The order of results in this section is:

relation to other operations,

particular operations and relations,

Cartan formulas,

computations in π_*S,

remarks.

In order not to interrupt the main flow of ideas, we have deferred a number of remarks until the end of the section.

Throughout this section let $E_r(X,Y)$ be the ordinary mod p Adams spectral sequence converging to $[X,Y]_*$, and let $E_r(S,\mathcal{D})$ be the spectral sequence of IV §6 based on ordinary mod p homology. Let $\mathcal{D}$ be the sequence

$$\mathcal{D} = \{D_pS^n \cdots \to D_p^iS^n \to \cdots \to D_p^1S_n \to D_p^0S^n\}.$$

From the spectral sequence $E_r(S,\mathcal{D})$ we obtain an isomorphism between an associated graded of $\pi_*D_pS^n$ and $E_\infty(S,\mathcal{D})$:

$$E^0(\pi_*D_pS^n) \cong E_\infty(S,\mathcal{D}).$$

Write $E^0(\alpha)$ for the image in $E_\infty^{s,*}(S,\mathcal{D})$ of an element $\alpha \in \pi_*D_pS^n$ of filtration s. By IV.7.5, $E_2(S,\mathcal{D})$ is free over $E_2(S,S)$ on generators e_i corresponding to the cells of D_pS^n. By 2.9 below, a more convenient basis over $E_2(S,S)$ is given by the elements

$$\beta^\varepsilon P^j = (-1)^j \nu(n) e_{jq-\varepsilon-n(p-1)}$$

where $\varepsilon = 0$ or 1 ($\varepsilon = 0$ if $p = 2$), $q = 2(p-1)$ ($q = 1$ if $p = 2$), $jq-\varepsilon \geq n(p-1)$ and ν is the function defined in IV.2.4 ($\nu = 1$ if $p = 2$). Thus, $E^0(\alpha)$ can be written as a linear combination of the $\beta^\varepsilon P^j$ with coefficients in $E_2(S,S)$. Recall the operation $\alpha^*: \pi_n Y \to \pi_N Y$ associated to each element $\alpha \in \pi_N D_p S^n$.

Relation of the α^* to other operations

Proposition 1.1. If $\iota: S^{np} \to D_p S^n$ is the natural map then $\iota^*(x) = x^p$ and

$$E^0(\iota) = \begin{cases} P^n & p = 2 \\ P^j & p > 2 \text{ and } n = 2j \\ 0 & p > 2 \text{ and } n \text{ odd} \end{cases}$$

Propoosition 1.2. Let $h: \pi_* \to H_*$ be the Hurewicz homomorphism. If $E^0(\alpha) = \beta^\varepsilon P^j$ then $h \circ \alpha^* = \beta^\varepsilon Q^j \circ h$, where $\beta^\varepsilon Q^j$ is the Dyer-Lashof operation defined in III.1.

If $E^0(\alpha) = \sum a_{j,\varepsilon} \beta^\varepsilon P^j$, with each $a_{j,\varepsilon} \in E_2(S,S)$ and $\bar{x} \in E_2(S,Y)$, we let $E^0(\alpha)(\bar{x}) = \sum a_{j,\varepsilon} \beta^\varepsilon P^j(\bar{x})$.

Proposition 1.3. (Kahn, Milgram) If $x \in \pi_n Y$ is detected by $\bar{x} \in E_2(S,Y)$, then $\alpha^*(x)$ is detected by $E^0(\alpha)(\bar{x})$.

To see the relation to Toda brackets, suppose we have compressed α into the np+i skeleton $D_p^i S^n$ and that it projects to $\tilde{a}$ on the top cell S^{np+i}. Let $D_p^{i-1}(x) = D_p(x)|D_p^{i-1}S^n$ and let $c_i \in \pi_{np+i-1} D_p^{i-1} S^n$ be the attaching map of the np+i cell.

Proposition 1.4. $\alpha^*(x) \in \langle \tilde{a}, c_i, \xi D_p^{i-1}(x) \rangle$. The set of all such $\alpha^*(x)$ is a coset of $\xi D_p^{i-1}(x) \circ \pi_N D_p^{i-1} S^n$.

Note: We will frequently find further that $E^0(\alpha) = a\beta^\varepsilon P^j$ where $i = jq-\varepsilon-n(p-1)$ and $(-1)^j \nu(n) a$ detects $\tilde{a}$. Then

$$E^0(\alpha)(\bar{x}) = E^0(\alpha^*(\bar{x})) = a\beta^\varepsilon P^j(\bar{x}),$$

so that α^* is detected by Toda brackets in essentially the same fashion as by Steenrod operations in $E_2(S,Y)$.

Particular operations and relations

Hereafter, if $\theta \in E_\infty(S,\mathcal{D})$ and $x \in \pi_n Y$, let $\theta(x) = \{\alpha^*(x) | E^0(\alpha) = \theta\}$. Clearly, the indeterminacy in $\theta(x)$, defined to be

$$\mathrm{Ind}(\theta(x)) = \{\alpha^*(x) - \beta^*(x) | E^0(\alpha) = \theta = E^0(\beta)\},$$

is the set of values of all homotopy operations on x whose corresponding element in $E_\infty(S,\mathcal{D})$ has higher filtration than does θ.

Proposition 1.5 (Kahn, Milgram): The following are equivalent:

(i) $\beta^\varepsilon P^j$ acts on $\pi_n Y$

(ii) $e_i \in E_\infty(S,\mathcal{D})$, $i = jq-\varepsilon-n(p-1)$

(iii) $D_p^i S^n$ is reducible

(iv) if $p = 2$ then $n \equiv -i-1 \quad (2^{\phi(i)})$;
if $p > 2$ then $\varepsilon = 0$ and $n = 2j$,
or $\varepsilon = 1$ and $j \equiv 0 \quad (p^{\psi(i)})$.

The functions ϕ and ψ are defined in 2.5 and 2.11 below.

Definition 1.6. If $p = 2$, let $\beta_0 = 2$, $\beta_1 = \eta$, $\beta_2 = \nu$ and let β_j be a generator of Im J in dimension $8a+2^b-1$, where $j = 4a+b$ and $0 \le b \le 3$. If $p > 2$, let $\alpha_0 = p$, and let α_j be a generator of Im J in dimension $jq-1$.

Theorem 1.7 (Toda, Barratt, Mahowald, Cooley): Let $p = 2$. If $x \in \pi_n Y$ and $j = 4a+b$, $0 \le b \le 3$, then

$$\beta_j \circ x^2 = 0 \quad \text{if} \quad n \equiv 2^j - 8a - 2^b - 1 \quad (2^{j+1})$$

and $\beta_j \circ P^{n+1}(x) = \alpha x^2$ for some $\alpha \in \pi_{8a+2^b}S$ if $n \equiv 0 \quad (2)$ and $n \equiv 2^j - 8a - 2^b - 2 \quad (2^{j+1})$.

Theorem 1.8. Let $p > 2$ and $x \in \pi_n Y$. Let $\varepsilon_p(a)$ denote the exponent of p in the prime factorization of a. If $n = 2k-1$ then

$$\alpha_j \circ \beta P^k x = 0 \qquad \text{if } j = 0 \text{ or } j > 0 \text{ and } \varepsilon_p(k+j) = j-1.$$

If $n = 2k$ then

$$\alpha_j \circ \beta P^{k+1} x = \alpha x^p \quad \text{for some } \alpha \ \pi_{(j+1)q-2}S \quad \text{if } j = 0 \text{ or } j > 0 \text{ and } \varepsilon_p(k+j+1) = j-1.$$

<u>Theorem 1.9</u>. The operations listed in Tables 1.1 and 1.3 exist on π_n and satisfy the relations listed in Tables 1.2 and 1.4. In Tables 1.1 and 1.3 the columns labelled "indeterminacy" list generators for the indeterminacy of each operation, and the columns labelled "τ_{p*}" list the values of

$$\tau_{p*}:\pi_N D_p S^n \to \pi_N S^{np} \cong \pi_{N-np}S$$

thereby indicating the deviation from additivity of the given operation (by IV.7.4).

TABLE 1.1

Operations on π_n for $p > 2$

n	operations	indeterminacy	τ_{p*}
$n = 2k-1$	βP^k	0	0
	$h_0 P^k$	0	0
	$g_1 P^k$	0	0
$n = 2k-1$, $k \equiv -1$ (p)	βP^{k+1}	$h_0 P^k$	0
$n = 2k-1$	$h_0\beta P^{k+1}$	$\alpha_1\beta P^k$	0
$k \equiv -2$ (p)	βP^{k+2}	$g_1 P^k$ and $h_0 P^{k+1}$ (if it exists)	0
$n = 2k$	P^k	0	$p!$
	βP^{k+1}	$\alpha_1 P^k$	multiple of α_1
	$h_0 P^{k+1}$	$\alpha_2 P^k$	multiple of α_2
$n = 2k$, $k \equiv -2$ (p)	βP^{k+2}	$h_0 P^{k+1}$ and $\alpha_2 P^k$	multiple of α_2

TABLE 1.2

Relations among operations on π_n for $p > 2$

n	relations
$n = 2k-1$	$p\beta P^k = ph_0P^k = pg_1P^k = 0$ $(k+1)\alpha_1\beta P^k = 0$
$n = 2k-1$ $k \equiv -1\ (p)$	$p\beta P^{k+1} = -h_0P^k$ $\alpha_1\beta P^{k+1} \equiv 0 \mod \alpha_2\beta P^k$
$n = 2k-1$ $k \equiv -2\ (p)$	$ph_0\beta P^{k+1} \equiv 0 \mod \alpha_2\beta P^k$
$n = 2k$	$k\alpha_1P^k = p\beta P^{k+1}$ $(k+2)\alpha_1\beta P^{k+1} = 0$
$n = 2k$ $k \equiv -2\ (p)$	$p\beta P^{k+2} \equiv -h_0P^{k+1} \mod \alpha_2P^k$

TABLE 1.3

Operations on π_n for p = 2

n	operations	indeterminacy	τ_{2*}
$n \equiv 0$ (4)	P^n	$2P^n$	2
	P^{n+1}	ηP^n	η
	P^{n+3}	$2P^{n+3}, \nu P^n$	multiple of ν
	h_1P^{n+2}	$2h_1P^{n+2}, \nu P^n$	multiple of ν
$n \equiv 1$ (4)	P^n	0	0
	h_1P^{n+1}	$\eta^2 P_n$	0 or η^2
	P^{n+2}	$2P^{n+2}$	0 or η^2
	h_1P^{n+5}	$2h_1P^{n+5}, \nu^2 P^n$	0 or ν^2
	$h_1^2P^{n+4}$	$2h_1^2P^{n+4}, \nu^2 P^n$	0 or ν^2
	$h_1^3P^{n+3}$	$2h_1^3P^{n+3}, \nu^2 P^n$	0 or ν^2
$n \equiv 1$ (8)	P^{n+6}	$2P^{n+6}$	0 or ν^2
$n \equiv 2$ (4)	P^n	$2P^n$	2
	P^{n+1}	ηP^n	0
	h_1P^{n+4}	$2h_1P^{n+4}$	0
	$h_1^2P^{n+3}$	$2h_1^2P^{n+3}$	0
	$h_1^3P^{n+2}$	$2h_1^3P^{n+2}$	0
	h_2P^{n+3}	$\nu^2 P^n$	0 or ν^2
$n \equiv 2$ (8)	P^{n+5}	$2P^{n+5}$	0
$n \equiv 3$ (4)	P^n	0	0
	h_1P^{n+1}	0	0 or η^2
	h_1P^{n+3}	$2h_1P^{n+3}$	0
	$h_1^2P^{n+2}$	$\eta^2 h_1 P^{n+1}$	0
	h_2P^{n+2}	0	0
$n \equiv 3$ (8)	P^{n+4}	$2P^{n+4}$	0

TABLE 1.4

Relations among operations for p = 2

$n \equiv 0\ (4)$	$2P^{n+1} = 0$ $2h_1P^{n+2} = \eta^2P^{n+1}$	
$n \equiv 0\ (8)$	$2P^{n+3} = h_1P^{n+2}$ $\eta P^{n+3} = 0$ $2\nu P^{n+3} = \nu h_1P^{n+2} = 0$	
$n \equiv 4\ (8)$	$2P^{n+3} = h_1P^{n+2} + \nu P^n$ $\eta P^{n+3} = \nu P^{n+1}$ $\nu h_1P^{n+2} = \nu^2P^n$	
$n \equiv 1\ (4)$	$2P^n = 0$ $2h_1P^{n+1} = \eta^2P^n$ $2P^{n+2} = h_1P^{n+1}$ $\eta h_1P^{n+1} = 0$	$2h_1P^{n+5} = h_1^2P^{n+4}$ $2h_1^2P^{n+4} = h_1^3P^{n+3}$ $2h_1^3P^{n+3} = 0$
$n \equiv 1\ (8)$	$\eta P^{n+2} = 0$ $2\nu P^{n+2} = 0$ $2P^{n+6} = h_1P^{n+5}$	
$n \equiv 5\ (8)$	$\eta P^{n+2} = \nu P^n$ $\nu P^{n+2} = 0$	
$n \equiv 2\ (4)$	$2P^{n+1} = \eta P^n$ $\eta P^{n+1} = 0$ $4\nu P^n = 0$	$2h_1P^{n+4} = h_1^2P^{n+3}$ $2h_1^2P^{n+3} = h_1^3P^{n+2}$ $2h_1^3P^{n+2} = 0$
$n \equiv 2\ (8)$	$2P^{n+5} = h_1P^{n+4}$ $\eta P^{n+5} = h_2P^{n+3}$	
$n \equiv 6\ (8)$	$\nu P^{n+1} = 0$ $2h_2P^{n+3} = \nu^2P^n$ $\eta h_1P^{n+4} \equiv 0 \mod \nu^2P^n$	

$n \equiv 3\ (4)$	$2P^n = 0$	$2h_1P^{n+3} = h_1^2P^{n+2}$
	$\eta P^n = 0$	$2h_1^2P^{n+2} = \eta^2h_1P^{n+1}$
	$2h_1P^{n+1} = 0$	$2h_2P^{n+2} = 0$
$n \equiv 3\ (8)$	$2P^{n+4} = h_1P^{n+3}$	
	$\eta P^{n+4} = h_2P^{n+2}$	
	$\eta h_2P^{n+2} = \nu^2P^n$	
$n \equiv 7\ (8)$	$\nu P^n = 0$	
	$\eta h_1P^{n+3} = 0$	
	$\eta h_2P^{n+2} = 0$	

Cartan Formulas

For later computations we need the Cartan formulas for the first operation above the pth power.

Proposition 1.10. Let $p = 2$, $x \in \pi_n Y$, $y \in \pi_m Y$. Assume $n+m$ is even. Then

$$P^{n+m+1}(xy) = \begin{cases} P^{n+1}(x)y^2 + x^2P^{m+1}(y) + c_{n,m}\eta x^2y^2 & n \equiv m \equiv 0 \ (2) \\ S_{n,m}(x,y) & n \equiv 3 \ (4) \text{ or } m \equiv 3 \ (4) \\ S_{n,m}(x,y) + c_{n,m}\eta x^2y^2 & n \equiv m \equiv 1 \ (4) \end{cases}$$

where $S_{n,m}:\pi_n \times \pi_m \to \pi_{2(n+m)+1}$ is an operation such that

$$E^0(S_{n,m}) = P^nP^{m+1} + P^{n+1}P^m$$

and $$2S_{n,m}(x,y) = \begin{cases} \eta x^2y^2 & n \equiv m \equiv 1 \ (4) \\ 0 & n \equiv 3 \ (4) \text{ or } m \equiv 3 \ (4) \ , \end{cases}$$

and where $c_{n,m}$ is an integer depending only on n and m.

Proposition 1.11. Let $p > 2$, $x \in \pi_n Y$ and $y \in \pi_m Y$. Then

(i) if $n = 2j$ and $m = 2k$,

$$\beta P^{j+k+1}(xy) = \beta P^{j+1}(x)y^p + x^p\beta P^{k+1}(y) + d_{n,m}\alpha_1 x^py^p$$

where $d_{n,m}$ is an integer depending only on n and m.

(ii) if $n = 2j$ and $m = 2k-1$,

$$\beta P^{j+k}(xy) = x^p\beta P^k(y)$$

(iii) if $n = 2j-1$ and $m = 2k-1$,

$$\beta P^{j+k}(xy) = S_{j,k}(x,y)$$

where $S_{j,k} : \pi_{2j-1}Y \times \pi_{2k-1}Y \to \pi_{2(j+k)p-3}Y$ is an operation such that $E^0(S_{j,k}) = \beta P^j \cdot P^k + P^j \cdot \beta P^k$ and $pS_{j,k}(x,y) = 0$.

Computations

Our final results contain extensions to all H_∞ ring spectra of classical results about π_*S due to Toda, Barratt, Mahowald, Gray and Milgram, as well as some low dimensional calculations at the prime 2.

Let $\doteq$ denote equality up to multiplication by a unit.

Proposition 1.12. If $p = 2$ then $P^1(2) = \eta$.

Proposition 1.13. If $p > 2$ then $\beta P^1(p) \doteq \alpha_1$ and $\beta P^{p-1}(\alpha_1) = \beta_1$.

Combined with the Cartan formulas 1.10 and 1.11, these yield the following results.

Proposition 1.14. Let $x \in \pi_n Y$ and $n = 2j$. If $p = 2$ then $P^{n+1}(2x) = \eta x^2$. If $p > 2$ then $\beta P^{j+1}(px) \doteq \alpha_1 x^p$ and $\beta P^{j+p-1}(\alpha_1 x) = \beta_1 x^p$. The indeterminacy of each is 0.

Corollary 1.15. Let $x \in \pi_n Y$. If $p = 2$, $n \not\equiv 1$ (4) and $2x = 0$, then $\eta x^2 = 0$. If $p > 2$ and $px = 0$ then $\alpha_1 x^p = 0$. If $p > 2$ and $\alpha_1 x = 0$ then $\beta_1 x^p = 0$. In particular, $\alpha_1 \beta_1^p = 0$.

In the next proposition, the statement "$aP^j(x) = y \bmod A$" means that A is the indeterminacy of aP^j when applied to x. If the indeterminacy is not mentioned, it is 0.

Proposition 1.16. The following hold in π_*S localized at 2.

(i) $P^1(\eta) = \eta^2$

(ii) $P^3(\nu) = \nu^2$, $h_1P^4(\nu) = \eta\sigma$ or $\bar{\nu}$, $h_1^2P^5(\nu) = 0$.

(iii) P^3, h_1P^4, h_1P^6, $h_1^2P^5$, and h_2P^5 annihilate 2ν and 4ν.

(iv) P^6, P^7, $h_1^2P^9$, and $h_1^3P^8$ annihilate ν^2.

(v) $P^7(\sigma) = \sigma^2$, $h_1P^8(\sigma) = \eta^*$ or $\eta^* + \eta\rho$,

$$h_1P^{10}(\sigma) = \nu^* \bmod \langle 2\nu^*\rangle + \langle \eta\bar{\mu}\rangle,$$

$$h_1^2P^9(\sigma) = 2\nu^* \bmod \langle 4\nu^*\rangle + \langle \eta\bar{\mu}\rangle.$$

(vi) $P^7(2\sigma) = 0$, $h_1P^8(2\sigma) = 0$, $h_2P^9(2\sigma) = 0$,

$$h_1P^{10}(2\sigma) = 2\nu^* \bmod \langle 4\nu^*\rangle + \langle \eta\bar{\mu}\rangle,$$

$$h_1^2P^9(2\sigma) = 4\nu^* \bmod \langle \eta\bar{\mu}\rangle.$$

(vii) P^7, h_1P^8, $h_1^2P^9$ and h_2P^9 annihilate 4σ,

$$h_1P^{10}(4\sigma) = 4\nu^* \mod \langle\eta\bar{\mu}\rangle .$$

Remarks: These are listed by the result to which they refer.

(1.4): The indeterminacy of the Toda bracket $\langle\tilde{a}, c_i, \xi D_p^{i-1}(x)\rangle$ in Proposition 1.4 is $\xi D_p^{i-1}(x) \circ \pi_N D_p^{i-1} S^n + (\pi_{np+i} Y) \circ \tilde{a}$, while the indeterminacy of $\alpha^*(x)$ is only $\xi D_p^{i-1}(x) \circ \pi_N D_p^{i-1} S^n$. This reflects the fact that $\alpha^*(x)$ uses the canonical null homotopy $D_p^i(x)$ of $D_p^{i-1}(x) \circ c_i$, whereas the Toda bracket allows any null homotopy of $\xi D_p^{i-1}(x) \circ c_i$.

(1.8): Since π_{pq-2} is the first nonzero homotopy group of S in a dimension congruent to -2 mod q, we get

$$\alpha_j \beta P^{k+1} x = 0$$

for $j < p-1$ satisfying the hypotheses of (1.8).

(1.9): (i) In the range of dimensions listed, the operations and relations given in Tables 1.1 through 1.4 generate all the operations and relations over $\pi_* S$. For examples, when $n \equiv 0$ (4) and $p = 2$:

(a) ηP^n and $\eta^2 P^n$ are nonzero operations because the relations listed do not force them to be 0;

(b) the relation $4h_1P^{n+2} = 0$ follows from the listed relation

$$2h_1P^{n+2} = \eta^2 P^{n+1},$$

and is therefore omitted;

(c) the redundant operation h_1P^{n+2} is included because the relation

$$2P^{n+3} = h_1P^{n+2}$$

which makes it redundant reflects a universally hidden extension:

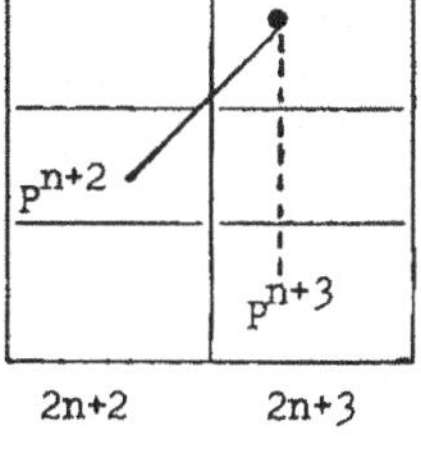

$h_0P^{n+3} = 0$ in E_∞ and $2P^{n+3}x$ is detected by $h_1P^{n+2}x$.

(ii) The operations of degree n+3 for $n \equiv 0$ (4) and p = 2 are particularly interesting. If $n \equiv 0$ (8) then by [59] $\pi_{2n+3}D_2S^n = Z_8 \oplus Z_8$. It is generated by νP^n and P^{n+3} with relations

$$2P^{n+3} = h_1P^{n+2}$$

and

$$4P^{n+3} = 2h_1P^{n+2} = \eta^2P^{n+1}.$$

If $n \equiv 4$ (8) then [59] gives $\pi_{2n+3}D_2S^n = Z_4 \oplus Z_{16}$ and it is generated by h_1P^{n+2} (of order 4) and P^{n+3} (of order 16) with relations

$$2h_1P^{n+2} = \eta^2P^{n+1}$$

$$2P^{n+3} = h_1P^{n+2} + \nu P^n$$

$$4P^{n+3} = \eta^2P^{n+1} + 2\nu P^n$$

$$8P^{n+3} = 4\nu P^n.$$

(iii) Entries in the τ_{p*} column such as "0 or η^2" indicate that we have not calculated τ_{p*}. Such entries simply list the elements of π_*S in the relevant dimension. Even this limited information is useful in Proposition 1.16.

<u>(1.10) and (1.11)</u>: Let $\psi: \mathcal{A} \to \mathcal{A} \otimes \mathcal{A}$ be the diagonal of the Steenrod algebra $(\psi(P^n) = \sum P^i \otimes P^{n-i})$. If

$$E^0(\alpha) = \sum a_iA_i, \quad a_i \in E_2(S,S), \quad A_i \in \mathcal{A}$$

then

$$E^0(\delta_*(\alpha)) = \sum a_i\psi(A_i).$$

This defines $\delta_*(\alpha)$ and, hence, the formula for $\alpha^*(xy)$, modulo higher filtration in $E_\infty(S,\mathcal{D})$.

<u>(1.15)</u>: This proof that $\alpha_1\beta_1^p = 0$ differs from Toda's in that Toda views the product in π_*S as composition and studies $D_p(S^n \cup_p e^{n+1})$ while we view it as the smash product and study $D_pS^n \wedge D_pS^m$. Toda shows that

$$D_p(S^n \cup_p e^{n+1}) \supset S^{np} \cup_{\alpha_1} e^{np+q}$$

and

$$D_p(S^n \cup_{\alpha_1} e^{n+q}) \supset S^{np} \cup_{\beta_1} e^{np+pq-1}.$$

Thus, if px = 0 or $\alpha_1x = 0$ then $\alpha_1x^p = 0$ or $\beta_1x^p = 0$, respectively. The proof given in 1.15 uses the values of the operations on p and α_1, rather than the structure of D_p of their cofibers.

Segal [49] saw that the Cartan formula for homotopy operations should provide a proof that $\alpha_1\beta_1^p = 0$, but his explicit formulas were incorrect.

There is still another proof that $\alpha_1\beta_1^p = 0$ which uses virtually none of the machinery of homotopy operations, but does require that we have calculated enough of π_*S to know that the p^2q-3 stem is either 0 or Z_p. Given this, the relation

$$-\alpha_1\beta_1^p = p\beta P^{p^2-p}(\beta_1)$$

from Table 1.2 implies that $\alpha_1\beta_1^p = 0$.

Remark 1.17: This is a quick survey of results on homotopy operations which are not included here. Toda [106] shows thAt the extended powers propagate several relations. For example, if $\langle\alpha_1,p,x\rangle = 0$ then $\beta_s x^p = 0 \bmod \alpha_1$ for $1 < s < p$. As corollaries he shows that $\beta_2\beta_1^p = 0$ and the β_s are nilpotent, foreshadowing Nishida's proof, a few years later, that all positive dimensional elements of π_*S are nilpotent.

Gray [36] obtained results similar to 1.15 using homotopy operations which are associativity or commutativity obstructions for ring spectra.

Oka and Toda [92] have extensive information on the cell structure of $D_p(S^n\cup_p e^{n+1})$ which they use, in particular, to show that $\gamma_1 \neq 0$.

Milgram [80] also uses extended powers $D_2(S^n\cup_{2^i} e^{n+1})$ to define homotopy operations which can be iterated to yield infinite families of elements in π_*S, presumably related to the elements detected by K-theory.

Cooley, in his thesis [30], uses extended powers to compute some Toda brackets and to derive 1.7 as well as the relation $\varepsilon x^2 = 0$ if $x \in \pi_n$, $n \equiv 2,3,7$ (8), which is not in 1.7.

Milgram [79 and 81] computes the Coker J part of the operations on π_8S and π_9S using Steenrod operations in $E_2(S,S)$.

§2. Extended powers of spheres

In this section we collect the results on extended powers of spheres which are needed to prove the results of §1. They will also be essential to our results on differentials in the next chapter. First, we recall the values of the K and J groups of lens spaces. Then, we identify the spectra $D_\pi^i S^n$, π cyclic, as the suspension spectra of stunted lens spaces and determine when they are stably reducible or coreducible. Also, we show that, after localizing at p, D_pS^n is a wedge summand of $D_\pi S^n$, which gives a simple cell structure to D_pS^n.

Throughout this section, let p be a prime, let $\pi \subset \Sigma_p$ be the p-Sylow subgroup generated by the p-cycle $(1\ 2\cdots p)$, and let W^k be the k-skeleton of a contractible π or Σ_p free CW complex W. (Definitions 2.1 and 2.7 provide the π free CW complexes which we shall use most frequently.)

The results for p = 2 are analogous to the results for odd primes, but are sufficiently simpler that we state them separately. We begin with odd primes.

Definition 2.1. Let $p > 2$ and let $\rho = \exp(2\pi i/p)$. Let π act on the unit sphere $S^{2k+1} \subset \mathbb{C}^{k+1}$ by letting a generator of π send (z_i) to (ρz_i). Let

$$\tilde{L}^{2k+1} = S^{2k+1}/\pi,$$

$$\tilde{L}^{2k} = \{[z_0,\ldots,z_k] \in \tilde{L}^{2k+1} \mid z_k^p \text{ is real and } \geq 0\},$$

and

$$\tilde{L}_n^{n+k} = \tilde{L}^{n+k}/\tilde{L}^{n-1},$$

where $[z_0,\ldots,z_k]$ denotes the equivalence class of $(z_0,\ldots,z_k)$ and $\tilde{L}^{2k-1}$ is embedded in $\tilde{L}^{2k}$ by setting $z_k = 0$. We call $\tilde{L}_n^{n+k}$ a stunted lens space.

Each representation of π on $\mathbb{C}^{k+1}$ without trivial subrepresentations yields a free π action on S^{2k+1} and a corresponding lens space S^{2k+1}/π. Since they are all stably equivalent we have simply chosen our favorite. Note, however, that the others reappear briefly in the proof of Proposition 2.4.

It is easy to see that $\tilde{L}^n - \tilde{L}^{n-1}$ is an open n cell. Thus $\tilde{L}_n^{n+k}$ has one cell in each dimension between n and n+k inclusive. Note that $\tilde{L}_1^n = \tilde{L}^n$ and $\tilde{L}_0^n = (\tilde{L}^n)^+$, the union of $\tilde{L}^n$ and a disjoint basepoint.

Since $\tilde{L}^\infty = S^\infty/\pi$ is a $K(\pi,1)$, $H^*(\tilde{L}^\infty;Z_p) = E\{x\} \otimes P\{\beta x\}$, with $|x| = 1$, and the Steenrod operations are specified by

$$P^i(x^\varepsilon(\beta x)^j) = \binom{j}{i}x^\varepsilon(\beta x)^{j+i(p-1)}.$$

The isomorphisms

$$H^i\tilde{L}_n^{n+k} \longrightarrow H^i\tilde{L}^{n+k} \longleftarrow H^i\tilde{L}^\infty$$

for $n \leq i \leq n+k$ then determine $H^*\tilde{L}_n^{n+k}$ as an $\mathcal{A}_p$ module.

Definition 2.2. Let $p > 2$ and let π act on C by multiplication by ρ. Let $\xi \in KU(\tilde{L}^{2k+1})$ be the bundle

$$S^{2k+1} \times_\pi C \longrightarrow S^{2k+1} \times_\pi \{0\} = \tilde{L}^{2k+1},$$

let $\zeta_i = r(\xi^i) \in KO(\tilde{L}^{2k+1})$ where $r: KU \to KO$ forgets complex structure, let

$\zeta = J(\zeta_1) \in J(\tilde{L}^{2k+1})$, and let $\sigma = \xi - 1_{\mathbf{C}} \in \widetilde{KU}(\tilde{L}^{2k+1})$. Let ξ, ζ_i, ζ and σ also denote the restrictions of these elements to $\tilde{L}^{2k}$.

We collect some results from [47], [48] and [58] in the following theorem.

Theorem 2.3. Let $\tilde{L}^{2k} \to \tilde{L}^{2k+1}$ be the inclusion and let $\langle x \rangle$ denote the cyclic group generated by x.

(i) $i^*: KU(\tilde{L}^{2k+1}) \to KU(\tilde{L}^{2k})$ is an isomorphism and

$$\widetilde{KU}(\tilde{L}^{2k}) = \langle\sigma\rangle \oplus \langle\sigma^2\rangle \oplus \cdots \oplus \langle\sigma^{p-1}\rangle$$

(ii) $r: \widetilde{KU}(\tilde{L}^{2k}) \to \widetilde{KO}(\tilde{L}^{2k})$ is an epimorphism,

$$\widetilde{KO}(\tilde{L}^{2k+1}) = \widetilde{KO}(\tilde{L}^{2k}) \oplus \widetilde{KO}(S^{2k+1}),$$

and i^* is projection onto the first summand under this isomorphism.

(iii) $\tilde{J}(\tilde{L}^{2k}) = \langle Jr(\sigma)\rangle = \langle \zeta - 2\rangle$ and has order $p^{[k/(p-1)]}$,

$$\tilde{J}(\tilde{L}^{2k+1}) = \tilde{J}(\tilde{L}^{2k}) \oplus \tilde{J}(S^{2k+1})$$

and i^* is projection onto the first summand under this isomorphism. Also, $J(\zeta_i) = \zeta$ for $i = 1,2,\ldots,p-1$.

Proof. This is all in [47], [48] and [58] except $J(\zeta_i) = \zeta$, which follows from the Adams conjecture:

$$J(\zeta_i) = Jr\xi^i = Jr\psi^i\xi = Jr\xi = \zeta. \quad //$$

The extended powers $D^k_\pi S^n$ are suspension spectra of Thom spaces of complex bundles over $\tilde{L}^k = W^k/\pi$. Thus Theorem 2.3 ensures us that the following theorem (proved in [81]) identifies all such spectra. Note that its proof does not require p to be a prime.

Theorem 2.4. If $s \geq 0$, the Thom complex of $r + s\zeta$ over $\tilde{L}^k$ satisfies

$$\Sigma^\infty T(r + s\zeta) = \Sigma^\infty \Sigma^r \tilde{L}^{2s+k}_{2s}.$$

Proof. The contribution of the trivial r dimensional fibration is obvious and may be ignored. We will actually prove a much more precise result. If α is an n-dimensional representation of π, we let $R^n(\alpha)$ and $S^{n-1}(\alpha)$ denote R^n and S^{n-1} with π action given by α. If the action is free on S^{n-1} we obtain a closed manifold $L(\alpha) = S^{n-1}(\alpha)/\pi$. If α and β are two such representations of dimension n and k respectively, let $\alpha|L(\beta)$ be the bundle

$$S^{k-1}(\beta) \times_\pi R^n(\alpha) \longrightarrow S^{k-1}(\beta) \times_\pi \{0\} = L(\beta).$$

We claim that there is a homeomorphism

$$T(\alpha|L(\beta)) \cong L(\beta \oplus a)/L(\alpha),$$

where $L(\alpha)$ is embedded in $L(\beta \oplus a)$ as the last n coordinates. This will imply Theorem 2.4 for odd k (since $L(\beta)$ is odd dimensional, p being odd). The even case will follow by removing the top cell on each side, since the homeomorphism will be cellular if we give the Thom complex $T(\alpha|L(\beta))$ the natural cell structure compatible with that of $L(\beta)$.

To establish the claim, let $f: S^{k-1}(\beta) \times_\pi R^n(\alpha) \to S^{n+k-1}(\beta \oplus \alpha)/\pi$ be induced by the natural inclusion $S^{k-1}(\beta) \times R^n(\alpha) \to R^{n+k}(\beta \oplus \alpha) - \{0\}$ followed by the radial retraction $R^{n+k} - \{0\} \to S^{n+k-1}$. It is easy to check that f is one-to-one and maps onto everything except the copy of $L(\alpha)$ embedded as the last n coordinates. Just as easily, one sees that f sends the zero section of $\alpha|L(\beta)$ to the embedding of $L(\beta)$ as the first k coordinates. It follows that $\alpha|L(\beta)$ is the normal bundle of this embedding $L(\beta) \to L(\beta \oplus \alpha)$ and that its Thom complex is $L(\beta \oplus \alpha)/L(\alpha)$. //

The fact that $\zeta \in J(\tilde{L}^k)$ has finite order enables us to define stunted lens spectra in positive and negative dimensions.

Definition 2.5. Let $\psi(k) = [k/2(p-1)]$. If n is any integer, $\varepsilon = 0$ or 1, and $k \geq \varepsilon$, let

$$\tilde{L}_{2n+\varepsilon}^{2n+k} = \Sigma^{2(n-r)}\Sigma^{\infty}\tilde{L}_{2r+\varepsilon}^{2r+k}$$

for $r \equiv n \ (p^{\psi(k)})$ such that $r \geq 0$.

The following result shows that the spectrum $\tilde{L}_n^{n+k}$ is well-defined up to equivalence in $\bar{h}\mathcal{S}$. Recall that an n-dimensional complex X is reducible if $X/X^{n-1} \simeq S^n$ and the projection $X \to S^n$ has a right inverse. Dually, an (n-1)-connected complex X is coreducible if $X^n = S^n$ and the inclusion $S^n \to X$ has a left inverse. Let $W = S^\infty$, let $q: W \to \tilde{L}^\infty$ be the quotient map and let $W^k = q^{-1}(\tilde{L}^k)$. Then we may define $D_\pi^k X = W^k \ltimes_\pi X^{(p)}$.

Theorem 2.6. Let S^n be the p-local n-sphere spectrum. Then

(i) $D_\pi^k S^n \simeq \Sigma^n \tilde{L}_{n(p-1)}^{n(p-1)+k}$.

(ii) $\tilde{L}_{2n}^{2n+k}$ is coreducible iff $n \equiv 0 \ (p^{\psi(k)})$, while $\tilde{L}_{2n+1}^{2n+k}$ is coreducible iff $k = 1$.

(iii) If $\varepsilon = 0$ or 1, $k \geq \varepsilon$ and $n \equiv r\ (p^{\psi(k)})$ then

$$\tilde{L}^{2n+k}_{2n+\varepsilon} \simeq \Sigma^{2(n-r)}\tilde{L}^{2r+k}_{2r+\varepsilon}.$$

(iv) $\tilde{L}^{b}_{a}$ and $\tilde{L}^{-a-1}_{-b-1}$ are (-1) dual spectra.

(v) If $\varepsilon = 0$ or 1 and $k \geq \varepsilon$ then $\tilde{L}^{2n+k}_{2n+\varepsilon}$ is reducible iff either $k = \varepsilon$ or k is odd and $2n+k+1 \equiv 0\ (p^{\psi(k)})$.

Proof. If $n \geq 0$ then $D^k_\pi S^n = W^k \times_\pi S^{n(p)} = \Sigma^\infty T(n\gamma_k)$, where γ_k is the restriction to $\tilde{L}^k$ of the bundle over $\tilde{L}^\infty = B\pi$ induced by the regular representation of π. Since $\gamma_k = 1 + \zeta_1 + \cdots + \zeta_m$, $J(n\gamma_k) = n + nm\zeta$ (where $2m = p-1$). By Theorem 2.4,

$$\Sigma^\infty T(n\gamma_k) = \Sigma^n\tilde{L}^{n(p-1)+k}_{n(p-1)}.$$

If $n < 0$ then, by [Equiv, VI.5.3 and 5.4]

$$W^k \ltimes_\pi S^{n(p)} = W^k \ltimes_\pi (\Sigma^n S)^{(p)} \simeq \Sigma^{-N}\Sigma^\infty T(N + n\gamma_k)$$

for sufficiently large N, and since $J(n\gamma_k) = n + nm\zeta$, we find that

$$W^k \ltimes_\pi S^{n(p)} \simeq \Sigma^{-N}\Sigma^\infty T(N+n + nm\zeta) \simeq \Sigma^n\tilde{L}^{n(p-1)+k}_{n(p-1)}$$

by Definition 2.5 and Theorems 2.4 and 2.3.(iii). This proves (i).

By Theorem 2.4, $\tilde{L}^{2n+k}_{2n} = \Sigma^\infty T(n\zeta|\tilde{L}^k)$. By [15], $\Sigma^\infty T(n\zeta)$ is coreducible if and only if $\mathfrak{J}(n\zeta) = 0$, so the first half of (ii) follows by Theorem 2.3.(iii). For the second part of (ii) we need only note that the Bockstein is nonzero on H^{2n+1} if $k > 1$.

To prove (iii), note that $\mathfrak{J}(n\zeta) = \mathfrak{J}(r\zeta)$ if $n \equiv r\ (p^{\psi(k)})$ by Theorem 2.3.(iii).

To prove (iv), first consider $\tilde{L}^{2n+k}_{2n}$ with k odd. By Theorem 2.4, $\tilde{L}^{2n+k}_{2n} = \Sigma^\infty T(n\zeta|\tilde{L}^k)$. Since k is odd, $\tilde{L}^k$ is a closed manifold. By considering the fibration

$$S^1 \to \tilde{L}^k \to CP^{[k/2]},$$

we see that the tangent bundle of $\tilde{L}^k$ is $([k/2] + 1)\zeta - 1$. Atiyah's duality theorem [15, Theorem 3.3] implies that the (-1) dual of $\tilde{L}^{2n+k}_{2n}$ is $\Sigma^\infty T(1 - (n+[k/2] + 1)\zeta) = \tilde{L}^{-2n-1}_{-2n-k-1}$. To prove (iv) for the other three possible combinations of odd or even top and bottom cells, we use the duality between inclusion of the bottom cell of a complex and projection onto the top cell of its dual.

Finally, (v) follows from (ii) and (iv) by the duality between reductions and coreductions. //

Now we present the analogs of 2.1 through 2.6 for D_pS^n instead of $D_\pi S^n$. Since the transfer splits D_pS^n off as a wedge summand of $D_\pi S^n$, we can use this as a short-cut to the results we need. Let $X_{(p)}$ denote the p-localization of a spectrum or space X. The following result is proved in [7].

Proposition 2.7. There is a CW spectrum L with one cell in each nonnegative dimension congruent to 0 or -1 modulo 2(p-1), such that $L \simeq (\Sigma^\infty B\Sigma_p^+)_{(p)}$.

Definition 2.8. Let L^k be the k-skeleton of L and let $L_n^{n+k} = L^{n+k}/L^{n-1}$ if $n > 0$. If $n < 0$, $\varepsilon = 0$ or 1, and $k \geq \varepsilon$, let $L_{2n+\varepsilon}^{2n+k} = \Sigma^{2(n-r)} L_{2r+\varepsilon}^{2r+k}$ for $r \equiv n\ (p^{\psi(k)})$ such that $r \geq 0$.

Note that n and k are not uniquely determined by L_n^{n+k} as they are by $\tilde{L}_n^{n+k}$. For example, $L_1^q = L_2^q = \cdots = L_{q-1}^q$, where $q = 2(p-1)$, since L has no cells in dimensions $1,2,\ldots,q-2$.

Theorem 2.9. Let S^n be the p-local n-sphere spectrum and let $q = 2(p-1)$. Then

(i) $D_pS^{2j} \simeq \Sigma^{2j}L_{jq}^\infty$ and $D_pS^{2j-1} \simeq \Sigma^{2j-1}L_{jq-1}^\infty$. The maps $D_\pi S^n \to D_pS^n$ and $\tilde{L}_n^{n+k} \to L_n^{n+k}$ induced by the inclusion $\pi \subset \Sigma_p$ are projections onto wedge summands.

(ii) L_{jq}^{jq+k} is coreducible iff $j \equiv 0\ (p^{\psi(k)})$, while L_{jq-1}^{jq+k} is coreducible iff $k = -1$.

(iii) If $\varepsilon = 0$ or 1 and $i \equiv j\ (p^{\psi(k+2\varepsilon)})$ then

$$L_{jq-\varepsilon}^{jq+k} \simeq \Sigma^{(j-i)q} L_{iq-\varepsilon}^{iq+k}.$$

(iv) If ε and δ are 0 or 1 then $L_{jq-\varepsilon}^{iq-\delta}$ is (-1) dual to $L_{-iq+\delta-1}^{-jq+\varepsilon-1}$.

(v) If $\varepsilon = 0$ or 1 then $L_{jq-\varepsilon}^{jq+k}$ has a reducible jq+k cell iff either $k = \varepsilon = 0$ or $k = iq-1$ and $i+j \equiv 0\ (p^{i+\varepsilon-1})$.

Note: Part (i) shows that bottom dimensions of the form $jq-\varepsilon$, $\varepsilon = 0$ or 1, are more natural in this context than $jq+\varepsilon$. This accounts for the exponent $\psi(k+2\varepsilon)$ in (iii), where $\psi(k)$ might be expected.

Proof. By the remark preceding the theorem, the first statement in (i) can be abbreviated to $D_pS^n = \Sigma^n L^\infty_{n(p-1)}$. The transfer $(\Sigma^\infty B\Sigma_p)_{(p)} \to \Sigma^\infty B\pi$ splits off L^∞ and $L^\infty_{n(p-1)}$ as wedge summands of $\tilde{L}^\infty$ and $\tilde{L}^\infty_{n(p-1)}$ respectively. Similarly, the transfer splits off D_pS^n as a wedge summand of $D_\pi S^n$. The maps

$$D_pS^n \xrightarrow{t_1} D_\pi S^n \simeq \Sigma^n \tilde{L}^\infty_{n(p-1)} \xrightarrow{i_1} \Sigma^n L^\infty_{n(p-1)}$$

and

$$\Sigma^n L^\infty_{n(p-1)} \xrightarrow{t_2} \Sigma^n \tilde{L}^\infty_{n(p-1)} \simeq D_\pi S^n \xrightarrow{i_2} D_pS^n ,$$

where t_1 and t_2 are transfers, and i_1 and i_2 are induced by the inclusion $\pi \subset \Sigma_p$ are inverse equivalences because their composites induce isomorphisms in mod p homology. This proves (i). Now (ii)-(v) follows from 2.6 and (i). //

The preceding theorem does not assert that $W^k \ltimes_{\Sigma_p} S^{n(p)} \simeq \Sigma^n L^{n(p-1)+k}_{n(p-1)}$ where W^k is the k-skeleton of a contractible free Σ_p space, because this is not true. In general, $W^k \ltimes_{\Sigma_p} S^{n(p)}$ will have homology in dimension np+k which goes to 0 in D_pS^n and in $\Sigma^n L^{n(p-1)+k}_{n(p-1)}$. Since we are only interested in homology which is nonzero in D_pS^n, $\Sigma^n L^{n(p-1)+k}_{n(p-1)}$ is more useful to us than is $W^k \ltimes_{\Sigma_p} S^{n(p)}$.

Therefore we will let $D^k_pS^n = \Sigma^n L^{n(p-1)+k}_{n(p-1)}$, rather than $W^k \ltimes_{\Sigma_p} S^{n(p)}$.

The preceding theorem also shows that we may ignore the distinction between L^{n+k}_n and $\tilde{L}^{n+k}_n$ without harm. We used $\tilde{L}^{n+k}_n$ and $D_\pi S^n$ as a stepping stone to information about D_pS^n because J theory only gives information about coreducibility of Thom complexes, and we need Atiyah's S-duality theorem to convert this to information about reducibility. The S-duality theorem of Atiyah only applies to Thom complexes of bundles over manifolds so cannot be used on bundles over the skeleta of $B\Sigma_p$, or over the even skeleta of $B\pi$. Conveniently, the odd skeleta of $B\pi$ are manifolds (if we use a lens space for $B\pi$). To obtain analogous information about D_rS^n for nonprime r, a similar technique works. First, we split D_rS^n off of $D_\tau S^n$ using the transfer, where $\tau \subset \Sigma_r$ is a p-Sylow subgroup. Then the structure of τ (a Cartesian product of iterated wreath products of π) suggests manifolds mapping to $B\tau$ which we can use just as the odd skeleta of $B\pi$ are used here.

We now turn to the analogs of 2.1 through 2.6 for p = 2.

Definition 2.10. Let $n \geq 0$, let $\pi = \Sigma_2$ act antipodally on S^n and let

$$P^n = S^n/\pi$$

$$\text{and} \quad P_n^{n+k} = P^{n+k}/P^{n-1} .$$

We call P_n^{n+k} a stunted projective space. Let ξ in $KO(P^n)$ be the canonical real line bundle and let $\lambda = \xi - 1 \in KO(P^n)$.

Remarks. (1) If $p = 2$ we will agree to let L^n and $\tilde{L}^n$ mean P^n and let L_n^{n+k} and $\tilde{L}_n^{n+k}$ mean P_n^{n+k} so that uniform statements of results for all primes can be given. The P^n and P_n^{n+k} notation will still appear frequently because many of the results are not the same for even and odd primes.

(2) It is easy to see that $P^n - P^{n-1}$ is an open n-cell so that P_n^{n+k} has one cell in each dimension between n and n+k inclusive. Since $P^\infty = S^\infty/Z_2$ is a $K(Z_2,1)$, $H^*(P^\infty;Z_2) = P\{x\}$ with $|x| = 1$ and

$$Sq^i x^j = \binom{j}{i} x^{i+j}.$$

The isomorphisms

$$H^i P_n^{n+k} \to H^i P^{n+k} \leftarrow H^i P^\infty$$

for $n \leq i \leq n+k$ thus determine $H^* P_n^{n+k}$ as an $\mathcal{A}_2$ module.

Theorem 2.11. Let $\phi(n)$ be the number of integers j congruent to 0,1,2, or 4 mod 8 such that $0 < j \leq n$. Then $\widetilde{KO}(P^n) = \langle\lambda\rangle$ and has order $2^{\phi(n)}$. Furthermore,

$$J: KO(P^n) \to J(P^n)$$

is an isomorphism.

Proof. $KO(P^n)$ is computed in [1]. The computations there and the Adams conjecture imply the last statement. //

Theorem 2.12. If $s \geq 0$ the Thom complex of $r+s\xi$ over P^n satisfies

$$\Sigma^\infty T(r + s\xi) = \Sigma^\infty P_s^{s+n}.$$

Proof. The proof of Proposition 2.4 can easily be adapted to prove this as well.

As for odd primes, we can now define stunted projective spectra starting and ending in any positive or negative dimensions.

Definition 2.13. For $k \geq 0$ and any n let

$$P_n^{n+k} = \Sigma^{n-r}\Sigma^\infty P_r^{r+k}$$

for any $r \equiv n \ (2^{\phi(k)})$, $r \geq 0$.

The following result shows that P_n^{n+k} is well defined up to equivalence in $\overline{h}\mathcal{S}$. Let S^k have the antipodal action of π. We define $D_2^k X = S^k \ltimes_\pi X^{(2)}$.

Theorem 2.14. Let S^n be the 2-local n-sphere spectrum. Then

(i) $D_2^k S^n \simeq \Sigma^n P_n^{n+k}$

(ii) P_n^{n+k} is coreducible if and only if $n \equiv 0 \ (2^{\phi(k)})$

(iii) If $n \equiv m \ (2^{\phi(k)})$ then $P_n^{n+k} \simeq \Sigma^{n-m} P_m^{m+k}$

(iv) P_a^b and P_{-b-1}^{-a-1} are (-1) dual spectra

(v) P_n^{n+k} is reducible if and only if $n+k+1 \equiv 0 \ (2^{\phi(k)})$.

Proof (i) follows for $n \geq 0$ from Theorem 2.12 once we observe that the regular representation γ_k is $1 + \xi$. For $n < 0$ we have

$$D_2^k S^n = D_2^k(\Sigma^n S) \simeq \Sigma^{-N}\Sigma^\infty T(N + n\gamma_k)$$

by VI.5.3 and VI.5.4 of [Equiv], for sufficiently large N. Hence $D_2^k S^n \simeq \Sigma^n P_n^{n+k}$ for $n < 0$ also, again by 2.12.

Parts (ii) through (v) follow exactly as in 2.6. In (iv) we use the fact that P^n is a closed manifold with tangent bundle $(n+1)\xi - 1$. //

The last results in this section identify the top dimensional component of any attaching map of $D_p S^n$ by combining Theorems 2.6 and 2.14 with Milnor's result on Thom complexes of sphere bundles over suspensions. First we must define the maps under consideration. As in §1, $q = 2(p-1)$ and $\varepsilon = 0$ or 1 ($q = 1$ and $\varepsilon = 0$ if $p = 2$).

Definition 2.15. Define a function v_p by

$$v_p(n) = \max\{v \mid L_{n-v+1}^n \text{ is reducible}\}.$$

Let $v = v_p(n)$ and define $a_p(n) \in \pi_{v-1}S$ to be Σ^{v-n} of the composite

$$S^{n-1} \longrightarrow L^{n-v} \longrightarrow S^{n-v}$$

in which the first map is a lift of the ataching map of the n cell and the second is projection onto the top cell of L^{n-v}.

The indeterminacy in the definition of $a_p(n)$ is the kernel of the homomorphism induced on π_{n-1} by the inclusion of the bottom cell of L^{n-1}_{n-v} .

We will often omit the subscript p for typographical simplicity. The notations v and a are intended to be mnemonic: v stands for "vector field number" and a stands for "attaching map". Actually, v is not quite the vector field number as defined by Adams [1]; $v_2(n)$ is $\rho(n-1)$ in Adams' notation. The function v_p tells us how far we can compress each of the attaching maps of L^∞. The attaching map of the n cell factors through L^{n-v} if and only if L^n_{n-v+1} is reducible. Thus, it factors through L^{n-v} but not through L^{n-v-1}, where $v = v_p(n)$. By the definition of $v_p(n)$, $a_p(n)$ is nonzero. We obtain a good hold on v_p and a_p from the following two lemmas. Let $\varepsilon_p(j)$ be the exponent of p in the prime factorization of j.

<u>Proposition 2.16.</u> If $p > 2$ then, with $q = 2(p-1)$,

$$v_p(jq-\varepsilon) = \begin{cases} 1 & \varepsilon = 0 \\ q(1 + \varepsilon_p(j)) & \varepsilon = 1 . \end{cases}$$

If $p = 2$ then $v_2(j) = 8a + 2^b$, where $\varepsilon_2(j+1) = 4a + b$ and $0 \leq b \leq 3$.

<u>Proposition 2.17.</u> If $v_p(n) = 1$ then $a_p(n)$ is the map of degree p. If $v_p(n) > 1$ then $a_p(n) \otimes 1$ generates $\operatorname{Im} J \otimes Z_{(p)}$ in dimension $v_p(n)-1$.

<u>Proof of 2.16.</u> Theorem 2.14.(v) shows that $v_2(j)$ is the maximum s such that $\varepsilon_2(j+1) = \phi(s-1)$. The formula for $v_2(j)$ follows easily from this. Theorem 2.9.(v) shows that if $p > 2$ then $v_p(jq) = 1$ while $v_p(jq-1)$ is the maximum s such that $\varepsilon_p(jq) = \psi(s-1)$. The formula for $v_p(jq-\varepsilon)$ follows immediately. //

<u>Proof of 2.17.</u> Let $n = jq-\varepsilon$, $v = v_p(n)$ and $a = a_p(n)$. We wish to construct a map of cofiber sequences

$$\begin{array}{ccccccc} S^{n-1} & \longrightarrow & L^{n-1}_{n-v} & \longrightarrow & L^n_{n-v} & \longrightarrow & S^n \\ \| & & \uparrow b & & \uparrow & & \| \\ S^{n-1} & \xrightarrow{a} & S^{n-v} & \longrightarrow & Ca & \longrightarrow & S^n \end{array}$$

where $Ca = S^{n-v} \cup_a e^n$, b is the inclusion of the bottom cell, and $a \otimes 1$ generates $\operatorname{Im} J \otimes Z_{(p)}$. By S-duality and Theorems 2.9.(iv) and 2.14.(iv), it is equivalent to construct a map of cofiber sequences

$$(*) \qquad \begin{array}{ccccccc} S^{-n} & \longleftarrow & L^{v-n-1}_{-n} & \longleftarrow & L^{v-n-1}_{-n-1} & \longleftarrow & S^{-n-1} \\ \| & & \downarrow b^* & & \downarrow & & \| \\ S^{-n} & \xleftarrow{a} & S^{v-n-1} & \longleftarrow & Ca & \longleftarrow & S^{-n-1} \end{array}$$

in which b^* is the collapse onto the top cell and a is as before. The lemma is trivial when $v=1$ so we may assume $v>1$ and hence, that n is odd. Let γ be the bundle $-(n+1)\xi$ if $p=2$ and $-j(p-1)\zeta$ if $p>2$ over L^v. Then $L^{v-n-1}_{-n-1}=T(\gamma)$. By the definition of v, γ is trivial over L^{v-1} but not over L^v. This implies $\gamma=\pi^*\nu$ where $\pi: L^v \to L^v/L^{v-1}=S^v$ is the collapsing map and $0\neq\nu$ $KO(S^v)$. By [85], $T(\nu)$ has attaching map $J(\nu)$. Thus, the inclusions of the fiber S^{-n-1} into $T(\gamma)$ and $T(\nu)$ induce a map (*) of cofiber sequences with $a=J(\nu)$. Since v is greater than 1, it is even when $p>2$ by 3.2. Thus, 2.3.(iii) and 2.9.(i) when $p>2$, and 2.11 when $p=2$, imply that the kernel of $\mathfrak{J}(L^v)\to\mathfrak{J}(L^{v-1})$ is Z_p. Hence $\mathfrak{J}(\gamma)$ generates it, being nonzero. Since $\pi^*(a)=\mathfrak{J}(\gamma)$, $a\in\mathfrak{J}(S^v)$ must generate $\mathfrak{J}(S^v)\otimes Z_{(p)}$. //

In the notation of 1.6, Propositions 2.16 and 2.17 are summarized by the equations

$$a_2(j) \doteq \beta_{\varepsilon_2(j+1)}$$

$$a_p(jq) = p$$

and

$$a_p(jq-1) = \alpha_{1+\varepsilon_p(j)}$$

where $\doteq$ denotes equality up to multiplication by a unit of $Z_{(p)}$.

§3. Proofs for section 1 and other calculations

This section primarily consists of proofs of results of §1 with the additional necessary results (3.1-3.4) interspersed. Note, however, that the spectral sequence charts in Figures 3.1 to 3.9 can be very useful in conjunction with Theorem 1.10 since they show where in the Adams spectral sequence the elements detecting the results of homotopy operations must lie.

Proof of 1.1. $\iota^*(x)=x^p$ by IV.7.3.(iii). Clearly, $E^0(\iota)=e_0\otimes\iota_n^p=e_0$, so the second statement is immediate from the definition:

$$\beta^\varepsilon P^j = (-1)^j\nu(n)e_{jq-\varepsilon-n(p-1)}.$$

Proof of 1.2. Recall from III §1 that the homology operations are defined by

$$Q^jx = \xi_*(e_{j-n}\otimes x^2) \quad \text{if } p=2,$$

and

$$\beta^\varepsilon Q^jx = \xi_*((-1)^j\nu(n)e_{jq-\varepsilon-n(p-1)}\otimes x^p) \quad \text{if } p>2.$$

To prove 1.2 we simply calculate. If $p=2$ and $E^0(\alpha)=P^j$ then

$$\begin{aligned} h\alpha^*(x) &= [\alpha^*(x)]_*(\iota_N) \\ &= \xi_* D_p(x)_* \alpha_*(\iota_N) \\ &= \xi_* D_p(x)_*(e_{j-n} \otimes \iota_n^2) \\ &= \xi_*(e_{j-n} \otimes h(x)^2) \\ &= Q^j h(x). \end{aligned}$$

The proof is essentially the same when $p > 2$. //

Proof of 1.3. This is just the naturality of the spectral sequence $E_r(S, \mathcal{D})$. //

Proof of 1.4. Consider the following commutative diagram, in which the row is the cofiber sequence of c_i and α' is a lift of α to $D_p^i S^n$.

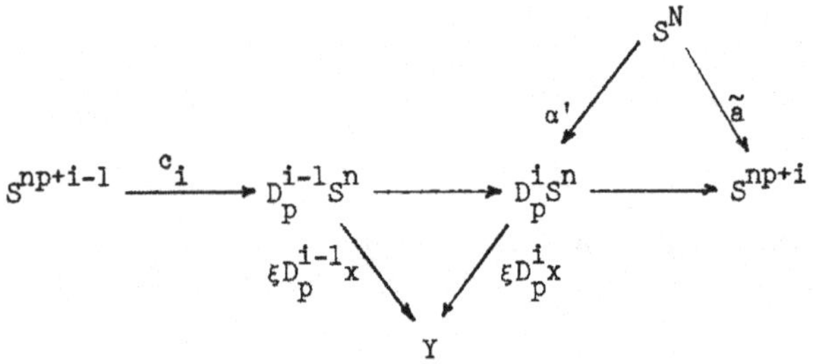

Clearly $\alpha^*(x) = \xi D_p(x)\alpha = \xi D_p^i(x)\alpha'$ and this lies in the Toda bracket $\langle \tilde{a}, c_i, \xi D_p^{i-1}(x)\rangle$. If α and β both lift to $D_p^i S^n$ and project to $\tilde{a}$ on S^{np+i}, then $\alpha - \beta$ lifts to $D_p^{i-1}S^n$ so that $\alpha^*(x) - \beta^*(x)$ is in $\xi D_p^{i-1}(x) \circ \pi_N D_p^{i-1} S^n$. Conversely, if $\gamma \in \pi_N D_p^{i-1} S^n$ then $\alpha + \gamma$ also lifts to $D_p^i S^n$ and projects to α' on S^{np+i}. //

Proof of 1.5. By definition, $\beta^{\epsilon} P^j$ is defined on π_n if and only if e_i is a permanent cycle in $E_\infty(S, \mathcal{D})$. Thus (i) and (ii) are equivalent. Let $\mathcal{D}_i$ be $\mathcal{D}$ truncated at the np+i cell. The map of spectral sequences $E_r(S, \mathcal{D}_i) \to E_r(S, S^{np+i})$ induced by the projection $D_p^i S^n \to S^{np+i}$ sends e_i to the identity map of S^{np+i}. If $D_p^i S^n$ is reducible then there is a map back which splits $E_r(S, S^{np+i})$ off $E_r(S, \mathcal{D}_i)$, forcing e_i to be a permanent cycle. Conversely, if e_i is a permanent cycle then any map detecting it will be a reduction. Thus (ii) and (iii) are equivalent. Finally, (iii) and (iv) are equivalent by Theorems 2.6.(v), 2.9.(v) and 2.14.(v). //

Proof of 1.7. To show $\beta_j \circ x^2 = 0$, where $\beta_j \in \pi_{v-1}S$, we need only show that P_{n+1}^{n+v} is reducible and P_n^{n+v} is not, since this implies that the n+v cell is attached only to the n cell of P_n^{n+v}, and Proposition 2.17 implies that the attaching map is a generator of Im J in $\pi_{v-1}S$. If $j = 4a + b$ then $v = 8a + 2^b$, so 2.14.(v) implies that n must satisfy

$$n + 8a + 2^b \equiv -1 \quad (2^j)$$

$$\text{and} \qquad n + 8a + 2^b \not\equiv -1 \ (2^{j+1}).$$

To show $\beta_j \circ P^{n+1}x$ is a multiple of x^2, we must show that P_{n+1}^{n+v+1} is not reducible, but P_{n+2}^{n+v+1} is reducible, for then the top cell will be attached to the cells carrying x^2 and $P^{n+1}x$. The rest of the proof is the same as in the first case. //

Proof of 1.8. To show that $\alpha_j \circ \beta P^k x = 0$, for $x \in \pi_n Y$ and $n = 2k-1$, is trivial when $j = 0$. Simply note that L_{kq-1}^{kq} is a mod p Moore spectrum. When $j > 0$ we must show $L_{kq}^{(k+j)q-1}$ is reducible, while $L_{kq-1}^{(k+j)q-1}$ is not. By 2.9.(v) we need $k+j \equiv 0 \ (p^{j-1})$ but $k+j \not\equiv 0 \ (p^j)$.

When $n = 2k$, the relation $\alpha_j \circ \beta P^{k+1}x = \alpha \circ x^p$ for some α is also trivial when $j = 0$. We need only note that L_{kq+q-1}^{kq+q} is a mod p Moore spectrum. For $j > 0$, we must show that $L_{kq+q}^{(k+j)q+q-1}$ is reducible, but $L_{kq+q-1}^{(k+j)q+q-1}$ is not. By 2.9.(v) we must have $k+j+1 \equiv 0 \ (p^{j-1})$ but $k+j+1 \not\equiv 0 \ (p^j)$. //

When $n = 2k$, if we try to show $\alpha_j \circ x^p = 0$ by this technique we find we must assume $k+j \equiv 0 \ (p^{j-1})$ and $k+j \not\equiv 0 \ (p^{j-1})$, so that no information is available.

Before we compute the first few homotopy groups of D_pS^n (and hence the first few homotopy operations), we describe the attaching maps of the first few cells. Exact definitions of the maps used in the following proposition can be found in the proof.

Proposition 3.1. Let $p = 2$.

(i) If $n \equiv 1 \ (4)$ then $P_n^{n+3} \simeq S^n \cup_2 e^{n+1} \vee S^{n+2} \cup_{\tilde{\eta}+2} e^{n+3}$

(ii) If $n \equiv 2 \ (4)$ then $P_n^{n+3} \simeq S^n \vee S^{n+1} \cup_{\eta+2} e^{n+2} \cup_\eta e^{n+3}$

(iii) If $n \equiv 3 \ (4)$ then $P_n^{n+3} \simeq S^n \cup_2 e^{n+1} \cup_\eta e^{n+2} \cup_2 e^{n+3}$

(iv) If $n \equiv 0 \ (4)$ then $P_n^{n+3} \simeq S^n \vee S^{n+1} \cup_2 e^{n+2} \vee S^{n+3}$.

Proof. Much of the structure of P_n^{n+3} is determined by Sq^1 and Sq^2 in $H^*P_n^{n+3}$. We will assume this information and fill in the rest. Suppose $n \equiv 0$ (4). Then 2.14 implies P_n^{n+3} is both reducible and coreducible, so only the middle two cells are attached. When $n \equiv 1$ (4), collapsing the bottom cell of the previous case yields $P_n^{n+2} \simeq S^n \cup_2 e^{n+1} \vee S^{n+2}$. Computing Sq^1 and Sq^2 shows e^{n+3} is attached to S^{n+2} by a map of degree 2, and is attached to the Moore spectrum by a map which projects to η on S^{n+1}. This projection induces an epimorphism

$$Z_4 = \pi_{n+2}(S^n \cup_2 e^{n+1}) \longrightarrow \pi_{n+2}S^{n+1} = Z_2.$$

Therefore, the attaching map is a generator $\tilde{\eta}$ of $\pi_{n+2}(S^n \cup_2 e^{n+1})$.

When $n \equiv 2$ (4), we start with $P_n^{n+2} \simeq S^n \vee S^{n+1} \cup_{\eta+2} e^{n+2}$. The long exact homotopy sequence of $S^n \vee S^{n+1} \to P_n^{n+2}$ shows that the inclusion $S^{n+1} \to P_n^{n+2}$ induces an isomorphism on π_{n+2}. Since Sq^2 is nonzero on $H^{n+1}P_n^{n+3}$, the n+3 cell is attached by the map

$$S^{n+2} \xrightarrow{\ \eta\ } S^{n+1} \longrightarrow P_n^{n+2},$$

which we also call η.

Finally, when $n \equiv 3$ (4), we start with $P_n^{n+2} \simeq S^n \cup_2 e^{n+1} \cup_\eta e^{n+2}$. The map $P_n^{n+2} \to S^{n+1} \vee S^{n+2}$ which collapses the bottom cell, induces on π_{n+2} a monomorphism

$$\pi_{n+2}P_n^{n+2} = Z_2 \oplus Z \rightarrowtail Z_2 \oplus Z = \pi_{n+2}S^{n+1} \oplus \pi_{n+2}S^{n+2}$$

which sends (a,b) to (a,2b). Computing Sq^1 and Sq^2 shows that the attaching map of the n+3 cell is $(0,1) \in \pi_{n+2}P_n^{n+2}$, which projects to the map of degree 2 on S^{n+2}. We simply call this map 2. //

Proposition 3.2. Let $p > 2$.

(1) $$L_{jq}^{jq+2q-1} \simeq S^{jq} \vee S^{jq+q-1} \bigcup_{-j\alpha_1+p} e^{jq+q} \bigcup_{-(j+2)\alpha_1} e^{jq+2q-1}.$$

(2) $$L_{jq-1}^{jq+q} \simeq S^{jq-1} \cup_p e^{jq} \bigcup_{-(j+1)\alpha_1} e^{jq+q-1} \bigcup_{-j\alpha_1+p} e^{jq+q}.$$

Proof. Recall that the first three nonzero homotopy groups of S localized at p are $\pi_0 = Z$, $\pi_{q-1} = Z_p$ generated by α_1, and $\pi_{2q-1} = Z_p$ generated by α_2. Thus $L_{jq}^{jq+q-1} = S^{jq} \vee S^{jq+q-1}$ is the only possibility. Computing β and P^1 in $H^*L_{jq}^{jq+q}$ shows that $L_{jq}^{jq+q} \simeq S^{jq} \vee S^{jq+q-1} \bigcup_{-j\alpha_1+p} e^{jq+q}$. Finally, the long exact homotopy sequence of $S^{jq} \vee S^{jq+q-1} \to L_{jq}^{jq+q}$ shows that the inclusion of S^{jq+q-1} induces an isomorphism of $\pi_{jq+2q-2}$. Thus the attaching map of the jq+2q-1 cell factors through S^{jq+q-1} and is determined to be $-(j+2)\alpha_1$ by computing P^1.

Collapsing the bottom cell and redefining j we find that $L^{jq+q-1}_{jq-1} \simeq S^{jq-1} \cup_p e^{jq} \underset{-(j+1)\alpha_1}{\cup} e^{jq+q-1}$. The long exact homotopy sequence of $S^{jq-1} \to L^{jq+q-1}_{jq-1}$ shows that the attaching map of the jq+q cell is determined by its projections onto S^{jq} and S^{jq+q-1}. Computing P^1 and β shows these to be $-j\alpha_1$ and p respectively. //

Diagrams of the cohomology with Sq^1 and Sq^2 or β and P^1 indicated are convenient mnemonic devices. For p = 2 we have

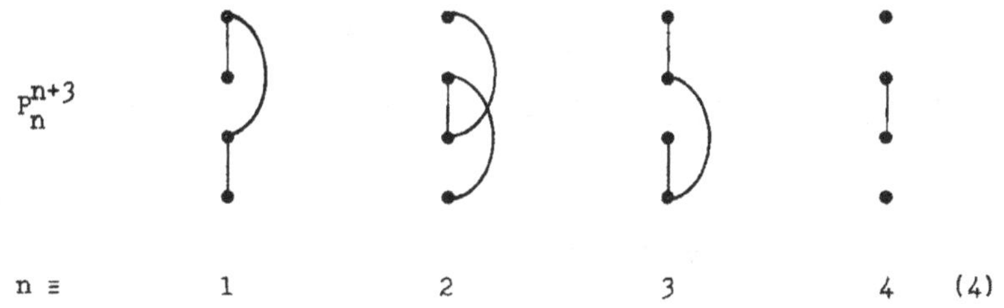

For p > 2, we have

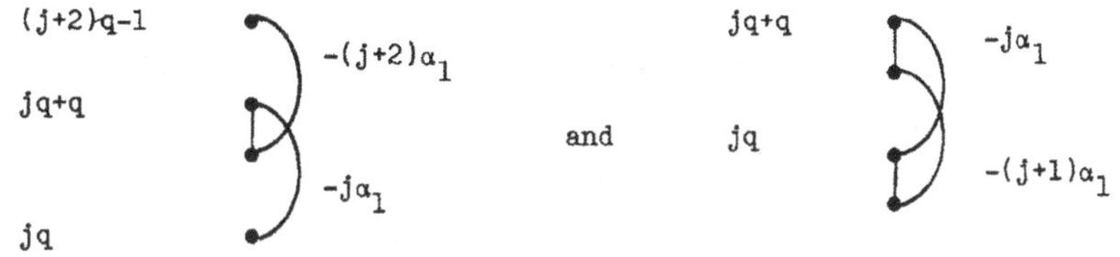

We can also think of these diagrams as indicating cells by dots and attaching maps by lines, and this is how we have labelled the diagrams for p > 2.

The spectral sequence $E_r(S, \wp)$ will enable us to glean a maximal amount of information from Propositions 3.1 and 3.2. We begin with p = 2. Recall, from [66], the initial segment of the HZ_2 Adams spectral sequence for π_*S.

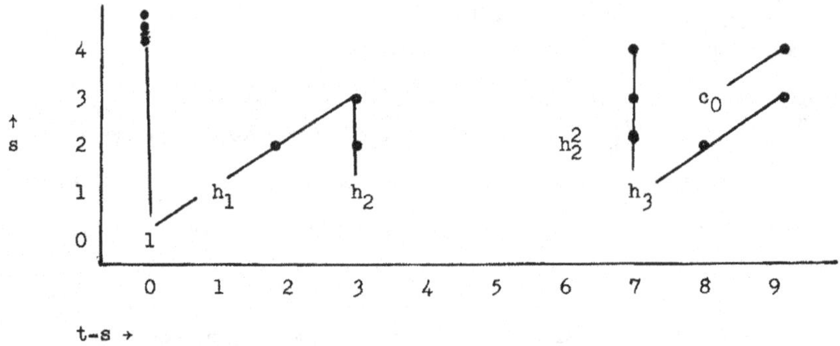

Vertical lines represent multiplication by h_0, detecting the map of degree 2, and diagonals represent multiplication by h_1, detecting η. We shall only use the first 8 stems ($t-s \leq 8$). Let $\mathcal{P}$ be the sequence

$$P_n^{n+8} \longleftarrow P_n^{n+7} \longleftarrow \cdots \longleftarrow P_n^{n+1} \longleftarrow P_n^n .$$

(Omitting the Σ^n from $D_2^i S^n = \Sigma^n P_n^{n+i}$ means a class in $E_r(S,\mathcal{P})$ will have stem degree equal to the amount by which the corresponding homotopy operation raises degrees.)

Proposition V.7.5 says that $E_2(S,\mathcal{P})$ is free over $E_2(S,S)$ on generators in each degree from n to n+k. Write x(i) for the element of $E_2(S,\mathcal{P})$ which is $x \in E_2(S,S)$ in the i summand, if $i \geq n$. Let x(i) mean 0 if $i < n$.

<u>Theorem 3.3</u>. In $E_r(S,\mathcal{P})$, for $t-s \leq 6$,

$$d_2 x(i) = h_0 x(i-1) \quad \text{if } i \equiv 0 \quad (2),$$

$$d_3 x(i) = h_1 x(i-2) \quad \text{if } i \equiv 0,1 \quad (4),$$

and $$d_5 x(i) = h_2 x(i-4) \quad \text{if } i \equiv 0,1,2,3 \quad (8).$$

In the same range, $E_\infty(S,\mathcal{P})$ is given by Figures 3.1 through 3.4.

<u>Note</u>: Dotted vertical lines indicate "hidden extensions". That is, they represent multiplications by 2 which cause an increase of more than 1 in filtration. Similarly, dotted diagonals indicate the effect of multiplication by η when this causes an increase in filtration of more than 1. See the proof of 1.9 for their derivation.

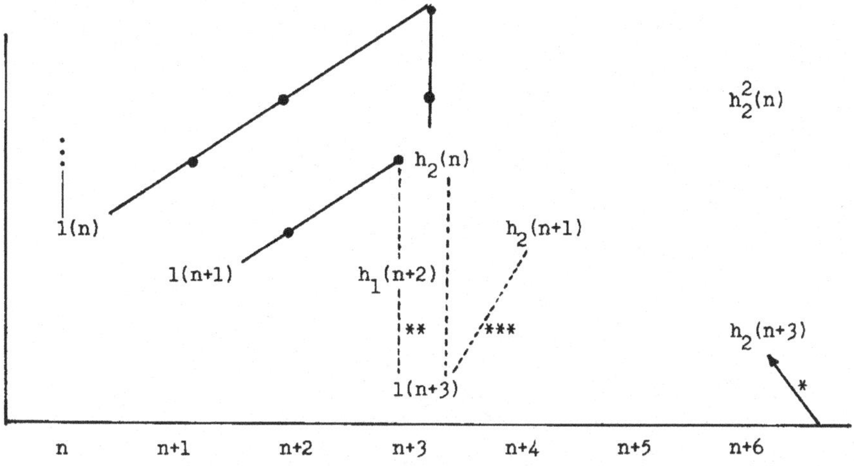

$n \equiv 0\ (4)$ *) hit by $d_5(1(n+7))$ iff $n \equiv 4\ (8)$

**) 2 times $1(n+3)$ is $h_1(n+2)$ if $n \equiv 0\ (8)$ and it is "$h_1(n+2) + h_2(n)$" if $n \equiv 4\ (8)$

***) if $n \equiv 4\ (8)$

Figure 3.1

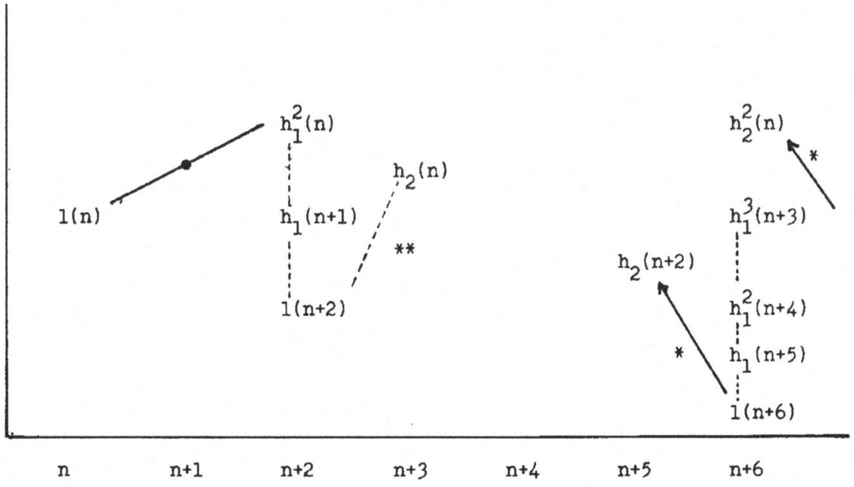

$n \equiv 1\ (4)$ *) differential iff $n \equiv 5\ (8)$

**) if $n \equiv 5\ (8)$

Figure 3.2

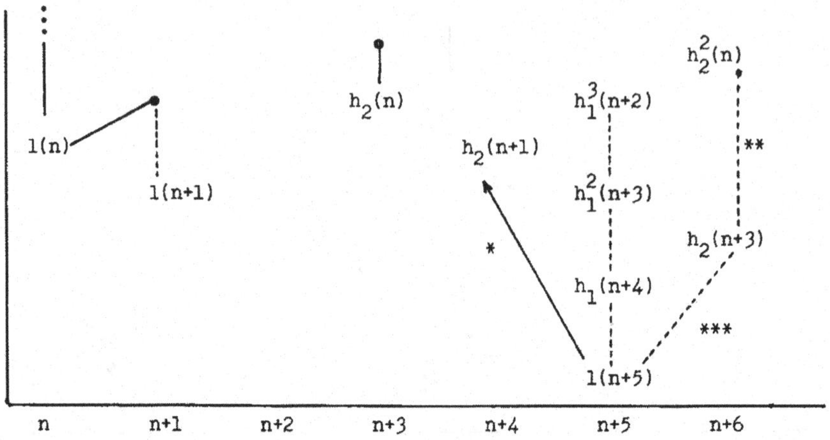

$n \equiv 2\ (4)$ *) differential iff $n \equiv 6\ (8)$

**) if $n \equiv 6\ (8)$

Figure 3.3 ***) if $n \equiv 2\ (8)$

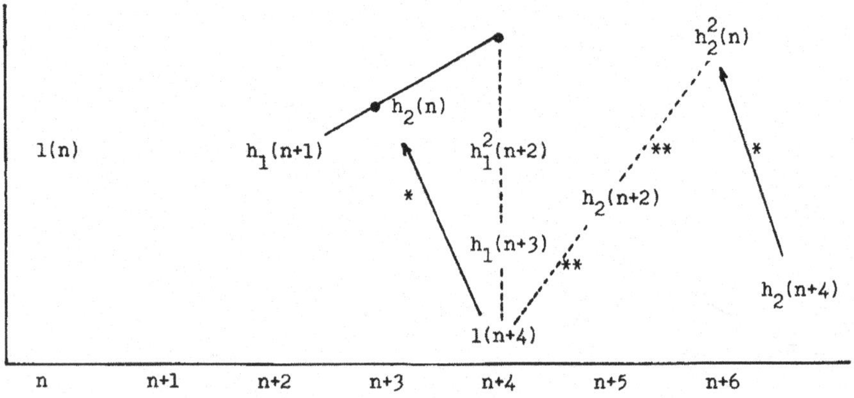

$n \equiv 3\ (4)$ *) differential if $n \equiv 7\ (8)$

**) if $n \equiv 3\ (8)$

Figure 3.4

<u>Proof of 3.3</u>: The differentials listed correspond to attaching maps which can be detected by Sq^1, Sq^2 and Sq^4, and they hold in the spectral sequences for $\mathcal{B}'$, $\mathcal{B}''$ and $\mathcal{B}'''$ below

$\mathcal{B}'$ $$S^{i-1} \cup_2 e^i \leftarrow S^{i-1} \leftarrow *$$

$\mathcal{B}''$ $$S^{i-2} \cup_\eta e^i \leftarrow S^{i-2} = S^{i-2} \leftarrow *$$

$\mathcal{B}'''$ $$S^{i-4} \cup_\nu e^i \leftarrow S^{i-4} = S^{i-4} = S^{i-4} = S^{i-4} \leftarrow *$$

The differential $d_2x(i) = h_0x(i-1)$ if $i \equiv 0$ (2) is immediate, since $1(i) \notin E_2$ and by dimensional considerations $d_21(i) = h_0(i-1)$ is the only possible d_2 on $1(i)$. The module structure over $E_2(S,S)$ now gives $d_2x(i) = h_0x(i-1)$.

The d_3 differential is slightly more complicated. There are two cases. If $i \equiv 1$ (4) then the i cell is not attached to the i-1 cell, but is attached to the i-2 cell by η; $d_31(i) = h_1(i-2)$ follows as for d_2, and this implies $d_3x(i) = h_1x(i-2)$. If $i \equiv 0$ (4) then $1(i) \notin E_3$ since $d_21(i) = h_0(i-1)$. However, the map of spectral sequences induced by $\mathcal{C} \to \mathcal{P}''$

$$\begin{array}{ccccc}
\mathcal{C} & S^{i-2}\vee S^{i-1}\cup_{\eta+2}e^i & \longleftarrow & S^{i-2}\vee S^{i-1} & \longleftarrow & S^{i-2} \\
\downarrow & \downarrow & & \downarrow & & \| \\
\mathcal{P}'' & S^{i-2}\cup_\eta e^i & \longleftarrow & S^{i-2} & = & S^{i-2}
\end{array}$$

shows that elements of $E_3(S,\mathcal{C})$ must satisfy $d_3x(i) = h_1x(i-2) + k$ where k is the kernel of $E_3(S,\mathcal{C}) \to E_3(S,\mathcal{P}'')$, that is, k must have the form $y(i-1)$. By inspection k must be 0 in the dimensions considered. Now, by truncating $\mathcal{P}$ at the i cell, then collapsing the i-3 skeleton we can compare $E_3(S,\mathcal{P})$ to $E_3(S,\mathcal{C})$. Again we have $d_3x(i) = h_1x(i-2) + k$, where k is now a sum of elements coming from the i-3 cell or below. The first possibility is when $n \equiv 0$ (4). We must decide between $d_3h_1(n+4) = h_1{}^2(n+2)$ and $d_3h_1(n+4) = h_1{}^2(n+2) + h_2(n+1)$. Let P^n, P^{n+1}, h_1P^{n+2}, and P^{n+3} denote elements detected by $1(n)$, $1(n+1)$, $h_1(n+2)$, and $1(n+3)$, respectively. Comparing with Mahowald's calculations [59], we find that $2 \circ P^{n+3} = h_1P^{n+2}$ or $h_1P^{n+2} + \nu \circ P^n$, depending on n mod 8. Composing with η yields $\eta \circ h_1P^{n+2} = 0$. But if $d_3h_1(n+4)$ were $h_1^2(n+2) + h_2(n+1)$ we would have $\eta \circ h_1P^{n+2} = \nu \circ P^{n+1}$. Therefore we must have $d_3h_1(n+4) = h_1{}^2(n+2)$. The same argument, with minor variations, finishes all the d_3 differentials.

Finally, the d_5 differentials follow by similar comparisons with $E_5(S,\mathcal{P}''')$. In all but one case, there is nothing in filtrations less than or equal to the filtration of $h_2x(i-4)$ so the comparison with $E_5(S,\mathcal{P}''')$ is sufficient. The one remaining case is when $n \equiv 1$ (4). Here $h_1^3(n+3)$ lies between $h_2(n+4)$ and $h_2^2(n)$. Since the n+4 cell is not attached to the n+3 cell, the $d_5h_2(n+4) = h_2^2(n)$ is right here also.

There are no further possible differentials by inspection. The hidden extensions here are all evident from Mahowald's computation in [59] of the Adams spectral sequence of P_n^∞. //

Note. The spectral sequence $E_r(S,\mathcal{P})$ has far more hidden extensions than $E_r(S,P_n^\infty)$ since the cells are spread apart in $E_r(S,\mathcal{P})$ whereas they all occur in the same filtration in $E_r(S,P_n^\infty)$. By IV.7.6, the same hidden extensions occur among the elements generated by the $\beta^\varepsilon P^j x$ for a fixed x.

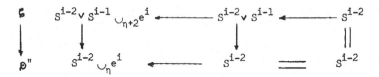

<u>Proof of 1.9 when $p = 2$</u>: A permanent cycle $x(i)$ corresponds to an operation xP^i. Thus, Table 1.3 is simply a list of the elements of $E_\infty(S,\mathcal{B})$, omitting most of those which are multiples by elements of π_*S of other elements of $E_\infty(S,\mathcal{B})$. The indeterminacy of an operation consists of those elements in the same stem and higher filtration, so it too can be read off Figures 3.1 through 3.4. With the exception of $\tau_{2*}(P^n)$ and $\tau_{2*}(P^{n+1})$, the values of τ_{2*} listed are the only elements of π_*S in the relevant dimension. Since $\pi_{2n}D_2S^n = Z_2$ when n is odd, $\tau_{2*}(P^n) = 0$ in this case. When n is even, $\iota:S^{2n} \to D_2S^n$ induces an isomorphism of π_{2n}. By II.1.10, the composite $\iota\tau_2:D_2S^n \to D_2S^n$ is multiplication by 2 on $H_{2n} \cong \pi_{2n}$. Thus $\tau_{2*}(P^n) = 2$. To calculate $\tau_{2*}(P^{n+1})$, first suppose $n \equiv 2\ (4)$. By Theorem 3.3, $\pi_{2n+2}D_2S^n = 0$. Therefore, $\eta P^{n+1} = 0$ and hence $\eta\tau_{2*}(P^{n+1}) = 0$. This forces $\tau_{2*}(P^{n+1})$ to be 0, not η. When $n \equiv 0\ (4)$, Theorem 3.3 gives $\pi_{2n+1}D_2S^n = Z_2 \oplus Z_2$ with generators P^{n+1} and ηP^n. By II.2.8, $\tau_{2*}(P^{n+1})$ is not zero and hence must be η.

Determining the relations in Table 1.4 amounts to determining the π_*S module structure of $\pi_*D_2S^n$. The indeterminacy of the operations in Table 1.3 induces a similar indeterminacy in the relations of Table 1.4. The relations are to be interpreted as asserting equality modulo the sum of the indeterminacies of the two sides. Thus, in order to prove that they hold, we need only show that they hold for some choice of representatives. The E_∞ terms in Theorem 3.3 force the following thirteen relations:

$$2P^n = 0 \qquad n \equiv 1,3\ (4)$$

$$\eta h_1P^{n+1} = 0 \qquad n \equiv 1\ (4)$$

$$2\nu P^{n+2} = 0 \qquad n \equiv 1\ (8)$$

$$\nu P^{n+2} = 0 \qquad n \equiv 5\ (8)$$

$$4\nu P^n = 0 \qquad n \equiv 2\ (4)$$

$$\eta P^{n+1} = 0 \qquad n \equiv 2\ (4)$$

$$\nu P^{n+1} = 0 \qquad n \equiv 6\ (8)$$

$$\left.\begin{aligned} \eta P^n &= 0 \\ 2h_1P^{n+1} &= 0 \\ 2h_2P^{n+2} &= 0 \end{aligned}\right\} \qquad n \equiv 3\ (4)$$

$$\left.\begin{aligned} \nu P^n &= 0 \\ \eta h_1P^{n+3} &= 0 \\ \eta h_2P^{n+2} &= 0 \end{aligned}\right\} \qquad n \equiv 7\ (8)$$

Another eighteen relations follow by considering the attaching maps given in Proposition 3.1, the spectral sequences in Theorem 3.3 and the reducibility and coreducibility given in Theorem 2.14. These are

$$\left.\begin{array}{l} 2P^{n+1} = 0 \\ 2h_1P^{n+2} = \eta^2P^{n+1} \end{array}\right\} \quad n \equiv 0\ (4)$$

$$\left.\begin{array}{l} \eta P^{n+3} = 0 \\ 2P^{n+3} = h_1P^{n+2} \\ 2\nu P^{n+3} = \nu h_1P^{n+2} = 0 \end{array}\right\} \quad n \equiv 0\ (8)$$

$$\left.\begin{array}{l} 2P^{n+3} = h_1P^{n+2} + \nu P^n \\ \eta P^{n+3} = \nu P^{n+1} \\ \nu h_1P^{n+2} = \nu^2P^n \end{array}\right\} \quad n \equiv 4\ (8)$$

$$2P^{n+2} = h_1P^{n+1} \quad n \equiv 1\ (4)$$

$$\left.\begin{array}{l} \eta P^{n+2} = 0 \\ 2P^{n+6} = h_1P^{n+5} \end{array}\right\} \quad n \equiv 1\ (8)$$

$$\eta P^{n+2} = \nu P^n \quad n \equiv 5\ (8)$$

$$2P^{n+1} = \eta P^n \quad n \equiv 2\ (4)$$

$$\left.\begin{array}{l} 2P^{n+5} = h_1P^{n+4} \\ \eta P^{n+5} = h_2P^{n+3} \end{array}\right\} \quad n \equiv 2\ (8)$$

$$\eta h_1P^{n+4} \equiv 0 \bmod \nu^2P^n \quad n \equiv 6\ (8)$$

$$\left.\begin{array}{l} 2P^{n+4} = h_1P^{n+3} \\ \eta P^{n+4} = h_2P^{n+2} \end{array}\right\} \quad n \equiv 3\ (8)$$

For example, when $n \equiv 0\ (8)$, the attaching map of the n+4 cell gives $2P^{n+3} = h_1P^{n+2}$. Then $2\nu P^{n+3} = \nu h_1P^{n+2}$ must be either 0 or ν^2P^n by the E_∞ term in Figure 3.1. But P_n^{n+7} is coreducible, so ν^2P^n is impossible. Similarly, when $n \equiv 4\ (8)$, the attaching map of the n+4 cell gives $2P^{n+3} = h_1P^{n+2} + \nu P^n$. (Note that, since P_n^{n+3} is coreducible, νP^n need not be considered a part of the indeterminacy of $2P^{n+3}$ or h_1P^{n+2}.) Thus $2\nu P^{n+3} = \nu h_1P^{n+2} + \nu^2P^n$. But νP^{n+3} is either 0 or ν^2P^n by the E_∞ term in Figure 3.1. Thus $2\nu P^{n+3} = 0$ and hence $\nu h_1P^{n+2} = -\nu^2P^n = \nu^2P^n$.

Four more relations come from the fact that $\pi_{n+2}(S^n \cup_2 e^{n+1}) \approx Z_4$, so that the composite of 2 and a map which projects to η on S^{n+1}, lifts to η^2 on S^n. These are

$$\left.\begin{array}{l} 2h_1P^{n+1} = \eta^2P^n \\ 2h_1P^{n+5} = h_1^2P^{n+4} \end{array}\right\} \quad n \equiv 1\ (4)$$

$$2h_1P^{n+4} = h_1^2P^{n+3} \quad n \equiv 2\ (4)$$

$$2h_1P^{n+3} = h_1^2P^{n+2} \quad n \equiv 3\ (4)$$

The relations

$$2h_1^2P^{n+4} = h_1^2P^{n+3} \qquad n \equiv 1\ (4)$$
$$2h_1^3P^{n+3} = 0$$
$$2h_1^2P^{n+3} = h_1^3P^{n+2} \qquad n \equiv 2\ (4)$$
$$2h_1^2P^{n+2} = h_1^3P^{n+1} \qquad n \equiv 3\ (4)$$
$$\eta h_2P^{n+2} = \nu^2P^n \qquad n \equiv 3\ (8)$$

are the only possibilities consistent with Mahowald's calculations [59] (note that these are not hidden extensions in his spectral sequence).

Finally, the relation $2h_2P^{n+3} = \nu^2P^n$ when $n \equiv 6\ (8)$ follows by comparison with the spectral sequence for the cofiber of the inclusion $P_{n+1}^{n+2} \to P_n^{n+4}$. In the cofiber, $2P^{n+3} = \nu P^n$ is obvious from the attaching maps. //

Now consider the odd primary case. Recall, from [55], that, in degrees less than pq-2, the HZ_p Adams spectral sequence has elements

	$a_0^i \in E_2^{i,i}$	detecting p^i, $i = 0,1,2,\ldots,$
	$h_0 \in E_2^{1,q}$	detecting $\alpha_1 \in \pi_{q-1}$,
and	$g_{i-1} \in E_2^{i,iq+i-1}$	detecting $\alpha_i \in \pi_{iq-1}$, for $2 \le i \le p$.

Let $\mathcal{B}$ be the sequence

$$L_{n(p-1)}^{n(p-1)+ps} \longleftarrow L_{n(p-1)}^{n(p-1)+ps-1} \longleftarrow \cdots \longleftarrow L_{n(p-1)}^{n(p-1)+1} \longleftarrow L_{n(p-1)}^{n(p-1)}$$

Since $L_{n(p-1)}^{\infty}$ has cells only in dimensions n(p-1) and greater which are congruent to 0 or -1 mod q, $E_2(S,\mathcal{B})$ is free over $E_2(S,S)$ on generators in those degrees. Write $x(j,\varepsilon)$ for the element of $E_2(S,\mathcal{B})$ which is x $E_2(S,S)$ in the $jq-\varepsilon$ summand, if $jq-\varepsilon \ge n(p-1)$. We agree to let $x(j,\varepsilon) = 0$ if $jq-\varepsilon < n(p-1)$.

<u>Theorem 3.4</u>. In $E_r(S,\mathcal{B})$, $d_2(x(j,0)) = a_0x(j,1)$ and

$$d_{2p-1}(x(j,1)) = -jh_0x(j-1,1).$$

In low dimensions $E_{2p}(S,\mathcal{B})$ is given by Figures 3.5 through 3.9.

<u>Notes</u>: (1) The dotted arrows to the left represent possible d_{2p} differentials which we have not computed. This is why the theorem only claims to give $E_{2p}(S,\mathcal{B})$. The indicated d_{2p} is the only possible remaining differential in the range listed. This is true for dimensional reasons except when $n = 2k-1$ and $k \equiv -2\ (p)$. Here the

possibility that $d_{4p-2}(1(k+2,1))$ is nonzero is excluded by the fact that $L_{kq-1}^{(k+2)q-1}$ is reducible when $k \equiv -2$ (p) by Theorem 2.9.(v).

(2) Dashed vertical lines represent hidden extensions. Precisely, if x and y are detected by $\overline{x}$ and $\overline{y}$, the notation

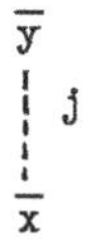

means that $px \equiv jy$ modulo higher filtrations. Of course, if j is 0 this means the extension is trivial. We replace j by a question mark if we have not settled the extension.

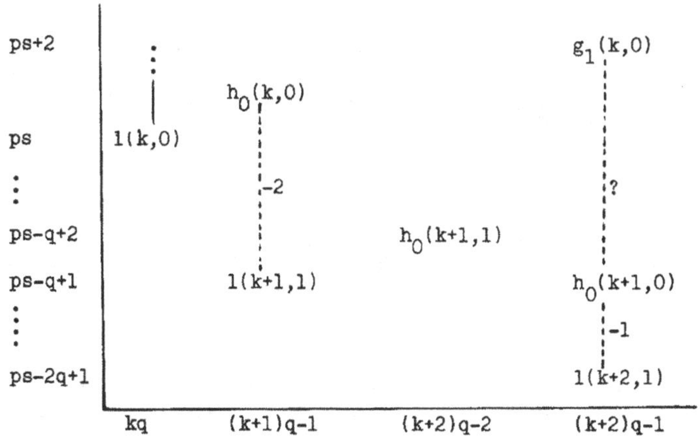

Figure 3.5 $n = 2k$, $k \equiv -2$ (p)

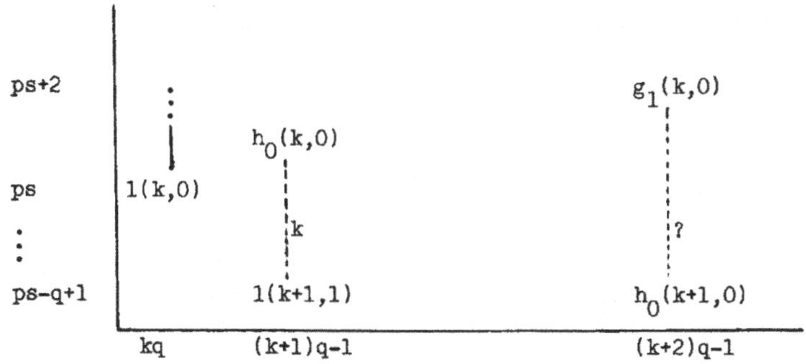

Figure 3.6 $n = 2k$, $k \not\equiv -2$ (p)

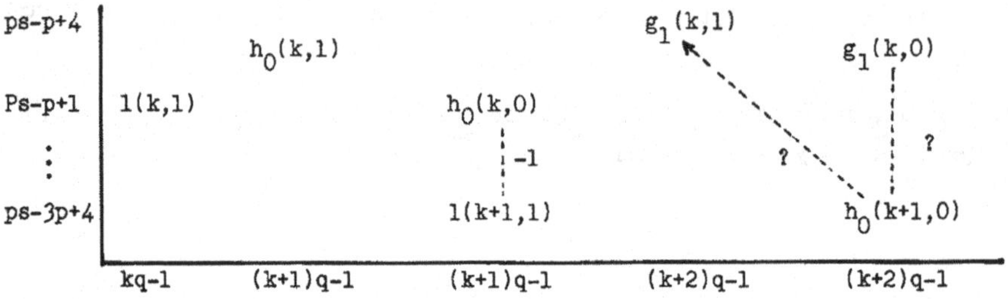

Figure 3.7 $n = 2k-1,\ k \equiv -1\ (p)$

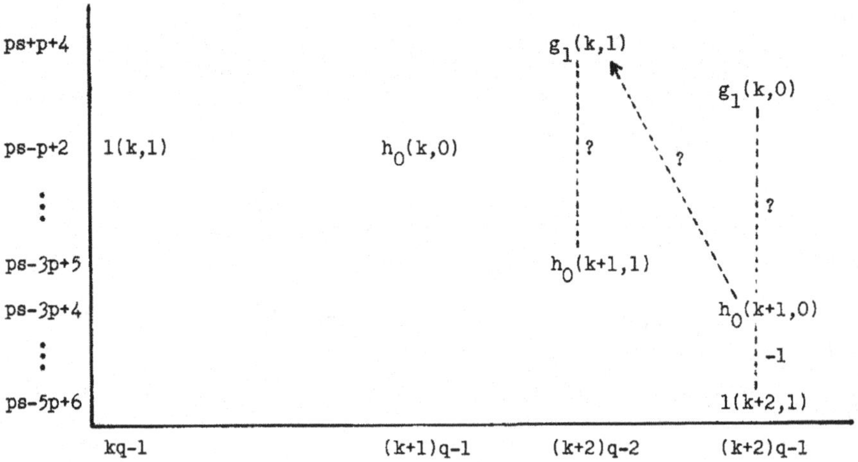

Figure 3.8 $n = 2k-1,\ k \equiv -2\ (p)$

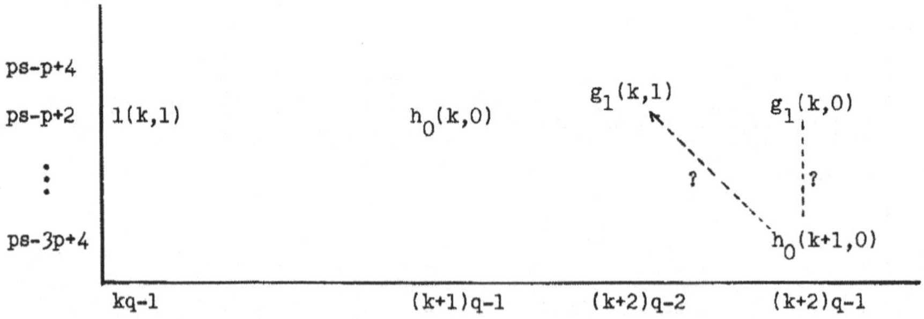

Figure 3.9 $n = 2k-1,\ k \not\equiv -1 \text{ or } -2\ (p)$

Proof of 3.4. The differentials follow from the attaching maps in Proposition 3.2 just as 3.3 follows from 3.1. Applying them gives the values of $E_{2p}(S,\mathcal{P})$ listed in Figures 3.5 through 3.9. The indicated hidden extensions all come from the attaching maps of the even cells of $L^{\infty}_{n(p-1)}$. //

Proof of 1.9 when $p > 2$: A permanent cycle $x(j,\varepsilon)$ corresponds to a homotopy operation $x\beta^{\varepsilon}P^{j}$. Thus Table 1.1 is a list of those elements in Figures 3.5 through 3.9 which must be permanent cycles by Theorem 3.4. The indeterminacy is obtained from Figures 3.4 through 3.9 as for $p = 2$. The values of τ_{p*} listed are the only elements of $\pi_* S$ in the relevant dimensions, except for $\tau_{p*}(P^k) = p!$, which follows from II.1.10.

The relations in Table 1.2 are all determined by the attaching maps from Proposition 3.2. //

Proof of 1.10. By IV. 7.3.(v), to determine $P^{n+m+1}(xy)$ we must calculate the image of $P^{n+m+1}\in\pi_{2(n+m)+1}D_2S^{n+m}$ under $\delta_*:\pi_*D_2S^{n+m}\to\pi_*(D_2S^n\wedge D_2S^m)$. We need only consider

$$P^{n+m+2}_{n+m}\to P^{n+2}_{n}\wedge P^{m+2}_{m}$$

for dimensional reasons. If $\mathcal{B}_{n,m}$ is the skeletal filtration of $P^{n+2}_{n}\wedge P^{m+2}_{m}$, then $E_2(S,\mathcal{B}_{n,m})$ is generated over $E_2(S,S)$ by elements $1(j,k)$ with $n \le j \le n+2$ and $m \le k \le m+2$ corresponding to the cells of P^{n+2}_{n} and P^{m+2}_{m} in an obvious fashion. The attaching maps of P^{n+2}_{n} and P^{m+2}_{m} determine the differentials in low dimensions from which we get $E_\infty(S,\mathcal{B}_{n,m})$. The extension questions in $\pi_{2(n+m)+1}$ are also determined by P^{n+2}_{n} and P^{m+2}_{m} when $n \equiv m \equiv 0$ (2). When $n \equiv m \equiv 1$ (2) we need the fact that the top cell of the smash product of two mod 2 Moore spaces is attached to the bottom cell by η, to settle the extension question. We conclude that if $n \equiv m \equiv 0$ (2) then $\pi_{2(n+m)+1}$ is generated by $P^{n+1}P^m$, P^nP^{m+1}, and ηP^nP^m with relations

$$2P^{n+1}P^m = \begin{cases} 0 & n \equiv 0 \ (4) \\ \eta P^nP^m & n \equiv 2 \ (4) \end{cases}$$

and

$$2P^nP^{m+1} = \begin{cases} 0 & m \equiv 0 \ (4) \\ \eta P^nP^m & m \equiv 2 \ (4) . \end{cases}$$

If $n \equiv m \equiv 1$ (2) then $\pi_{2(n+m)+1}$ is generated by an element we call $S_{n,m}$ which is detected by $1(n+1,m) + 1(n,m+1)$ with the relation

$$2S_{n,m} = \begin{cases} 0 & n \equiv 3 \text{ or } m \equiv 3 \ (4) \\ \eta P^nP^m & n \equiv m \equiv 1 \ (4) . \end{cases}$$

From the image of $S_{n,m}$ in $E_\infty(S, \ {}_{n,m})$ we can see that

$$E^0(S_{n,m}) = P^{n+1}P^m + P^nP^{m+1}.$$

Finally $\delta_*(P^{n+m+1})$ is determined modulo the kernel of the Hurewicz homomorphism by commutativity of the following diagram, in which the isomorphisms are Thom isomorphisms

$$\begin{array}{ccc}
\pi_*D_2S^{n+m} & \xrightarrow{\delta_*} & \pi_*D_2S^n \wedge D_2S^m \\
\downarrow h & & \downarrow h \\
H_*D_2S^{n+m} & \xrightarrow{\delta_*} & H_*D_2S^n \wedge D_2S^m \\
\wr\| & & \wr\| \\
H_*BZ_2 & \xrightarrow{\Delta_*} & H_*(BZ_2 \times BZ_2)
\end{array}$$

Since ηP^nP^m generates the kernel of the Hurewicz homomorphism we are done. //

Proof of 1.11. The commutative diagram above shows that the Hurewicz homomorphism must map the Cartan formula for a homotopy operation into the Cartan formula for its Hurewicz image. Case (i), $n = 2j$ and $m = 2k$, follows by an argument formally identical to, but easier than, the proof of 1.10 when $n \equiv m \equiv 0$ (2). Case (ii) is immediate from the homology Cartan formula because in this case we're in the Hurewicz dimension. Case (iii) follows just as in the proof of 1.10 when $n \equiv m \equiv 3$ (4). //

Proof of 1.12. In $E_2(S,S)$, $Sq^1(h_0) = h_1$ by [3]. Therefore, $P^1(2) = \eta$. //

Proof of 1.13. By definition $\beta P^1(p)$ is a unit times the composite

$$S^{2p-3} \xrightarrow{\beta P^1} D_pS \xrightarrow{D_p(p)} D_pS \xrightarrow{\xi} S,$$

where βP^1 is the inclusion of the 2p-3 cell. By II.1.8, $D_p(p) \equiv \iota_p\tau_p$ mod p, and by II.2.8, $\tau_p \circ \beta P^1 \neq 0$. Since $\xi\iota_p = 1$, the composite and hence $\beta P^1(p)$ are nonzero. The fact that $\beta P^{p-1}(\alpha_1) = \beta_1$ follows from the fact that in the Adams spectral sequence, $\beta P^{p-1}(h_0) = b_1^1$ using the notation of [66]. The latter can be computed directly from the definition of βP^{p-1} using the definitions

$$h_0 = [\xi_1], \quad b_1^1 = \sum_{i+j = p-1} \frac{1}{p}(i,j)[\xi_1^{pi}|\xi_1^{pj}]$$

in the bar construction. Alternatively, we may refer to Liulevicius' computation [55, pp. 26, 30] using [66, II-6.6] to translate it into our notation. //

Proof of 1.14. This is now immediate:

$$P^{n+1}(2x) = P^1(2)x^2 + 4P^{n+1}(x) + 4c_{0n}\eta x^2$$
$$= \eta x^2$$

since $2P^{n+1}(x)$ is either 0 or ηx^2 by 1.10. Similarly,

$$\beta P^{j+1}(px) = \beta P^1(p)x^p + p^p\beta P^{j+1}(x) + d_{0n}\alpha_1 p^p x^p$$
$$= \beta P^1(p)x^p + jp^{p-1}\alpha_1 P^j(x)$$
$$\doteq \alpha_1 x^p$$

since $p\beta P^{j+1}(x) = j\alpha_1 P^j(x)$. Finally $\beta P^{j+p-1}(\alpha_1 x) = x^p\beta P^{p-1}(\alpha_1) = x^p\beta_1$. The indeterminacy is always zero because where it is not automatically zero it is $4\eta x^2$ or $p^p\alpha_1 x^p$. //

Proof of 1.15. If $p = 2$ then $\eta x^2 = 0$ by Theorem 1.10 when $n \equiv 3$ (4) (even if $2x \neq 0$) while $0 = P^{n+1}(2x) = \eta x^2$ by Proposition 1.14 when $n \equiv 0$ (2). If $p > 2$ then $x^p = 0$ if n is odd, while if $n = 2j$, Proposition 1.14 implies that $0 = \beta P^{j+1}(px) \doteq \alpha_1 x^p$ and $0 = \beta P^{j+p-1}(\alpha_1 x) = \beta_1 x^p$. When $x = \beta_1$ the second of these formulas is $\alpha_1\beta_1^p = 0$. //

Proof of 1.16. Several of the computations follow from $P^n(x) = x^2$ if $x \in \pi_n$, others from $\pi_4 = \pi_5 = \pi_{12} = \pi_{13} = 0$. Similarly, several indeterminacies are zero from Theorem 1.10 or because they lie in filtrations which are 0. We will prove the remainder of the results.

Since $P^4(h_2) = h_3$, $h_1P^4(\nu)$ is detected by h_1h_3 so is either $\eta\sigma$ or $\bar{\nu}$. By 1.10, $h_1^2P^5(\nu) = 2h_1P^6(\nu) = 0$ since $2\pi_{10} = 0$. Similarly, $h_1P^4(2\nu) = 0$ by calculating Steenrod operations in Ext. Since $\tau_{2*}(h_1P^6) = 0$, we get $h_1P^6(2\nu) = 2h_1P^6(\nu) = 0$, and since $\tau_{2*}(h_2P^5) = 0$, we get $h_2P^5(2\nu) = 2h_2P^5(\nu) = 0$. By 1.10, $h_1^2P^5(2\nu) = 2h_1P^6(2\nu) = 0$ also. The operations on 4ν can all be calculated from the additivity rule $\alpha^*(4\nu) = 2\alpha^*(2\nu) + \tau_{2*}(\alpha)(2\nu)^2 = 2\alpha^*(2\nu)$.

Since $2\pi_{17} = 0$, the relations $h_1^2P^9(\nu^2) = 2h_1P^{10}(\nu^2)$ and $h_1^3P^8(\nu^2) = 2h_1^2P^9(\nu^2)$ force these elements to be 0 mod 0.

Since $P^8(h_3) = h_4$, $h_1P^8(\sigma)$ is detected by h_1h_4 so must be η^* or $\eta^* + \eta\rho$. Since $2h_1^2P^9 = \eta^2h_1P^8$ and $\eta^2h_1P^8(\sigma)$ is detected by $h_1^3h_4 = h_0^2h_2h_4$, it follows that $h_1^2P^9(\sigma)$ is detected by $h_0h_2h_4$. Since $2h_1P^{10} = h_1^2P^9$ it follows that $h_1P^{10}(\sigma)$ is detected by h_2h_4. Thus $h_1P^{10}(\sigma) = \nu^* \text{ or } \nu^* + \eta\bar{\mu}$ modulo $\langle 2\nu^* \rangle$, which is its indeterminacy, and similarly for $h_1^2P^9(\sigma)$.

Since $P^7(2\sigma) = 4\sigma^2 = 0$, we have

$$h_1P^8(2\sigma) = 2h_1P^8(\sigma) + \left\{\begin{matrix} 0 \\ \text{or} \\ \eta^2 \end{matrix}\right\} \sigma^2 = 0 + 0 = 0.$$

The remaining operations are additive except for

$$h_1P^8(4\sigma) = 2h_1P^8(2\sigma) + \left\{\begin{matrix} 0 \\ \text{or} \\ \eta^2 \end{matrix}\right\} 4\sigma^2 = 0 + 0 = 0 \; . \qquad //$$

CHAPTER VI

THE ADAMS SPECTRAL SEQUENCE of H_∞ RING SPECTRA

by Robert R. Bruner

In this chapter we show how to use an H_∞ ring structure on a spectrum Y to produce formulas for differentials in the Adams spectral sequence of π_*Y. We shall confine attention to the Adams spectral sequence based on mod p homology, although it is clear that similar results will hold in generalized Adams spectral sequences as well.

The differentials have two parts. The first is the reflection in the Adams spectral sequence of relations in homotopy like those in Chapter V. For example, when $x \in \pi_n Y$ and $n \equiv 1$ (4), there is no homotopy operation $P^{n+1}x$ since the n+1 cell of P_n^∞ is attached to the n cell by a degree 2 map. In the Adams spectral sequence there is a Steenrod operation $Sq^{n+1}\bar{x}$ and a differential $d_2 Sq^{n+1}\bar{x} = h_0 Sq^n \bar{x} = h_0 \bar{x}^2$. Therefore $h_0 \bar{x}^2 = 0$ in E_∞. This in itself only implies that $2x^2$ has filtration greater than that of $h_0\bar{x}^2$ in the Adams spectral sequence, but by examining its origin as a homotopy operation we see that $2x^2 = 0$. Thus, the formulas we produce for differentials are most effective when combined with the results about homotopy operations in Chapter V. The differential $d_2 Sq^{n+3}\bar{x} = h_0 Sq^{n+2}\bar{x}$, still assuming $n \equiv 1$ (4), is a perfect illustration of this. The corresponding relation in homotopy is $2P^{n+2}x = h_1 P^{n+1}x$ where $h_1 P^{n+1}$ is an indecomposable homotopy operation detected by $h_1 Sq^{n+1}$ in the Adams spectral sequence. The differential on $Sq^{n+3}\bar{x}$ represented geometrically is the sum of maps representing $h_0 Sq^{n+2}\bar{x}$ and $h_1 Sq^{n+1}\bar{x}$, but since $h_1 Sq^{n+1}\bar{x}$ has filtration one greater

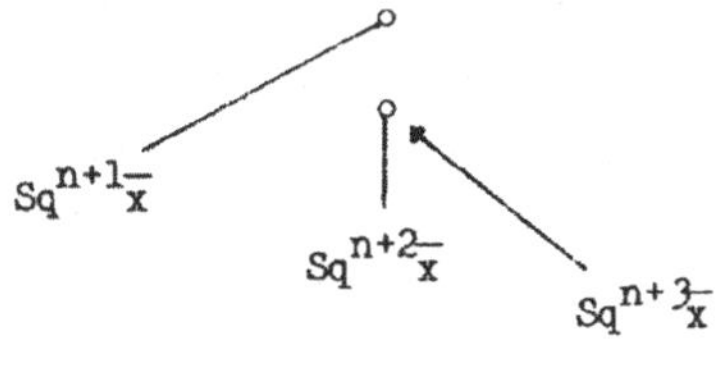

than does $h_0 Sq^{n+2}\bar{x}$, it does not appear in the differential. This reflects a hidden extension in the Adams spectral sequence: $2P^{n+2}x$ appears to be 0 in the Adams spectral sequence (i.e. $h_0 Sq^{n+2}\bar{x} = 0$ in E_∞) only because of the filtration shift. In fact, $2P^{n+2}x = h_1 P^{n+1}x$. The moral of this is just the obvious fact mentioned above: the differentials should not be considered in isolation but should be combined with the homotopy operations of Chapter V. Further examples will be given in section 1.

The second part of the differentials arises when we consider Steenrod operations on elements that are not permanent cycles. If x in filtration s survives until E_r we can make x into a permanent cycle by truncating the spectral sequence at filtration s+r. Thus the differentials of the type just discussed apply to x until we get to E_r. However, by analyzing the contribution of $d_r x$ we can show that it will not affect the differentials on $\beta^{\varepsilon}P^j x$ until E_{pr-p+1} where it contributes $\beta^{\varepsilon}P^j d_r x$. Thus the differentials of the first type apply far beyond the range in which we are justified in pretending that x is a permanent cycle. (To be precise we should note that $d_r x$ can occasionally affect differentials on $\beta^{\varepsilon}P^j x$ through a term containing $x^{p-1}d_r x$ in E_{r+1}.)

The first results of this type were established by D. S. Kahn [45] who showed that the H_∞ ring map $\xi_2: W \ltimes_{Z_2} S^{(2)} \to S$ (obtained through coreductions of stunted projective spaces) could be filtered to obtain maps representing the results of Steenrod operations in $\mathrm{Ext}_A(Z_2,Z_2)$ and that some differentials were implied by this. Milgram [81] extended Kahn's work to the odd primary case and introduced the spectral sequence of IV.6 which is by far the most effective tool for computing the first part of the differential. His work was confined to the range in which it is possible to act as if one is operating on a permanent cycle. Nonetheless he was able to use the resulting formulas for differentials to substantially shorten Mahowald and Tangora's calculation [61] of the first 45 stems at the prime 2 and to catch a mistake in their calculation. The next step was taken by Makinen [62], who showed how to incorporate the contribution of $d_r x$ in the differentials on $Sq^j x$ for $p = 2$. Unfortunately, he apparently did not apply his formulas to the known calculations of the stable stems, for one of his most interesting formulas (published in 1973),

$$d_3 Sq^j x = h_1 Sq^{j-2} x + Sq^j d_2 x \quad \text{if } n \equiv 1 \ (4),$$

combined with Milgram's calculation of Steenrod operations [81], implies that $d_3 e_1 = h_1 t$, contradicting Theorem 8.6.6 of Mahowald and Tangora [61]. This application was left for the author to discover in 1983. Note that the differential is out of Milgram's range since a nonzero $d_2 x$ prevents us from calculating $d_3 Sq^j x$ unless we incorporate terms involving $d_2 x$. The argument in [61] that e_1 is a permanent cycle is an intricate one, involving the existence of various Toda brackets, while the proof that $d_3 Sq^j x = h_1 Sq^{j-2} x + Sq^j d_2 x$ if $n \equiv 1$ (4) is relatively straightforward. This appears to be convincing evidence that the H_∞ structure in the form of Steenrod operations in Ext is a powerful computational tool.

One other piece of related work is the thesis of Clifford Cooley [30]. He obtains formulas similar to Milgram's [61] by using the spectral sequence connecting homomorphism for a cofiber sequence of stunted projective spaces to reduce them to d_1's which he gets from a lambda algebra resolution of the cohomology of the appropriate stunted projective space. Calculating differentials this way or by the spectral sequence of IV.6 is probably a matter of indifference. The most interesting aspect of Cooley's thesis is that he works unstably, examining the interaction of the Steenrod operations and the EHP sequence. As in all other earlier work on this subject he views the H_∞ ring structure in terms of coreductions of stunted projective spaces. The interaction of the Steenrod operations and the EHP sequence had been discovered by William Singer [97] using the algebraic EHP sequence obtained from the lambda algebra.

In the work at hand, we extend the ideas of Makinen to the odd primary case to obtain comprehensive formulas for the first nontrivial differential on $\beta^{\varepsilon}P^{j}x$, which we state in §1. These apply to the mod p Adams spectral sequence of any H_∞ ring spectrum. The remainder of §1 consists of calculations using these formulas in the Adams spectral sequence of a sphere, including the differential discussed above. These are intended to illustrate especially the interaction between the homotopy operations and the differentials, specifically to obtain better formulas in particular cases than hold in general. One of these is $d_3r = h_1d_0^2$, which forces h_4^2 to be a permanent cycle. This is the shortest proof we know of this fact.

In §§2 and 3 we describe the natural Σ_p equivariant cell decomposition of $(\Sigma X)^{(p)}$ and use it to relate extended powers of X and of ΣX.

In §4 we start the proof of the formulas in §1, using the results of §§2 and 3. We also prove that the geometry splits naturally into three cases, which we deal with one at a time in the remaining §§5-7.

1. Differentials in the Adams spectral sequence

In this section we state our theorems concerning differentials, explain some of the subtleties involved in understanding what they are really saying, and calculate some examples in order to illustrate their use and demonstrate their power.

Localize everything at p. Let Y be an H_∞ ring spectrum. Let $E_r^{s,n+s}(S,Y) \Rightarrow \pi_nY$ be the Adams spectral sequence based on ordinary mod p homology. We shall adopt the following shorthand notation for differentials. If A is in filtration s and B_1 and B_2 are in filtrations $s+r_1$ and $s+r_2$ respectively, then

$$d_*A = B_1 \dotplus B_2$$

means that $d_iA = 0$ for $i < \min(r_1,r_2)$ and

$$d_{r_1}A = B_1 \qquad \text{if } r_1 < r_2$$

$$d_rA = B_1 + B_2 \qquad \text{if } r_1 = r = r_2\text{, and}$$

$$d_{r_2}A = B_2 \qquad \text{if } r_1 > r_2$$

Note. This does not mean that this differential is necessarily nonzero. Nor does it mean that if B_1 happens to be 0, then $d_{r_2}A = B_2$ regardless of whether $r_2 > r_1$ or not. More likely, B_1 is zero because it comes from a map which lifts to filtration $s+r_1+1$ or more and, hence, B_1 could conceivably lead to a nonzero $d_{r_1+1}A$. The point is that you can't tell what B_1 is contributing to the differential if all you know is that it is zero in filtration $s+r_1$. However, when we explicitly state that $T_p = 0$ in Theorem 1.2 we mean that it is to be treated as having filtration ∞.

The geometry behind the formula $d_*A = B_1 \dotplus B_2$ will make it clear exactly what the formula can and cannot tell you. The formula means that for some $r_0 > \max(r_1,r_2)$, A is represented by a map whose boundary splits into a sum $\overline{B}_1 + \overline{B}_2 + \overline{B}_0$, where each $\overline{B}_i$ lifts to filtration $s+r_i$, and where $\overline{B}_1$ and $\overline{B}_2$ represent B_1 and B_2 respectively. It is irrelevant what $\overline{B}_0$ represents because $\overline{B}_1 + \overline{B}_2$ lies in a lower filtration. This is fortunate, since in general $\overline{B}_0$ is very complicated. In particular cases however, we can often analyze $\overline{B}_0$ in order to get more complete information about d_*A. For examples of this, see Proposition 1.17(ii) (the formula $d_3r_0 = h_1d_0^2$) and Proposition 1.6.

Two remaining points about the formula are best made using examples. The formulas we will shortly prove say that, under appropriate circumstances,

$$d_*Sq^j x = Sq^j d_r x \dotplus \overline{a}xd_r x$$

and

$$d_*Sq^j d_r x = \overline{a}(d_r x)^2$$

where $\overline{a} \in E_\infty(S,S)$. The algebra structure also implies that

$$d_r(\overline{a}xd_r x) = \overline{a}(d_r x)^2.$$

If the filtration of $Sq^j x$ is s, then the filtration of $Sq^j d_r x$ is $s+2r-1$, while that of $\overline{a}xd_r x$ is $s+r+f+k$ (f is the filtration of $\overline{a}$ and k will be defined shortly). The three ways these differentials can combine are illustrated below

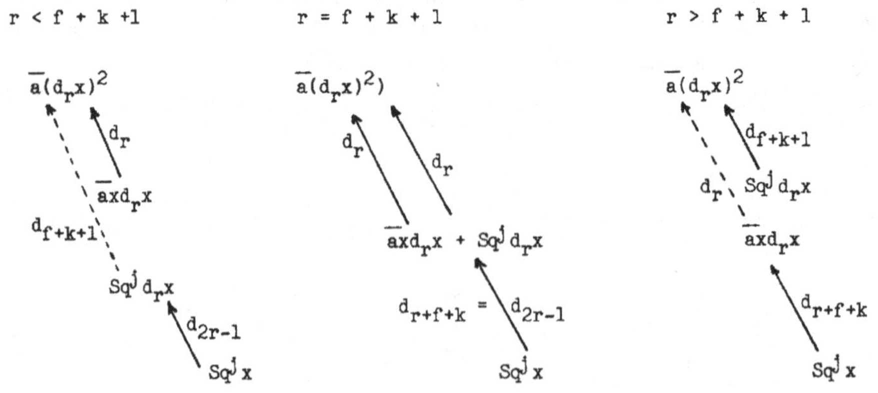

Taken individually, the terms $Sq^j d_r x$ and $\bar{a}xd_r x$ do not always appear to survive long enough for $Sq^j x$ to be able to hit them. For example, when $r > f+k+1$, the differential $d_{r+f+k}Sq^j x = \bar{a}xd_r x$ is preceded by the differential $d_r(\bar{a}xd_r x) = \bar{a}(d_r x)^2$, which would have prevented $\bar{a}xd_r x$ from surviving until E_{r+k+f}, had it not happened that a still earlier differential ($d_{f+k+1}Sq^j d_r x = \bar{a}(d_r x)^2$) had already hit $\bar{a}(d_r x)^2$. This is completely typical. The formula $d_* A = B_1 \dot{+} B_2$, as used here, carries with it the claim that the right-hand side will survive long enough for this differential to occur, and even shows the "coconspirator" which will make this possible when it seems superficially false.

The other point illustrated by this example occurs when $Sq^j d_r x$ and $xd_r x$ are permanent cycles and $r > f+k+1$. Then the differential $d_{r+k+f}Sq^j x = \bar{a}xd_r x$ reflects a hidden extension: $\bar{a}(xd_r x)$ is zero in E_∞ because of a filtration shift. It is actually detected by $Sq^j d_r x$. Relations among homotopy operations typically cause such phenomena. Note that the cell which carries $Sq^j x$ is also the cell which produces the relation in homotopy. In a suitably relative sense this is the meaning of all differentials in the Adams spectral sequence ("relative" because the terms in a relation corresponding to a differential will typically be relative homotopy classes which do not survive to E_∞ to become absolute homotopy classes).

We can now state our main theorems. Assume given $x \in E_r^{s,n+s}$ and consider the element $\beta^\varepsilon P^j x$ (as usual, $\varepsilon = 0$ and $P^j = Sq^j$ if $p = 2$). Let

$$k = \begin{cases} j-n & p = 2 \\ (2j-n)(p-1)-\varepsilon & p > 2, \end{cases}$$

so that $\beta^\varepsilon P^j x \in E_2^{ps-k,p(n+s)}$, which lies in the $k+np$ stem. Using the functions v_p and a_p of V.2.15, V.2.16 and V.2.17 we define $v = v_p(k+n(p-1))$ and $a = a_p(k+n(p-1)) \in \pi_{v-1}S$. Recall that a is the top component of an attaching map of a stunted lens space after the attaching map has been compressed into the lowest possible skeleton. Let

$$\bar{a} \in E_\infty^{f,f+v-1}(S,S)$$

detect a (this defines f as well). Recall that $a_0 \in E_\infty^{1,1}$ detects the map of degree p when $p > 2$.

Theorem 1.1. There exists an element $T_p \in E_2^{**}(S,Y)$ such that

(i) if $p = 2$ then $d_* Sq^j x = Sq^j d_r x \dot{+} T_2$,

(ii) if $p > 2$ then

$$d_{r+1}P^j x = d_{r+1}x^p = a_0 x^{p-1} d_r x \quad \text{if } 2j = n,$$

$$d_2 P^j x = a_0 \beta P^j x \quad \text{if } 2j > n, \text{ and}$$

$$d_* \beta P^j x = -\beta P^j d_r x \dot{+} T_p .$$

Theorem 1.2.

$$T_2 = \begin{cases} 0 & v > k+1 \text{ or } 2r-2 < v < k \\ \bar{a}xd_r x & v = k+1 \\ \bar{a}Sq^{j-v}x & v = k \text{ or } (v < k \text{ and } v \le 10) \end{cases}$$

If $p > 2$ then

$$T_p = \begin{cases} 0 & v > k+1 \text{ or } pr-p < v < k \\ (-1)^e\, \bar{a}x^{p-1}d_r x & v = k+1 \\ (-1)^{e-1}\, \bar{a}\beta P^{j-e-1}x & v < k \text{ and } v \le pq. \end{cases}$$

where e is the exponent of p in the prime factorization of j.

Note. When $p > 2$, k and v have opposite parity so that $v = k$ never occurs.

Theorems 1.1 and 1.2 give complete information on the first possible nonzero differential except when

$$pq < v < \min(k, pr-p+1) \quad \text{if } p > 2,$$

or

$$10 < v < \min(k, 2r-1) \quad \text{if } p = 2.$$

The sketch of the proof given in Section 4 should make it clear what the obstruction is in these cases. We do have some partial information which we collect in the following theorem.

Theorem 1.3. If $p > 2$ and $v > q$ then $d_i\beta P^j x = 0$ if $i < v+2 \le pr-p+1$, while $d_{pr-p+1}\beta P^j x = -\beta P^j d_r x$ if $v + 2 > pr-p+1$. If $p = 2$ and $v > 8$ then $d_i Sq^j x = 0$ if $i < v+2 \le 2r-1$, while $d_{2r-1}Sq^j x = Sq^j d_r x$ if $v+2 > 2r-1$.

To apply these results we must know the values of the Steenrod operations in $E_2 = \mathrm{Ext}_{\mathcal{A}}(Z_p, H_*Y)$. For our examples we will concentrate primarily on $p = 2$ and $Y = S^0$, since this is a case in which there are many nontrivial examples. We cannot resist also showing how useful the Steenrod operations are in the purely algebraic task of determining the products in Ext.

We begin with the elements $h_n \in E_2^{1,2^n-1}$ dual to the Sq^{2^n}. Parts (i) and (iii) of the following propositon may also be found in [88].

Proposition 1.4. (i) (Adams [3]) $Sq^{2^n}h_n = h_{n+1}$ and $Sq^{2^n-1}h_n = h_n^2$.

(ii) (Adams [2]) $h_n h_{n+1} = 0$, $h_{n+1}^3 = h_n^2 h_{n+1}$ and $h_n h_{n+2}^2 = 0$.

(iii) (Novikov [91]) $h_n^2 h_{n+3}^2 = 0$, $h_0^{2^n} h_{n+2}^2 = 0$ and, if $n > 0$, $h_0^{2^n} h_n = 0$.

Proof $Sq^{2^n-1}h_n = h_n^2$ because the first operation is always the square. If we let $S: \mathrm{Ext}^{s,*} \to \mathrm{Ext}^{s,*}$ be Sq^{n+s} on $\mathrm{Ext}^{s,n+s}$, then Proposition 11.10 of [68] shows that in the cobar construction $S[x_1|\cdots|x_j] = [x_1^2|\cdots|x_j^2]$. Since h_n is represented by $[\xi_1^{2^n}]$, it follows that $Sq^{2^n}h_n = S(h_n) = h_{n+1}$. For dimensional reasons, the Cartan formula reduces to $S(xy) = S(x)S(y)$. Thus, to show (ii) we need only show $h_0h_1 = 0$, $h_1^3 = h_0^2h_2$, and $h_0h_2^2 = 0$. These occur in such low dimensions that they may be checked "by hand". In fact, only the first and third must be done this way since $Sq^2(h_0h_1) = h_0^2h_2 + h_1^3$. The relation $h_n^2h_{n+3}^2 = 0$ follows similarly from $h_0^2h_3^2 = Sq^8(h_0h_2^2) = 0$. The only nonzero operation on h_{n+2}^2 is $Sq^{2^{n+3}}h_{n+2}^2 = h_{n+3}^2$ since (ii) implies that $h_{n+2}^4 = h_{n+2}(h_{n+1}^2h_{n+3}) = 0$. The relation $h_0^{2^n}h_{n+2}^2 = 0$ then follows by induction from $h_0^2h_3^2 = 0$. Finally, $h_0^{2^n}h_n = 0$ follows by induction from $h_0^2h_1 = 0$ since

$$Sq^{2^n}(h_0^{2^n}h_n) = h_0^{2^{n+1}}h_{n+1}\,.$$

As is well known, the preceding proposition implies the Hopf invariant one differentials.

Corollary 1.5. $d_2h_{n+1} = h_0h_n^2$ for all $n > 0$.

Proof. By Theorems 1.1 and 1.2 we find that

$$d_*h_{n+1} = d_*Sq^{2^n}h_n = Sq^{2^n}d_2h_n \dotplus h_0h_n^2$$

so that

$$d_2h_{n+1} = h_0h_n^2$$

since $Sq^{2^n}d_2h_n$ is in filtration 4. (It follows, of course, that $Sq^{2^n}d_2h_n = Sq^{2^n}h_0h_{n-1}^2 = h_0^2h_n^2$.)

The next result shows how we may use the relation with homotopy operations to get stronger results than the differentials themselves give.

Proposition 1.6. h_1h_4 and h_2h_4 are permanent cycles.

Proof. Since $h_1h_4 = Sq^9(h_0h_3)$, it is carried by the 9-cell of P_7^9. The attaching map is η, to the 7-cell, and hence its boundary is $\eta(2\sigma)^2 = 0$. Similarly, $h_2h_4 = Sq^{10}(h_1h_3)$, so h_2h_4 is carried by the 10-cell of $P_8^{10} \simeq S^8 \vee (S^9 \cup_2 e^{10})$. The 9-cell carries $P^9(\eta\sigma)$, which has order 2 by the Cartan formula in Theorem V.1.10. Thus, the boundary of the 10-cell maps to 0 and h_2h_4 is a permanent cycle.

Before turning to other families of elements we should note that the Hopf invariant one differentials of Corollary 1.5 account for only a few of the non-trivial differentials on the $h_0^i h_{n+1}$. In fact, Proposition 1.4 implies $d_2 h_0^i h_{n+1} = h_0^{i+1} h_n^2$ is 0 if $i+1 \geq 2^{n-2}$. On the other hand, $h_0^i h_{n+1} \neq 0$ for $i < 2^{n+1}$, and from the known order of Im J, there must be higher differentials on many of the $h_0^i h_{n+1}$ which survive to E_3. It seems difficult to determine these higher differentials in terms of the Steenrod operations, though Milgram [81] has indicated that it may be possible with a sufficiently good hold on the chain level operations. More disappointing is the fact that it doesn't seem possible to propagate these higher differentials. That is, even if we accept as given a differential like $d_3 h_0 h_4 = h_0 d_0$, we don't seem to get any information on $d_3 h_0^3 h_5$.

The operation we call S in Proposition 1.4 will be very useful so we collect its properties before proceeding.

Proposition 1.7. If $S = Sq^{n+s}: Ext^{s,n+s} \to Ext^{s,2(n+s)}$ then

(i) $S[x_1|\cdots|x_k] = [x_1^2|\cdots|x_k^2]$ in the cobar construction

(ii) $S(xy) = S(x)S(y)$

(iii) $Sq^j Sx = SSq^{j-n-s}x$

(iv) $S\langle x_0, x_1, \ldots, x_n\rangle \subset \langle Sx_0, Sx_1, \ldots, Sx_n\rangle$

Proof. (i) is Proposition 11.10 of [68], while (ii) and (iii) are immediate from the Cartan and Adem relations since all the other terms must be 0 for dimensional reasons. Part (iv) is proved in [78].

For our remaining sample calculations we will explore the consequences of the squaring operations on the elements c_0, d_0, e_0 and f_0. The key elements we will be concerned with are collected in Table 1.1 along with Massey product representations. With the exception of f_0 and y_0, the Massey products have no indeterminacy.

s	n = t-s	Name	Massey product
3	8	c_0	$\langle h_1,h_0,h_2^2\rangle$
4	14	d_0	$\langle h_0,h_2^2,h_0,h_2^2\rangle$
4	17	e_0	$\langle h_0^2,h_3^2,h_1,h_o\rangle$
4	18	f_0	$\langle h_0^2,h_3^2,h_2\rangle$
4	20	g_1	-------
6	30	r_0	$\langle h_0^2,h_3^2,h_3^2,h_0^2\rangle$
7	35	m_0	$\langle h_2,h_1,r_0\rangle$
6	36	t_0	------
5	37	x_0	$\langle h_3,h_4,d_0\rangle$
6	38	y_0	$\langle h_0^4,h_4^2,h_3\rangle$

TABLE 1.1

Also, note that the elements Mahowald and Tangora call r,m,t,x and y, we are calling r_0, m_0, t_0, x_0 and y_0. The reason for the subscript will be apparent from the following definition.

Definition 1.8. If $i \geq 0$ and $a \in \{c,d,e,f,g,r,m,t,x,y\}$, let $a_0 = a$ and

$$a_{i+1} = Sa_i.$$

Applying Proposition 1.7(iv) we find immediately that

$$c_i \in \langle h_{i+1},h_i,h_{i+2}^2\rangle$$

$$d_i \in \langle h_i,h_{i+2}^2,h_i,h_{i+2}^2\rangle$$

$$e_i \in \langle h_i^2,h_{i+3}^2,h_{i+1},h_i\rangle$$

$$f_i \in \langle h_i^2,h_{i+3}^2,h_{i+2}\rangle$$

$$r_i \in \langle h_i^2,h_{i+3}^2,h_{i+3}^2,h_i^2\rangle$$

$$m_i \in \langle h_{i+2},h_{i+1},r_i\rangle$$

$$x_i \in \langle h_{i+3},h_{i+4},d_i\rangle$$

and

$$y_i \in \langle h_i^4,h_{i+4}^2,h_{i+3}\rangle \; .$$

However, we shall not make any use of these Massey product representations here.

From the calculations of Mukohda [88] or Milgram [81] we collect the values of the Steenrod operations on c_0, d_0, e_0 and f_0. The following abbreviation will be very convenient: if $x \in \mathrm{Ext}^{s,n+s}$ let $Sq^*(x) = (Sq^n x, Sq^{n+1}, \ldots, Sq^{n+s}x) = (x^2, \ldots, Sx)$

<u>Theorem 1.9.</u>

$$Sq^* c_0 = (c_0^2, h_0 e_0, f_0, c_1)$$

$$Sq^* d_0 = (d_0^2, 0, r_0, 0, d_1)$$

$$Sq^* e_0 = (e_0^2, m_0, t_0, x_0, e_1)$$

$$Sq^* f_0 = (0, h_3 r_0, y_0, 0, f_1)$$

The indeterminacy in the Massey product representations of f_0 and y_0 suggests that we should define them by the squaring operations above:

$$f_0 = Sq^{10} c_0 \quad \text{and} \quad y_0 = Sq^{20} f_0.$$

Applying Proposition 1.7.(iii) we immediately obtain the following corollary.

<u>Corollary 1.10.</u>

$$Sq^* c_i = (c_i^2, h_i e_i, f_i, c_{i+1})$$
$$Sq^* d_i = (d_i^2, 0, r_i, 0, d_{i+1})$$
$$Sq^* e_i = (e_i^2, m_i, t_i, x_i, e_{i+1})$$
$$Sq^* f_i = (0, h_{i+3} r_i, y_i, 0, f_{i+1}).$$

Before computing the differentials that this corollary implies, it will be useful to obtain a number of relations in Ext. This also gives us an opportunity to illustrate how powerful the Steenrod operations are in propagating relations. The relations we will assume known are all calculated by Tangora [103] by means of the May spectral sequence. In general, this technique only yields relations modulo terms of lower weight. However, the particular relations we need do not suffer from this ambiguity, since there are no terms of lower weight in their bidegree.

<u>Proposition 1.11</u> (i) $h_0 c_0 = 0$, $h_2 c_0 = 0$, $h_3 c_0 = 0$, $h_0 c_1 = 0$, $h_1 f_0 = 0$, $h_1 r_0 = 0$, $h_1 m_0 = 0$.

(ii) $c_0^2 = h_1^2 d_0$, $h_2 d_0 = h_0 e_0$, $h_1 e_0 = h_0 f_0$, $h_2 e_0 = h_0 g_1$, $h_0^2 d_0 = P^1 h_2^2$, $h_2 t_0 = c_1 g_1$.

(iii) $h_0^6 r_0 = 0$, $h_4 f_0 = 0$, $h_3 d_0^2 = 0$, $h_2 d_1 = h_4 g_1$, $h_0^6 x_0 = 0$, $h_2 m_0 = h_0^2 y_0$, $h_0^2 f_1 = h_1^2 e_1$.

These relations are grouped as follows: (i) holds because the relevant bidegree is 0 or is not annihilated by h_0, as multiples of h_1 must be; (ii) follows from [103] since, again by [103], there are no elements of lower weight in the given bidegrees; (iii) now follows either by applying Steenrod operations to relations in (i) and (ii) or by the same argument as (ii). (The point is that the relations in (iii) are dependent on those in (i) and (ii) under the action of the Steenrod algebra.)

Corollary 1.12. (i) $h_ic_i = 0$, $h_{i+2}c_i = 0$, $h_{i+3}c_i = 0$, $h_{i-1}c_i = 0$, $h_{i+1}f_i = 0$, $h_{i+1}r_i = 0$, $h_{i+1}m_i = 0$.

(ii) $c_i^2 = h_{i+1}^2 d_i$, $h_{i+2}d_i = h_ie_i$, $h_{i+1}e_i = h_if_i$, $h_{i+2}e_i = h_ig_{i+1}$, $h_{i+2}t_i = c_{i+1}g_{i+1}$.

(iii) $h_{i+4}f_i = 0$, $h_{i+3}d_i^2 = 0$, $h_{i+1}d_i = h_{i+3}g_i$, $h_{i+2}m_i = h_i^2y_i$, $h_{i-1}^2f_i = h_i^2e_i$.

Proof These are immediate from Proposition 1.11 since S is a ring homomorphism by Proposition 1.7(ii).

A comparison of the preceding proposition and corollary will show that if we view the periodicity operator as a Massey product

$$P^rx = \langle h_{r+2}, h_0^{2^{r+1}}, x\rangle ,$$

then we have only Milgram's theorem (Proposition 1.7.(iv)) to use in calculating $S(P^rx)$, and this generally leaves us with too much indeterminacy. For example, $P^1h_1h_3 = c_0^2$ so $S(P^1h_1h_3) = Sc_0^2 = c_1^2$. On the other hand, $S(P^1h_1h_3) = S\langle h_3, h_0^4, h_1h_3\rangle \in \langle h_4, 0, h_2h_4\rangle = 0$ modulo indeterminacy which is divisible by h_4. Of course, since $c_1^2 \neq 0$, it follows that $h_2h_4g = c_1^2$ since $h_4(h_2g)$ is the only possible nonzero element divisible by h_4. This example shows that to calculate $S(P^rx)$, we need another representation of P^rx. It also shows that the Massey product representation can lead to useful information (although in this case the product $h_2h_4g = c_1^2$ was already true in the associated graded). Accordingly, we provide the following formula for the interaction of the Sq^i and the periodicity homomorphisms P^r.

Proposition 1.13. Let $Sq_i = Sq^{t-i}: Ext^{s,t} \to Ext^{s+i,2t}$. Modulo the ideal generated by $\{h_{r+1}^2, h_{r+2}, Sq_0x, \ldots, Sq_ix\}$ we have

$$Sq_i P^{r-1}x = \begin{cases} 0 & i < 2^r \\ P^r Sq_{i-2^r}x + \langle h_{r+1}^2, h_0^{2^{r+1}}, Sq_{i-2^r-1}x\rangle & i \geq 2^r. \end{cases}$$

If $i = 0$, the indeterminacy (of $Sq_0 = S$) is generated by h_{r+2} and Sq_0x.

Proof. This is a special case of Milgram's general result [78], which, for three-fold Massey products says

$$Sq_i\langle a,b,c\rangle \subset \left\langle (Sq_i a,\dots,Sq_0 a), \begin{pmatrix} Sq_0 b & & \\ \vdots & \ddots & \\ Sq_i b & \cdots & Sq_0 b \end{pmatrix}, \begin{pmatrix} Sq_0 c \\ \vdots \\ Sq_i c \end{pmatrix} \right\rangle,$$

since $Sq_0 h_0^n = h_1^n = 0$ for $n \geq 4$, $Sq_n h_0^n = h_0^{2n}$, and $Sq_i h_0^n = 0$ otherwise.

Corollary 1.14. $\langle h_4, h_0^8, h_3^2\rangle = P^2h_3^2 = h_0^4 r_0$ with no indeterminacy.

Proof. By Proposition 1.11, $P^1h_2^2 = h_0^2 d_0$. By Theorem 1.9 we have $Sq^{16}h_0^2 d_0 = h_0^4 r_0 + h_1^2 d_0^2 = h_0^4 r_0$, since $h_1 d_0^2$ must be divisible by h_0 so $h_1^2 d_0^2 = 0$. By Proposition 1.13, $Sq^{16}P^1h_2^2 = Sq_4 P^1 h_2^2 = P^2h_3^2$ with indeterminacy generated by h_3^2 and h_4. For dimensional reasons the indeterminacy is 0.

Combining Proposition 1.11 with Theorem 1.9 we can produce a number of relations in Ext which do not hold in the associated graded calculated by Tangora.

Proposition 1.15.

(i)	$h_0 r_0 = s_0$	and hence	$h_i r_i = s_i$
(ii)	$h_3 r_0 = h_1 t_0 + h_0^2 x_0$	and hence	$h_{i+3} r_i = h_{i+1} t_i + h_i^2 x_i$
(iii)	$h_2 e_0^2 = h_0^4 x_0$	and hence	$h_{i+2} e_i^2 = 0$ if $i > 0$
(iv)	$h_2^2 d_1 = h_1 x_0$	and hence	$h_{i+1}^2 d_i = h_i x_{i-1}$
(v)	$h_1 y_0 = h_2 t_0$	and hence	$h_{i+1} y_i = h_{i+2} t_i$
(vi)	$h_2 x_0 = 0$	and hence	$h_{i+2} x_i = 0$
(vii)	$h_1 f_1 = h_0^2 c_2$	and hence	$h_i f_i = h_{i-1}^2 c_{i+1}$
(viii)	$h_2 y_0 = 0$	and hence	$h_{i+2} y_i = 0$
(ix)	$h_3 x_0 = h_0^2 g_2$	and hence	$h_{i+3} x_i = h_i^2 g_{i+2}$

Note. Mahowald and Tangora [61] found (i)-(iii) by other techniques. Barratt, Mahowald and Tangora [20] also found (iv), (vii), and (ix) by other techniques. Milgram [81] found (i) and (ii) by using the Steenrod operations. Mukohda [88] found (iv)-(vi) and (ix), partly by using the Steenrod operations and the cobar construction, and partly by means of a minimal resolution.

Proof. Given (ii), (i) follows because $h_0h_3r_0 = h_0^3x_0 \neq 0$, from which it follows that $h_0r_0 \neq 0$. The only possibility is $h_0r_0 = s_0$. To prove (ii), apply Sq^{20} to the relation $h_2d_0 = h_0e_0$. To prove (iii), apply Sq^{19} to the relation $h_1e_0 = h_0f_0$ and use the fact that $h_1m_0 = 0$. To prove (iv), apply Sq^{21} to the relation $h_2d_0 = h_0e_0$ and use the fact that $h_0^2e_1 = 0$. To prove (v), apply Sq^{21} to the relation $h_1e_0 = h_0f_0$ and use (iv) to show that $h_1^2x_0 = h_1(h_2^2d_1) = 0$. To prove (vi), apply Sq^{22} to the relation $h_1e_0 = h_0f_0$ to show that $h_2x_0 = h_1^2e_1 + h_0^2f_1$, and apply Proposition 1.11.(iii) to show that this is 0. For (vii), we apply Sq^{22} to $h_0c_1 = 0$. Similarly, Sq^{21} applied to $h_1f_0 = 0$ yields (viii). Finally, (ix) follows by applying Sq^{24} to the relation $h_2e_0 = h_0g_1$ to get $h_0^2g_2 = h_3x_0 + h_2^2e_1$, and noting that $h_2^2e_1 = h_2(h_1f_1) = 0$. The calcultion of $Sq^{24}(h_0g_1)$ is possible because $Sq^{24}g_1 = g_2$ by definition, while $Sq^{23}g_1 = 0$ for dimensional reasons.

Now we examine the differentials implied by the squaring operations in the c_i, d_i, e_i and f_i families. The results we obtain for $t-s \geq 45$ are all new. In the range $t-s \leq 45$ they are due to May [66], Maunder [65], Mahowald and Tangora [61], Milgram [81] and Barratt, Mahowald and Tangora [20] with the exception of $d_3e_1 = h_1t$, which is new and corrects a mistake in [20]. As noted by Milgram [81] the proofs using Steenrod operations are usually far simpler and more direct than the original proofs. In addition, when they replace proofs which relied on prior knowledge of the relevant homotopy groups we obtain independent verification of the calculation of those homotopy groups.

If $x \in E_r^{s,n+s}$, let us write $x \in (s,n)$ or $x \in (s,n)_r$ for convenience. Theorems 1.1, 1.2 and 1.3 imply that

$$d_*Sq^jx = Sq^jd_rx \doteq \begin{cases} 0 & v > k+1 \text{ or } 2r-2 < v < k \\ \bar{a}xd_rx & v = k+1 \\ \bar{a}Sq^{j-v}x & v = k \text{ or } (v < k \text{ and } v \leq 10) \end{cases}$$

where $k = j-n$, $v = 8a + 2^b$ if $j+1 = 2^{4a+b}(\text{odd})$, and $\bar{a}$ detects a generator of Im J in $\pi_{v-1}S^0$.

We start with a general observation about families $\{a_i\}$ with $a_{i+1} = S(a_i)$. If $a_i \in (s,n_i)$ then

$$n_i + s = 2(n_{i-1} + s) = 2^i(n_0 + s).$$

If N is the integer such that $2^{N-1} < s+2 \leq 2^N$ then the differentials on the elements $Sq^j a_i$ depend on the congruence class of n_i modulo 2^N. Clearly, $n_i \equiv -s$ modulo 2^N if $i \geq N$. Thus, the differentials on all but the first N members of such a family follow a pattern which depends only on the filtration in which the family lives.

Consider the c_i family. We have $c_0 \in (3,8)_\infty$, so in general $c_i \in (3,2^i \cdot 11-3)$.

Proposition 1.16. (i) $c_1 \in E_\infty$ while $d_2 c_i = h_0 f_{i-1}$ for $i \geq 2$

(ii) $d_2 f_0 = h_0^2 e_0$, $f_1 \in E_5$, and $d_3 f_i = h_1 y_{i-1}$ for $i \geq 2$

(iii) $d_3 c_i^2 = h_0^2 h_{i+2} r_{i-1}$ for $i \geq 2$

Note. We will show shortly that $d_2 h_0 y_{i-1} = h_0^2 h_{i+2} r_{i-1}$. This, together with (iii) implies that $d_3 c_i^2 = 0$.

Corollary 1.17. $d_2 e_0 = c_0^2$ and $\nu\theta_4 \neq 0$, where θ_4 is the Arf invariant one element detected by h_4^2.

Proof. Since $c_0 \in (3,8)_\infty$, $Sq^* c_0 = (c_0^2, h_0 e_0, f_0, c_1)$ is carried by $\Sigma^8 P_8^{11} = S^{16} \vee (S^{17} \cup_2 e^{18}) \vee S^{19}$. Therefore $c_1 \in E_\infty$ and $d_2 f_0 = h_0^2 e_0$. Applying Proposition 1.11 we find that $d_2 h_1 e_0 = d_2 h_0 f_0 = h_0^3 e_0 = h_1^3 d_0 = h_1 c_0^2$, from which it follows that $d_2 e_0 = c_0^2$.

Since $c_1 \in (3,19)_\infty$, $Sq^* c_1 = (c_1^2, h_1 e_1, f_1, c_2)$ is carried by $\Sigma^{19} P_{19}^{23} = (S^{38} \cup_2 e^{39} \cup_\eta e^{40}) \cup_2 e^{41}$. Therefore $d_2 c_2 = h_0 f_1$ and $d_3 f_1 = h_1 c_1^2 = h_1 h_2^2 d_1 = 0$, so that $f_1 \in E_5$ for dimensional reasons. Since $c_2 = \langle h_3, h_2, h_4^2 \rangle$ and $c_2 \notin E_\infty$, the Toda bracket $\langle \sigma, \nu, \theta_4 \rangle$ does not exist. We shall show in the next proposition that $h_4^2 \in E_\infty$ so that θ_4 exists. Since $\sigma\nu = 0$, it follows that $\nu\theta_4 \neq 0$.

Now assume for induction that $d_2 c_i = h_0 f_{i-1}^2$ and that $i \geq 2$. We can arrange the relevant information in the following table.

j (mod 4)	$Sq^j c_i$	$Sq^j (h_0 f_{i-1}^2)$	k	v	$\bar{a}$
1	c_i^2	$h_0^2 h_{i+2} r_{i-1}$	0	2	h_1
2	$h_i e_i$	$h_0^2 y_{i-1} + h_1 h_{i+2} r_{i-1}$	1	1	h_0
3	f_i	$h_1 y_{i-1}$	2	≥ 4	-
4	c_{i+1}	$h_0^2 f_i$	3	1	h_0

It follows that $d_3c_i^2 = h_0^2h_{i+2}r_{i-1}$, $d_2h_ie_i = h_0c_i^2$, $d_3f_i = h_1y_{i-1}$ and $d_2c_{i+1} = h_0f_i$. This completes the inductive step and finishes the proof of Propositon 1.16 and Corollary 1.17. Note that we have omitted $d_2h_ie_i$ from the statement of the proposition because it will follow from our calculation of d_2e_i below.

Proposition 1.18. (i) $d_2k = h_0d_0^2$

(ii) $d_3r_0 = h_1d_0^2$ and $h_4^2 \in E_\infty$

(iii) $r_i \in E_3$ for $i \geq 1$

(iv) $d_i \in E_3$ for $i \geq 1$

Note. Mahowald and Tangora show [61] that d_1 is actually in E_∞, not just E_3. Also, the proof given here that $h_4^2 \in E_\infty$ is much simpler than the proof in [61].

Proof. Since $d_0 \in (4,14)_\infty$, $Sq^*d_0 = (d_0^2, 0, r_0, 0, d_1)$ is carried by $\Sigma^{14}P_{14}^{18}$, which has attaching maps as shown

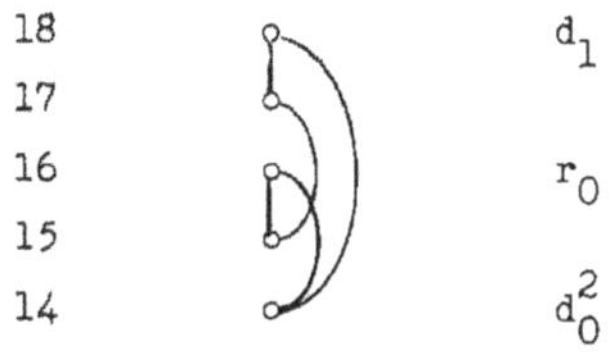

Now $d_3h_0h_4 = h_0d_0$ implies $h_0d_0^2 = 0$ in E_4. The only possibility is that $d_2k = h_0d_0^2$. This implies that $2\pi_{29} = 0$. Since the boundary of the 16 cell carries $h_1d_0^2$ plus twice something, we get $d_3r_0 = h_1d_0^2$. Nothing is left for h_4^2 to hit, so $h_4^2 \in E_\infty$. Finally, $d_2(d_1) = h_0 \cdot 0 = 0$ so $d_1 \in E_3$. Now assume for induction that $i \geq 1$ and $d_i \in E_3$. The terms $Sq^jd_3d_i$ in the differentials on Sq^jd_i will not contribute until E_5, so will not affect the proof of (iii) and (iv). Since $Sq^*d_i = (d_i^2,0,r_i,0,d_{i+1})$ we find that $d_2r_i = h_0 \cdot 0 = 0$ and $d_2(d_{i+1}) = h_0 \cdot 0 = 0$, proving (iii) and (iv) and completing the induction.

Proposition 1.19. (i) $d_2m_0 = h_0e_0^2$, $t_0 \in E_{11}$ and $d_3e_1 = h_1t_0$

(ii) $e_1^2 \in E_5$, $d_5m_1 = Sq^{39}h_1t_0$, $d_2t_1 = h_0m_1$, $d_3x_1 = h_1m_1$ and $d_2e_2 = h_0x_1$.

(iii) If $i \geq 2$ and $n = 2^i \cdot 21 - 4$ then $d_3e_i^2 = h_0^2e_ix_{i-1} + Sq^nh_0x_{i-1}$, $d_3m_i = Sq^{n+1}h_0x_{i-1}$, $d_2t_i = h_0m_i$, $d_3x_i = Sq^{n+3}h_0x_{i-1}$, and $d_2e_i = h_0x_{i-1}$.

Proof. By Corollary 1.17, $d_2e_0 = c_0^2$. The information needed to calculate the differentials on the Sq^je_0 is most conveniently presented in a table.

j	$Sq^j e_0$	k	v	$\bar{a}$	$Sq^j c_0^2$	conclusion
17	e_0^2	0	2	h_1	0	$d_3 e_0^2 = 0$
18	m_0	1	1	h_0	$h_0^2 e_0^2$	$d_2 m_0 = h_0 e_0^2$
19	t_0	2	4	h_2	0	$d_3 t_0 = 0$
20	x_0	3	1	h_0	0	$d_2 x_0 = h_0 t_0 = 0$
21	e_1	4	2	h_1	0	$d_3 e_1 = h_1 t_0$

We omit $d_3 e_0{}^2$ and $d_2 x_0 = 0$ from the proposition because they also follow simply for dimensional reasons. Similarly, since t_0 is in E_4 it must be in E_{11} for dimensional reasons. Thus (i) is proved.

Since $d_3 e_1 = h_1 t_0$, the term $Sq^j h_1 t_0$ will contribute to $d_5 Sq^j e_1$ if $Sq^j e_1$ lives that long. Again, the information is most conveniently organized into a table.

j	$Sq^j e_1$	k	v	$\bar{a}$	conclusion
38	e_1^2	0	1	h_0	$d_4 e_1^2 = h_0 e_1 h_1 t_0 = 0$
39	m_1	1	8	h_3	$d_5 m_1 = Sq^{39} h_1 t_0$
40	t_1	2	1	h_0	$d_2 t_1 = h_0 m_1$
41	x_1	3	2	h_1	$d_3 x_1 = h_1 m_1$
42	e_2	4	1	h_0	$d_2 e_2 = h_0 x_1$

All of (ii) follows immediately . Now assume for induction that $d_2 e_i = h_0 x_{i-1}$ and $i \geq 2$. Again we organize the information in tabular form. Let $n = 2^i \cdot 21 - 4$ so that $e_i \in (4,n)_2$.

j	$Sq^j e_i$	k	v	$\bar{a}$	conclusion
n	e_i^2	0	1	h_0	$d_3 e_i^2 = h_0^2 e_i x_{i-1} + Sq^n h_0 x_{i-1}$
n+1	m_i	1	2	h_1	$d_3 m_i = Sq^{n+1} h_0 x_{i-1}$
n+2	t_i	2	1	h_0	$d_2 t_i = h_0 m_i$
n+3	x_i	3	4	h_2	$d_3 x_i = Sq^{n+3} h_0 x_{i-1}$
n+4	e_{i+1}	4	1	h_0	$d_2 e_{i+1} = h_0 x_i$

This establishes (iii) and completes the induction.

Note that three of the 5 entries in the above table satisfy $\bar{v} = k+1$. The corresponding differentials therefore contain terms of the form $\bar{a}xd_rx$, specifically $\bar{a}h_0e_ix_{i-1}$ in this instance.

Only one of the differentials on the Sq^jf_i is interesting.

Proposition 1.20. For all $i \geq 0$, $d_2y_i = h_0h_{i+3}r_i$.

Proof. The terms in d_*Sq^jx involving d_rx do not contribute to d_2Sq^jx. If $n = 2^i \cdot 22 - 4$ so that $f_i \in (4,n)$ then $Sq^{n+1}f_i = h_{i+3}r_i$ and $Sq^{n+2}f_i = y_i$. Since $n+2$ is even the proposition follows immediately.

This completes our sampler. We have calculated only about one fourth of the differentials found by Mahowald and Tangora, but they include some of the most difficult. The remaining differentials follow more or less directly from those calculated here just as in Mahowald and Tangora's original paper [61].

2. Extended Powers of Cells

In order to study Steenrod operations on elements of the Adams spectral sequence which are not permanent cycles, we need a relative version of the extended power construction. The extended power functor $E\pi \ltimes_\pi X^{(p)}$, for $\pi \subset \Sigma_p$, factors as the composite of the functors

$$X \longmapsto X^{(p)}$$

and

$$Y \longmapsto E\pi \ltimes_\pi Y$$

If we replace X by a pair (X,A) then $X^{(p)}$ is replaced by a length p+1 filtration $X^{(p)} \supset \cdots \supset A^{(p)}$ of π spectra and we may apply $E\pi \ltimes_\pi (?)$ to this termwise. The resulting diagram is the relativization which we need. While the formalism applies to any pair (X,A), we will confine attention to pairs (CX,X), where CX is the cone on X, both for notational simplicity and because the pth power of such a pair has special properties which we shall exploit. In particular, note that Lemma 2.4 is the geometric analog of the fact that a trivial one-dimensional representation splits off the permutation representation of $\pi \subset \Sigma_p$ on R^p. Most of this section is devoted to this fact and its consequences.

An element $x \in E_r^{s,n+s}(X,Y)$ can be represented by a map of pairs

$$(CX,X) \longrightarrow (Y_s,Y_{s+r}).$$

Extended powers of (CX,X) can be used to construct a map representing $\beta^\epsilon P^jx$. The final bit of the section establishes the facts about extended powers which will enable us to construct and analyze such a map.

We shall work first in the category of based π-spaces and based π-maps and the homotopy category of based π-spaces and π-homotopy classes of based π-maps with weak equivalences inverted. The results are then transferred to the category of π-spectra by small smash products, desuspensions, and colimits.

Let I be the unit interval. We choose 0 as the basepoint, justifying our choice by the resulting simplicity of the formulas in the proof of Lemma 2.4. For a space or spectrum X, let $CX = X \wedge I$. The isomorphism $X \cong X \wedge \{0,1\}$ and the cofibration $\{0,1\} \subset I$ induce a cofibration $X \to CX$ with cofiber ΣX.

Definition 2.1. For a space X, define a Σ_p-space $\Gamma_i(X)$ by

$$\Gamma_i(X) = \{c_1 \wedge \cdots \wedge c_p \in (CX)^{(p)} \mid \text{at least } i \text{ of the } c_j \text{ lie in } X\}.$$

If X is a spectrum, define a Σ_p spectrum $\Gamma_i(X) = X^{(p)} \wedge \Gamma_i(S^0)$.

Lemma 2.2. (i) For a space X, $\Gamma_i(X)$ is naturally and Σ_p equivariantly homeomorphic to $X^{(p)} \wedge \Gamma_i(S^0)$.

(ii) $\Gamma_i(\Sigma^\infty X) \cong \Sigma^\infty \Gamma_i(X)$ if X is a space.

(iii) $\Gamma_{i+1}(X) \to \Gamma_i(X)$ is a Σ_p-cofibration.

(iv) $\Gamma_i(X)/\Gamma_{i+1}(X)$ is equivalent to the wedge of all $(i,p-i)$ permutations of $X^{(i)} \wedge (\Sigma X)^{(p-i)}$. In particular, if (p) is the permutation representation of Σ_p on $\mathbb{R}^p$ then $\Gamma_0(X)/\Gamma_1(X) \cong (\Sigma X)^{(p)} \cong \Sigma^{(p)} X^{(p)}$ and $\Gamma_p(X) \cong X^{(p)}$.

(v) $\Gamma_1(X) \cong \Sigma^{p-1} X^{(p)}$ as Σ_p spaces or spectra, where S^{p-1} has the Σ_p action inherited from the p-cell $\Gamma_0(S^0) = I^{(p)}$.

Proof. (i) follows immediately from the shuffle map

$$(x_1 \wedge t_1) \wedge \cdots \wedge (x_p \wedge t_p) \longmapsto (x_1 \wedge \cdots \wedge x_p) \wedge (t_1 \wedge \cdots \wedge t_p).$$

(ii) is a consequence of the commutation of Σ^∞ and smash products.

(iii) follows for spectra if it holds for spaces. By (i) it holds for spaces if it holds for S^0. For S^0, it follows because $\Gamma_i(S^0)$ is the (p-i) skeleton of a CW decomposition of $\Gamma_0(S^0) = I^{(p)}$.

Similarly, (iv) holds in general if it holds for S^0, for which it is immediate.

(v) follows from the fact that $\Gamma_1(S^0)$ is the boundary of the p-cell $\Gamma_0(S^0)$.

Remark 2.3: We will complete what we have begun in (iv) and (v) above in Lemma 3.5, which shows that

$$\Gamma_i(X) \simeq \bigvee_{(p-i,i-1)} \Sigma^{np-i}X^{(p)}.$$

The next lemma is the key result of this section. Let I and S^1 have trivial Σ_p actions so that if X is a Σ_p space or spectrum then $CX = X \wedge I$ and $\Sigma X = X \wedge S^1$ are also.

Lemma 2.4. There are natural equivariant equivalences $\Gamma_0(X) \simeq C\Gamma_1(X)$ and $\Sigma\Gamma_1(X) \simeq (\Sigma X)^{(p)}$ such that the triangle

$$\begin{array}{ccc} & \subset & C\Gamma_1(X) \\ \Gamma_1(X) & & \wr\| \\ & \subset & \Gamma_0(X) \end{array}$$

commutes.

Proof. By definition and by 2.2(i) we may assume $X = S^0$. We define a Σ_p homeomorphism $\Gamma_0(S^0) \to C\Gamma_1(S^0)$ by

$$t_1 \wedge \cdots \wedge t_p \longrightarrow \left(\frac{t_1}{t} \wedge \cdots \wedge \frac{t_p}{t}\right) \wedge t$$

where $t = \max\{t_i\}$. The inverse homeomorphism is given by

$$(t_1 \wedge \cdots \wedge t_p) \wedge t \longmapsto tt_1 \wedge tt_2 \wedge \cdots \wedge tt_p \ .$$

Commutativity of the triangle is immediate. The equivalence $\Sigma\Gamma_1(X) \simeq (\Sigma X)^{(p)}$ follows since $\Sigma\Gamma_1(X) \simeq C\Gamma_1(X)/\Gamma_1(X) \simeq \Gamma_0(X)/\Gamma_1(X) \simeq (\Sigma X)^{(p)}$, the latter equivalence by 2.2(iv).

Lemma 2.5. For any $\pi \subset \Sigma_p$ and any π-free π space W, there are natural equivalences

$$W \ltimes_\pi \Gamma_0(X) \simeq C(W \ltimes_\pi \Gamma_1(X))$$

and

$$\Sigma(W \ltimes_\pi \Gamma_1(X)) \simeq W \ltimes_\pi (\Sigma X)^{(p)}$$

such that the following triangle commutes.

$$\begin{array}{ccc} & \nearrow & W \ltimes_\pi \Gamma_0(X) \\ W \ltimes_\pi \Gamma_1(X) & & \wr\| \\ & \searrow & C(W \ltimes_\pi \Gamma_1(X)) \end{array}$$

Proof. By Lemma 2.4, $W \ltimes_\pi \Gamma_0(X) \simeq W \ltimes_\pi(\Gamma_1(X) \wedge I)$ and by I.1.2.(ii) $W \ltimes_\pi(\Gamma_1(X) \wedge I) \simeq (W \ltimes_\pi \Gamma_1(X)) \wedge I = C(W \ltimes_\pi \Gamma_1(X))$. The second equivalence follows similarly. Commutativity of the triangle follows from naturality with respect to $\{0,1\} \subset I$.

In the remainder of this section we shall restrict attention to the special case of interest in section 4. The general case presents no additional difficulties but is notationally more cumbersome.

Let $\pi \subset \Sigma_p$ be cyclic of order p and let $W = S^\infty$ with the cell structure which makes $C_* W \cong \mathcal{W}$, the usual $Z[\pi]$ resolution of Z. Let W^k be the k-skeleton of W. As in V.2, W^k/π is the lens space $\tilde{L}^k$, and, by I.1.3.(ii), if $\Gamma_i = \Gamma_i(S^{n-1})$ then $W^k \ltimes_\pi \Gamma_i / W^{k-1} \ltimes_\pi \Gamma_i \simeq \Sigma^k \Gamma_i$. By Lemmas 2.2 and 2.5 we then have the following corollary of Theorems V.2.6 and V.2.14.

Corollary 2.6:
$$W^k \ltimes_\pi \Gamma_p \simeq \Sigma^{n-1} \tilde{L}^{(n-1)(p-1)+k}_{(n-1)(p-1)}$$
and
$$W^k \ltimes_\pi \Gamma_1 \simeq \Sigma^{n-1} \tilde{L}^{n(p-1)+k}_{n(p-1)} .$$

Now note that Lemma 2.5 also implies that $W^k \ltimes_\pi \Gamma_1 \cup W^{k-1} \ltimes_\pi \Gamma_0$ is the cofiber of the inclusion $W^{k-1} \ltimes_\pi \Gamma_1 \to W^k \ltimes_\pi \Gamma_1$. By Corollary 2.6 or by Lemma 2.2 and I.1.3.(ii) it follows that
$$W^k \ltimes_\pi \Gamma_1 \cup W^{k-1} \ltimes_\pi \Gamma_0 \simeq S^{np+k-1}.$$

To get this equivalence in a maximally useful form, first consider a more general situation. In order to analyze the Barratt-Puppe sequence of a map $a:A \to X$ one constructs the diagram below.

(2.1)

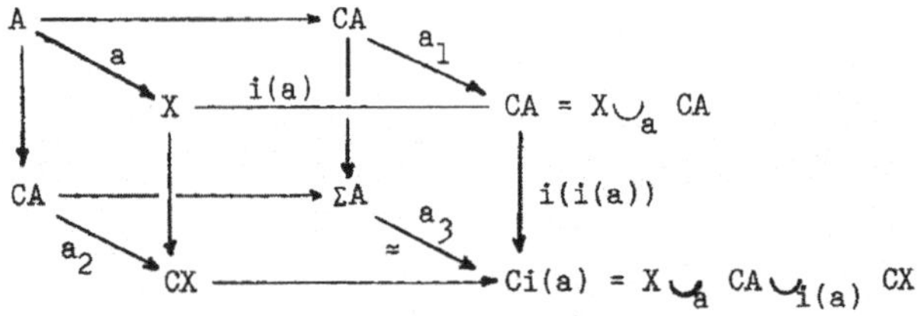

In diagram (2.1) the front and back squares are pushouts, a_3 is an equivalence, $a_2 = Ca = a \wedge 1$, a_1 is the obvious natural inclusion, and the maps a, i(a), and $a_3^{-1} i(i(a))$ are the beginning of the cofiber sequence of a. The following obvious fact about such diagrams will be used repeatedly.

Lemma 2.7. Let $B \to Y$ be a cofibration and let $\pi : Y \to Y/B$ be the natural map. For any map
$$f:(Ci(a),X) \to (Y,B),$$
we have $\pi f a_3 = \overline{fa_1} - \overline{fa_2}$ in $[\Sigma A, Y/B]$, where $\overline{fa_i}$ is the map $\Sigma A \to Y/B$ induced by $(fa_i, fa):(CA,A) \to (Y,B)$.

Proof. The only question is whether we should get $\overline{fa_1} - \overline{fa_2}$ or its negative. We choose $\overline{fa_1} - \overline{fa_2}$ for consistency with the Barratt-Puppe sequence signs. The point is that a_3 is a homotopy inverse to the map from $Ci(a)$ to ΣA which collapses CX, and the orientations on the two cones are determined by this fact.

Returning to the special case which prompted these generalities, let $a: S^{np+k-2} \to W^{k-1} \ltimes_\pi \Gamma_1$ be the attaching map of the top cell of $W^k \ltimes_\pi \Gamma_1$. Then diagram (2.1) becomes diagram (2.2) below.

(2.2)

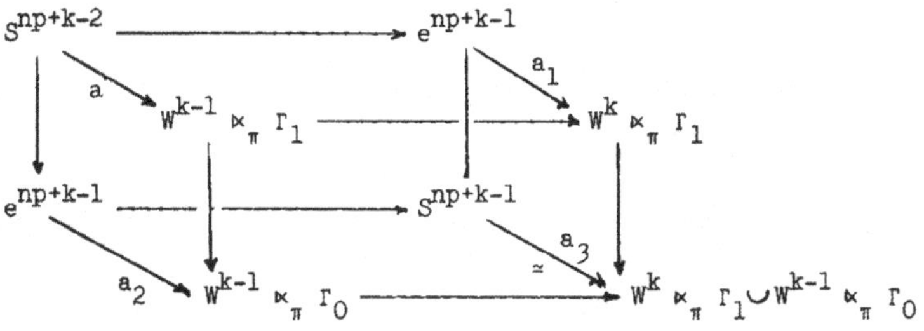

Corollary 2.8. Let $B \to Y$ and $\pi: Y \to Y/B$ be as in Lemma 2.7. For any map $f: (W^k \ltimes_\pi \Gamma_1 \cup W^{k-1} \ltimes_\pi \Gamma_0, W^{k-1} \ltimes_\pi \Gamma_1) \to (Y,B)$ we have $\pi f a_3 = \overline{fa_1} - \overline{fa_2}$ in $\pi_{np+k-1}(Y/B)$.

Let $v = v_p(n(p-1)+k)$ in the notation of Definition V.2.15, so that $a \in \pi_{np+k-2} W^{k-1} \ltimes_\pi \Gamma_1$ factors through $W^{k-v} \ltimes_\pi \Gamma_1$. Then we may replace the front face of diagram (2.2) by

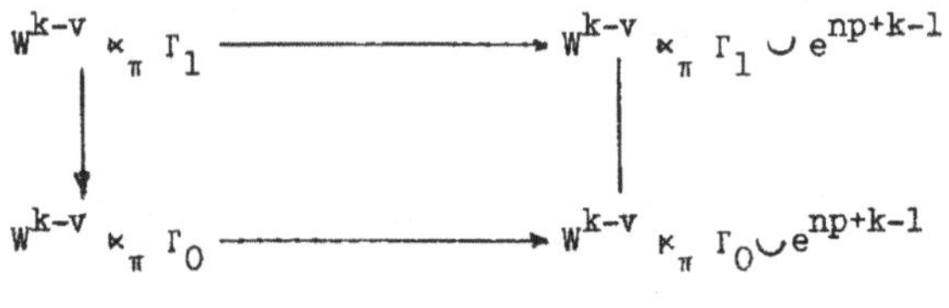

in which the np+k-1 cell is attached by a lift of a. This gives us a version of Corollary 2.8 in which f need only map $W^{k-v} \ltimes_\pi \Gamma_1$ into B and the map $\overline{fa_2}$ factors through $W^{k-v} \ltimes_\pi \Gamma_0$.

§3. Chain Level Calculations

In this section we define and study certain elements in the cellular chains of $W \ltimes_\pi \Gamma_0(S^{n-1})$. In sections 5-7 they will be used to investigate the homotopy groups of various pairs of subspaces of $W \ltimes_\pi \Gamma_0(S^{n-1})$. Here we use them to determine the effect in homology of a compression (lift) of the natural map $W^k \ltimes_\pi \Gamma_p(S^{n-1}) \to W^k \ltimes_\pi \Gamma_1(S^{n-1})$.

Let $\Gamma_i = \Gamma_i(S^{n-1})$. Give $e^n = C(S^{n-1})$ the cell structure with one n-cell x and one (n-1)-cell dx. Let $C_*(?)$ denote cellular chains and $C_*(?;R) = C_*(?) \otimes R$. Then $C_*\Gamma_0 = \langle x,dx\rangle^p$, the p-fold tensor product of copies of $C_*(e^n) = \langle x,dx\rangle$, and

$$C_i\Gamma_j = \begin{cases} C_i\Gamma_0 & i \leq np-j \\ 0 & i > np-j \end{cases} .$$

We shall find it convenient to omit the tensor product sign in writing elements of $C_*\Gamma_j$, so that, for example, $x^{p-1}dx$ denotes $x \otimes x \otimes \cdots \otimes x \otimes dx$. Let $W = S^\infty$ with the usual π-equivariant cell structure. Then C_*W is the minimal resolution $\mathcal{W}$ of Z over $Z[\pi]$. Let

$$\mathcal{W}(k)_j = \begin{cases} \mathcal{W}_j & j \leq k \\ 0 & j > k \end{cases}$$

so that $\mathcal{W}(k) = C_*(W^k)$, where W^k is the k-skeleton of W. Then by I.2.1, $C_*(W^k \ltimes_\pi \Gamma_i) \cong \mathcal{W}(k) \otimes_\pi C_*\Gamma_i$.

Let α be the p-cycle $(1\ 2\ \cdots\ p)$ in $\pi \subset \Sigma_p$, and let π and Σ_p act on $C_*\Gamma_i$ by permuting factors. Following [68, Theorem 3.1] we define elements $t_i \in C_*\Gamma_0$ as follows. Define a contracting homotopy for $C_*\Gamma_0$ by $s(ax) = 0$ and $s(adx) = (-1)^{|a|}ax$.

<u>Definition 3.1.</u> If $p = 2$, let $t_0 = dx^2$, $t_1 = xdx$, and $t_2 = x^2$. If $p > 2$, let $N = 1 + \alpha + \alpha^2 + \cdots + \alpha^{p-1}$. Let

$$t_0 = dx^p,\quad t_1 = dx^{p-1}x,$$
$$t_{2i} = s((\alpha^{-1} - 1)t_{2i-1}),\quad \text{and}$$
$$t_{2i+1} = s(Nt_{2i}).$$

<u>Lemma 3.2.</u> (i) If $p = 2$ then $d(t_2) = (\alpha + (-1)^n)t_1$ and $d(t_1) = t_0$.

(ii) If $p > 2$ then $d(t_1) = t_0$,

$$d(t_{2i}) = (\alpha^{-1} - 1)t_{2i-1}$$
$$\text{and}\quad d(t_{2i+1}) = Nt_{2i} \quad \text{if } i > 0.$$

(iii) If $p > 2$ then $t_p = (-1)^{mn} m! x^p$ and

$$t_{p-1} = m! x^{p-1} dx + (m-1)!(\alpha^{-1} - 1) Q x^{p-1} dx$$

where $\quad m = (p-1)/2 \quad$ and $\quad Q = (\alpha+1) \sum_{i=1}^{m} i\alpha^{2i}.$

Proof. (i) and (ii) are easy calculations, by induction on i for $d(t_{2i})$ and $d(t_{2i+1})$ using $(\alpha^{-1}-1)N = 0 = N(\alpha^{-1}-1)$ and $ds + sd = 1$.

In [68, Theorem 3.1] it is shown that $t_p = (-1)^{mn} m! x^p$ and that $t_{p-1} = (m-1)! P x^{p-1} dx$, where $P = \alpha + \alpha^3 + \cdots \alpha^{p-2}$. Since $P = m + (\alpha^{-1} - 1)Q$, (iii) follows.

Lemma 3.3. If $p = 2$, then in $C_*(W^{i+1} \ltimes_\pi \Gamma_1)$

$$e_{i+1} \otimes dx^2 \sim \begin{cases} (-1)^i e_i \otimes d(x^2) & n \not\equiv i \ (2) \\ (-1)^i e_i \otimes d(x^2) - 2e_i \otimes xdx & n \equiv i \ (2) \end{cases}$$

Proof. We have $d(e_i) = (\alpha + (-1)^i) e_{i-1}$ and $d(x^2) = dx\,x + (-1)^n x\,dx$. Therefore

$$\begin{aligned} d(e_{i+1} \otimes xdx) &= (\alpha + (-1)^{i+1}) e_i \otimes xdx + (-1)^{i+1} e_{i+1} \otimes dx^2 \\ &= e_i \otimes dx\,x + (-1)^{i+1} e_i \otimes xdx + (-1)^{i+1} e_{i+1} \otimes dx^2 , \end{aligned}$$

from which we obtain

$$\begin{aligned} e_{i+1} \otimes dx^2 &\sim (-1)^i e_i \otimes dx\,x - e_i \otimes xdx \\ &= (-1)^i e_i \otimes d(x^2) - (1 + (-1)^{i+n}) e_i \otimes xdx . \end{aligned}$$

Lemma 3.4. Let $p > 2$. If i is odd then, in $C_*(W^{i+p-1} \ltimes_\pi \Gamma_i)$,

$$e_{i+p-1} \otimes dx^p \sim (-1)^{mn+m} m! e_i \otimes d(x^p).$$

If i is even then, in $C_*(W^{i+p-1} \ltimes_\pi \Gamma_1)$,

$$e_{i+p-1} \otimes dx^p \sim (-1)^{mn+m} m!\, e_i \otimes d(x^p) - p \sum_{j=1}^{p-1} (-1)^{[j/2]} e_{i+p-j-1} \otimes t_j .$$

Hence, for any i,

$$e_{i+p-1} \otimes dx^p \sim (-1)^{mn+m} m!\, e_i \otimes d(x^p)$$

in $C_*(W^{i+p-1} \ltimes_\pi \Gamma_1, Z_p)$.

Proof. By Lemma 3.1 and the definition of $\mathcal{W}$ we find that if i is even then

$$d(e_{i+p-j} \otimes t_j) = \begin{cases} N(e_{i+p-j-1} \otimes t_j + e_{i+p-j} \otimes t_{j-1}) & j \text{ odd, } j \neq 1 \\ T(e_{i+p-j-1} \otimes t_j - e_{i+p-j} \otimes t_{j-1}) & j \text{ even} \\ Ne_{i+p-2} \otimes t_1 + e_{i+p-1} \otimes t_0 & j = 1 \end{cases}$$

and if i is odd then

$$d(e_{i+p-j} \otimes t_j) = \begin{cases} Te_{i+p-j-1} \otimes t_j - Ne_{i+p-j} \otimes t_{j-1} & j \text{ odd, } j \neq 1 \\ Ne_{i+p-j-1} \otimes t_j + Te_{i+p-j} \otimes t_{j-1} & j \text{ even} \\ Te_{i+p-2} \otimes t_1 - e_{i+p-1} \otimes t_0 & j = 1 , \end{cases}$$

where $N = 1 + \alpha + \alpha^2 + \cdots + \alpha^{p-1}$ and $T = \alpha - 1$.

Suppose i is odd. We define

$$c = \sum_{j=1}^{m} (-1)^{j-1}(e_{i+p-2j+1} \otimes t_{2j-1} - e_{i+p-2j} \otimes t_{2j}).$$

A routine calculation then shows that

$$d(c) = -e_{i+p-1} \otimes t_0 + (-1)^m e_i \otimes Nt_{p-1} ,$$

and hence, by Lemma 3.2.(ii) and (iii)

$$e_{i+p-1} \otimes t_0 \sim (-1)^m e_i \otimes Nt_{p-1} = (-1)^m e_i \otimes d(t_p) = (-1)^{mn+m} m! \; e_i \otimes d(x^p).$$

This establishes the result for odd i.

Now suppose i is even. We define

$$c = \sum_{j=1}^{m} (-1)^{j-1}(Me_{i+p-2j} \otimes t_{2j} + e_{i+p-2j+1} \otimes t_{2j-1})$$

where $M = \alpha^{p-2} + 2\alpha^{p-3} + \cdots + (p-2)\alpha + (p-1)$. One easily checks that $N = TM + p = MT + p$. A routine calculation then shows that

$$d(c) = e_{i+p-1} \otimes t_0 + p\sum_{j=1}^{m} (-1)^{j-1}(e_{i+p-2j} \otimes t_{2j-1} - e_{i+p-2j-1} \otimes t_{2j})$$
$$- (-1)^m e_i \otimes Nt_{p-1} ,$$

from which the result follows for even i by Lemma 3.2.(ii) and (iii) just as for odd i.

In order to prove the compression result (Lemma 3.6) we need to show that, ignoring the Σ_p action, $\Gamma_i(x)$ is just a wedge of suspensions of $X^{(p)}$.

Lemma 3.5. In $\overline{h}\mathcal{J}$ or $\overline{h}\mathcal{S}$, $\Gamma_i(X) \simeq \bigvee_{(p-i,i-1)} \Sigma^{np-i}X^{(p)}$.

Proof. By Definition 2.1 and Lemma 2.2.(i) we may assume $X = S^0$. Again let $\Gamma_i = \Gamma_i(S^0)$. Since $\Gamma_0 = e^{np}$ is contractible, $C_*\Gamma_0$ is exact. It follows that $C_*\Gamma_i$ is exact except in dimension np-i and that

$$H_k\Gamma_i = \begin{cases} 0 & k \neq np-i \\ \ker(C_{np-i}\Gamma_0 \to C_{np-i-1}\Gamma_0) & k = np-i \end{cases}$$

Thus $H_{np-i}\Gamma_i$ is free abelian, being a subgroup of the free abelian group $C_{np-i}\Gamma_0$. By the Hurewicz and Whitehead theorems Γ_i is a wedge of np-i spheres. Splitting $C_*\Gamma_0$ into short exact sequences shows that

$$\operatorname{rank} H_{np-i}\Gamma_i + \operatorname{rank} H_{np-i-1}\Gamma_{i+1} = \operatorname{rank} C_{np-i}\Gamma_0 = (p-i,i).$$

(Recall $(a,b) = (a+b)!/a!b!$). Since $H_{np-1}\Gamma_1$ has rank 1 by Lemma 2.2(v), we see by induction on i that

$$\operatorname{rank} H_{np-i}\Gamma_i = (p-i,i-1).$$

We are now prepared to prove the key result.

Lemma 3.6. The natural inclusion $W^{i+1} \ltimes_\pi \Gamma_{j+1} \to W^{i+1} \ltimes_\pi \Gamma_j$ is homotopic to a map $e: W^{i+1} \ltimes_\pi \Gamma_{j+1} \to W^i \ltimes_\pi \Gamma_j$. In integral homology $e = ee \cdots e: W^{i+p-1} \ltimes_\pi \Gamma_p \to W^i \ltimes_\pi \Gamma_1$ satisfies

(i) $\quad e_*(e_{i+p-1} \otimes (dx)^p) = (-1)^{mn+m} m! e_i \otimes d(x^p) \quad$ if $p > 2$ and i is odd,

(ii) $\quad e_*(e_{i+1} \otimes (dx)^2) = (-1)^i e_i \otimes d(x^2) \quad$ if $p = 2$ and $n \not\equiv i$ (2),

where we denote homology classes by representative cycles. In mod p homology, (i) and (ii) hold for all i and n. In integral homology $e: W^{p-1} \ltimes_\pi \Gamma_p \to W^0 \ltimes_\pi \Gamma_2 \simeq \Gamma_2$ satisfies

(iii) $\quad e_*(e_{p-2} \otimes (dx)^p) = (-1)^{m-1} Te_0 \otimes t_{p-2} \quad$ if $p > 2$.

Proof. The map compresses because $W^{i+1} \ltimes_\pi \Gamma_{j+1}$ is np+i-j dimensional while $W^{i+1} \ltimes_\pi \Gamma_j / W^i \ltimes_\pi \Gamma_j \simeq \bigvee S^{np+i-j+1}$ by the preceding lemma. In order to evaluate e_*, first assume $p > 2$ and consider the commutative triangle,

$$\begin{array}{ccc} W^{i+p-1} \ltimes_{\pi} \Gamma_p & \longrightarrow & W^{i+p-1} \ltimes_{\pi} \Gamma_1 \\ & \searrow^{e} & \uparrow \\ & & W^{i} \ltimes_{\pi} \Gamma_1 \end{array}$$

in which the unlabelled maps are the natural inclusions. In mod p homology the vertical map is an isomorphism, so it suffices to note that $e_{i+p-1} \otimes dx^p \sim (-1)^{mn+m} m! e_i \otimes d(x^p)$ by 3.4. Now assume i is odd. The vertical map is the quotient map $Z \to Z_p$, and the mod p case implies e_* is correct up to a multiple of p. The indeterminacy of the lift from $W^{i+1} \ltimes_\pi \Gamma_1$ to $W^i \ltimes_\pi \Gamma_1$ consists of maps

$$W^{i+p-1} \ltimes_\pi \Gamma_p \xrightarrow{c} S^{np+i-1} \xrightarrow{b} S^{np+i-1} \xrightarrow{a} W^i \ltimes_\pi \Gamma_1$$

in which c is projection onto the top cell, b is arbitrary, and a is the attaching map of the np+i cell. On integral homology c_* is the identity and a_* is multiplication by p. Thus it is possible to choose the lift e such that e_* is as stated in integral homology. (This is a general fact about maps obtained by cellular approximation, but we only need it here so do not bother with the general statement.)

The argument for $p = 2$ is exactly analogous to that just given.

§4. Reduction to three cases

In this section we start with an overview of the proof, then establish notations which we shall use in the remainder of this chapter, and finally start the proof of Theorems 1.1, 1.2 and 1.3 by showing that it splits into three parts and by proving some results which will be used in all three.

If $\Gamma_j = \Gamma_j(S^{n-1})$ as in Section 2, we would like to prove Theorems 1.1, 1.2 and 1.3 by doing appropriate calculations in a spectral sequence $E_r(S, \mathcal{D})$ where $\mathcal{D}$ is an inverse sequence constructed from the $W^i \ltimes_{\Sigma_p} \Gamma_j$'s. However, there are technical difficulties which have prevented this. If a proof can be constructed along these lines, it should immediately imply that T_p (see Theorem 1.2) is a linear combination of $\beta^\delta P^{j-i}x$ and $x^{p-k}(d_r x)^k$ for various δ, i and k, with coefficients in $E_2(S,S)$. The coefficient of the lowest filtration term would be $\bar{a}$, and the determination of the other coefficients would give complete information on the first possible nonzero differential on $\beta^\varepsilon P^j x$.

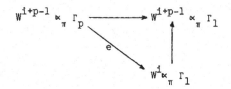

The proof we give runs as follows. The spectrum $W \ltimes_{\Sigma_p} \Gamma_j$ is a wedge summand of $W \ltimes_\pi \Gamma_j$, $\pi \subset \Sigma_p$ cyclic of order p. In a very convenient abuse of notation, we will write $D^i\Gamma_j$ for the np + i-j skeleton of this summand. There is a homotopy equivalence of (e^{k+np}, S^{k+np-1}) with $(D^k\Gamma_0, D^{k-1}\Gamma_0 \cup D^k\Gamma_1)$. The element $\beta^\varepsilon P^j x$ is represented by a map of $(D^k\Gamma_0, D^{k-1}\Gamma_0 \cup D^k\Gamma_1)$ into the Adams resoluton of our H_∞ ring spectrum Y. Thus, we must study lifts of the boundary $D^{k-1}\Gamma_0 \cup D^k\Gamma_1$ in order to compute $d_*\beta^\varepsilon P^j x$. Since $D^k\Gamma_1$ is homotopy equivalent to the stunted lens space $\Sigma^n L_{n(p-1)}^{n(p-1)+k}$ and $D^k\Gamma_0$ is the cone on $D^k\Gamma_1$, $D^{k-1}\Gamma_0 \cup D^k\Gamma_1 \simeq D^k\Gamma_1/D^{k-1}\Gamma_1 \simeq S^{k+np-1}$. Now $D^{k+p-1}\Gamma_p$ is also a stunted lens space and the natural inclusion $D^{k+p-1}\Gamma_p \to D^{k+p-1}\Gamma_1$ factors through $D^k\Gamma_1$ (Lemma 3.6). The resulting map $D^{k+p-1}\Gamma_p \to D^k\Gamma_1$ is equivalent to the cofiber of the inclusion of the bottom cell of $D^{k+p-1}\Gamma_p$. Thus $D^k\Gamma_1/D^{k-1}\Gamma_1 \simeq D^{k+p-1}\Gamma_p/D^{k+p-2}\Gamma_p$. The top cell of $D^{k+p-1}\Gamma_p$ carries the element $\beta^\varepsilon P^j d_r x$ and this is where this term comes from. The other term comes in because we are given a map of $D^{k-1}\Gamma_0 \cup D^k\Gamma_1$, not $D^k\Gamma_1/D^{k-1}\Gamma_1$, into the Adams resolution. Thus we must find another cell whose boundary is the same as the boundary of the top cell of $D^k\Gamma_1$ or $D^{k+p-1}\Gamma_p$, and we must lift it until it detects an element in homotopy or until it has filtration higher than that of $\beta^\varepsilon P^j d_r x$. Since $D^i\Gamma_0 \equiv CD^i\Gamma_1$, we can simply cone off the attaching map of the top cell of $D^k\Gamma_1$ as long as this cell is nontrivially attached. This produces the terms $\bar{a}P^{j-v}x$, $\bar{a}\beta P^{j-e-1}x$ and $a_0\beta P^j x$. If the top cell of $D^k\Gamma_1$ is unattached, the top cell of $D^{k+p-1}\Gamma_p$ may still be attached to the cell $D^{p-2}\Gamma_p$. There is a nullhomotopy of this cell in Γ_1 which carries $x^{p-1}d_r x$. This is the source of the terms $\bar{a}x^{p-1}d_r x$. Finally, when the top cell of $D^{k+p-1}\Gamma_p$ is unattached, it carries the entire boundary.

There are two complications to the above picture. First, the map $D^{k+p-1}\Gamma_p \to D^k\Gamma_1$ is a lift of the natural inclusion $D^{k+p-1}\Gamma_p \to D^{k+p-1}\Gamma_1$ and does not commute with the maps into the Adams resolution until we pass to a lower filtration. This necessitates extra work at some points. Second, the attaching map ataches the top cell to the whole lens space, not just to the cell carrying $P^{j-v}x$ or $\beta P^{j-e-1}x$. As the filtration of $\bar{a}$ increases, the possibility arises that a piece of the attaching map which attaches to a lower cell will show up in a lower filtration than the term $\bar{a}P^{j-v}x$ or $\bar{a}\beta P^{j-e-1}x$. This possibility accounts for the cases in which we do not have complete information.

Now let us establish notation to be used in this and the remaining sections. As in section 1 we assume given a p-local H_∞ ring spectrum Y and an element $x \in E_r^{s,n+s}(S,Y)$, the E_r term of the ordinary Adams spectral sequence converging to π_*Y. We wish to describe the first nontrivial differential on $\beta^\varepsilon P^j x$ in terms of x and $d_r x$. (Here $\varepsilon = 0$ if $p = 2$.) Recall from §1 the definition

$$k = \begin{cases} j-n & p = 2 \\ (2j-n)(p-1) - \varepsilon & p > 2 \end{cases}$$

Let

$$Y \simeq Y_0 \leftarrow Y_1 \leftarrow Y_2 \leftarrow \cdots$$

be an Adams resolution of Y and let

$$Y^{(p)} \simeq Y_0^{(p)} = F_0 \leftarrow F_1 \leftarrow F_2 \leftarrow \cdots$$

be its pth power as in IV.4. Represent x by a map $(e^n, S^{n-1}) \to (Y_s, Y_{s+r})$ and let $\Gamma_i = \Gamma_i(S^{n-1})$ be the ith filtration of $\Gamma_0 = e^{np}$ as in Definition 2.1. Recall that the spectrum $W \ltimes_{\Sigma_p} \Gamma_i$ is a wedge summand of $W \ltimes_\pi \Gamma_i$ where $\pi \subset \Sigma_p$ is cyclic of order p. In the remainder of this chapter, $D^k\Gamma_i$ will denote the np+k-i skeleton of this summand. Let us use ξ generically to denote the composites

$$\xi_{k,ps+ir}(1 \ltimes x^p): D^k\Gamma_i \to W^k \ltimes_\pi \Gamma_i \to W^k \ltimes_\pi F_{ps+ir} \to Y_{ps+ir-k},$$

the maps of pairs and unions constructed from them, and their composites with the maps $Y_{j+t} \to Y_j$. We will use the following consequence of Lemma 3.6 repeatedly. Recall that e is defined in Lemma 3.6.

<u>Lemma 4.1</u>. The following diagram commutes.

$$\begin{array}{ccc}
D^{k+p-1}\Gamma_p & \xrightarrow{e} & D^k\Gamma_1 \\
 & & \downarrow \xi \\
\downarrow \xi & & Y_{ps+r-k} \\
 & & \downarrow \\
Y_{ps+pr-k-p+1} & \longrightarrow & Y_{ps+r-k-1}
\end{array}$$

<u>Proof</u>. In the diagram below, the triangle commutes because $r \geq 1$ and the quadrilateral commutes by Lemma 3.6.

$$\begin{array}{ccc}
D^{k+1}\Gamma_{j+1} & \longrightarrow & D^k\Gamma_j \\
\downarrow & & \downarrow \\
Y_{ps+(j+1)r-k-1} & & Y_{ps+jr-k} \\
\downarrow & & \downarrow \\
Y_{ps+(j+1)r-k-2} & \longrightarrow & Y_{ps+jr-k-1}
\end{array}$$

The lemma follows by composing the diagrams for $j = 1,2,\ldots,p-1$.

In IV.2 we constructed a chain homomorphism $\phi: \mathcal{W} \otimes \mathcal{C}^p \to \mathcal{C}$, where is the cobar construction, which we used to construct Steenrod operations, and in IV.5 we showed that ξ induces such a homomorphism. In particular, Definition IV.2.4 says

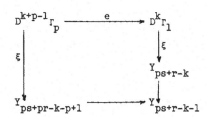

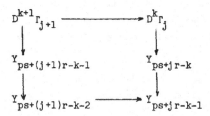

$$\beta^{\varepsilon}P^{j}x = (-1)^{j}\nu(n)\Phi_*(e_k \otimes x^p) \qquad p > 2$$

and
$$Sq^{j}x = \Phi_*(e_k \otimes x^2) \qquad p = 2.$$

The following relative version of Corollary IV.5.4 gives us maps which represent these elements. In it we let $\mathcal{C}$ be the cobar construction $C(Z_p, \mathcal{A}_p, H_*Y)$ so that $\mathcal{C}_{s,n+s} \cong \pi_n(Y_s/Y_{s+1}) \cong \pi_n(Y_s, Y_{s+1})$ and let $\mathcal{W} = C_*(W)$ so that $\mathcal{W}_k = C_k(W) \cong \pi_k(W^k/W^{k-1}) \cong \pi_k(W^k, W^{k-1})$.

Lemma 4.2. If $e \in \mathcal{W}_k$ is represented by $e \in \pi_k(W^k, W^{k-1})$ then $\Phi_*(e \otimes x^p)$ is represented by the composite

$$\begin{array}{ccc}
(e^{np+k}, S^{np+k-1}) & \xrightarrow{\ \Phi_*(e\otimes x^p)\ } & (Y_{ps-k}, Y_{ps-k+1}) \\
\wr\!\wr & & \\
(e^k \ltimes \Gamma_0, e^k \ltimes \Gamma_1 \cup S^{k-1} \ltimes \Gamma_0) & & \\
\Big\downarrow e \ltimes 1 & & \Big\uparrow \xi \\
(W^k \ltimes \Gamma_0, W^k \ltimes \Gamma_1 \cup W^{k-1} \ltimes \Gamma_0) & & \\
\Big\downarrow u & & \\
(W^k \ltimes_\pi \Gamma_0, W^k \ltimes_\pi \Gamma_1 \cup W^{k-1} \ltimes_\pi \Gamma_0) & \xrightarrow[1 \ltimes_\pi x^p]{} & (W^k \ltimes_\pi F_{ps}, W^k \ltimes_\pi F_{ps+r} \cup W^{k-1} \ltimes_\pi F_{ps})
\end{array}$$

where u is the passage to orbits map.

Note: If $e \in \mathcal{W}_k$ is a $Z[\pi]$ generator (e.g. $e = \alpha^i e_k$ for some i) then the vertical composite in the diagram is an equivalence by the same argument which was used to construct diagrams (2.1) and (2.2).

Proof. This is simply the relative version of Corollary IV.5.4. The natural isomorphism $\pi_*(X,A) \cong \pi_*(X/A)$ for cofibrations $A \to X$ enable one to pass freely between this version and the absolute version of IV.5.4.

We shall refer to the boundary of the map in Lemma 4.2 so frequently that we give it a name.

Definition 4.3. Let $\partial\Phi \in \pi_{np+k-1}Y_{ps-k+1}$ be the restriction to S^{np+k-1} of the map $\Phi_*(e_k \otimes x^p)$ of Lemma 4.2. Let $\iota \in \pi_{np+k-1}(D^k\Gamma_1 \cup D^{k-1}\Gamma_0)$ be the map with Hurewicz image

$$(-1)^k e_k \otimes d(x^p) + \begin{cases} 0 & k=0 \text{ or } k \text{ odd, } p > 2 \\ 0 & k+n \text{ odd, } p = 2 \\ pe_{k-1} \otimes x^p & 0 \neq k \text{ even, } p > 2 \\ (-1)^k 2e_{k-1} \otimes x^2 & k+n \text{ even, } p = 2 \end{cases}$$

Lemma 4.4. (i) $\partial\Phi = \xi_*(\iota)$

(ii) ι is an equivalence

(iii) Orienting the top cell of $D^k\Gamma_1$ correctly, the homotopy class ι contains the map a_3 of diagram (2.2).

Proof (i) holds because we are in the Hurewicz dimension of $D^k\Gamma_1 \cup D^{k-1}\Gamma_0 \simeq S^{np+k-1}$ so the Hurewicz image of ι is sufficient to determine ι, and its Hurewicz image is the boundary of the cell $e_k \otimes x^p$. Statement (ii) is immediate from the Hurewicz isomorphism, and statement (iii) is immediate from the fact that a_3 is an equivalence.

The differentials on $\beta^\varepsilon P^j x$ are given by the successive lifts of $(-1)^j \nu(n)\partial\Phi$ when $p > 2$, and of $\partial\Phi$ when $p = 2$. Corollary 2.8 and the discussion following it show that the attaching maps of lens spaces, and hence elements of Im J, enter into the question of lifting this boundary. In the remainder of this section we establish various facts about the numerical relations between the filtrations and dimensions involved, the last of which will enable us to split our proof into three very natural special cases.

Lemma 4.5. If $p > 2$, the generator of Im J in dimension $jq-1$ has filtration $\leq j$. If $p = 2$ the generator of Im J in dimension $8a+\varepsilon$ $(\varepsilon = 0,1,3,7)$ has filtration $\leq 4a+\varepsilon$ if $\varepsilon \neq 7$, and $\leq 4a+4$ if $\varepsilon = 7$.

Proof. The vanishing theorem for $\mathrm{Ext}_{A_p}(Z_p,Z_p)$ says that $\mathrm{Ext}^{st} = 0$ if $0 < t-s < U(s)$, where $U(s) = qs-2$ if $p > 2$ and

$$U(4a+\varepsilon) = \begin{cases} 8a-1 & \varepsilon = 0 \\ 8a+1 & \varepsilon = 1 \\ 8a+2 & \varepsilon = 2 \\ 8a+3 & \varepsilon = 3 \end{cases}$$

if $p = 2$ by [4] and [56]. First suppose $p > 2$. The Im J generator in dimension $jq-1$ is detected by an element of $\mathrm{Ext}^{s,t}$ where $t-s = jq-1$. Hence $jq-1 \geq U(s) = sq-2$, which implies $j \geq s$. Now, suppose $p = 2$. A trivial calculation shows that if

$s > 4a + \varepsilon$, $\varepsilon = 0,1,3,4$, then $U(s) > 8a + \varepsilon$ if $\varepsilon \neq 4$, $8a + 7$ if $\varepsilon = 4$. This immediately implies the lemma.

We apply this to prove the following three lemmas. As in §1 let v be $v_p(k + n(p-1))$, and let f be the Adams filtration of the generator of Im J in $\pi_{v-1}S^0$.

Lemma 4.6. Assume $p > 2$. If $v = k+1$ and $f \geq r-1$ then $pr-p-k+1 < 2r-1$.

Proof. Equivalently, we must show $k > (p-2)(r-1)$. By Lemma 4.5

$$f \leq \frac{v}{q} = \frac{k+1}{q} .$$

Thus $k+1 \geq qf \geq q(r-1)$ and hence it is sufficient to show that $q(r-1) - 1 > (p-2)(r-1)$. This is immediate since $r > 1$.

Lemma 4.7. Either $\min\{pr-p+1, v+f\} < v+r-1$ or $r = p = 2$ and $v = 1$ or 2.

Proof. Suppose $p > 2$. Then $f \leq v/q$. If $pr-p+1 \geq v+r-1$ then $v \leq (p-1)(r-1) + 1$ and hence

$$f \leq \frac{r-1}{2} + \frac{1}{q} < r-1.$$

Now suppose $p = 2$. We must show that if $r \geq v$ then $f < r-1$. It suffices to show $f < v-1$. This follows from Lemma 4.5 except when $v = 1,2$, or 4. In these cases $f = 1$ so the lemma holds when $v = 4$. If $v = 1$ or 2 then $f < r-1$ unless $r = 2$. This completes the lemma.

Lemma 4.8. Exactly one of the following holds:

(a) $v > k + p-1$,

(b) $v = k+1$ and if $p > 2$ then n is even,

(c) $v \leq k$.

Proof. There is nothing to prove if $p = 2$, so assume $p > 2$. We must show that if $k < v \leq k+p-1$ then $v = k+1$ and n is even. Recall that $k = (2j-n)(p-1)-\varepsilon$ and $v = v_p(k+n(p-1)) = v_p(2j(p-1)-\varepsilon)$. If $\varepsilon = 0$ then $v = 1$. Hence $k = 0$ and $n = 2j$ so that (b) holds as required. If $\varepsilon = 1$ then $v = q(1 + \varepsilon_p(j))$. Dividing the inequalities $k < v \leq k+p-1$ by $p-1$ yields

$$2j-n- \frac{1}{p-1} < 2(1+\varepsilon_p(j)) \leq 2j-n - \frac{1}{p-1} + 1$$

which has only one solution: $2(1 + \varepsilon_p(j)) = 2j-n$. Hence n is even and $v = q(1+\varepsilon_p(j)) = (2j-n)(p-1) = k+1$.

Lemma 4.8 is a consequence of the splitting of the mod p lens space into wedge summands, the summand of interest to us being the Σ_p extended power of a sphere. To see the relation, recall that v tells us how far we can compress the attaching map of the top cell of $W^k \ltimes_\pi \Gamma_1 \simeq \Sigma^{n-1} L_{n(p-1)}^{n(p-1)+k}$. When $v \leq k$, it compresses to $W^{k-v} \ltimes_\pi \Gamma_1$ and no further. When $v > k$ it is not attached to $W^k \ltimes_\pi \Gamma_1$. However, recall that there are equivalences

$$\begin{array}{ccc} W^{k+p-1} \ltimes_\pi \Gamma_p & \simeq & \Sigma^{n-1} L_{(n-1)(p-1)}^{n(p-1)+k} \\ \downarrow & & \downarrow \\ W^k \ltimes_\pi \Gamma_1 & \simeq & \Sigma^{n-1} L_{n(p-1)}^{n(p-1)+k} \end{array}$$

by Corollary 2.6, and that the top cell of $W^k \ltimes_\pi \Gamma_1$ is the image of the top cell of $W^{k+p-1} \ltimes_\pi \Gamma_p$ by Lemma 3.6. When $v > k$ this cell compresses to $W^{p-2} \ltimes_\pi \Gamma_p$. The first possibility is that it goes no further, and in this case the wedge summand of the lens space we are interested in has cells in dimensions n(p-1) and n(p-1)-1 so that n must be even. By the splitting of the lens space into wedge summands, the next possibility is $v = k+p-1$, which would have the top cell of $W^{k+p-1} \ltimes_p \Gamma_p$ attached to the bottom cell. In fact this cannot happen because the attaching map is in Im J and thus is not in an even stem. So $v > k+p-1$ is the only possibility if $v > k+1$, and this says that top cells of $W^{k+p-1} \ltimes_\pi \Gamma_p$ and $W^k \ltimes_\pi \Gamma_1$ are unattached. This "geometry" explains why the differentials on $\beta^\varepsilon P^j x$ are so different in these three cases. We shall start with the simplest of the three cases, and proceed to the most complicated.

§5. Case (a): $v > k+p-1$

Since $v > k+p-1 \geq 1$, it follows that $\varepsilon = 1$ if $p > 2$. Thus Theorems 1.1 and 1.2 say that

$$d_{2r-1} P^j x = P^j d_r x \qquad \text{if } p = 2$$

and

$$d_{pr-p+1} \beta P^j x = -\beta P^j d_r x \quad \text{if } p > 2.$$

Theorem 1.3 follows automatically from these facts, so these are what we shall establish.

By Lemma 4.1, the following diagram commutes.

$$\begin{array}{ccccc}
D^{k+p-1}\Gamma_p & \xrightarrow{\ e\ } & D^k\Gamma_1 & \longrightarrow & D^k\Gamma_1 \cup D^{k-1}\Gamma_0 \\
\Big\downarrow \xi & & \Big\downarrow \xi & & \Big\downarrow \xi \\
 & & E_{ps+r-k} & & \\
 & & \Big\downarrow & & \\
E_{ps+pr-k-p+1} & \longrightarrow & E_{ps+r-k+1} & \longrightarrow & E_{ps-k+1}
\end{array}$$

Because $v > k+p-1$, the top cell of $D^{k+p-1}\Gamma_p$ is not attached (Corollary 2.6 and Definition V.2.15). Thus there exists a reduction $\rho \in \pi_{np+k-1}(D^{k+p-1}\Gamma_p)$ whose Hurewicz image is $e_{k+p-1} \otimes dx^p$ (it is easy to check that $e_{k+p-1} \otimes dx^p$ generates H_{np+k-1}). Also, $v > k+p-1 \geq 1$ immplies that k is odd if $p > 2$ and that $k+n$ is odd if $p = 2$ by Proposition V.2.16. Combining Lemmas 3.6 and 4.4 we find that $\xi_*(\rho)$ is a lift of $\partial\Phi$ when $p = 2$, and of $(-1)^{mn+m-1}m!\partial\Phi$ when $p > 2$. Applying Lemma 4.2 or Corollary IV.5.4 we see that $\xi_*(\rho)$ represents $\Phi_*(e_{k+p-1} \otimes dx^p)$. Thus, if $p = 2$ we have

$$d_{2r-1}P^j x = \xi_*(\rho) = \Phi_*(e_{k+1} \times dx)^2 = P\, d^j x_r.$$

If $p > 2$, we have

$$d_{pr-p+1}\beta P^j x = (-1)^j \nu(n)(-1)^{mn+m-1}\frac{1}{m!}\xi_*(\rho)$$

$$= (-1)^{mn+m-1}(\nu(n)/m!\,\nu(n-1))\beta P^j d_r x.$$

It is easy to check that $\nu(n)/m!\,\nu(n-1) \equiv (-1)^{mn+m} \bmod p$ so that $d_{pr-p+1}\beta P^j x = -\beta P^j d_r x$.

§6. Case (b): $v = k+1$

We will begin by considering $p = 2$. Theorems 1.1 and 1.2 say that

$$\begin{array}{ll}
d_{2r-1}P^j x = P^j d_r x & \text{if } 2r-1 < r+f+k, \\
d_{2r-1}P^j x = P^j d_r x + \bar{a} x d_r x & \text{if } 2r-1 = r+f+k, \text{ and} \\
d_{r+f+k}P^j x = \bar{a} x d_r x & \text{if } 2r-1 > r+f+k.
\end{array}$$

Since the filtration f of $\bar{a}$ is positive and $r \geq 2$, Theorem 1.3 follows from Theorems 1.1 and 1.2.

Let $N = k+2n-1$ and let $C_2 \in \pi_N(D^{k+1}\Gamma_2, \Gamma_2)$ be the top cell of $D^{k+1}\Gamma_2$ with its boundary compressed as far as it will go. Then the Hurewicz image

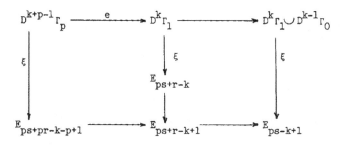

$h(C_2) = e_{k+1} \otimes dx^2$ and $\partial C_2 = a = a_2(k+n) \in \pi_{N-1}\Gamma_2 \cong \pi_k S^0$. Since $\Gamma_2 \simeq S^{2n-2}$ and $\Gamma_1/\Gamma_2 \simeq S^{2n-1} \vee S^{2n-1}$ by Lemma 2.2, the Hurewicz homomorphisms in

$$\begin{array}{ccc} \pi_{2n-1}(\Gamma_1,\Gamma_2) & \xrightarrow{\ h\ } & H_{2n-1}(\Gamma_1,\Gamma_2) \\ \Big\downarrow \partial & & \Big\downarrow \partial \\ \pi_{2n-2}\Gamma_2 & \xrightarrow{\ h\ } & H_{2n-2}\Gamma_2 \end{array}$$

are isomorphisms. Let $R \in \pi_{2n-1}(\Gamma_1,\Gamma_2)$ satisfy $h(R) = x\,dx = e_0 \otimes x\,dx$ in the notation of §3. Then $\partial R \in \pi_{2n-2}\Gamma_2$ is an equivalence since $h(\partial R) = dx^2 = e_0 \otimes dx^2$. Let a also denote $(Ca,a) \in \pi_N(e^{2n-1},S^{2n-2})$. Let i be the natural inclusion $i:(\Gamma_1,\Gamma_2) \to (D^{k-1}\Gamma_0,\Gamma_2)$ if $k > 0$ and let $i = 1:(\Gamma_1,\Gamma_2) \to (\Gamma_1,\Gamma_2)$ if $k = 0$. Let eC_2 denote $(e,1)_*(C_2) \in \pi_N(D^k\Gamma_1,\Gamma_2)$.

Lemma 6.1: $\partial\Phi = \xi_*(eC_2 \cup iRa)$ in $\pi_N Y_{2s-k+1}$.

Proof. First note that $eC_2 \cup iRa$ is defined since $\partial C_2 = \partial(iRa) = a \in \pi_{N-1}\Gamma_2$. By Lemma 4.4, $\partial\Phi = \xi_*(eC_2 \cup iRa)$ will follow if $eC_2 \cup iRa \in \pi_N(D^k\Gamma_1 \cup D^{k-1}\Gamma_0)$ has Hurewicz image $(-1)^k e_k \otimes d(x^2)$, since $v_2(k+n) = k+1$ implies that either $k+n$ is odd or $k = 0$. If $k \neq 0$ then $\pi: D^k\Gamma_1 \cup D^{k-1}\Gamma_0 \to D^k\Gamma_1/D^{k-1}\Gamma_1$ is an equivalence and Lemma 2.7 says that $\pi(eC_2 \cup iRa) = \overline{eC_2} \in \pi_N D^k\Gamma_1/D^{k-1}\Gamma_1$ since iRa factors through $D^{k-1}\Gamma_1$. Then $h(\overline{eC_2}) = e_* h(C_2) = (-1)^k e_k \otimes d(x^2)$ by Lemma 3.6 (since $k+n$ is odd) and we are done. If $k = 0$ then n is even, since $v_2(n) = 1$, and $eC_2 \cup Ra \in \pi_{2n-1}\Gamma_1$. Also, $a = -2 \in \pi_{2n-2}S^{2n-2}$ since $h(\partial C_2) = d(e_1 \otimes dx^2) = (\alpha-1)e_0 \otimes dx^2 =$ $= -2e_0 \otimes dx^2$. To compute $h(eC_2 \cup Ra)$, project to Γ_1/Γ_2 since $H_{2n-1}\Gamma_1 \to H_{2n-1}\Gamma_1/\Gamma_2$ is the monomorphism which sends $e_0 \otimes d(x^2)$ to $e_0 \otimes xdx + e_0 \otimes dx\,x$. By Lemma 2.7, $\pi(eC_2 \cup Ra): S^{2n-1} \to \Gamma_1 \to \Gamma_1/\Gamma_2$ equals $\overline{eC_2} - \overline{Ra}$ so

$$\begin{aligned} h(\pi(eC_2 \cup Ra)) &= h(\overline{eC_2}) - h(\overline{Ra}) \\ &= e_*(e_1 \otimes dx^2) + 2e_0 \otimes xdx \\ &= e_0 \otimes (dx)x - e_0 \otimes xdx + 2e_0 \otimes xdx \\ &= e_0 \otimes (dx)x + e_0 \otimes xdx. \end{aligned}$$

Therefore $h(eC_2 \cup Ra) = e_0 \otimes d(x^2)$ and we're done, proving Lemma 6.1.

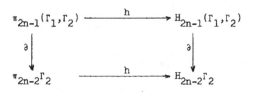

Since $\xi_*\partial C_2 \in \pi_*Y_{2s+2r}$, $\xi_*(eC_2 \cup iRa) = \xi_*(eC_2) - \xi_*(iRa)$ in $\pi_*(Y_{2s-k+1}, Y_{2s+2r})$. By Lemma 4.1 (or 3.6), $\xi_*(eC_2)$ and ξ_*C_2 have the same image in $\pi_*(Y_{2s-k+1}, Y_{2s+2r})$. Since $h(C_2) = e_{k+1} \otimes dx^2$, $\xi_*C_2 \in \pi_*(Y_{2s-k+2r-1}, Y_{2s+2r})$ represents $P^j d_r x$ by Lemma 4.2. Similarly, $h(R) = e_0 \otimes x\, dx$ implies that $\xi_*R \in \pi_*(Y_{2s+r}, Y_{2s+2r})$ represents xd_rx, and hence $\xi_*(Ra) \in \pi_*(Y_{2s+r+f}, Y_{2s+2r})$ represents $\bar{a}xd_rx$. This completes case (b) when $p = 2$.

When $p > 2$ (and $v = k+1$) we will treat $k = 0$ and $k > 0$ separately. First suppose $k = 0$. Then $v = 1$, $n = 2j$ and $\varepsilon = 0$. Also, $f = 1$, $\bar{a} = a_0 \in E_\infty^{1,1}(S,S)$ and $a \in \pi_0 S$ is the map of degree p. Thus, we must show

$$d_{r+1}x^p = a_0 x^{p-1} d_r x.$$

Heuristically this is exactly what one would expect from the fact that $d_r x^p = p(x^{p-1}d_r x)$. That this is too casual is shown by the fact that we have just proved (for $p = 2$) that

$$d_3 x^2 = h_0 x d_2 x + P^n d_2 x.$$

The extra term arises because when we lift the map representing $2xd_2x$ to the next filtration, we find also the map representing $P^n d_2 x$ which we added in order to replace $xd_2x + (d_2x)x$ by $2xd_2x$. Thus, our task for $p > 2$ is to show the analogous elements can always be lifted to a higher filtration than that in which $a_0 x^{p-1} d_r x$ lies. The following lemma will do this for us.

Lemma 6.2. There exists elements

$$C_1 \in \pi_{np-1}\Gamma_1 \qquad Y \in \pi_{np-1}(D^1\Gamma_2, \Gamma_2 \cup D^1\Gamma_3)$$
$$X \in \pi_{np-1}(\Gamma_1, \Gamma_2) \qquad Z \in \pi_{np-1}(D^2\Gamma_3, D^1\Gamma_3 \cup D^2\Gamma_4)$$

such that

$$C_1 = pX + pY + Z \quad \text{in} \quad \pi_{np-1}(D^1\Gamma_1 \cup D^2\Gamma_2, \Gamma^2 \cup D^1\Gamma_3 \cup D^2\Gamma_4),$$
$$h(C_1) = e_0 \otimes d(x^p), \text{ and}$$
$$h(X) = e_0 \otimes x^{p-1}dx.$$

Proof. Since $np-1$ is the Hurewicz dimension of all the spectra or pairs of spectra involved, we may define C_1, X, Y and Z by their Hurewicz images. Thus C_1 and X are given, and we let

$$h(Y) = \frac{1}{m} e_1 \otimes Qd(x^{p-1})dx - \frac{1}{m!} e_1 \otimes t_{p-2}, \text{ and}$$
$$h(Z) = -\frac{1}{m!} e_2 \otimes Nt_{p-3}.$$

As in section 3, $N = \sum \alpha^i$ and $Q = (\alpha+1)\sum_{i=1}^{m} i\alpha^{2i}$. We also let $M = \sum i\alpha^{p-i-1}$ and note that $M(\alpha-1) = N-p$. Define

$$C = \frac{1}{m!}(Me_1 \otimes t_{p-1} + e_2 \otimes t_{p-2}) + \frac{p}{m} e_1 \otimes Qx^{p-1}\, dx$$

in $C_*(D^1\Gamma_1 \cup D^2\Gamma_2, \Gamma_1 \cup D^1\Gamma_2 \cup D^2\Gamma_3)$. By Lemma 3.2 it follows that

$$d(C) = h(C_1) - ph(X) - ph(Y) - h(Z)$$

which shows that $C_1 = pX + pY + Z$.

By Lemmas 4.4 and 6.2, $\partial\Phi \in \pi_*Y_{ps+1}$ is the image of $\xi_*C_1 \in \pi_*Y_{ps+r}$. Lemma 6.2 also implies that

$$\xi_*C_1 = p\xi_*X + p\xi_*Y + \xi_*Z$$

in $\pi_*(Y_{ps+r-1}, Y_{ps+2r})$. Since $\xi_*Y \in \pi_*(Y_{ps+2r-1}, Y_{ps+2r})$ and $\xi_*Z \in \pi_*(Y_{ps+3r-2}, Y_{ps+3r-1})$ it follows that $\xi_*C_1 = p\xi_*X$ in $\pi_*(Y_{ps+r-1}, Y_{ps+2r})$ and that $\partial\Phi = p\xi_*X$ in $\pi_*(Y_{ps+1}, Y_{ps+2r})$. Lemma 4.2 implies that $\xi_*X \in \pi_*(Y_{ps+r}, Y_{ps+2r})$ represents $x^{p-1}d_rx$ and hence $p\xi_*X$ lifts to $\pi_*(Y_{ps+r+1}, Y_{ps+2r})$ where it represents $a_0x^{p-1}d_rx$. Finally, IV.3.1 implies

$$d_{r+1}P^jx = d_{r+1}x^p = a_0x^{p-1}d_rx.$$

Now suppose that $k > 0$. Then $v = k+1$ is greater than 1 and hence congruent to 0 mod $2(p-1)$ by V.2.16. Also by V.2.16, $\varepsilon = 1$ and $k = (2j-n)(p-1)-\varepsilon$ is therefore odd. Lemma 4.4 then implies $\partial\Phi = \xi_*(\iota)$ with $h(\iota) = -e_k \otimes d(x^p)$. The next three lemmas describe the pieces into which we will decompose $\partial\Phi$. In the first we define an element of π_{np-1} of the cofiber of $e: D^{p-2}\Gamma_p \to \Gamma_1$, which we think of as an element of a relative group $\pi_{np-1}(\Gamma_1, D^{p-2}\Gamma_p)$. In order to specify the image of such an element under the Hurewicz homomorphism, we use the cellular chains of the cofiber in the guise of the mapping cone of $e_*: C_*D^{p-2}\Gamma_p \to C_*\Gamma_1$. That is, we let

$$C_i(\Gamma_1, D^{p-2}\Gamma_p) = C_i\Gamma_1 \oplus C_{i-1}D^{p-2}\Gamma_p$$

with $d(a,b) = (d(a) - e_*(b), -d(b))$.

<u>Lemma 6.3</u>. There exists $R \in \pi_{np-1}(\Gamma_1, D^{p-2}\Gamma_p)$ such that

(i) $h(R) = ((-1)^{m-1}e_0 \otimes t_{p-1}, e_{p-2} \otimes t_0) \in H_*(\Gamma_1, D^{p-2}\Gamma_p)$

(ii) $h(\partial R) = e_{p-2} \otimes t_0 = e_{p-2} \otimes (dx)^p$, and

(iii) $\partial R \in \pi_{np-2}D^{p-2}\Gamma_p$ is an equivalence.

Proof. Since $d(e_0 \otimes t_{p-1}) = Te_0 \otimes t_{p-2}$ by Lemma 3.2 and $e_*(e_{p-2} \otimes t_0) = (-1)^{m-1}Te_0 \otimes t_{p-2}$ by Lemma 3.6.(iii), and since $d(e_{p-2} \otimes t_0) = 0$, it follows that $((-1)^m e_0 \otimes t_{p-1}, e_{p-2} \otimes t_0)$ is a cycle of $(\Gamma_1, D^{p-2}\Gamma_p)$. Since $\Gamma_1 \simeq S^{np-1}$ and $D^{p-2}\Gamma_p \simeq S^{np-2}$, the Hurewicz homomorphism is onto and R satisfying (i) exists. Now (ii) is obvious since the boundary homomorphism simply projects onto the second factor. Part (iii) is immediate from the fact that $e_{p-2} \otimes t_0$ generates $H_{np-2}D^{p-2}\Gamma_p$.

Now we split R into a piece we want and another piece modulo Γ_2.

Lemma 6.4. There exist $X \in \pi_{np-1}(\Gamma_1, \Gamma_2)$ and $Y \in \pi_{np-1}(D^1\Gamma_2, \Gamma_2)$ such that

(i) $h(X) = (-1)^{m-1} m! e_0 \otimes x^{p-1}dx$, and

(ii) $(i,e)_*(R) = i_*X + j_*Y$ in $\pi_*(D^1\Gamma_1, \Gamma_2)$ where $i:\Gamma_1 \to D^1\Gamma_1$, $j:D^1\Gamma_2 \to D^1\Gamma_1$ and $e:D^{p-2}\Gamma_p \to \Gamma_2$.

Proof. We are working in the Hurewicz dimension of all the pairs involved so it suffices to work in homology. We define X by (i) and define Y by

$$h(Y) = (-1)^{m-1}(m-1)! e_1 \otimes Qd(x^{p-1})dx.$$

On cellular chains, the map $(i,e):(\Gamma_1, D^{p-2}\Gamma_p) \to (D^1\Gamma_1, \Gamma_2)$ induces the homomorphism

$$C_k\Gamma_1 \oplus C_{k-1}D^{p-2}\Gamma_p \longrightarrow C_k\Gamma_1 \xrightarrow{i_*} C_kD^1\Gamma_1 \longrightarrow C_kD^1\Gamma_1/C_k\Gamma_2$$

in which the unlabelled maps are the obvious quotient maps. Thus, denoting equivalence classes by representative elements,

$$\begin{aligned} h((i,e)_*R) &= (-1)^{m-1}e_0 \otimes t_{p-1} \\ &= (-1)^{m-1}m! e_0 \otimes x^{p-1}dx + (-1)^{m-1}(m-1)! Te_0 \otimes Qx^{p-1}dx \end{aligned}$$

by Lemma 3.2. Since

$$d(e_1 \otimes Qx^{p-1}dx) = Te_0 \otimes Qx^{p-1}dx - e_1 \otimes Qd(x^{p-1})dx,$$

it follows that $h((i,e)_*R) = h(i_*X + j_*Y)$.

In our last lemma we split $\partial\Phi$ into two pieces modulo $D^{p-2}\Gamma_p$. Let $N = k+np-1$.

Lemma 6.5. If $v = k+1$ and $k > 0$, and if $C_p \in \pi_N(D^{k+p-1}\Gamma_p, D^{p-2}\Gamma_p)$ is the top cell ($h(C_p) = e_{k+p-1} \otimes dx^p$) with its boundary compressed as far as possible, then $\partial C_p =$

∂Ra in $\pi_{N-1}D^{p-2}\Gamma_p$ and

$$\partial\Phi = (-1)^{m-1}\frac{1}{m!}\xi_*(eC_p \cup iRa) \quad \text{in} \quad \pi_*Y_{ps-k+1} .$$

Proof. Since $v = k+1$, the attaching map of the top cell factors through $D^{p-2}\Gamma_p$. Since ∂R is an equivalence by Lemma 6.3.(iii), the definition of $a = a_p(k+n(p-1))$ ensures that $\partial C_p = (\partial R)a = \partial Ra$. Now $D^k\Gamma_1 \cup D^{k-1}\Gamma_0 \simeq D^k\Gamma_1/D^{k-1}\Gamma_1$ and, since $k > 0$, Ra factors through $\Gamma_1 \subset D^{k-1}\Gamma_1$. Hence, in $H_*(D^k\Gamma_1 \cup D^{k-1}\Gamma_0)$,

$$\begin{aligned} h(eC_p \cup iRa) &= h(eC_p) \\ &= e_*(e_{k+p-1} \otimes dx^p) \\ &= (-1)^m m! e_k \otimes d(x^p) \end{aligned}$$

by Lemma 3.6 (since k is odd and n is even). By Lemma 4.4, it follows that $\partial\Phi = (-1)^{m-1}\frac{1}{m!}\xi_*(eC_p \cup iRa)$.

We are now ready to prove Theorems 1.1, 1.2, and 1.3 in this remaining case ($p > 2$, $v = k+1$, and $k > 0$). We must show that

$$d_*\beta P^j x = -\beta P^j d_r x \dotplus (-1)^e \bar{a}\, x^{p-1} d_r x.$$

By Lemma 6.5, $d_*\beta P^j x$ is obtained by lifting

$$(-1)^j \nu(n)\partial\Phi = (-1)^{j+m-1}\nu(n)\frac{1}{m!}\xi_*(eC_p \cup iRa)$$

from $\pi_*(Y_{ps-k+1})$ to the highest filtration possible. Since $\xi_*(eC_p)$ and $\xi_*(iRa)$ have common boundary in $Y_{ps+pr-p+2}$, $\xi_*(eC_p \cup iRa) = \xi_*(eC_p) - \xi_*(iRa)$ in $\pi_*(Y_{ps-k+1}, Y_{ps+pr-p+2})$. By naturality of ξ, $\xi_*(iRa)$ is the image of

$$\xi_* Ra \in \pi_*(Y_{ps+r}, Y_{ps+pr-p+2})$$

and by Lemma 4.1, $\xi_*(eC_p)$ is the image of

$$\xi_* C_p \in \pi_*(Y_{ps+pr-k-p+1}, Y_{ps+pr-p+2}).$$

Lemma 6.4 implies that $\xi_* R = \xi_* X$ in $\pi_*(Y_{ps+r-1}, Y_{ps+2r-1})$ since $\xi_* Y$ is in filtration 2r-1 or higher. (Note that since ∂R is mapped into Γ_2 by e in 6.4.(ii), Lemma 4.1 forces us to work modulo filtration 2r-1, the filtration into which ξ maps $D^1\Gamma_2$.) Thus

$$\xi_*(eC_p \cup iRa) = \xi_* C_p - \xi_* Xa \quad \text{in} \quad \pi_*(Y_{ps-k+1}, Y_{ps+2r-1}),$$

and, since $\bar{a}$ has filtration f, $\xi_* Xa$ comes from $\pi_*(Y_{ps+r+f}, Y_{ps+2r})$. By Lemma 4.6, either r+f or pr-k-p+1 is less than 2r-1, so that at least one of $\xi_* C_p$ and $\xi_* Xa$ is nontrivial in $\pi_*(Y_{ps-k+1}, Y_{ps+2r-1})$ in general. Since $h(C_p) = e_{k+p-1} \otimes dx^p$ and $h(X) = (-1)^{m-1} m! e_0 \otimes x^{p-1}dx$, Lemma 4.2 implies that

$\xi_* C_p$ represents $(-1)^j \frac{1}{\nu(n-1)} \beta P^j d_r x$, and

$\xi_* Xa$ represents $(-1)^{m-1} m! \, \bar{a} x^{p-1} d_r x$.

It then follows that

$$
\begin{aligned}
d_* \beta P^j x &= (-1)^j \nu(n) \partial\Phi \\
&= (-1)^{j+m-1} \nu(n) \frac{1}{m!} (\xi_* C_p - \xi_* Xa) \\
&= (-1)^{m-1} \frac{\nu(n)}{\nu(n-1)} \frac{1}{m!} \beta P^j d_r x - (-1)^j \nu(n) \bar{a} x^{p-1} d_r x \\
&= - \beta P^j d_r x \dotplus (-1)^e \bar{a} x^{p-1} d_r x
\end{aligned}
$$

since $\nu(n)/\nu(n-1) \equiv (-1)^m m! \pmod p$ and since $v = k+1$ implies $2(e+1)(p-1) = (2j-n)(p-1)$ so that $n = 2(j-e-1)$ and hence

$$-(-1)^j \nu(n) = (-1)^{j+1}(-1)^{j-e-1} = (-1)^e.$$

This completes case (b).

§7. Case (c): $v \leq k$.

In this case the boundary $\partial\Phi$ splits into a piece which represents the same operation (P^j or $\beta^\varepsilon P^j$) on $d_r x$ and another piece which is an operation of lower degree applied to x times an attaching map of a stunted lens space. We begin with the lemma needed to identify this latter piece exactly. Recall the spectral sequence of IV.6, and recall the notations established in §1.

Lemma 7.1. Let $\alpha \in \pi_{k+np-1} D^{k-v} S^{n(p)}$ be the attaching map of the top cell of $D^k S^{n(p)}$ and let f be the filtration of $\rho_*(\alpha) = a_p(k+n(p-1))$, where $\rho: D^{k-v} S^{n(p)} \to S^{k+np-v}$ is projection onto the top cell. Let $\mathcal{D}$ be the sequence

$$D^{k-v} S^{n(p)} \leftarrow D^{k-v-1} S^{n(p)} \leftarrow \cdots \leftarrow S^{n(p)}.$$

In the spectral sequence $E_r(S, \mathcal{D})$ the following hold:

(a) $1 \leq \mathrm{filt}(\alpha) \leq f$,

(b) if $\mathrm{filt}(\alpha) = f$ then α is detected by

$$\bar{a} e_{k-v} + \sum_{i=0}^{k-v-1} c_i e_i$$

for some $c_i \in E_2(S,S)$,

(c) if $p = 2$ and $v \leq 10$ or $p > 2$ and $v \leq pq$ then $\mathrm{filt}(\alpha) = f$ and α is detected by $\bar{a}e_{k-v}$.

Proof. (a) Since $\alpha_* = 0$ in mod p homology, $\mathrm{filt}(\alpha) > 0$. Note that this fact (applied to all the attaching maps of $D^{k-v}S^{n(p)}$) ensures that the spectral sequence can be constructed. Since ρ induces a homomorphism from $E_r(S,\mathcal{B})$ to $E_r(S,S)$, and $\rho_*(\alpha)$ has filtration f, α must have filtration $\leq f$.

(b) By IV.6.1(i), every element has the form

$$\sum_{i=0}^{k-v} c_i e_i$$

for some c_i. If $\mathrm{filt}(\alpha) = f$ then the element detecting α projects to $\bar{a}$ in the Adams spectral sequence of the top cell. Hence $c_{k-v} = \bar{a}$. (In fact this argument shows that if $c_{k-v} \neq 0$ then $\mathrm{filt}(\alpha) = f$ and $c_{k-v} = \bar{a}$.)

(c) Under the stated hypothesis, $\bar{a}e_{k-v}$ is the only element of filtration $\leq f$ in degree $k+np-1$.

To prove Theorems 1.1, 1.2 and 1.3, let us first assume that $v = 1$. Then k is even and $\varepsilon = 0$ if $p > 2$, and $k+n$ is even if $p = 2$. Theorems 1.1 and 1.2 say that

$$d_2 P^j x = h_0 P^{j-1} x \qquad \text{if } p = 2, \text{ and}$$

$$d_2 P^j x = a_0 \beta P^j x \qquad \text{if } p > 2.$$

Theorem 1.3 follows from Theorems 1.1 and 1.2 in this case. The first step is to split the element ι of Definition 4.3 into two pieces. Recall that

$$h(\iota) = (-1)^k (e_k \otimes d(x^p) + pe_{k-1} \times x^p).$$

Lemma 7.2: If $k \geq v = 1$ and $C_1 \in \pi_{k+np-1}(D^k\Gamma_1, D^{k-1}\Gamma_1)$ is the top cell, oriented so that $h(C_1) = (-1)^k e_k \otimes d(x^p)$, there exists $A \in \pi_{k+np-1}(D^{k-1}\Gamma_0, D^{k-1}\Gamma_1)$ such that

$$h(A) = (-1)^{k-1} pe_{k-1} \otimes x^p$$

and $$\iota = C_1 \cup A \in \pi_{k+np-1}(D^k\Gamma_1 \cup D^{k-1}\Gamma_0).$$

Proof. Let $N = k+np-1$. To see that A exists, consider the boundary maps and Hurewicz homomorphisms

$$\begin{array}{ccccc}
\pi_N(D^{k-1}\Gamma_0, D^{k-1}\Gamma_1) & \xrightarrow[\partial]{\cong} & \pi_{N-1}D^{k-1}\Gamma_1 & \xleftarrow[\partial]{} & \pi_N(D^k\Gamma_1, D^{k-1}\Gamma_1) \\
\downarrow h & & \downarrow h & & \downarrow h \cong \\
H_N(D^{k-1}\Gamma_0, D^{k-1}\Gamma_1) & \xrightarrow[\partial]{\cong} & H_{N-1}D^{k-1}\Gamma_1 & \xleftarrow[\partial]{} & H_N(D^k\Gamma_1, D^{k-1}\Gamma_1)
\end{array}$$

The isomorphisms are isomorphisms because $D^{k-1}\Gamma_0 \simeq *$ by Lemma 2.4 and because $D^k\Gamma_1/D^{k-1}\Gamma_1 \simeq S^{k+np-1}$. Certainly A exists satisfying $\partial A = \partial C_1$. It follows that

$$\partial(h(A)) = \partial(h(C_1)) = \partial((-1)^{k-1}pe_{k-1} \otimes x^p),$$

showing that $h(A) = (-1)^{k-1}pe_{k-1} \otimes x^p$.

To show that $\iota = C_1 \cup A$, it is enough to show $h(\iota) = h(C_1 \cup A)$, since $D^k\Gamma_1 \cup D^{k-1}\Gamma_0 \simeq S^{k+np-1}$. With $N = k+np-1$, note that $H_N D^{k-1}\Gamma_1 = 0$. This implies that the homomorphism

$$H_N D^k\Gamma_1 \cup D^{k-1}\Gamma_0 \xrightarrow{i_*} H_N(D^k\Gamma_1 \cup D^{k-1}\Gamma_0, D^{k-1}\Gamma_1)$$

is injective, so that we need only show $i_*h(\iota) = i_*h(C_1 \cup A)$. By Lemma 2.7, $i_*h(C_1 \cup A) = h(C_1) - h(A)$ and the result follows.

We now have $\partial\Phi = \xi_*\iota = \xi_*(C_1 \cup A) = \xi_*C_1 - \xi_*A$ modulo $Y_{ps+r-k+1}$ since $\xi_*(D^{k-1}\Gamma_1) \subsetneq Y_{ps+r-k+1}$. Applying Lemma 7.1 we find that ξ_*A represents $(-1)^{k-1}a_0\Phi_*(e_{k-1} \otimes x^p)$ in $\pi_*(Y_{ps-k+2}, Y_{ps+r-k+1})$ (with $a_0 = h_0$ if $p = 2$). Sorting out the constants, we find using Definition IV.2.4 that $-\xi_*A$ contributes $a_0\beta P^j x$, if $p > 2$, and $h_0 P^{j-1}x$, if $p = 2$, to the differential on $P^j x$. Thus, it remains only to show that ξ_*C_1 is in a higher filtration than ξ_*A.

<u>Lemma 7.3</u>. If i_1 and i_2 are the maps

$$\begin{array}{ccc} (D^k\Gamma_1, D^{k-1}\Gamma_1) & \xrightarrow{\ i_1\ } & \\ \downarrow & & (D^{k+1}\Gamma_1, D^{k-1}\Gamma_1 \cup D^{k+1}\Gamma_2) \\ (D^k\Gamma_1, D^{k-1}\Gamma_1 \cup D^k\Gamma_2) & \xrightarrow{\ i_2\ } & \end{array}$$

then there exists X such that $i_{1*}C_1 = p(i_{2*}X)$.

<u>Proof</u>. Since $k+np-1$ is the Hurewicz dimension of the domain and codomain of i_2, it suffices to work in homology. First suppose $p > 2$. We let $h(X) = e_k \otimes x^{p-1}dx$, which is obviously a cycle modulo $D^{k-1}\Gamma_1 \cup D^k\Gamma_2$. Then, in the codomain of i_1 and i_2 we have

$$\begin{aligned} e_k \otimes d(x^p) &= e_k \otimes Nx^{p-1}dx \\ &= Te_k \otimes Mx^{p-1}dx + pe_k \otimes x^{p-1}dx \\ &\sim e_{k+1} \otimes M^{-1}d(x^{p-1})dx + pe_k \otimes x^{p-1}dx \\ &\equiv p\, e_k \otimes x^{p-1}dx, \end{aligned}$$

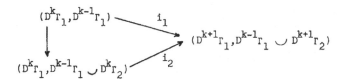

where $N = \sum \alpha^i$, $T = \alpha - 1$, and $M = \sum_{1}^{p-1} i\alpha^{p-i-1}$. The homology is due to $d(e_{k+1} \otimes Mx^{p-1}dx)$ and the congruence holds modulo $D^{k+1}\Gamma_2 \cup D^{k-1}\Gamma_1$. This implies that $i_{1*}C_1 = pi_{2*}X$.

Now suppose $p = 2$. We again let $h(X) = e_k \otimes xdx$ and again this is obviously a cycle. By Lemma 3.3 we have

$$\begin{aligned}(-1)^k e_k \otimes d(x^2) &\sim e_{k+1} \otimes dx^2 + 2e_k \otimes xdx \\ &\equiv 2e_k \otimes xdx,\end{aligned}$$

where the congruence holds modulo $D^{k+1}\Gamma_2 \cup D^{k-1}\Gamma_1$. This implies that $i_{1*}C_1 = 2i_{2*}X$.

We can now finish the proof of Theorems 1.1-1.3 for $v = 1$. By Lemma 7.3, the image of ξ_*C_1 in $\pi_*(Y_{ps-k+1}, Y_{ps-k+r+1})$ is zero, since it is the image of ξ_*pX, with $\xi_*X \in \pi_*(Y_{ps-k+r}, Y_{ps-k+r+1})$ so that $\xi_*pX \in \pi_*(Y_{ps-k+r+1}, Y_{ps-k+r+1}) = 0$. Thus the entire differential is given by $-\xi_*A$ and we are done.

Now suppose $1 < v \leq k$. Then, since $v = v_p(k+n(p-1))$, Lemma V.2.16 implies that $k+n$ is odd if $p = 2$ and that k is odd and $\varepsilon = 1$ if $p > 2$. Also, by Definition 4.3, $h(\iota) = (-1)^k e_k \otimes d(x^p)$. Let $N = k+np-1$.

<u>Lemma 7.4</u>. If $C_p \in \pi_N(D^{k+p-1}\Gamma_p, D^{k+p-1-v}\Gamma_p)$ is the top cell, oriented so that $h(C_p) = e_{k+p-1} \otimes dx^p$, then there exists $A \in \pi_N(D^{k-v}\Gamma_0, D^{k-v}\Gamma_1)$ such that $\partial A = e_*\partial C_p$ and $\iota \in \pi_N(D^k\Gamma_1 \cup D^{k-1}\Gamma_0)$ is the image of

$$\begin{Bmatrix} (-1)^{k+mn+m}\frac{1}{m!}(eC_p \cup A) & p > 2 \\ & \\ eC_2 \cup A & p = 2 \end{Bmatrix} \in \pi_N(D^k\Gamma_1 \cup D^{k-v}\Gamma_0)$$

<u>Proof</u>. To see that A exists consider the following diagram, whose upper square commutes and whose lower square anticommutes.

$$\begin{array}{ccc}
\pi_{N-1}D^{k+p-1-v}\Gamma_p & \xleftarrow{\ \partial\ } & \pi_N(D^{k+p-1}\Gamma_p, D^{k+p-1-v}\Gamma_p) \\
\downarrow e_* & & \cong \downarrow (e,e)_* \\
\pi_{N-1}D^{k-v}\Gamma_1 & \xleftarrow{\ \partial\ } & \pi_N(D^k\Gamma_1, D^{k-v}\Gamma_1) \\
\cong \uparrow \partial & (-1) & \cong \uparrow \partial \\
\pi_N D^{k-v}\Gamma_0/\Gamma_1 & \xleftarrow[\ \partial\ ]{} & \pi_{N+1}(D^k\Gamma_0/\Gamma_1, D^{k-v}\Gamma_0/\Gamma_1)
\end{array}$$

The isomorphisms are isomorphisms because $D^k\Gamma_0 \simeq * \simeq D^{k-v}\Gamma_0$ by Lemma 2.4 and (e,e) is an equivalence by Lemma 3.6. Thus, we may define $A = \partial^{-1}e_*\partial C_p$. To see that ι is the image of the claimed elements, it suffices to work in homology, as in Lemma 7.2. Here, $h(eC_p \cup A) = e_*h(C_p) - h(A) = e_*h(C_p)$ since $H_{N-1}D^{k-v}\Gamma_1 = 0$ for dimensional reasons. By hypothesis, $h(C_p) = e_{k+p-1} \otimes dx^p$, so

$$h(eC_p \cup A) = \begin{cases} (-1)^{mn+m} m!\, e_k \otimes d(x^p) & p > 2 \\ (-1)^k e_k \otimes d(x^2) & p = 2 \end{cases}$$

by Lemma 3.6. Comparing this with $h(\iota) = (-1)^k e_k \otimes d(x^p)$ finishes the proof.

Now,

$$d_*\beta^\varepsilon P^j x = \begin{cases} (-1)^j \nu(n)\xi_*\iota & p > 2 \\ \xi_*\iota & p = 2 \end{cases}$$

so, up to a scalar multiple, our differential is $\xi_*(eC_p \cup A) \in \pi_N Y_{ps-k+1}$. By Corollary 2.8 and Lemma 4.1 we find that

$$\begin{aligned} \xi_*(eC_p \cup A) &= \xi_*eC_p - \xi_*A \quad \text{in } \pi_N(Y_{ps-k+1}, Y_{ps-k+r+v}) \\ &= \xi_*C_p - \xi_*A \quad \text{in } \pi_N(Y_{ps-k+1}, Y_{ps-k+r+v-1}). \end{aligned}$$

It follows from the definition of C_p that ξ_*C_p lifts to $\pi_*(Y_{ps-k+pr-p+1}, Y_{ps-k+r+v})$. By Lemma 4.2, ξ_*C_p represents $\Phi_*(e_{k+p-1} \otimes dx^p)$, which equals $\beta^\varepsilon P^j d_r x$ up to a scalar multiple. When $p = 2$ this shows that ξ_*C_2 contributes $P^j d_r x$ to $d_*P^j x$. When $p > 2$, the coefficient of $\beta P^j d_r x$ is

$$(-1)^{2j+k+mn+m} \frac{\nu(n)}{\nu(n-1)} \frac{1}{m!} \equiv -1 \pmod p.$$

The congruence follows from the definition of ν, $\nu(2a+b) = (-1)^a (m!)^b$ if $b = 0$ or 1, and the congruence $(m!)^2 \equiv (-1)^{m-1} \pmod p$. This almost proves Theorem 1.1, with T_p consisting of $-\xi_*A \in \pi_N(Y_{ps-k+1}, Y_{ps-k+r+v})$ plus a possible "error term" in $\pi_N(Y_{ps-k+r+v-1}, Y_{ps-k+r+v})$ coming from the use of Lemma 4.1 above. "Almost" because this decomposition is only valid modulo filtration ps-k+r+v and we must still show that either $\beta^\varepsilon P^j d_r x$ or T_p will be a filtration lower than this in order to finish the proof of Theorem 1.1. To do this, we must identify ξ_*A. Referring to the diagram in the proof of Lemma 7.4, the element C_p in the upper right corner goes to A in the lower left corner if we follow the top and left arrows, while it goes to

$$\begin{cases} (-1)^{k+mn+m} m!\alpha & p > 2 \\ \alpha & p = 2, \end{cases}$$

where α is the attaching map of the cell $e_k \otimes x^p$, if we follow the bottom and right arrows. Since the lower square anticommutes and since k is odd if $p > 2$, it follows that

$$A = \begin{cases} (-1)^{mn+m} m!\alpha & p > 2 \\ -\alpha & p = 2. \end{cases}$$

Applying Lemma 7.1(a) we see that $\xi_* A$ has filtration less than or equal to ps-k+v+f. Lemma 4.7 implies that, unless $r = p = 2$ and $v = 1$ or 2, one of $\xi_* C_p$ and $\xi_* A$ will occur in a filtration less than ps-k+v+r-1. Thus Theorem 1.1 is proved unless $r = p = v = 2$ (since $v = 1$ has already been dealt with). Applying the rest of Lemma 7.1 we find that

$$\xi_* A = \begin{cases} (-1)^{mn+m} m!\ \bar{a}\ \Phi_*(e_{k-v} \otimes x^p) & p > 2 \\ -\bar{a}\ \Phi_*(e_{k-v} \otimes x^2) & p = 2 \end{cases}$$

if $v = k$ (since $D^{k-v}\Gamma_0/\Gamma_1 \simeq S^{n(p)}$ has only one cell in this case) or if $p = 2$ and $v \leq 10$ or if $p > 2$ and $v \leq pq$. Combining constants, we find that $T_2 = \bar{a}P^{j-v}x$ and that $T_p = (-1)^{e-1}\bar{a}\beta P^{j-e-1}x$ if $p > 2$ (recall that $e = \varepsilon_p(j)$). The constant in the odd primary case comes from the fact that $v = v_p(k+n(p-1)) = v_p(2j(p-1) - 1) = 2(p-1)(1+e)$ by V.2.16, so $k-v = (2(j-e-1) - n)(p-1) - 1$. This completes the proof of Theorem 1.2 except when $r = p = v = 2$ (as noted above) or when $pr-p < v < k$. In the latter case, Lemma 7.1.(a) still ensures us that

$$\begin{aligned} \text{filt}(\xi_* A) &\geq ps-k + v+1 \\ &> ps-k + pr - p+1 \\ &= \text{filt}(\xi_* C_p). \end{aligned}$$

Hence the term contributed to $d_*\beta^{\varepsilon}P^j x$ by $\xi_* C_p$ appears alone in this case. This completes the proof of Theorem 1.2 except when $r = p = v = 2$. Deferring the latter case until the end, we shall now prove Theorem 1.3. If $p = 2$ we may assume $v > 8$, while if $p > 2$ we may assume $v > q$. The attaching map α of Lemma 7.1 must then have filtration 2 or more. This is so because

(i) all but the top two cells are in filtration 2 or more,

(ii) the next to top cell component is the product of a positive dimensional element of $E_2(S,S)$ (since $v > 0$) and a cell in filtration 1, so has filtration at least 2,

(iii) the top cell component is a permanent cycle (being the image of the permanent cycle α), hence has filtration at least 2 by the nonexistence of Hopf invariant one elements in dimension $v-1$.

This implies that $\xi_* A$ has filtration $ps-k + v+2$ or more. Since $\xi_* C_p$ has filtration $ps-k + pr - p+1$ and $\partial\Phi$ splits into these pieces modulo filtration $ps-k + r + v-1$, we have $d_i \beta^\varepsilon P^j x = 0$ if

$$i \le \min\{v+1, pr-p, v+r-2\}$$
$$= \min\{v+1, pr-p\} ,$$

the equality holding because $v+r-2 < v+1$ implies $r = 2$, so that $pr-p = p < v = v+r-2$ by our assumption on v. This proves Theorem 1.3.

It remains only to prove Theorems 1.1 and 1.2 when $r = p = v = 2$. Together, they say $d_3 P^j x = P^j d_2 x + h_1 P^{j-2} x$. Let $N = k+2n-1$ and let $C_1 \in \pi_N(D^k\Gamma_1, D^{k-2}\Gamma_1)$ and $C_2 \in \pi_N(D^{k+1}\Gamma_2, D^{k-1}\Gamma_2)$ be the top cells, oriented so that $h(C_1) = (-1)^k e_k \otimes d(x^2)$ and $h(C_2) = e_{k+1} \otimes dx^2$.

Lemma 7.5. There exists $A \in \pi_N(D^{k-2}\Gamma_0, D^{k-2}\Gamma_1)$ such that $\partial A = \partial C_1$ and $\iota = C_1 \cup A$ in $\pi_N(D^k\Gamma_1 \cup D^{k-1}\Gamma_0)$.

Proof. Since $D^{k-2}\Gamma_0 \simeq *$ we may define $A = \partial^{-1}\partial C_1$

$$\pi_N(D^k\Gamma_1, D^{k-2}\Gamma_1) \xrightarrow{\partial} \pi_{N-1} D^{k-2}\Gamma_1 \xleftarrow[\cong]{\partial} \pi_N(D^{k-2}\Gamma_0, D^{k-2}\Gamma_1).$$

Clearly, $h(A) = 0$, so $h(C_1 \cup A) = h(C_1) = h(\iota)$. Thus $\iota = C_1 \cup A$.

It follows that

$$\partial\Phi = \xi_*\iota = \xi_*(C_1 \cup A) = \xi_* C_1 - \xi_* A \in \pi_N(Y_{2s-k+1}, Y_{2s-k+4}).$$

As before, we wish to replace $\xi_* C_1$ by $\xi_* C_2$ plus an error term which we can ignore. The following lemma is what we need in order to do this.

Lemma 7.6. Let
$$i_1 : D^{k-2}\Gamma_1 \to D^{k-1}\Gamma_2 \cup D^{k-2}\Gamma_1,$$
$$i_2 : D^{k-1}\Gamma_2 \to D^{k-1}\Gamma_2 \cup D^{k-2}\Gamma_1,$$
and
$$j : D^{k-1}\Gamma_1 \to D^k\Gamma_1$$

be the natural inclusions. Then there exists $X \in \pi_N(D^{k-1}\Gamma_1, D^{k-1}\Gamma_2 \cup D^{k-2}\Gamma_1)$ with positive filtration in the Adams spectral sequence, such that in $\pi_N(D^k\Gamma_1, D^{k-1}\Gamma_2 \cup D^{k-2}\Gamma_1)$

$$(1,i_1)_*C_1 = (e,i_2)_*C_2 + (j,1)_*X$$

Proof. Since $\rho:(D^k\Gamma_1, D^{k-2}\Gamma_1 \cup D^{k-1}\Gamma_2) \to (D^k\Gamma_1, D^{k-1}\Gamma_1)$ is the cofiber of $(j,1)$, we need only show $\rho_*(1,i_1)_*C_1 = \rho_*(e,i_2)_*C_2$ in order to establish the existence of X satisfying

$$(1,i_1)_*C_1 = (e,i_2)_*C_2 + (j,1)_*X.$$

The filtration of X is necessarily positive because

$$D^{k-1}\Gamma_1/D^{k-1}\Gamma_2 \cup D^{k-2}\Gamma_1 \simeq \bigvee S^{N-1}$$

by I.1.3 and Lemma 2.2. Since N is the Hurewicz dimension of $(D^k\Gamma_1, D^{k-1}\Gamma_1)$ it suffices to show $h(\rho_*(e,i_2)_*C_2) = h(\rho_*(1,i_1)_*C_1)$. This is immediate from Lemma 3.6.

With Lemma 7.6 we can now finish the proof of Theorems 1.1 and 1.2. The element ξ_*X is in $\pi_N(Y_{2s-k+3}, Y_{2s-k+4})$, but since X has filtration greater than 0, $\xi_*X = 0$ in $\pi_N(Y_{2s-k+3}, Y_{2s-k+4})$. Thus $\xi_*C_1 = \xi_*(1,i_1)_*C_1 = \xi_*(e,i_2)_*C_2$ in $\pi_N(Y_{2s-k+2}, Y_{2s-k+4})$. By Lemma 4.1, $\xi_*(e,i_2)_*C_2 = \xi_*C_2$ in $\pi_N(Y_{2s-k+1}, Y_{2s-k+4})$, and ξ_*C_2 lifts to $\pi_N(Y_{2s-k+3}, Y_{2s-k+4})$ where it represents $P^j d_2 x$ by Lemma 4.2. Finally, ξ_*A also lifts to $\pi_N(Y_{2s-k+3}, Y_{2s-k+4})$ where it represents $h_1 P^{j-2}x$ by Lemma 7.1. Thus

$$d_3 P^j x = P^j d_2 x + h_1 P^{j-2} x.$$

CHAPTER VII

H_∞ RING SPECTRA VIA SPACE-LEVEL HOMOTOPY THEORY

J. E. McClure

Our main goal in this chapter is to show that the spectrum KU representing periodic complex K-theory has an H_∞ structure. The existence of such a structure is important since it will allow us to develop a complete theory of Dyer-Lashof operations in K-theory, including the computation of $K_*(QX)$; this program is carried out in chapter IX. Of course, we already know that the connective spectrum kU has an H_∞ structure since it has an E_∞ structure by [71, VIII. 2.1]. However, it is not known whether KU has an E_∞ structure, and the distinction between kU and KU is crucial for our work in chapter IX. We therefore require a new method for constructing H_∞ ring spectra.

As usual, the case of ordinary ring spectra provides a useful analogy. The easiest way to give KU a ring structure is to use Whitehead's original theory of spectra [108]. We use the term "prespectrum" for a spectrum in the sense of Whitehead [108, p. 240], reserving the term "spectrum" for the stricter definition of I§1. The Bott periodicity theorem for BU gives rise at once to a prespectrum ([108, p. 241]; more work is needed in order to get a spectrum), and the tensor product of vector bundles gives this prespectrum a ring structure in the sense of [108, p. 270]. Now the Whitehead category is not equivalent to the stable category $\bar{h}\mathcal{S}$, but it is a quotient of it, and one can lift structures in this category to $\bar{h}\mathcal{S}$ as long as certain $\lim^1$ terms vanish. These $\lim^1$ terms do vanish for KU and we obtain the desired ring structure.

In order to carry this through for H_∞ structures we must give the Bott prespectrum a "Whitehead" H_∞ structure (which is fairly easy) and show how to lift it to $\bar{h}\mathcal{S}$ (which is considerably more difficult). Our main concern in this chapter is with the lifting process, which is called the cylinder construction and denoted by Z. We begin in Sections 1 and 2 by giving a careful development of the cases already mentioned, namely the passage from prespectra to spectra and from ring prespectra to ring spectra. Our account is based on that in [67] and [71, II §3] but is adapted to allow generalization to the H_∞ case to which we turn next. In section 3 we give a general result allowing construction of maps $D_\pi E \to F$ in $\bar{h}\mathcal{S}$ from prespectrum-level data. Although the basic idea is similar to that of section 2 this situation requires new hypotheses and methods. Section 4 is a digression which gives a convenient sufficient condition for the vanishing of the $\lim^1$ terms encountered in sections 1, 2, and 3. In section 5 we define H_∞ structures on prespectra (for technical reasons, these are called H_∞^d structures) and show that they lift to H_∞ structures in $\bar{h}\mathcal{S}$ when the relevant $\lim^1$ terms vanish. In section 6 we observe that spectra obtained in this way actually have H^d_∞ structures as defined in I.4.3 and that there is in fact an "approximate equivalence" between H^d_∞ structures on spectra and prespectra. Section 7 gives the application to K-theory. The necessary H^d_∞ structure on the Bott prespectrum is obtained from the E_∞ structure on kU; a more elementary construction not depending on E_∞ theory (but still using the results of this chapter) will be given in VIII §4. Section 8 gives a technical result which is used in section 3. Except for section 8 and one place in section 1 we use only the formal properties of $\overline{h\mathcal{A}}$ and D_π given in I§1 and I§2.

This chapter and the next are a revised version of my Ph. D. dissertation. I would like to take this opportunity to thank my advisor Peter May for his warm support and encouragement both in the course of this work and in the years since. I would also like to thank my colleagues Gaunce Lewis and Anne Norton, my friend Deborah Harrold, my parents, and a person who wishes to remain anonymous for their no less valuable support. However, the views expressed in these chapters are my own and do not necessarily reflect their opinions.

§1. The Whitehead category and the stable category

In this section we describe the relation between the Whitehead category, denoted $\overline{w\mathcal{P}}$, and the stable category $\overline{h\mathcal{S}}$. The results are well-known, but we give them in some detail in order to fix notation and because we need particularly precise statements for our later work.

We begin by defining $\overline{w\mathcal{P}}$. An object T, called a prespectrum, is a sequence of spaces T_i (for $i \geq 0$) and maps $\sigma_i : \Sigma T_i \to T_{i+1}$ in $\overline{h\mathcal{J}}$ (see I§1; the use of $\overline{h\mathcal{J}}$ here is technically convenient but could be avoided by systematic use of CW-approximations). If the adjoints $\tilde{\sigma}_i : T_i \to \Omega T_{i+1}$ are weak equivalences we call T an Ω-prespectrum. A morphism $f : T \to U$ is a sequence of maps $f_i : T_i \to U_i$ such that $f_{i+1} \circ \sigma_i \simeq \sigma_i \circ \Sigma f_i$ in $\overline{h\mathcal{J}}$. This should be compared with the much stricter definition of morphism in $\overline{h\mathcal{S}}$ given in I§1; it is precisely because morphisms in $\overline{w\mathcal{P}}$ are defined in terms of homotopy that this category is a useful intermediate step between space-level and spectrum-level homotopy theory. The set of maps in $\overline{w\mathcal{P}}$ from T to U is denoted $[T,U]_w$. If U is an Ω-prespectrum then this set is an abelian group and is equal to the inverse limit $\lim_i [T_i, U_i]$ with respect to the maps

$$[T_{i+1}, U_{i+1}] \xrightarrow{\Omega} [\Omega T_{i+1}, \Omega U_{i+1}] \xrightarrow{\tilde{\sigma}_i^*} [T_i, \Omega U_{i+1}] \xrightarrow{(\tilde{\sigma}_i)_*^{-1}} [T_i, U_i].$$

There is an evident forgetful functor $z : \overline{h\mathcal{S}} \to \overline{w\mathcal{P}}$. Although there is no useful functor in the other direction, there is an "approximately functorial" construction Z, called the cylinder construction. This can be defined in several essentially equivalent ways (see I§6 of the sequel). For our purposes it is easiest to define

$$ZT = \operatorname*{Tel}_i \Sigma^{-i}\Sigma^\infty T_i,$$

where the telescope is taken with respect to the maps

$$\Sigma^{-i}\Sigma^\infty T_i \cong \Sigma^{-i}\Sigma^{-1}\Sigma\Sigma^\infty T_i \cong \Sigma^{-i-1}\Sigma^\infty \Sigma T_i \longrightarrow \Sigma^{-i-1}\Sigma^\infty T_{i+1} .$$

We write θ_i for the inclusion $\Sigma^\infty T_i \to \Sigma^i ZT$. If $f:T \to U$ is any map in $\overline{w\mathcal{P}}$ there exists a map $F:ZT \to ZU$ induced by f in the sense that the diagram

$$\begin{array}{ccc}
\Sigma^{-i}\Sigma^\infty T_i & \xrightarrow{\Sigma^{-i}\Sigma^\infty f_i} & \Sigma^{-i}\Sigma^\infty U_i \\
\Big\downarrow{\scriptstyle \Sigma^{-i}\theta_i} & & \Big\downarrow{\scriptstyle \Sigma^{-i}\theta_i} \\
ZT & \xrightarrow{\quad F \quad} & ZU
\end{array}$$

commutes for all $i \geq 0$. Unfortunately, this map is not in general unique. To clarify the situation consider the Milnor $\lim^1$ sequence

$$0 \longrightarrow \lim{}^1\, [\Sigma^{1-i}\Sigma^\infty T_i, ZU] \longrightarrow [ZT,ZU] \longrightarrow \lim[\Sigma^{-i}\Sigma^\infty T_i, ZU] \longrightarrow 0.$$

Clearly, the map induced by f is unique if and only if the $\lim^1$ term vanishes. We shall use the notation Zf for this map when this condition is satisfied (and not otherwise). We have $Z(f \circ g) = Zf \circ Zg$ whenever all three are defined.

The $\lim^1$ term just mentioned is only the first of many which will arise in our work. For applications we wish to know when they vanish. This question will be considered in detail in §4; for the moment we simply remark that for the cases of interest to us (namely Bott spectra and certain bordism spectra) all relevant $\lim^1$ terms do in fact vanish.

Although Z is not a functor, it has several useful properties. In fact, one may think of the pair (z,Z) as an "approximate adjoint equivalence" between $\overline{h\mathcal{S}}$ and the full subcategory of Ω-prespectra in $\overline{w\mathcal{P}}$. The following result makes this precise.

<u>Theorem 1.1.</u> For each $T \in \overline{w\mathcal{P}}$ and $E \in \overline{h\mathcal{S}}$ there exists maps $\kappa:T \to zZT$ and $\lambda:ZzE \to E$ with the following properties.

(i) κ is natural in the sense that $zZf \circ \kappa = \kappa \circ f$ whenever Zf is defined.

(ii) κ is an equivalence whenever T is an Ω-prespectrum.

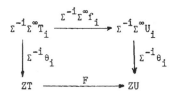

(iii) λ is natural in the sense that $f \circ \lambda = \lambda \circ Zzf$ whenever Zzf is defined.

(iv) λ is an equivalence for all $E \in \overline{h}\mathcal{S}$.

(v) $z\lambda \circ \kappa$ is the identity map of zE.

(vi) The map $\tau : [ZT,E] \to [T,zE]_w$ defined by $\tau f = zf \circ \kappa$ is an isomorphism whenever $\lim^1 E^{i-1}T_i = 0$.

(vii) The map Zf, whenever it is defined, is uniquely determined by the equation $\tau(Zf) = \kappa \circ f$.

The rest of this section gives the proof of 1.1. In order to construct κ and λ we need an alternative description of the i-th space functor from $\overline{h}\mathcal{S}$ to $\overline{h}\mathcal{T}$.

<u>Lemma 1.2</u>. There is a natural equivalence $E_i \simeq \Omega^\infty\Sigma^i E$. If θ_i' denotes the adjoint map $\Sigma^\infty E_i \to \Sigma^i E$ then the following diagrams commute.

$$(1)\quad \begin{array}{ccc} \Sigma E_i & \simeq & \Sigma\Omega^\infty\Sigma^i E \\ \downarrow{\scriptstyle \sigma_i} & & \downarrow \\ E_{i+1} & \simeq & \Omega^\infty\Sigma^{i+1}E \end{array} \qquad\qquad (2)\quad \begin{array}{ccc} \Sigma^\infty\Sigma E_i & \cong & \Sigma\Sigma^\infty E_i \\ \downarrow{\scriptstyle \Sigma^\infty\sigma_i} & & \downarrow{\scriptstyle \Sigma\theta_i'} \\ \Sigma^\infty E_{i+1} & \xrightarrow{\theta_{i+1}'} & \Sigma^{i+1}E \end{array}$$

For the proof see I§7 of the sequel. The fact that such an equivalence exists should not be surprising since it is well-known that the reduced E-cohomology groups E^iX of a based space X can be defined either as $[\Sigma^\infty X, \Sigma^i E]$ or as $[X,E_i]$. The diagrams of Lemma 1.2 (which are adjoints of each other) simply say that one obtains the same suspension isomorphism with either of these two definitions.

Given $T \in \overline{w}\mathcal{P}$ we can now define $\kappa : T \to zZT$ by letting the i-th component $\kappa_i : T_i \to (ZT)_i$ be the composite

$$T_i \longrightarrow \Omega^\infty\Sigma^\infty T_i \xrightarrow{\Omega^\infty\theta_i} \Omega^\infty\Sigma^i ZT \simeq (ZT)_i .$$

We note for later use that the following diagram commutes.

$$(3)\quad \begin{array}{ccc} \Sigma^\infty T_i & \xrightarrow{\Sigma^\infty\kappa_i} & \Sigma^\infty(ZT)_i \\ & {\scriptstyle \theta_i}\searrow \quad \swarrow{\scriptstyle \theta_i'} & \\ & \Sigma^i ZT & \end{array}$$

The verification that κ is in fact a $\overline{w}\mathcal{P}$-map is a routine diagram chase using diagram (1) above. It is clear that κ satisfies 1.1(i); in fact it has the stronger property that $zF \circ \kappa = \kappa \circ f$ whenever $F : ZT \to ZU$ is induced by f. For part (ii) we first compute

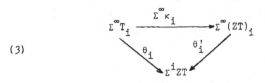

$$\pi_k(ZT)_i = \pi_{k-i}ZT = \operatorname*{colim}_j \pi_{k-i+j}\Sigma^\infty T_j$$
$$= \operatorname*{colim}_j \operatorname*{colim}_\ell \pi_{k-i+j+\ell}\,\Sigma^\ell T_j\ .$$

A cofinality argument shows that the inclusion of $\operatorname*{colim}_j \pi_{k-i+j}\, T_j$ in the last group is an isomorphism. If T is an Ω-prespectrum, then the inclusion

$$\pi_i T_k \to \operatorname*{colim}_j \pi_{k-i+j}\, T_j$$

is an isomorphism and the result follows.

Next we define $\lambda: ZzE \to E$ to be any map obtained by passage to the telescope from the maps

$$\Sigma^{-i}\theta_i' : \Sigma^{-i}\Sigma^\infty E_i \to E.$$

Part (v) is immediate, and (iv) follows from (ii) and (v). For (iii) it suffices, by the definition of Zzf, to show that $\lambda^{-1} \circ f \circ \lambda : ZzE \to ZzE'$ is induced by zf, i.e., that the diagram

$$\begin{array}{ccccccc}
\Sigma^{-i}\Sigma^\infty E_i & & \xrightarrow{\Sigma^{-i}\Sigma^\infty f_i} & & & & \Sigma^{-i}\Sigma^\infty E_i' \\
\downarrow{\scriptstyle \Sigma^{-i}\theta_i} & & & & & & \downarrow{\scriptstyle \Sigma^{-i}\theta_i} \\
ZzE & \xrightarrow{\lambda} & E & \xrightarrow{f} & E' & \xleftarrow{\lambda} & ZzE'
\end{array}$$

commutes for all $i \geq 0$. This in turn follows from the definition of λ and the naturality of θ_i'.

For part (vi) consider the $\lim^1$ sequence

$$0 \longrightarrow \lim{}^1[\Sigma\Sigma^{-i}\Sigma^\infty T_i, E] \longrightarrow [ZT,E] \xrightarrow{\bar{\tau}} \lim[\Sigma^{-i}\Sigma^\infty T_i, E] \longrightarrow 0.$$

The map $\bar{\tau}$ agrees with τ under the isomorphism

$$\lim[\Sigma^{-i}\Sigma^\infty T_i, E] \cong \lim[T_i, E_i] = [T, zE]_w$$

and the result follows.

Finally for (vii) we calculate

$$\tau(Zf) = zZf \circ \kappa = k \circ f.$$

The uniqueness follows from (vi).

§2. Pairings of spectra and prespectra.

In this section we give a multiplicative version of the results of §1 which in particular will allow us to produce a ring spectrum in $\overline{h\mathcal{S}}$ from suitable input in $\overline{w\mathcal{P}}$. Again the results are well-known.

For the rest of the chapter we fix an integer $d > 0$ and consider prespectra indexed on nonnegative multiples of d. This is convenient in the present section (for dealing with Bott spectra) and will be crucial in §3.

Let $E, E', F \in \overline{h\mathcal{S}}$. By a pairing of E and E' into F we mean simply a map $\phi: E \wedge E' \to F$. Although the category $\overline{w\mathcal{P}}$ has no smash product, a suitable prespectrum-level notion of pairing has been given by Whitehead [108, p. 255]; we recall it here.

Definition 2.1. Let $T, T', U \in \overline{w\mathcal{P}}$. A pairing $\psi: (T, T') \to U$ consists of a collection of maps

$$\psi_{i,j}: T_{di} \wedge T'_{dj} \to U_{d(i+j)}$$

such that the following diagram commutes in $\overline{h\mathcal{J}}$ for all $i, j \geq 0$.

$$\begin{array}{ccccc}
\Sigma^d T_{di} \wedge T'_{dj} & \xrightarrow{\sigma_i\ 1} & T_{d(i+1)} \wedge T'_{dj} & & \\
 & & & \searrow{\scriptstyle \psi_{i+1,j}} & \\
\wr\!\!\mid & (-1)^{dj} & & & \\
\Sigma^d(T_{di} \wedge T'_{dj}) & \xrightarrow{\Sigma^d \psi_{i,j}} & \Sigma^d U_{d(i+j)} & \xrightarrow{\sigma_{i+j}} & U_{d(i+j+1)} \\
\wr\!\!\mid & & & \nearrow{\scriptstyle \psi_{i,j+1}} & \\
T_{di} \wedge \Sigma^d T'_{dj} & \xrightarrow{1 \wedge \sigma_j} & T_{di} \wedge T'_{d(j+1)} & &
\end{array}$$

If $\phi: E \wedge E' \to F$ is a pairing in $\overline{h\mathcal{S}}$ and $f: \hat{E} \to E$, $f': \hat{E}' \to E'$, and $g: F \to \hat{F}$ are maps in $\overline{h\mathcal{S}}$ there is an evident pairing

$$g \circ \phi \circ (f \wedge f'): \hat{E} \wedge \hat{E}' \to \hat{F}.$$

Similarly, if $\psi: (T, T') \to U$ is a pairing in $\overline{w\mathcal{P}}$ and $f: \hat{T} \to T$, $f': \hat{T}' \to T'$, and $g: U \to \hat{U}$ are maps in $\overline{w\mathcal{P}}$ there is a composite pairing

$$g \circ \psi \circ (f, f'): (\hat{T}, \hat{T}') \to \hat{U}.$$

Next we show how to lift pairings from $\overline{w\mathcal{P}}$ to $\overline{h\mathcal{S}}$. If $\psi: (T, T') \to U$ is a pairing then $ZT \wedge ZT'$ is equivalent to

$$\mathrm{Tel}\ \Sigma^{-2di}\Sigma^{\infty}(T_{di} \wedge T'_{di})$$

and we can obtain an induced pairing $ZT \wedge ZT' \to ZU$ by passage to telescopes from the maps $\Sigma^{-2di}\Sigma^{\infty}\psi_{i,i}$. The induced pairing is unique if the group

$$\lim{}^1 (ZU)^{2di-1}(T_{di}\ \ T'_{di})$$

vanishes, and we denote it by $Z\psi$ when this condition is satisfied. Note that we now have two distinct, but analogous, meanings for the symbol Z, and we shall give another in section 3. There is no risk of confusion since the context will always indicate whether Z is being applied to a map in $\overline{w\mathcal{P}}$, a pairing, or an extended pairing as defined in section 3. Clearly we have

$$Zg \circ Z\psi \circ (Zf \wedge Zf') = Z(g \circ \psi \circ (f,f'))$$

whenever both sides are defined.

Next, given a pairing $\phi: E \wedge E' \to F$ in $\overline{h\mathcal{S}}$ we wish to define a pairing $z\phi:(zE,zE') \to zF$ (again, this use of the notation z is distinct from that in section 1). In contrast to section 1, it is inconvenient to do this directly from the definitions since the definition of $E \wedge E'$ is too complicated. Instead, we use the maps provided by Lemma 1.2. First let

$$\phi_{i,j}: \Sigma^{\infty}(E_{di} \wedge E'_{dj}) \to \Sigma^{d(i+j)}F$$

be the composite

$$\Sigma^{\infty}(E_{di} \wedge E'_{dj}) \simeq \Sigma^{\infty}E_{di} \wedge \Sigma^{\infty}E'_{dj} \xrightarrow{\theta'_i \wedge \theta'_j} \Sigma^{di}E \wedge \Sigma^{dj}E' \simeq \Sigma^{d(i+j)}E \wedge E' \longrightarrow \Sigma^{d(i+j)}F$$

Then the diagram

$$\begin{array}{ccccc}
\Sigma^{\infty}(\Sigma^{d}E_{di} \wedge E'_{dj}) & \longrightarrow & \Sigma^{\infty}(E_{d(i+1)} \wedge E'_{dj}) & \searrow^{\phi_{i+1,j}} & \\
\wr & & (-1)^{dj} & & \\
\Sigma^{\infty}(\Sigma^{d}(E_{di} \wedge E'_{dj})) & \simeq & \Sigma^{d}\Sigma^{\infty}(E_{di} \wedge E'_{dj}) & \xrightarrow{\Sigma^{d}\phi_{i,j}} & \Sigma^{d(i+j+1)}F \\
\wr & & & & \\
\Sigma^{\infty}(E_{di} \wedge \Sigma^{d}E'_{dj}) & \longrightarrow & \Sigma^{\infty}(E_{di} \wedge E'_{d(j+1)}) & \nearrow_{\phi_{i,j+1}} &
\end{array}$$

commutes by Lemma 1.2. We now define

$$(z\phi)_{i,j}: E_{di} \wedge E'_{dj} \to F_{d(i+j)}$$

to be the composite

$$E_{di} \wedge E'_{dj} \longrightarrow \Omega^{\infty}\Sigma^{\infty}(E_{di} \wedge E'_{dj}) \xrightarrow{\Omega^{\infty}\phi_{i,j}} \Omega^{\infty}\Sigma^{d(i+j)}F \simeq F_{d(i+j)} \,.$$

The fact that $z\phi$ is a pairing follows from the diagram above and another application of Lemma 1.2. We clearly have

$$z(g \circ \phi \circ (f \wedge f')) = zg \circ z\phi \circ (zf, zf').$$

Finally, given a pairing $\phi: ZT \wedge ZT' \to F$ we can define a pairing $\tau(\phi):(T,T') \to zF$ by $\tau(\phi) = z\phi \circ (\kappa,\kappa)$. In analogy with Theorem 1.1 we have

<u>Proposition 2.2</u> (i) If ψ is a pairing in $\overline{w\mathcal{P}}$ then $zZ\psi \circ (\kappa,\kappa) = \kappa \circ \psi$ whenever $Z\psi$ is defined.

(ii) If ϕ is a pairing in $\overline{h\mathcal{S}}$ then $\lambda \circ Zz\phi = \phi \circ (\lambda \ \ \lambda)$ whenever $Zz\phi$ is defined.

(iii) If $\lim^1 F^{2di-1}(T_{di} \wedge T'_{di}) = 0$ then τ is a one-to-one correspondence between pairings $ZT \wedge ZT' \to F$ and pairings $(T,T') \to zF$.

(iv) The pairing $Z\psi$, whenever it is defined, is uniquely determined by the equation $\tau(Z\psi) = \kappa \circ \psi$.

The proof is completely parallel to that of 1.1 and will be omitted.

As a special case we consider ring spectra and prespectra. Let S be the zero-sphere in $\overline{h\mathcal{S}}$ and let $\underline{S}$ be the prespectrum whose di-th term is S^{di} (with the evident structural maps). A <u>ring spectrum</u> is a spectrum E with maps $\phi: E \wedge E \to E$ and $e: S \to E$ satisfying the usual associativity, commutativity and unit axioms. Similarly, a <u>ring prespectrum</u> is a prespectrum T with a pairing $\psi:(T,T) \to T$ and a map $e:\underline{S} \to T$ satisfying associativity, commutativity and unit axioms. The unit axiom in this case is the commutativity of the following diagram in $\overline{h\mathcal{J}}$.

$$\begin{array}{ccccc}
S^{di} \wedge T_{dj} & \xrightarrow{e_{di} \wedge 1} & T_{di} \wedge T_{dj} & \xrightarrow{1 \wedge e_{dj}} & T_{di} \wedge S^{dj} \\
& \searrow^{\sigma} & \downarrow{\scriptstyle \psi_{i,j}} & \swarrow^{\sigma} & \\
& & T_{d(i+j)} & &
\end{array}$$

There are also evident notions of morphism for these structures. As a consequence of Proposition 2.2 we have the following.

<u>Corollary 2.3.</u> (i) If E is a ring spectrum then zE is a ring prespectrum. If f is a ring map in $\overline{h\mathcal{S}}$ then zf is a ring map in $\overline{w\mathcal{P}}$.

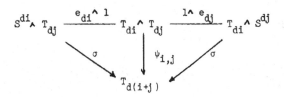

(ii) If T is a ring prespectrum with $\lim^1 (ZT)^{2di-1}(T_{di} \wedge T_{di}) = 0$ then ZT is a ring spectrum and $\kappa: T \to zZT$ is a ring map. If in addition $f: T \to T'$ is a ring map and

$$\lim{}^1 (ZT')^{2di-1}(T_{di} \wedge T_{di}) = \lim{}^1 (ZT')^{2di-1}(T'_{di} \wedge T'_{di}) = 0$$

then Zf is a ring map. If E is a ring spectrum and $\lim{}^1 E^{2di-1}(E_{di} \wedge E_{di}) = 0$ then $\lambda: ZzE \to E$ is a ring map.

§3. Extended pairings of spectra and prespectra

Let π be a fixed subgroup of Σ_j. In this section we generalize the results of section 2 by relating maps of the form $f: D_\pi E \to F$ in $\overline{h\mathcal{S}}$ to certain structures in $\overline{w\mathcal{P}}$ called extended pairings. This is our basic technical result, which will be applied in this chapter and the next to various problems in the theory of H_∞ ring spectra.

First we need a generalization of Definition 2.1. The difficulty is that, unlike the smash product, D_π does not commute with suspension. The situation becomes clearer when one realizes that $D_\pi \Sigma^d X$ is a relative Thom complex. For if p is the bundle

$$E\pi \times_\pi (R^d)^j \to B\pi$$

and p_X is the pullback of this bundle along the map

$$E\pi \times_\pi X^j \to B\pi,$$

then $D_\pi \Sigma^d X$ is the quotient $T(p_X)/T(p_*)$, where $*$ denotes the basepoint of X. The failure of D_π to commute with suspension arises from the fact that the bundle p is nontrivial. This suggests that we consider theories for which this bundle is at least orientable and replace the suspension isomorphisms which were implicitly present in section 2 with Thom isomorphisms. Note that the orientability of p with respect to a certain theory may well depend on the positive integer d.

Definition 3.1. Let F be a ring spectrum. A π-orientation for F is a map

$$\mu: D_\pi S^d \to \Sigma^{dj} F$$

such that the diagram

$$\begin{array}{ccc} (S^d)^{(j)} & \xrightarrow{\ \iota\ } & D_j S^d \\ \wr\!\wr & & \downarrow \mu \\ S^{dj} & \xrightarrow{\ \Sigma^{dj} e\ } & \Sigma^{dj} F \end{array}$$

commutes in $\overline{h\mathcal{S}}$. If U is a ring prespectrum, a π-orientation for U is a map

$$\nu: D_\pi S^d \to U_{dj}$$

such that the diagram

$$\begin{array}{ccc} (S^d)^{(j)} & \xrightarrow{\iota} & D_\pi S^d \\ \wr\| & & \downarrow \nu \\ S^{dj} & \xrightarrow{e_{dj}} & U_{dj} \end{array}$$

commutes in $\overline{h\mathcal{J}}$. A ring spectrum F or a ring prespectrum U with a fixed choice of π-orientation is called π-oriented. A ring map of π-oriented spectra or prespectra is π-oriented if it preserves the orientation.

It is now easy to give an analog for Definition 2.1. Recall the natural map δ defined in I§2.

Definition 3.2. Let T be a prespectrum and let (U,ν) be a π-oriented ring prespectrum. An extended pairing

$$\zeta:(\pi,T) \to (U,\nu)$$

is a sequence of maps

$$\zeta_i:D_\pi T_{di} \to U_{dij}$$

such that the following diagram commutes in $\overline{h\mathcal{J}}$ for all $i \geq 0$.

$$\begin{array}{ccccc} D_\pi(T_{di} \wedge S^d) & & \xrightarrow{\delta} & & D_\pi T_{di} \wedge D_\pi S^d \\ \downarrow D_\pi \sigma & & & & \downarrow \zeta_i \wedge \nu \\ D_\pi T_{d(i+1)} & & & & U_{dij} \wedge U_{dj} \\ & \searrow^{\zeta_{i+1}} & & \swarrow_{\psi_{ij,j}} & \\ & & U_{d(i+1)j} & & \end{array}$$

We shall usually suppress the orientation ν from the notation.

Definition 3.1 is general enough for our purposes, but it could be made more general by allowing U to be a module prespectrum over some π-oriented ring prespectrum. Everything which follows would work in this generality.

If $g:U \to U'$ is a π-oriented ring map and $f:T' \to T$ is any map in $\overline{w\mathcal{S}}$ we define the composite

$$g \circ \zeta \circ (\pi,f):(\pi,T') \to U'$$

by letting $(g \circ \zeta \circ (\pi,f))_i = g_{dji} \circ \zeta_i \circ D_\pi(f_{di})$. We also have composites in the π-variable: if ρ is a subgroup of π and U has a ρ-orientation consistent with its π-orientation then the maps

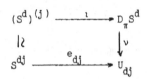

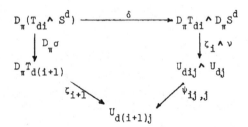

$$\zeta_i \circ \iota : D_\rho T_{di} \to U_{dij}$$

form an extended pairing denoted $\zeta \circ (\iota,1)$.

There is an evident stable version of 3.2: if F is a π-oriented ring spectrum we define an extended pairing from E to F to be a map $\xi : D_\pi E \to F$. We do not assume any relation between ξ and the orientation μ, but the presence of μ is necessary for the comparison with the prespectrum level. We can define composites $g \circ \xi \circ D_\pi f$ and $\xi \circ \iota$ as in the prespectrum case.

To complete the program of section 2 must show how to define $z\xi$ and $Z\zeta$ with suitable properties. Both of these will be defined by using a spectrum-level variant of the Thom homomorphism to which we turn next. If F is a π-oriented ring spectrum and $f : D_\pi E \to \Sigma^n F$ is any map we write $\Phi(f)$ for the composite

$$D_\pi \Sigma^d E \xrightarrow{\delta} D_\pi E \wedge D_\pi S^d \xrightarrow{f \wedge \mu} \Sigma^n F \wedge \Sigma^{dj} F \xrightarrow{\phi} \Sigma^{n+dj} F.$$

Since each class in $F^n(D_\pi E)$ is represented by some f we obtain a homomorphism

$$\Phi : F^n(D_\pi E) \to F^{n+dj}(D_\pi \Sigma^d E)$$

called the Thom homomorphism. We write $\Phi^{(i)}$ for the iterate $F^n(D_\pi E) \to F^{n+dij}(D_\pi \Sigma^{di} E)$. If $E = \Sigma^\infty X$ for some space X then it is easy to see that Φ is the relative Thom homomorphism for the bundle p_X and is therefore an isomorphism. Thus the following result should not be surprising.

Theorem 3.3. Φ is an isomorphism for every $E \in \overline{h\mathscr{S}}$.

The proof of this result, while not difficult, involves the definition of D_π and not just its formal properties and is deferred until section 8.

We can now define $z\xi$ for an extended pairing $\xi : D_\pi E \to F$. Give zF the orientation

$$z(\mu) : D_\pi S^d \longrightarrow \Omega^\infty \Sigma^\infty D_\pi S^d \simeq \Omega^\infty D_\pi S^d \longrightarrow \Omega^\infty \Sigma^{dj} F \simeq F_{dj} .$$

For each $i \geq 0$ let $(z\xi)_i$ be the composite

$$D_\pi E_{di} \longrightarrow \Omega^\infty D_\pi \Sigma^\infty E_{di} \xrightarrow{\Omega^\infty D_\pi \theta_{di}} \Omega^\infty D_\pi \Sigma^{di} E \xrightarrow{\Omega^\infty \Phi^{(i)} \xi} \Omega^\infty \Sigma^{dij} F \simeq F_{dij} .$$

The verification that $z\xi$ is in fact an extended pairing is completely similar to the analogous verification in section 2. Further, z is natural in the sense that $z(g \circ \xi \circ D_\pi f) = zg \circ z\xi \circ (\pi, zf)$ and $z(\xi \circ \iota) = z\xi \circ (\iota,1)$. Note that $z\xi$ depends not just on the map ξ but also on the orientation μ.

Unfortunately, $Z\zeta$ cannot be constructed directly as in sections 1 and 2. Instead we observe that we could have used 1.1(vi) and 2.2(iv) to define Zf and $Z\psi$ by means of the equations $\tau(Zf) = \kappa \circ f$ and $\tau(Z\psi) = \kappa \circ \psi$. If ξ is an extended pairing from ZT to F let $\tau(\xi)$ be the extended pairing

$$z\xi \circ (\pi,\kappa): (\pi,T) \to zF.$$

At the end of this section we shall prove

Theorem 3.4. If $\lim^1 F^{-1}(D_\pi \Sigma^{-di}\Sigma^\infty T_{di}) = 0$ then τ is a bijection between extended pairings $D_\pi ZT \to F$ and extended pairings $(\pi,T) \to zF$.

We can now define $Z\zeta$ for an extended pairing $\zeta:(\pi,T) \to U$ when the relevant $\lim^1$ terms vanish. Give ZU the π-orientation

$$Z(\nu):D_\pi S^d \simeq \Sigma^\infty D_\pi S^d \to \Sigma^\infty U_{dj} \to \Sigma^{dj} ZU.$$

and let $Z(\zeta)$ be $\tau^{-1}(\kappa \circ \zeta)$.

Corollary 3.5. (i) $zZ\zeta \circ (\pi,\kappa) = \kappa \circ \zeta$ whenever $Z\zeta$ is defined.

(ii) $Z(g \circ \zeta \circ (\pi,f)) = Zg \circ Z\zeta \circ D_\pi Zf$ and $Z(\zeta \circ (\iota,1)) = Z\zeta \circ \iota$ whenever both sides are defined.

(iii) $\lambda \circ Zz\xi = \xi \circ D_\pi\lambda$ whenever $Zz\xi$ is defined.

Proof of 3.5. (i) is the definition of $Z\zeta$. For the first equation in (ii) we calculate

$$\begin{aligned}\tau(Zg \circ Z\zeta \circ D_\pi Zf) &= zZg \circ zZ\zeta \circ (\pi,zZf) \circ (\pi,\kappa)\\ &= zZg \circ zZ\zeta \circ (\pi,\kappa) \circ (\pi,f)\\ &= zZg \circ \kappa \circ \zeta \circ (\pi,f)\\ &= \kappa \circ g \circ \zeta \circ (\pi,f)\\ &= \tau(Z(g \circ \zeta \circ (\pi,f)));\end{aligned}$$

the result follows by 3.4. The verification of the other equation in (ii) is similar. For part (iii) we have

$$\begin{aligned}\tau(\lambda^{-1} \circ \xi \circ D_\pi\lambda) &= z\lambda^{-1} \circ z\xi \circ (\pi,z\lambda) \circ (\pi,\kappa)\\ &= \kappa \circ z\xi = \tau(Z\xi)\end{aligned}$$

with the second equality following from 1.1(v); the result follows by 3.4.

Next we make some observations that will be important in sections 5 and 6. Part (iii) of our next result gives an alternate description of $Z\zeta$ which is similar to the definitions of Zf and $Z\psi$ in sections 1 and 2.

Corollary 3.6. Let $\xi : D_\pi ZT \to F$ be an extended pairing.

(i) $\tau(\xi)_i$ is the composite

$$D_\pi T_{di} \longrightarrow \Omega^\infty D_\pi \Sigma^\infty T_{di} \xrightarrow{\Omega^\infty D_\pi \theta_i} \Omega^\infty D_\pi \Sigma^{di} ZT \xrightarrow{\Omega^\infty \Phi^{(i)}\xi} \Omega^\infty \Sigma^{dij} F \simeq F_{dij}$$

(ii) If $\xi' : D_\pi ZT \to F$ is another extended pairing and τ is a bijection then $\xi = \xi'$ if and only if

$$\Phi^{(i)}\xi \circ D_\pi \theta_i = \Phi^{(i)}\xi' \circ D_\pi \theta_i$$

for all $i \geq 0$.

(iii) If $\zeta : (\pi, T) \to U$ is an extended pairing and $Z\zeta$ is defined then $Z\zeta$ is the unique map for which the following diagram commutes for all $i \geq 0$.

$$\begin{array}{ccc} D_\pi \Sigma^\infty T_{di} & \xrightarrow{D_\pi \theta_{di}} & D_\pi \Sigma^{di} ZT \\ \wr\| & & \\ \Sigma^\infty D_\pi T_{di} & & \Big\downarrow \Phi^{(i)} Z\zeta \\ \Big\downarrow \Sigma^\infty \zeta_i & & \\ \Sigma^\infty U_{dij} & \xrightarrow{\theta_{dij}} & \Sigma^{dij} ZU \end{array}$$

Proof of 3.6. Part (i) is immediate from the definition of τ and diagram (3) of section 1. Part (ii) follows at once from part (i). In part (iii) the commutativity follows from part (i) and the definition of $Z\zeta$, while the fact that $Z\zeta$ is the only such map follows from (ii).

Remark 3.7. Let D be a functor which is naturally equivalent to D_π for some π. More precisely, we assume that there are space and spectrum level functors , both called D and compatible with Σ^∞, and space and spectrum level equivalences $D \simeq D_\pi$ which are also compatible under Σ^∞; the cases of interest are $D_j \wedge D_k$ and $D_j D_k$. We can clearly carry through everything in this section with D_π replaced everywhere by D. The necessary maps

$$\delta : D(X \wedge Y) \to DX \wedge DY$$

and

$$\iota : X^{(j)} \to DX$$

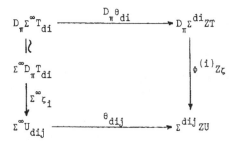

may be obtained from those for D_π by means of the given natural equivalence. Of course, D may already possess transformations δ and ι compatible with those for D_π; this is the case for $D = D_j \wedge D_k$ and $D = D_j D_k$. If π is a subgroup of $\rho \subset \Sigma_j$ and ι' denotes the composite

$$D \simeq D_\pi \xrightarrow{\iota} D_\rho$$

then (provided that ι' preserves the orientations) we can compose an extended pairing $\xi : D_\rho E \to F$ with ι' to get an extended pairing in the new sense from DE to F. Clearly z and Z will preserve such composites. The examples of interest for ι' are the maps $\alpha_{j,k}$ and $\beta_{j,k}$ defined in I§2.

We conclude this section with the proof of 3.4. If $\xi : D_\pi ZT \to F$ is an extended pairing we write $[\xi]$ for the element of $F^0 D_\pi ZT$ represented by ξ. Now D_π preserves telescopes by I.1.2(iii) so

$$D_\pi ZT \simeq \mathrm{Tel}\ D_\pi \Sigma^{-di} \Sigma^\infty T_{di} .$$

Hence the $\lim^1$ hypothesis implies

$$F^0 D_\pi ZT \cong \lim F^0 D_\pi \Sigma^{-di} \Sigma^\infty T_{di} .$$

The image of $[\xi]$ in the i-th term of the limit is $(D_\pi \Sigma^{-di} \theta_i)^* [\xi]$.

On the other hand if $\zeta : (\pi, T) \to zF$ is an extended pairing then each ζ_i represents an element $[\zeta_i] \in F^{dij} D_\pi T_{di}$, and Definition 3.2 says precisely that

$$\Phi[\zeta_i] = (D_\pi \sigma)^* [\zeta_{i+1}].$$

Hence the extended pairings $(\pi, T) \to zF$ are in one-to-one correspondence with the elements of

$$\lim F^{dij} D_\pi T_{di},$$

where the maps of the inverse system are the composites

$$F^{d(i+1)j} D_\pi T_{d(i+1)} \xrightarrow{(D_\pi \sigma)^*} F^{d(i+1)j} D_\pi \Sigma^d T_{di} \xrightarrow{\Phi^{-1}} F^{dij} D_\pi T_{di} .$$

Thus τ gives a map

$$\lim F^0 (D_\pi \Sigma^{-di} \Sigma^\infty T_{di}) \longrightarrow \lim F^{dij} (D_\pi T_{di}).$$

We claim this map is $\lim \Phi^{(i)}$, from which the result follows by 3.3. For by 3.6(i) and the naturality of Φ we have

$$[(\tau\xi)_i] = (D_\pi \theta_{di})^* \Phi^{(i)} [\xi] = \Phi^{(i)} ((D_\pi \theta_{di})^* [\xi]).$$

§4. A vanishing condition for $\lim^1$ terms

In order to apply the results of sections 1,2, and 3, one must have some way of showing that the relevant $\lim^1$ terms vanish. In this section, which is based on a paper of D. W. Anderson [10], we give a simple sufficient condition which is satisfied in our applications.

If F is a spectrum and X is a space we denote the F-cohomology Atiyah-Hirzebruch spectral sequence of X by $E_r(X;F)$. We say that the pair (X,F) is Mittag-Leffler (abbreviated M-L) if for each p and q there is an r with $E_r^{p,q}(X;F) = E_\infty^{p,q}(X;F)$; in particular this is true if the spectral sequence collapses.

Definition 4.1. A pair (T,F) with $T \in \overline{w\mathcal{P}}$ and $F \in \overline{h\mathcal{S}}$ is $\lim^1$-free if

(i) F and each T_{di} have finite type.

(ii) The pair (T_{di},F) is M-L for each $i \geq 0$.

(iii) If d is odd then $H^n(T_{di})$ and $\pi_n F$ are finite for all n. If d is even they are finite for odd n.

We say that $T \in \overline{w\mathcal{P}}$ is $\lim^1$-free if the pair (T,ZT) is.

The integer d in part (iii) is the one which was fixed at the beginning of section 2.

In practice it is easy to see whether a particular pair satisfies (i) and (iii). It is sometimes easier to deal with condition (ii) in the following equivalent form ([10, p. 291]).

Proposition 4.2. Suppose $E_2(X;F)$ has finite type. Then the pair (X,F) is M-L if and only if for each p and q the infinite cycles $Z_\infty^{p,q}(X;F)$ have finite index in $E_2^{p,q}(X;F)$.

Proof. Fix p and q. Let $C_r^{p,q}$ be the quotient of $E_r^{p,q}$ by its infinite cycles. If $Z_\infty^{p,q}$ has finite index in $E_r^{p,q}$ then $C_2^{p,q}$ is finite. Since $C_{r+1}^{p,q}$ is a subquotient of $C_r^{p,q}$ there must be an r_0 with $C_r^{p,q} = C_{r_0}^{p,q}$ for all $r \geq r_0$. But then clearly $C_{r_0}^{p,q} = 0$, hence $E_{r_0}^{p,q} = E_\infty^{p,q}$.

For the converse we recall that the rationalization $F \to F_{\mathbf{Q}}$ induces a rational isomorphism of E_2 terms. Since $F_{\mathbf{Q}}$ splits as a wedge of rational Eilenberg-Mac Lane spectra the spectral sequence $E_r(X;F_{\mathbf{Q}})$ collapses. Hence an element of infinite order in $E_r^{p,q}(X;F)$ cannot have as boundary another element of infinite order. It follows that $Z_r^{p,q}$ has finite index in $E_r^{p,q}$ and that the projection $Z_r^{p,q} \to E_{r+1}^{p,q}$ has finite kernel. But if $E_{r_0}^{p,q} = E_r^{p,q}$ then $C_{r_0}^{p,q} = 0$ and hence $C_2^{p,q}$ is finite as required.

Corollary 4.3. Suppose $E_r(X;F)$ and $E_r(X';F')$ have finite type. If $f:E_r(X;F) \to E_r(X';F')$ is a map of spectral sequences which induces a rational epimorphism in each bidegree of the E_2-terms, and if the pair (X,F) is M-L, then so is the pair (X',F').

As a consequence we get a way of generating new $\lim^1$-free pairs from known ones.

Corollary 4.4. Let (T,F) be a $\lim^1$-free pair and let $f:F \to F'$ and $g:T' \to T$ be maps inducing rational epimorphisms onto $\pi_* F'$ and $H^* T'_{di}$ for each i. If F' and each T'_{di} have finite type then the pair (T',F') is $\lim^1$-free.

Proof. The pair (T',F') clearly satisfies 4.1(iii), and it also satisfies 4.1(ii) since

$$f_* g^*_{di} : E_2(T_{di};F) \to E_2(T'_{di};F')$$

is a rational epimorphism in each bidegree.

In the remainder of this section we show that $\lim^1$ terms arising in previous sections do in fact vanish for $\lim^1$-free pairs. The reader willing to believe this can proceed to section 5.

By a filtered group we mean an abelian group A with a descending filtration

$$A = A^0 \supset A^1 \supset A^2 \supset \cdots .$$

A is complete if the map $A \to \lim A/A^n$ is an isomorphism (this includes the Hausdorff property), or equivalently if $\lim A^n = \lim^1 A^n = 0$. Filtered groups form a category whose morphisms are the filtration preserving maps.

Let $\{A_i\}_{i \geq 0}$ be an inverse system of filtered groups, and let A_i^n be the n-th filtration of A_i. Let $G^n A_i = A_i^n / A_i^{n+1}$. We need an algebraic fact ([10, Lemma 1.13]).

Proposition 4.5. Suppose that $\lim^1_i G^n A_i = 0$ for each n and that A_i is complete for each i. Then $\lim^1 A_i = 0$.

Using this we can prove the standard result about convergence of the Atiyah-Hirzebruch spectral sequence ([10, Theorem 2.1]). Recall that the skeletal filtration of $F^m X$ has as its n-th filtration the kernel of the restriction to the (n-1)-skeleton X(n-1). The associated graded groups of this filtration are the E_∞-term of the Atiyah-Hirzebruch spectral sequence.

Corollary 4.6. If the pair (X,F) is M-L then

(i) $\lim_n F^m X(n) = 0$ for each m,

(ii) The map $F^m X \to \lim_n F^m X(n)$ is an isomorphism, and

(iii) The skeletal filtration of $F^m X$ is complete.

Proof. Clearly (i) $\Longrightarrow$ (ii) $\Longrightarrow$ (iii) so we need only prove (i). Let $A_i = F^m X(i)$ with its skeletal filtration. This filtration is discrete, hence certainly complete, so by 4.5 it suffices to show $\lim_i^1 E_\infty^{p,q}(X(i);F) = 0$ for each p and q. Now the restriction

$$E_1^{p,q}(X;F) \to E_1^{p,q}(X(i);F)$$

is an isomorphism for $p \leq i$, hence the map

$$E_r^{p,q}(X;F) \to E_r^{p,q}(X(i);F)$$

is an isomorphism for $p \leq i-r+1$. Thus, if r_0 is such that $E_\infty^{p,q}(X;F) = E_{r_0}^{p,q}(X;F)$ we see that $E_\infty^{p,q}(X;F) \to E_\infty^{p,q}(X(i);F)$ is an isomorphism for $i \geq p+r_0-1$, so that $\lim_i^1 E_\infty^{p,q}(X(i);F) = 0$.

Now we can deal with the $\lim^1$ term of section 1.

Corollary 4.7. If the pair (T,F) is $\lim^1$-free then $\lim^1 F^{di-1}T_{di} = 0$.

Proof. Give $F^{di-1}T_{di}$ the skeletal filtration, which is complete by 4.6. Then each group of the associated graded is finite by 4.1(iii), hence the hypothesis of 4.5 is satisfied and we conclude that $\lim^1 F^{di-1}T_{di} = 0$.

Next we consider the relation with multiplicative structures.

Proposition 4.8. [10, p. 291] Suppose that F is a spectrum of finite type having the form ZU for a ring prespectrum U (in particular F may be a ring spectrum). If X and Y are spaces of finite type and the pairs (X,F) and (Y,F) are M-L, then so is $(X \wedge Y,F)$.

Proof. The hypothesis on F makes F-cohomology a ring-valued theory on spaces (but not necessarily on spectra). For each p and q the resulting product map

$$\bigoplus_{p'+p''=p} (E_2^{p',0}(X;F) \times E_2^{p'',q}(Y;F)) \to E_2^{p,q}(X \wedge Y;F)$$

is a rational epimorphism. Now $Z_\infty^{p',0}(X;F)$ and $Z_\infty^{p'',q}(Y;F)$ have finite index in $E_2^{p',0}(X;F)$ and $E_2^{p'',q}(Y;F)$ by 4.2, and the image of $Z_\infty^{p',0} \otimes Z_\infty^{p'',q}$ is contained in $Z_\infty^{p,q}(X \wedge Y;F)$. Hence $Z_\infty^{p,q}(X \wedge Y;F)$ has finite index in $E_2^{p,q}(X \wedge Y;F)$ and the result follows by 4.2.

This allows us to handle the $\lim^1$ term in section 2.

Corollary 4.9. If (T,F) and (T',F) are $\lim^1$-free and F has the form ZU for a ring prespectrum U then $\lim^1 F^{2di-1}(T_{di} \wedge T'_{di}) = 0$.

Proof. The skeletal filtration of $F^{2di-1}(T_{di} \wedge T'_{di})$ is complete by 4.6 and 4.8, and each group of the associated graded is finite by 4.1(iii). The result follows by 4.5.

We now consider extended powers.

Corollary 4.10. If X and F have finite type, F has the form ZU for a ring prespectrum U, and the pair (X,F) is M-L, then so is $(D_\pi X,F)$ for any $\pi \subset \Sigma_j$.

Proof. The transfer, which is a stable map from $D_\pi X$ to $X^{(j)}$, gives a rational epimorphism

$$E_2^{p,q}(X^{(j)};F) \to E_2^{p,q}(D_\pi X;F).$$

The result follows by 4.2 and 4.8.

Next we dispose of the $\lim^1$ term of section 3.

Corollary 4.11. If (T,F) is $\lim^1$-free and F is a π-oriented ring spectrum then $\lim^1 F^{-1} D_\pi \Sigma^{-di} \Sigma^\infty T_{di} = 0$.

Proof. The proof of 3.4 shows that the given inverse system is isomorphic to the inverse system $F^{dij-1} D_\pi T_{di}$ with structural maps $\Phi^{-1} \circ (D_\pi \sigma)^*$. Now the Thom isomorphism Φ preserves the skeletal filtration so we have a filtered inverse system of groups which are complete by 4.10. The associated graded groups are finite by 4.1(iii) and the proof of 4.10. The result follows by 4.5.

Finally, we record a result of Anderson which generalizes 4.6.

Proposition 4.12 [10, Corollary 2.4]. Suppose that X and F have finite type and (X,F) is M-L. If X is a countable CW-complex then the map

$$F^n X \to \lim_\alpha F^n X_\alpha ,$$

where $\{X_\alpha\}$ is the set of finite subcomplexes of X, is an isomorphism for each n.

§5. H_∞ ring spectra and prespectra

In this section we show that H_∞ ring spectra can be obtained by lifting the following structures in $\overline{w\mathcal{P}}$.

Definition 5.1. An H_∞^d ring prespectrum is a ring prespectrum U with maps

$$\zeta_{j,i}: D_j U_{di} \to U_{dij}$$

for all $i,j \geq 0$ such that each $\zeta_{1,i}$ is the identity map and the following diagrams commute in $\overline{h\mathcal{J}}$ for all $i,j,k \geq 0$.

$$\begin{array}{ccc} D_j U_{di} \wedge D_k U_{di} & \xrightarrow{\alpha} & D_{j+k} U_{di} \\ \downarrow{\scriptstyle \zeta_{j,i} \wedge \zeta_{k,i}} & & \downarrow{\scriptstyle \zeta_{j+k,i}} \\ U_{dij} \wedge U_{dik} & \xrightarrow{\psi} & U_{di(j+k)} \end{array} \qquad \begin{array}{ccc} D_j D_k U_{di} & \xrightarrow{\beta} & D_{jk} U_{di} \\ \downarrow{\scriptstyle D_j \zeta_{k,i}} & & \downarrow{\scriptstyle \zeta_{jk,i}} \\ D_j U_{dik} & \xrightarrow{\zeta_{j,ik}} & U_{dijk} \end{array}$$

$$\begin{array}{ccccc} D_k(U_{di} \wedge U_{dj}) & & \xrightarrow{\delta} & & D_k U_{di} \wedge D_k U_{dj} \\ \downarrow{\scriptstyle D_k \psi} & & & & \downarrow{\scriptstyle \zeta_{k,i} \wedge \zeta_{k,j}} \\ D_k U_{d(i+j)} & & & & U_{dik} \wedge U_{djk} \\ & \searrow{\scriptstyle \zeta_{k,i+j}} & & \swarrow{\scriptstyle \psi} & \\ & & U_{d(i+j)k} & & \end{array}$$

A ring map $f: U \to U'$ between H_∞^d ring prespectra is an H_∞^d ring map if $\zeta_{j,i} \circ D_j f_{di} = f_{dij} \circ \zeta_{j,i}$ for all $i,j \geq 0$.

The significance of the positive integer d in this definition is that a prespectrum may have an H_∞^d structure but not an $H_\infty^{d'}$ structure for $d' < d$. (Some examples of this phenomenon are given in the next section.) The third diagram in Definition 5.1 has no analog in the definition of H_∞ ring spectrum since in that situation the analog of the third diagram follows from the other two by (ii) and (iii) of I.3.4.

Definition 5.1 has several consequences. The first diagram implies the commutativity of

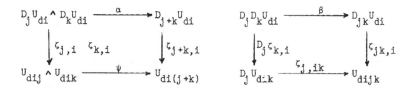

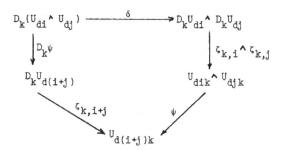

$$\begin{array}{ccc} (U_{di})^{(j)} & \xrightarrow{\ \iota\ } & D_j U_{di} \\ & {\scriptstyle\psi}\searrow \quad \swarrow{\scriptstyle\zeta_{j,i}} & \\ & U_{dij} & \end{array}$$

for all i and j. In particular the composite

$$\nu_j : D_j S^d \xrightarrow{D_j e_d} D_j U_d \xrightarrow{\zeta_{j,1}} U_{dj}$$

is a Σ_j-orientation for U. These orientations are consistent in the sense that the diagrams

$$(1) \quad \begin{array}{ccc} D_j S^d \wedge D_k S^d & \xrightarrow{\ \alpha\ } & D_{j+k} S^d \\ \downarrow{\scriptstyle \nu_j \wedge \nu_k} & & \downarrow{\scriptstyle \nu_{j+k}} \\ U_{dj} \wedge U_{dk} & \xrightarrow{\ \psi\ } & U_{d(j+k)} \end{array} \qquad (2) \quad \begin{array}{ccc} D_j D_k S^d & \xrightarrow{\ \beta\ } & D_{jk} S^d \\ \downarrow{\scriptstyle D_j \nu_k} & & \downarrow{\scriptstyle \nu_{jk}} \\ D_j U_{dk} & \xrightarrow{\ \zeta_{j,k}\ } & U_{djk} \end{array}$$

commute for all j and k. Now the unit diagram in the definition of a ring prespectrum and the third diagram in Definition 5.1 imply that for each fixed j the maps $\zeta_{j,i}$ give an extended pairing

$$\zeta_j : (\Sigma_j, U) \to (U, \nu_j).$$

<u>Theorem 5.2.</u> If U is a $\lim^1$-free H_∞^d ring prespectrum then the maps

$$Z(\zeta_j) : D_j ZU \to ZU$$

give ZU an H_∞ ring structure. If $f: U \to U'$ is an H_∞^d ring map and U,U' and the pair (U,ZU') are $\lim^1$-free then Zf is an H_∞ ring map.

The proof will occupy the rest of this section. We write F for ZU, ξ_j for $Z(\zeta_j)$ and ϕ for the multiplication $Z\psi$. Let μ_j be the orientation

$$Z(\nu_j) : D_j S^d \to \Sigma^{dj} ZU = \Sigma^{dj} F,$$

as defined after Theorem 3.4. First we claim that the μ_j are consistent in the following sense.

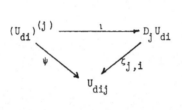

Lemma 5.3. The diagrams

$$(3)\qquad \begin{array}{ccc} D_jS^d \wedge D_kS^d & \xrightarrow{\ \alpha\ } & D_{j+k}S^d \\ \downarrow{\scriptstyle \mu_j \wedge \mu_k} & & \downarrow{\scriptstyle \mu_{j+k}} \\ \Sigma^{dj}F \wedge \Sigma^{dk}F & \xrightarrow{\ \Sigma^{d(j+k)}\phi\ } & \Sigma^{d(j+k)}F \end{array} \qquad\qquad (4)\qquad \begin{array}{ccc} D_jD_kS^d & \xrightarrow{\ \beta\ } & D_{jk}S^d \\ \downarrow{\scriptstyle D_j\mu_k} & & \downarrow{\scriptstyle \mu_{jk}} \\ D_j\Sigma^{dk}F & \xrightarrow{\ \phi^{(k)}(\xi_j)\ } & \Sigma^{djk}F \end{array}$$

commute for all $j,k \geq 0$.

Proof. For diagram (4) recall that μ_i is the composite $\theta_{di} \circ \Sigma^\infty \nu_i$, where θ_{di} is the natural map $\Sigma^\infty U_{di} \to \Sigma^{di}F$. Hence

$$\begin{aligned} \mu_{jk} \circ \beta &= \theta_{djk} \circ \Sigma^\infty(\nu_{jk} \circ \beta) \\ &= \theta_{djk} \circ \Sigma^\infty(\zeta_{j,k}) \circ \Sigma^\infty D_j \nu_k && \text{by diagram (2)} \\ &= \phi^{(k)}(\xi_j) \circ D_j\theta_{dk} \circ D_j\Sigma^\infty \nu_k && \text{by Corollary 3.6(iii)} \\ &= \phi^{(k)}(\xi_j) \circ D_j\mu_k . \end{aligned}$$

The proof for diagram (3) is similar.

Next we need another preliminary result.

Lemma 5.4. The diagram

$$\begin{array}{ccccc} D_k(F\wedge F) & & \xrightarrow{\ \delta\ } & & D_kF \wedge D_kF \\ \downarrow{\scriptstyle D_k\phi} & & & & \downarrow{\scriptstyle \xi_k \wedge \xi_k} \\ D_kF & & & & F\wedge F \\ & \searrow{\scriptstyle \xi_k} & & \swarrow{\scriptstyle \phi} & \\ & & F & & \end{array}$$

commutes for all $k \geq 0$.

In order to prove 5.4 we need the following variant of 3.6(ii).

Lemma 5.5. Let η_1 and η_2 be two maps

$$D_\pi(ZT \wedge ZT') \to F,$$

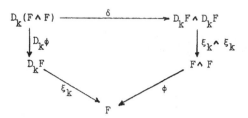

where F is a π-oriented ring spectrum and the pairs (T,F) and (T',F) are $\lim^1$-free. Then $\eta_1 = \eta_2$ if and only if the equation

$$\Phi^{(2i)}(\eta_1) \circ D_\pi(\theta_i \wedge \theta_i) = \Phi^{(2i)}(\eta_2) \circ D_\pi(\theta_i \wedge \theta_i) \tag{5}$$

holds for all $i \geq 0$.

Proof of 5.5. The composite isomorphism

$$F^0(D_\pi(ZT \wedge ZT')) \xrightarrow{\cong} \operatorname{Lim} F^0 D_\pi \Sigma^{-2di}(T_{di} \wedge T'_{di}) \xrightarrow{\lim \Phi^{(2i)}} \lim F^{2dij} D_\pi(T_{di} \wedge T'_{di})$$

takes η_1 to $\Phi^{(2i)}(\eta_1) \circ D_\pi(\theta_i \wedge \theta_i)$, and similarly for η_2.

Proof of 5.4. Let η_1 be the counterclockwise composite in the diagram and η_2 the clockwise composite. Consider the following diagram of spectra, where we have suppressed Σ^∞ to simplify the notation and the unlabeled arrows are all induced by maps θ_{di}.

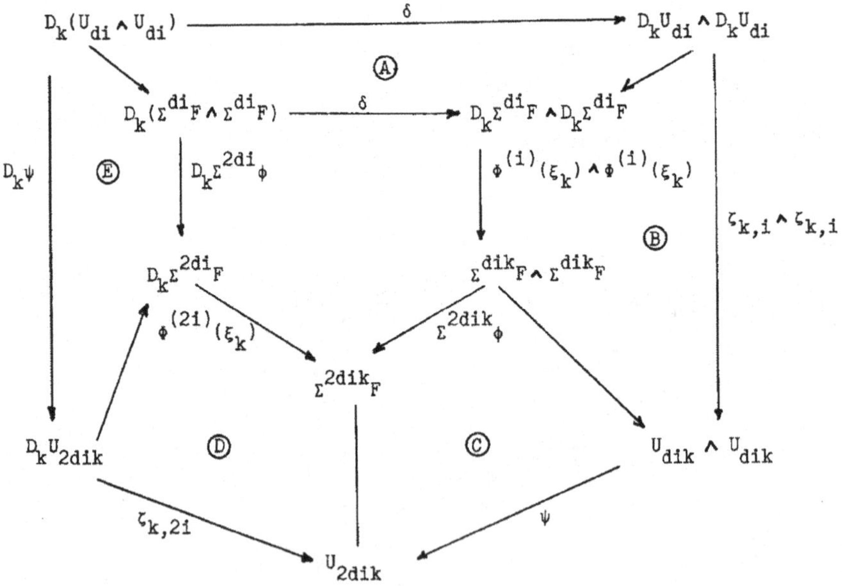

It is easy to see that the counterclockwise and clockwise composites in the inner pentagon are $\Phi^{(2i)}(\eta_1)$ and $\Phi^{(2i)}(\eta_2)$. To verify equation (5) it suffices to show that the outer pentagon and parts A, B, C, D and E commute. But the outer pentagon is the third diagram of Definition 5.1. Part A commutes by naturality of δ, parts C and E by definition of $\phi = Z\psi$, and parts B and D by 3.6(iii).

We now turn to the main part of the proof of 5.2. We shall show that the following diagram commutes; the other is similar.

(6)
$$\begin{array}{ccc} D_j D_k F & \xrightarrow{\ \beta\ } & D_{jk} F \\ \big\downarrow {\scriptstyle D_j \xi_k} & & \big\downarrow {\scriptstyle \xi_{jk}} \\ D_j F & \xrightarrow{\ \xi_j\ } & F \end{array}$$

We shall apply Remark 3.7 with $D = D_j D_k$. First orient $D_j D_k S^d$ using either of the two equal composites in diagram (4) of Lemma 5.3, and denote the associated Thom isomorphism by $\overline{\Phi}$. We write η_1 and η_2 for the counterclockwise and clockwise composites in diagram (6); these are extended pairings in the sense of Remark 3.7. By 3.6(ii) it suffices to show

(7)
$$\overline{\Phi}^{(i)} \eta_1 \circ D_j D_k \theta_i = \overline{\Phi}^{(i)} \eta_2 \circ D_j D_k \theta_i$$

for each $i \geq 0$. Consider the following diagram, where we have again suppressed Σ^∞ and the unlabeled arrows are all induced by maps θ_{di}.

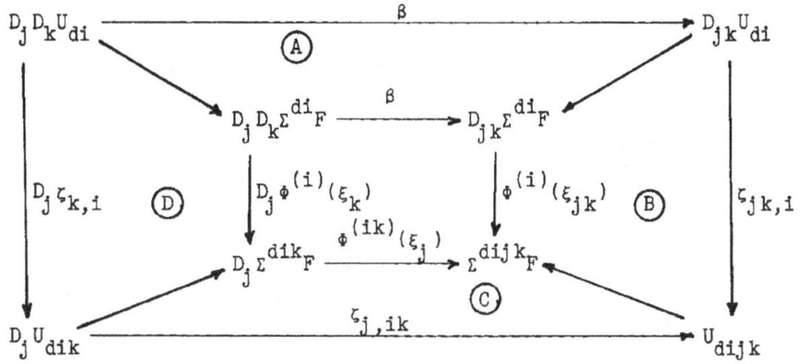

In the inner square the clockwise composite is clearly $\overline{\Phi}^{(i)}(\eta_2)$. Using Lemma 5.4 one can show that the counterclockwise composite is $\overline{\Phi}^{(i)}(\eta_1)$. To verify equation (7) we must show that the outer square and parts A, B, C and D commute. The outer square is the second diagram of Definition 5.1. Part A commutes by naturality of β and parts B,C, and D by 3.6(iii). This completes the proof.

§6. H_∞^d ring spectra.

Theorem 5.2 gives a useful relation between H_∞ structures in $\overline{h\mathscr{S}}$ and H_∞^d structures in $\overline{w\mathscr{P}}$. However, it does not provide a satisfactory analog for Corollary 2.3 since an arbitrary H_∞ ring spectrum F need not possess the Σ_j-orientations necessary to give an H_∞^d structure for zF. For example, if F = S then zF is not an H_∞^d prespectrum for <u>any</u> $d > 0$ (cf. Proposition 6.1). What is needed is a notion of H_∞ ring spectrum with built-in orientations. It turns out that the right objects to look at are H_∞^d ring spectra as defined in I.4.3.

If F is an H_∞ ring spectrum we say that a sequence of Σ_j-orientations is <u>consistent</u> if the diagrams of Lemma 5.3 commute. If F has an H_∞^d structure let μ_j be the composite

$$D_j S^d \xrightarrow{D_j \Sigma^d e} D_j \Sigma^d F \xrightarrow{\xi_{j,1}} \Sigma^{dj} F.$$

Then each μ_j is a Σ_j-orientation by I.4.4(iii) and an easy diagram chase shows that the μ_j are consistent. On the other hand, some H_∞ ring spectra do not even have Σ_2-orientations, and thus are certainly not H_∞^d. This is illustrated by our next result.

<u>Proposition 6.1.</u> (i) The sphere spectrum S is not an H_∞^d ring spectrum for any $d > 0$.

(ii) If F is an H_∞^d ring spectrum for d odd, then $\pi_* F$ has characteristic 2. If, in addition, F is connective and $\pi_0 F$ is augmented over Z_2 then F splits as a wedge of suspensions of HZ_2.

<u>Proof.</u> Let p^d be the bundle

$$E\Sigma_2 \times_{\Sigma_2} (R^d)^2 \to B\Sigma_2.$$

Then p^d is the d-fold Whitney sum of p^1 with itself, and p^1 is the sum of the Hopf bundle with a trivial bundle. The Thom complex of p^d is $D_2 S^d$, and so p^d is F-orientable if and only if F has a Σ_2-orientation (for the given value of d).

For (i) we recall (e.g. from [71, III.2.7]) that a bundle is S-orientable if and only if it is stably fibre-homotopy trivial. But p^d clearly has nontrivial Stiefel-Whitney classes for every $d \geq 1$.

(ii) Let $R = \pi_0 F$ and observe that F-orientability implies HR-orientability by virtue of the canonical map $F \to HR$. Consider the spectral sequence with

$$E_2^{p,q} = H^p(Z_2; H^q(S^d \quad S^d; R))$$

converging to $H^*(D_2 S^d; R)$. There is only one nonzero row and so $H^{2d}(D_2 S^d; R)$ is isomorphic to $H^0(Z_2; H^2(S^d \wedge S^d; R))$, which is the Z_2-fixed subgroup of

$H^{2d}(S^d \wedge S^d;R) \cong R$. But Z_2 acts on R as multiplication by -1, so we conclude that $H^{2d}(D_2S^d;R)$ is isomorphic to the 2-torsion subgroup of R. If on the other hand p^1 has an HR-orientation then $H^{2d}(D_2S^d;R) \cong R$, so that R must have characteristic 2. If in addition F is connective and R is augmented over Z_2 then the proof of Steinberger's splitting theorem III.4.1 gives the splitting of F.

Now let F be an H_∞^d ring spectrum. An easy diagram chase shows that the equation

$$\xi_{j,i} = \Phi^{(i)}(\xi_{j,0}):D_j\Sigma^{di}F \to \Sigma^{dij}F$$

holds for each i and j, where $\Phi^{(i)}$ is the Thom isomorphism determiend by the induced Σ_j-orientation of F. Thus the H_∞^d structure on F is uniquely determined by its underlying H_∞ structure and the set of induced Σ_j-orientations. Conversely, we have

Proposition 6.2. If F is an H_∞ ring spectrum with consistent Σ_j-orientations then the maps $\xi_{j,i}$ defined by $\xi_{j,i} = \Phi^{(i)}(\xi_j)$ give F an H_∞^d structure.

Using this, we can give a precise analog of 2.3.

Corollary 6.3 (i) If F is an H_∞^d ring spectrum then zF is an H_∞^d ring prespectrum. If f is an H_∞^d ring map in $\overline{h\mathcal{S}}$ then zf is an H_∞^d ring map in $\overline{w\mathcal{P}}$.

(ii) If U is a $\lim^1$-free H_∞^d ring prespectrum then ZU is an H_∞^d ring spectrum and $\kappa:U \to zZU$ is an H_∞^d ring map. If in addition $f:U \to U'$ is an H_∞^d ring map and U' and (U,ZU') are $\lim^1$-free then Zf is an H_∞^d ring map. If F is an H_∞^d ring spectrum and zF is $\lim^1$-free then $\lambda:ZzF \to F$ is an H_∞^d ring map.

Proof of 6.3. For part (i), the adjoint of the composite

$$\Sigma^\infty D_jF_{di} \cong D_j\Sigma^\infty F_{di} \xrightarrow{D_j\theta'_{di}} D_j\Sigma^{di}F \xrightarrow{\xi_{j,i}} \Sigma^{dij}F$$

is a map $\zeta_{j,i}:D_jF_{di} \to F_{dij}$. An easy diagram chase shows that the $\zeta_{j,i}$ satisfy Definition 5.1. Part (ii) is immediate from 5.2, 5.3 and 6.2.

The rest of this section gives the proof of 6.2. Let ω_j denote the composite

$$D_jS \xrightarrow{D_je} D_jF \xrightarrow{\xi_j} F$$

and let $\mu_j^{(i)} = \Phi^{(i)}\omega_j:D_jS^{di} \to \Sigma^{dij}F$; in particular $\mu_j^{(1)} = \mu_j$. Then $\xi_{j,i}$ is the composite

$$D_j\Sigma^{di}F \xrightarrow{\delta} D_jF\wedge D_jS^{di} \xrightarrow{\xi_j\wedge\mu_j^{(i)}} F\wedge\Sigma^{dij}F \xrightarrow{\Sigma^{dij}\phi} \Sigma^{dij}F.$$

It clearly suffices to show the commutativity of the following diagrams for all i,j,k.

$$(1)\quad \begin{array}{ccc} D_j S^{di} \wedge D_k S^{di} & \xrightarrow{\ \alpha\ } & D_{j+k} S^{di} \\ \Big\downarrow \mu_j^{(i)} \wedge \mu_k^{(i)} & & \Big\downarrow \mu_{j+k}^{(i)} \\ \Sigma^{dij} F \wedge \Sigma^{dik} F & \xrightarrow{\ \Sigma^{di(j+k)}\phi\ } & \Sigma^{di(j+k)} F \end{array} \qquad (2)\quad \begin{array}{ccc} D_j D_k S^{di} & \xrightarrow{\ \beta\ } & D_{jk} S^{di} \\ \Big\downarrow D_j \mu_k^{(i)} & & \Big\downarrow \mu_{jk}^{(i)} \\ D_j \Sigma^{dik} F & \xrightarrow{\ \phi^{(ik)}(\xi_j)\ } & \Sigma^{dijk} F \end{array}$$

$$(3)\quad \begin{array}{ccc} D_k(S^{di} \wedge S^{dj}) & \xrightarrow{\ \delta\ } & D_k S^{di} \wedge D_k S^{dj} \\ \Big\| & & \Big\downarrow \mu_k^{(i)} \wedge \mu_k^{(j)} \\ D_k S^{d(i+j)} & & \Sigma^{dik} F \wedge \Sigma^{djk} F \\ & \searrow^{\mu_k^{(i+j)}} \qquad \swarrow^{\Sigma^{d(i+j)k}\phi} & \\ & \Sigma^{d(i+j)k} F & \end{array}$$

In diagram (3) the clockwise composite is $\phi^{(j)}\mu_k^{(i)} = \phi^{(j)}\phi^{(i)}\omega_k = \phi^{(i+j)}\omega_k$. Hence the diagram commutes. Diagrams (1) and (2) commute when $i = 0$ since $e{:}S \to F$ is an H_∞ ring map. They commute when $i = 1$ by the consistency of the μ_j, and for $i \geq 1$ by induction. A similar induction shows that they will commute for all negative i if they do for $i = -1$. We prove commutativity of (2) when $i = -1$; the proof for (1) is similar. We apply Remark 3.7 with $D = D_j D_k$. Give $D_j D_k S^d$ either of the two equal orientations indicated in the second diagram of Lemma 5.3 and let Φ denote the associated Thom isomorphism. Let η_1 be the counterclockwise composite in diagram (2) and let η_2 be the clockwise composite. Clearly, we have $\Phi(\eta_2) = \omega_{jk} \circ \beta$, and since $\omega_{jk} \circ \beta = \xi_j \circ D_j\omega_k$ (this is the case $i = 0$ of diagram (2)) it suffices to show

$$\overline{\Phi}(\eta_1) = \xi_j \circ D_j\omega_k.$$

This is demonstrated by the following commutative diagram.

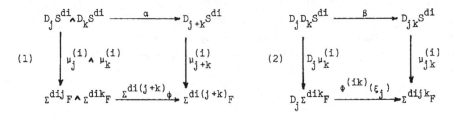

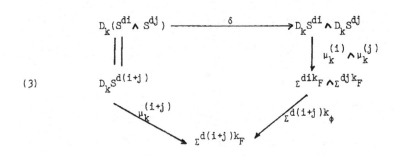

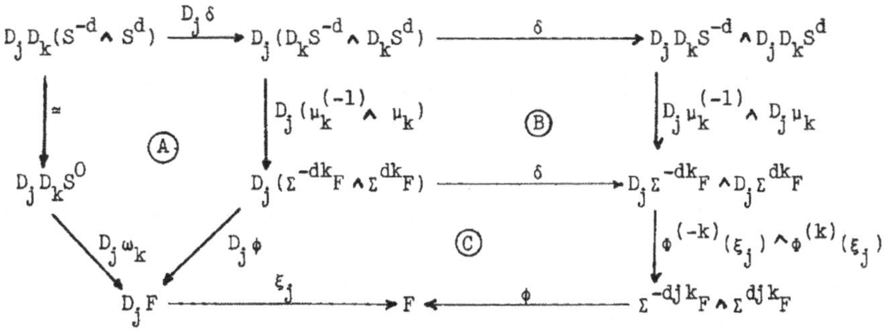

Here part Ⓐ is D_j applied to one case of diagram (3), part Ⓑ commutes by naturality of δ, and part Ⓒ follows from diagram (3) and the fact that ϕ is an H_∞ ring map (see parts (ii) and (iii) of I.3.4). This completes the proof.

§7. K-theory spectra

For our work in chapter IX with Dyer-Lashof operations in K-theory it will be essential to know that the spectrum KU representing periodic complex K-theory is an H_∞ ring spectrum. This is immediate from Corollary 6.3 once one has the necessary space-level input. We begin this section with a quick proof using as input the fact that the connective spectrum kU has an E_∞ ring structure. This in turn raises a consistency question which is settled in the remainder of the section. In VIII §4 we shall use Atiyah's power operations as input to give a more leisurely and elementary proof that KU is an H_∞ ring spectrum. Although we concentrate on the complex case in this section, everything goes through in the orthogonal case with the usual changes.

First recall from [71, VIII §2] that the spectrum kU representing connective complex K-theory is an E_∞ ring spectrum. Hence (as explained in I§4) it is an H_∞ ring spectrum. Throughout this section we will write ξ_j for the structural maps $D_j kU \to kU$. Now by I.3.9 the zero-th space of kU, which we denote by X, is an $H_{\infty 0}$ space with structural maps $D_jX \to X$ which will be denoted by ζ_j. The space X is of course equivalent to $BU \times Z$, and by Bott periodicity we can define an Ω-prespectrum $\mathcal{K}U$ with $\mathcal{K}U_{2i} = X$. We give $\mathcal{K}U$ an H_∞^2 structure by letting each map $D_j\mathcal{K}U_{2i} \to \mathcal{K}U_{2ij}$ be $\zeta_j : D_jX \to X$. We define KU to be $Z\mathcal{K}U$. At this point we need to know something about $\lim^1$ terms.

<u>Proposition 7.1.</u> $\mathcal{K}U$ and $\mathcal{K}O$ are $\lim^1$-free.

<u>Proof</u>. The pair $(\mathcal{K}U, KU)$ clearly satisfies 4.1(i) and (iii). Since $E_r(BU \times Z; KU)$ collapses for dimensional reasons it also satisfies 4.1(ii) and hence is $\lim^1$-

free. The result for $\mathcal{K}O$ follows from 4.4 by letting $f:KU \to KO$ be realification and $g:\mathcal{K}O \to \mathcal{K}U$ be complexification.

Now we can apply 6.3 to get

Theorem 7.2. KU is an H_∞^2 ring spectrum and KO is an H_∞^8 ring spectrum.

Remark 7.3. (i) We shall see in VIII§6 that the H_∞^8 structure of KO extends to an H_∞^4 structure.
(ii) It is shown in [71, VIII. 2.6 and VIII. 2.9] that the Adams operation ψ^k induces an E_∞ ring map of kU when completed away from k. We shall see in VIII§7 that ψ^k also induces an H_∞ ring map of $KU_{(p)}$ for p prime to k but that this is not an H_∞^2 ring map. Since the methods of the present section can only give H_∞^2 ring maps they cannot be applied directly to this question.

Next we wish to show that the H_∞ structure on KU is consistent with the original structure on kU. The point is that (as we shall see in a moment) kU inherits an H_∞ structure from that just given for KU, and we would like to know that the inherited structure is its original one. The proof will occupy the rest of this section.

First recall the n-connected-cover functors in $\overline{h\mathcal{S}}$ ([71, II.2.11]). We write c for the connective (i.e., -1-connected) cover functor. These functors have the usual property that any map from an n-connected spectrum lifts uniquely to the n-connected cover of its target ([71, II.2.10]). In particular, we have

Proposition 7.4. If F is an H_∞ ring spectrum then cF has a unique H_∞ structure for which the map $cF \to F$ is H_∞.

We shall prove

Proposition 7.5. There is an H_∞ ring map from kU (with its E_∞ structure) to cKU (with the H_∞ structure given by 7.2 and 7.4) which is an equivalence.

The analogous comparison of ring structures was given in]71, II§3].

First we observe that the iterated Bott map

$$B:\Sigma^{2i}kU \to kU$$

is equivalent to the (2i-1)-connected cover of kU. We can therefore define

$$\mu_j : D_j S^2 \to \Sigma^{2j} kU$$

to be the unique lift of the composite

$$D_j S^2 \xrightarrow{D_j \Sigma^2 e} D_j \Sigma^2 kU \xrightarrow{D_j B} D_j kU \xrightarrow{\xi_j} kU.$$

The μ_j are consistent Σ_j-orientations in the sense of 6.2 and hence kU is an H_∞^2 ring spectrum. It follows that zkU is an H_∞^2 ring prespectrum. We write

$$\eta_{j,i}: D_j(kU)_{2i} \to (kU)_{2ij}$$

for its structural maps.

Now define a map

$$\gamma: zkU \to \mathcal{K}U$$

by letting γ_{2i} be the composite

$$(zkU)_{2i} \simeq \Omega^\infty \Sigma^{2i} kU \xrightarrow{\Omega^\infty B} \Omega^\infty kU = X = (\mathcal{K}U)_{2i}.$$

We claim that γ is an H_∞^2 ring map. This is demonstrated by the commutativity of the following diagram.

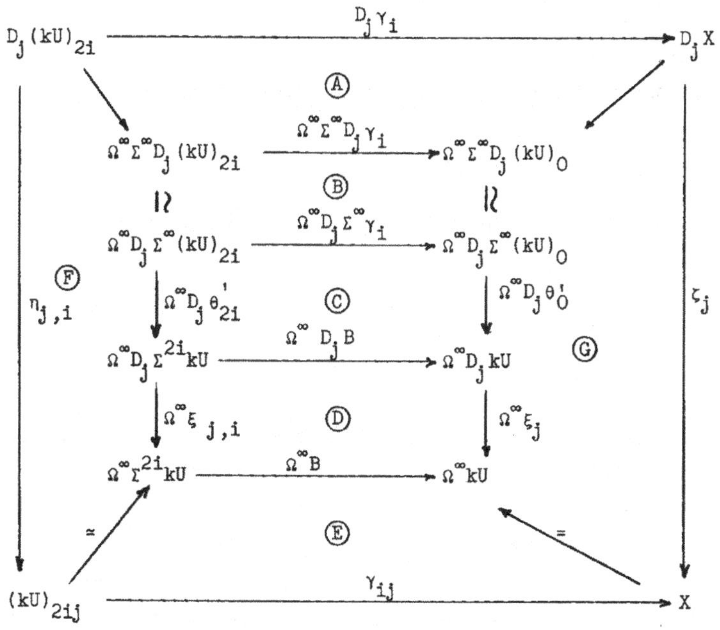

Parts F and G commute by definition of $\eta_{j,i}$ and ζ_j. Parts A and B commute by naturality, parts C and E by the definition of γ. Commutativity of part D follows from the definitiion of μ_i.

Next we need more $\lim^1$ information.

Proposition 7.6. zkU, zkO and the pairs (zkU,KU) and (zkO,KO) are $\lim^1$ free.

Proof. The Serre spectral sequence shows that the pairs (zkU,kU) and (zkU,KU) satisfy the finiteness requirement of 4.1(i) and (iii). Now by [10,4.3] and the proof of [10,3.13] (specifically the fifth line on p.301) we see that the pair $((kU)_{2i},kU)$ is M-L for each i and hence zkU is $\lim^1$-free. Since

$$E_2^{p,q}((kU)_{2i};KU) = E_2^{p,q}((kU)_{2i};kU)$$

for $q \leq 0$ it follows that $Z_\infty^{p,q}((kU)_{2i});KU)$ has finite index in $E_2^{p,q}((kU)_{2i};KU)$ for $q \leq 0$, hence for all q by Bott periodicity. Thus the pair (kU,KU) is $\lim^1$-free. The orthogonal case follows as in the proof of 7.1.

We can now define

$$\Gamma : kU \to KU$$

to be $Z_Y \circ \lambda^{-1}$, where Z and λ are as in §1. Then Γ is an H_∞^2 ring map by 6.3 and is clearly an equivalence of zeroth spaces. Hence the unique lift of Γ to cKU is an H_∞^2 ring map and an equivalence. This completes the proof of 7.5.

The fact that Γ is an H_∞^2 ring map, and thus preserves the orientations, has the following additional consequence which will be used in VIII §4.

Corollary 7.7. $\mu_j : D_j S^2 \to \Sigma^{2j} KU$ is the composite

$$D_j S^2 \xrightarrow{D_j \Sigma^2 e} D_j \Sigma^2 KU \xrightarrow{D_j B} D_j KU \xrightarrow{\xi_j} KU \xrightarrow{B^{-2j}} \Sigma^{2j} KU.$$

§8. A Thom isomorphism for spectra

In this section we prove Theorem 3.3. This is the only place in our work where we need the actual definition of D_π, instead of just its formal properties. We accordingly begin by giving a form of the definition; for a general discussion see the sequel.

Let $\mathcal{L}(j)$ be the space of linear isometries from $(R^\infty)^j$ to R^∞. Then $\mathcal{L}(j)$ is a free contractible π-space and hence there is a π-map $\chi : E\pi \to \mathcal{L}(j)$. Choose an increasing sequence W_i of finite π-subcomplexes of $E\pi$ with $\bigcup_i W_i = E\pi$. If $V \subset (R^\infty)^j$ is a finite-dimensional subspace then (since W_i is compact) the union

$$\bigcup_{w \in W_i} \chi(w)(V) \subset R^\infty$$

is contained in a finite-dimensional subspace. In particular, if we let A_i be the standard copy of R^{di} in R^∞ then there is a finite-dimensional subspace A_i' of R^∞ with

$$\chi(w)(A_i \oplus \cdots \oplus A_i) \subset A_i'$$

for every $w \in W_i$. Let a_i be the dimension of A_i'. We may assume that the A_i' form an increasing sequence, and we write B_i and B_i' for the orthogonal complements of A_i in A_{i+1} and of A_i' in A_{i+1}'.

Now consider the map from $W_i \times (A_i)^j$ to $W_i \times A_i'$ which takes $(w, x_1, \ldots, x_j)$ to $(w, \chi(w)(x_1 \oplus \cdots \oplus x_j))$. This gives an embedding of the trivial bundle

(1) $\quad W_i \times (A_i)^j \to W_i$

in the trivial bundle

(2) $\quad W_i \times A_i' \to W_i$.

The orthogonal complement is a nontrivial vector bundle over W_i. We let η_i be the associated sphere bundle (obtained by fibrewise one-point compactification). We write $S(\eta_i)$ and $T(\eta_i)$ for the total space and the Thom complex of η_i. If we let π act through permutations on $(A_i)^j$ and trivially on A_i' we obtain diagonal actions on the bundles (1) and (2) and hence on $S(\eta_i)$ and $T(\eta_i)$.

Next observe that the diagram of embeddings

$$\begin{array}{ccc}
W_i \times (A_i)^j & \longrightarrow & W_{i+1} \times (A_i)^j \\
\downarrow & & \downarrow \\
W_i \times A_i' & & W_{i+1} \times (A_{i+1})^j \\
\downarrow & & \downarrow \\
W_i \times A'_{i+1} & \longrightarrow & W_{i+1} \times A'_{i+1}
\end{array}$$

commutes. Hence there is a bundle map

$$\eta_i \oplus B_i' \to \eta_{i+1} \oplus (B_i)^j$$

covering the inclusion $W_i \to W_{i+1}$. The induced map

$$T(\eta_i) \wedge S^{B_i'} \longrightarrow T(\eta_{i+1}) \wedge (S^{B_i})^{(j)}$$

of Thom complexes is a π-map if we give each side the diagonal π-action; here S^{B_i} is the one-point compactification of B_i, etc.

Now let U be a prespectrum (indexed on multiples of d as usual). We define a new prespectrum U^χ indexed on the set $\{a_i\}$ as follows (we haven't previously considered prespectra indexed on sets like $\{a_i\}$, but everything in section 1 goes through with the obvious modifications). Let $(U^X)_{a_i}$ be the space

$$T(\eta_i) \wedge_\pi (U_{di})^{(j)}$$

with the structural maps σ indicated in the following diagram.

$$\begin{array}{ccccc}
\Sigma^{a_{i+1}-a_i} T(\eta_i) \wedge_\pi (U_{di})^{(j)} & \cong & (T(\eta_i) \wedge S^{B_i^!}) \wedge_\pi (U_{di})^{(j)} & \longrightarrow & (T(\eta_{i+1}) \wedge (S^{B_i})^{(j)}) \wedge_\pi (U_{di})^{(j)} \\
\downarrow \sigma & & & & \wr\!\! \parallel \\
T(\eta_{i+1}) \wedge_\pi (U_{d(i+1)})^{(j)} & \longleftarrow & T(\eta_{i+1}) \wedge_\pi (\Sigma^d U_{di})^{(j)} & \cong & T(\eta_{i+1} \;{}_\pi (S^{B_i} \wedge U_{di})^{(j)}
\end{array}$$

Finally, given $E \in \overline{h\mathcal{J}}$ we choose a prespectrum U with $ZU \simeq E$ (for example, we could let $U = zE$) and define

$$D_\pi E = Z(U^X) = \underset{i}{\mathrm{Tel}}\; \Sigma^{-a_i} \Sigma^\infty [T(\eta_i) \wedge_\pi (U_{di})^{(j)}].$$

This agrees up to weak equivalence with the more sophisticated definition given in the sequel, and in particular it does not depend on the choice of χ or U.

Now we can give the proof of 3.3. First we observe that the Thom isomorphism theorem holds in F-cohomology of <u>spaces</u> for any F-orientable bundle. This is well-known when the base space is finite-dimensional (see e.g. [71,III. 1.4]) and the general case follows since the Thom homomorphism induces a map of Milnor $\lim^1$ sequences. Similarly, the relative Thom isomorphism theorem holds for any F-oriented bundle over a pair (X,Y). For example, let U be a prespectrum, let

$$X = S(\eta_i) \times_\pi (U_{di})^j$$

and let Y be the subspace in which at least one coordinate is a point at ∞ or the basepoint of U_{di}. Note that X/Y is $(U^X)_{a_i}$. Let q be the pullback of the bundle

$$p: E\pi \times_\pi (R^d)^j \to B\pi$$

along the map

$$X = S(\eta_i) \times_\pi (U_{di})^j \to E\pi \times_\pi * = B\pi.$$

Then the relative Thom complex $T(q)/T(q|Y)$ is

$$T(\eta_i) \;{}_\pi (\Sigma^d U_{di})^{(j)} = (\Sigma^d U)^X_{a_i} .$$

Let δ_i denote the composite indicated in the following diagram.

$$\begin{array}{ccc}
(\Sigma^d U)^X_{a_i} = T(\eta_i)\wedge_\pi(\Sigma^d U_{di})^{(j)} & \xrightarrow{\Delta\wedge 1} & (T(\eta_i)\wedge E\pi^+)\wedge_\pi(\Sigma^d U_{di})^{(j)} \\
\downarrow{\scriptstyle \delta_i} & \swarrow & \\
U^X_{a_i}\wedge D_\pi S^d = [T(\eta_i)\wedge_\pi(U_{di})^{(j)}]\wedge[E\pi^+\wedge_\pi(S^d)^{(j)}] & &
\end{array}$$

If F is a π-oriented ring spectrum then the relative Thom isomorphism for q is the composite

$$F^n(U^X_{a_i}) \longrightarrow F^{n+dj}(U^X_{a_i}\quad D_\pi S^d) \xrightarrow{\delta_i^*} F^{n+dj}((\Sigma^d U)^X_{a_i}),$$

where the first map is multiplication by the π-orientation μ. We denote this composite by Φ_i.

Next, we note that if $E \simeq ZU$ then $\Sigma^d E \simeq Z(\Sigma^d U)$. It is shown in the sequel that the map

$$\delta : D_\pi \Sigma^d E \to D_\pi E \wedge D_\pi S^d$$

is obtained by passage to telescopes from the δ_i. We therefore have a map of Milnor $\lim^1$ sequences

$$\begin{array}{ccccccccc}
0 & \to & \lim\limits_i{}^1 F^{n+a_i-1}(U^X_{a_i}) & \longrightarrow & F^n D_\pi E & \to & \lim\limits_i F^{n+a_i}(U^X_{a_i}) & \to & 0 \\
 & & \downarrow{\scriptstyle \lim^1 \Phi_i} & & \downarrow{\scriptstyle \Phi} & & \downarrow{\scriptstyle \lim \Phi_i} & & \\
0 & \to & \lim\limits_i{}^1 F^{n+dj+a_i-1}((\Sigma^d U)^X_{a_i}) & \to & F^{n+dj} D_\pi \Sigma^d E & \to & \lim\limits_i F^{n+dj+a_i}((\Sigma^d U)^X_{a_i}) & \to & 0
\end{array}$$

The result follows by the five lemma.

We conclude this section with a technical fact which will be needed in VIII §6. Let $\zeta:(\pi,T) \to U$ be an extended pairing and suppose that the pair (T,ZU) is $\lim^1$-free. Then $Z\zeta$ exists and is clearly determined by the composites

$$T(\eta_i)\wedge_\pi(T_{di})^{(j)} = T^X_{a_i} \xrightarrow{\kappa} (D_\pi ZT)_{a_i} \xrightarrow{(Z\zeta)_{a_i}} (ZU)_{a_i}$$

for $i \geq 0$. It is natural to ask for an explicit description of the elements

$$z_i \in (ZU)^{a_i}(T(\eta_i)\wedge_\pi(T_{di})^{(j)})$$

represented by these composites. We shall give such a description by calculating the image of z_i under the relative Thom isomorphism

$$\Psi : (ZU)^{a_i}(T(\eta_i)\wedge_\pi(T_{di})^{(j)}) \longrightarrow (ZU)^{a_i+dij}(T(\eta_i)\wedge_\pi(\Sigma^{di}T_{di})^{(j)}).$$

Let $y_i \in (ZU)^{dij}(W_i^+ \wedge_\pi (T_{di})^{(j)})$ be represented by the composite

$$W_i^+ \wedge_\pi (T_{di})^{(j)} \hookrightarrow D_\pi T_{di} \xrightarrow{\zeta_i} U_{dij} \xrightarrow{\kappa} (ZU)_{dij}$$

and recall the homeomorphism

$$T(n_i) \wedge_\pi (\Sigma^{di} T_{di})^{(j)} \cong \Sigma^{a_i} W_i^+ \wedge_\pi (T_{di})^{(j)}.$$

Proposition 8.1. $\Psi z_i = \Sigma^{a_i} y$.

Proof. Write a for a_i. It will be shown in the sequel that the following diagram commutes for any space X.

$$\begin{array}{ccc} T(n_i) \wedge_\pi (\Sigma^{di} X)^{(j)} & \xrightarrow{\kappa_a} & (D_\pi \Sigma^\infty X)_a \\ \wr\!\!\! \| & & \wr\!\!\! \| \\ \Sigma^a (W_i^+ \wedge_\pi (T_{di})^{(j)}) \hookrightarrow \Sigma^a D_\pi T_{di} & \xrightarrow{\kappa_a} & (\Sigma^\infty D_\pi X)_a \end{array}$$

Letting $X = T_{di}$ gives the commutativity of the left square in the next diagram.

$$\begin{array}{ccccc} \Sigma^\infty(T(n_i) \wedge_\pi (\Sigma^{di} T_{di})^{(j)}) & \xrightarrow{\theta_a} & \Sigma^a D_\pi \Sigma^\infty T_{di} & \xrightarrow{\Sigma^a D_\pi \theta_{di}} & \Sigma^a D_\pi \Sigma^{di} ZT \\ \wr\!\!\! \| & & & & \\ \Sigma^\infty \Sigma^a (W_i^+ \wedge_\pi (T_{di})^{(j)}) & & \downarrow \simeq & & \downarrow \Sigma^a \phi(i) Z\zeta \\ \downarrow & & & & \\ \Sigma^\infty \Sigma^a D_\pi T_{di} & \xrightarrow{\simeq} & \Sigma^a \Sigma^\infty D_\pi T_{di} \xrightarrow{\Sigma^a \Sigma^\infty \zeta_i} \Sigma^a \Sigma^\infty U_{dij} & \xrightarrow{\Sigma^a \theta_{dij}} & \Sigma^{a+dij} ZU \end{array}$$

The right square commutes by Corollary 3.6(iii), and we therefore have equality of the two composites around the outside. But the counterclockwise composite is clearly $\Sigma^a y_i$, and the proof of Theorem 3.3 given in this section shows that the clockwise composite is Ψz_i. This completes the proof.

CHAPTER VIII

POWER OPERATIONS IN H_∞^d RING THEORIES

by J. E. McClure

It was shown in Chapter I that an H_∞^d ring structure on a spectrum E induces certain operations $\mathcal{P}_j$ in E-cohomology. In this chapter we investigate these operations in some important special cases, namely ordinary cohomology, K-theory, and cobordism.

In section 1 we collect the properties of the $\mathcal{P}_j$ and their internal variants P_j; most of these have already been shown in Chapter I. We also show that the results of Chapter VII allow one to construct an H_∞^d structure on E by giving space-level operations with certain properties. The section concludes with a brief account of a multiplicative transfer in E-cohomology which generalizes the norm map of Evens [35].

In section 2 we show that the general facts given in section 1 are strong enough to prove the usual properties of the Steenrod operations without any use of chain-level arguments. In section 3 we show that the same arguments applied to the spectrum $HZ_p \wedge X$ give the Dyer-Lashof operations in $H_*(X;Z_p)$ with all of their usual properties; in particular, we give new proofs of the Adem and Nishida relations which involve less calculation than the standard proofs.

In section 4 we show that the power operations in K-theory induced by the H_∞^d structures on KU and KO are precisely those defined by Atiyah [17]; this gives a rather concrete description of these H_∞^d structures. In section 5 we show that cobordism operations defined by tom Dieck in [31] lead to H_∞^d structures on the classical cobordism spectra which agree with their E_∞ structures; again, this fact gives a rather concrete homotopical description of the E_∞ structure. In section 6 we show that the Atiyah-Bott-Shapiro orientations are H_∞^d ring maps; it is still an open question whether they are E_∞ maps.

In section 7 we show that questions about H_∞^d ring maps simplify considerably when the spectra involved are p-local. We use this to show that the Adams operations are H_∞ ring maps (a fact which will be important in Chapter IX) and that the Adams summand of p-local K-theory is an H_∞^2 ring spectrum. We also give a sufficient condition for BP to be an H_∞^2 ring spectrum; however the question of whether it actually is an H_∞^2 ring spectrum remains open.

<u>Notation</u>. In chapters VIII and IX we shall write ΣX for $S^1 \wedge X$, instead of $X \wedge S^1$ as in chapters I-VII. We shall also use Σ to denote the suspension

isomorphism $\tilde{E}^n X \to \tilde{E}^{n+1} X$. In particular, if E is a ring spectrum the fundamental class in $\tilde{E}^n S^n$ will be denoted by $\Sigma^n 1$.

§1. General properties of power operations

Let E and F be spectra, let π be a subgroup of Σ_k, and let d be a fixed positive integer. By a power operation on $\overline{h\mathscr{S}}$ in the most general sense we mean simply a sequence $\mathscr{P}_\pi$ of natural transformations

$$E^{di}X \to F^{dik} D_\pi X,$$

one for each $i \in Z$, which are defined for all $X \in \overline{h\mathscr{S}}$. We shall also call $\mathscr{P}_\pi$ an (E,π,F) power operation when it is necessary to be more specific. In this section we consider the relation between power operations, extended pairings, and H_∞^d ring structures. In particular, we collect the properties of the canonical power operations associated to an H_∞^d ring structure and of the related internal operations.

The most important class of power operations for us will be the operations

$$\mathscr{P}_\pi : E^{di}X \to E^{dij} D_\pi X$$

determined by an H_∞^d ring structure on E. As usual, we abbreviate $\mathscr{P}_{\Sigma_j}$ by $\mathscr{P}_j$. Recall the definition from I§4: if $x \in E^{di}X$ is represented by $f: X \to \Sigma^{di}E$ then $\mathscr{P}_\pi x$ is represented by the composite

$$D_\pi X \xrightarrow{\iota} D_k X \xrightarrow{D_k f} D_k \Sigma^{di} E \xrightarrow{\xi_{k,i}} \Sigma^{dik} E.$$

Our first result collect the properties of these operations.

Proposition 1.1. Let E be an H_∞^d ring spectrum and let $x \in E^{di}X$, $y \in E^{dj}Y$, $\pi \subset \Sigma_k$,

(i) $\alpha^* \mathscr{P}_{j+k} x = (\mathscr{P}_j x)(\mathscr{P}_k x) \in E^{d(j+k)i}(D_j X \wedge D_k X)$.

(ii) $\beta^* \mathscr{P}_{jk} x = \mathscr{P}_j \mathscr{P}_k x \in E^{djki}(D_j D_k X)$

(iii) $\delta^*[(\mathscr{P}_\pi x)(\mathscr{P}_\pi y)] = \mathscr{P}_\pi(xy) \in E^{d(i+j)k} D_\pi (X \wedge Y)$.

(iv) $\iota^* \mathscr{P}_\pi x = x^k \in E^{dij}(X^{(k)})$

(v) If $1 \in E^0 S$ is the unit then $\mathscr{P}_\pi 1$ is the unit in $E^0(D_\pi S) = E^0(B\pi^+)$.

(vi) If X = Y and i = j then

$$\mathscr{P}_k(x+y) = \mathscr{P}_k x + \mathscr{P}_k y + \sum_{0<\ell<k} \tau^*_{\ell,k-\ell}[(\mathscr{P}_\ell x)(\mathscr{P}_{k-\ell} y)]$$

in $E^{dik} D_k X$, where

$$\tau_{\ell,k-\ell}:D_kX \longrightarrow D_\ell X \wedge D_{k-\ell}X$$

is the transfer defined in II.1.4.

(vii) If E is p-local then $\mathcal{P}_\pi x = \frac{1}{|\pi|}\tau_\pi^* x^k$ whenever $|\pi|$ is prime to p, where τ_π is the transfer $D_\pi X \to X^{(k)}$ of II.1.4.

(viii) If E is p-local then

$$\mathcal{P}_p(x+y) = \mathcal{P}_p x + \mathcal{P}_p y + \tau_p^*[\frac{1}{p!}((x+y)^p - x^p - y^p)].$$

<u>Proof</u>. (i), (ii), and (iii) are immediate from Definition I.4.3. Part (iv) follows from Remark I.4.4. Part (v) follows from I.3.4(i). Parts (vi) and (viii) were shown in II.2.1 and II.2.2, and part (vii) follows from the proof of the latter.

We shall also want to go in the other direction, that is, to start from a set of operations having certain properties and deduce the existence of an H_∞^d ring structure. Let E be a ring spectrum. We say that a set $\{\mathcal{P}_j\}_{j\geq 0}$ of (E,Σ_j,E) power operations is <u>consistent</u> if it satisfies 1.1(i), (ii), and (iii). Given a consistent set of operations $\mathcal{P}_j$ on E we can define maps

$$\xi_{j,i}:D_j\Sigma^{di}E \to \Sigma^{dij}E$$

by applying $\mathcal{P}_j$ to the classes represented by the identity maps $\Sigma^{di}E \to \Sigma^{di}E$. It is easy to see that the $\xi_{j,i}$ form an H_∞^d ring structure on E whose induced power operations are the given $\mathcal{P}_j$. On the other hand, two H_∞^d ring structures on E which determine the same power operations are clearly equal. Thus there is a one-to-one correspondence between H_∞^d ring structures on E and consistent sets of (E,Σ_j,E) power operations.

Next we consider a more general situation. Let π be a subgroup of Σ_k and let F be a π-oriented ring spectrum with orientation $\mu:D_\pi S^d \to \Sigma^{dk}F$ (see VII§3). The class in $F^{dk}(D_\pi S^d)$ represented by the orientation will also be denoted by μ. An (E,π,F) power operation $\mathcal{P}_\pi$ is <u>stable</u> if the equation

$$(1) \qquad \mathcal{P}_\pi(\Sigma^d x) = \delta^*(\mu \cdot \mathcal{P}_\pi x)$$

holds in $F^{d(i+1)k}(D_\pi\Sigma^d X)$ for all $x \in E^{di}X$. 1.1(iii) implies that the (E,π,E) power operations determined by an H_∞^d ring structure on E are stable. More generally, let $\xi:D_\pi E \to F$ be any map (in the terminology of VII§3, ξ is called an extended pairing). If $x \in E^{di}X$ is represented by $f:X \to \Sigma^{di}E$ define $\mathcal{P}_\pi x \in F^{dik}D_\pi X$ to be the element represented by the composite

$$D_\pi X \xrightarrow{D_\pi f} D_\pi\Sigma^{di}E \xrightarrow{\delta} (D_\pi S^d)^{(i)} \wedge D_\pi E \xrightarrow{\mu^{(i)}\wedge\,\xi} (\Sigma^{dk}F)^{(i)} \wedge F \xrightarrow{\phi} \Sigma^{dik}F,$$

where ϕ is the product map for F. Then $\mathcal{P}_\pi$ is a stable power operation. Conversely, given a stable operation $\mathcal{P}_\pi$ we obtain a map $\xi: D_\pi E \to F$ by applying $\mathcal{P}_\pi$ to the identity map $E \to E$. Clearly, this gives a one-to-one correspondence between maps $\xi: D_\pi E \to F$ and stable power operations. To sum up, we have shown

Proposition 1.2. (i) There is a one-to-one correspondence between consistent sets of (E,Σ_j,E) power operations and H_∞^d ring structures on E.

(ii) If F is a π-oriented ring spectrum and E is any spectrum, there is a one-to-one correspondence between stable (E,π,F) power operations and maps $\xi: D_\pi E \to F$.

For applications of 1.2 it is usually easiest to work with space-level instead of spectrum-level power operations. Our next result will allow us to reduce to this case. Let $\mathcal{C}$ be the homotopy category of finite CW complexes. Let $\{(E\pi)_\alpha\}_{\alpha\in A}$ be the set of finite π-subcomplexes of $E\pi$. By an (E,π,F) power operation on $\mathcal{C}$ we mean a sequence $\mathcal{P}_\pi$ of natural transformations

$$\tilde{E}^{di}X \to \lim_{\alpha} \tilde{F}^{dik}((E\pi)_\alpha^+ \wedge_\pi X^{(k)}),$$

one for each $i \in Z$, which are defined for all $X \in \mathcal{C}$. $\mathcal{P}_\pi$ is stable if it satisfies equation (1). A set $\{\mathcal{P}_j\}_{j>0}$ of (E,Σ_j,E) power operations on $\mathcal{C}$ is consistent if it satisfies 1.1(i),(ii) and (iii). Recall the cylinder construction Z from VII§1.

Proposition 1.3. (i) Let T be a prespectrum and suppose that each T_{di} has the homotopy type of a countable CW-complex. Let F be a ring spectrum. If the pair (T,F) is $\lim^1$-free in the sense of VII.4.1 then every stable (ZT,π,F) operation on $\mathcal{C}$ extends uniquely to a stable operation on $\overline{h\mathcal{S}}$.

(ii) Let E be a ring spectrum and suppose that each E_{di} has the homotopy type of a countable CW-complex and that zE is $\lim^1$-free. Then every consistent set $\{\mathcal{P}_j\}$ of (E,Σ_j,E) operations on $\mathcal{C}$ extends uniquely to a consistent set of operations on $\overline{h\mathcal{S}}$.

Proof. For part (i), let $\{X_{i,\beta}\}$ be the set of finite subcomplexes of T_{di} and let $x_{i,\beta} \in \tilde{E}^{di}X_{i,\beta}$ be the class of the inclusion map $X_{i,\beta} \to T_{di}$. The elements $\mathcal{P}_\pi(x_{i,\beta})$ determine an element of $\lim_{\alpha,\beta} \tilde{F}^{dik}((E\pi)_\alpha^+ \wedge_\pi X_{i,\beta})$ and hence of $\tilde{F}^{dik}D_\pi T_{di}$ by VII.4.10 and VII.4.12. It is easy to see that the maps $\zeta_i: D_\pi T_{di} \to F_{dik}$ representing these elements form an extended pairing of prespectra as defined in VII.3.2. Part (i) now follows from VII.3.4. For part (ii), a similar argument shows that the set $\{\mathcal{P}_j\}$ determines an H_∞^d ring structure on the prespectrum zE and the result follows from VII.6.3.

The definitions we have given are closely related to tom Dieck's axioms for "generalized Steenrod operations" [31]. Let E be a ring spectrum. In tom Dieck's terminology, a generalized Steenrod operation is what we have called an (E,π,E) power operation. His axioms P1 and P2 are 1.1(iv) and 1.1(ii) respectively. In particular, if $\mathcal{P}_\pi$ satisfies P1 then $\mathcal{P}_\pi\Sigma^d 1$ is a π-orientation for E. Axiom P3 is equation (1) above with $\mu = \mathcal{P}_\pi\Sigma^d 1$. Thus an operation satisfying P1 and P3 is stable in our sense (but not conversely). tom Dieck's final axiom P4 will also be of interest in what follows. If q is a vector bundle over X then $E\pi \times_\pi q^k$ is a vector bundle over $E\pi \times_\pi X^k$ whose Thom complex is homeomorphic to $D_\pi T(q)$. If ν is an E-orientation for q and $\mathcal{P}_\pi$ is an operation satisfying P1 then $\mathcal{P}_\pi(\nu)$ is clearly an E-orientation for $E\pi \times_\pi q^k$. Axiom P4 is the statement that E has canonical orientations for some class of vector bundles and that $\mathcal{P}_\pi$ takes the canonical orientation for q to that for $E\pi \times_\pi q^k$. This axiom will be satisfied in all of the particular cases considered in this chapter.

From now on we fix an H_∞^d ring spectrum E and let $\mathcal{P}_\pi$ denote the associated power operations. Let X be a space. Let Δ be the diagonal map

$$X \wedge B\pi^+ = X \wedge D_\pi S^0 \to D_\pi(X \wedge S^0) = D_\pi X$$

defined in II.3.1. We define the <u>internal power operation</u>

$$P_\pi : \tilde{E}^{di}X \to \tilde{E}^{dik}(X \wedge B\pi^+)$$

to be the composite

$$\tilde{E}^{di}X \xrightarrow{\mathcal{P}_\pi} \tilde{E}^{dik}D_\pi X \xrightarrow{\Delta^*} \tilde{E}^{dik}(X \wedge B\pi^+).$$

Since $X^+ \wedge B\pi^+ = (X \times B\pi)^+$ we obtain an unreduced operation

$$P_\pi : E^{di}X \to E^{dik}(X \times B\pi).$$

Our next result summarizes the properties of the unreduced operations; similar statements hold for the reduced ones.

<u>Proposition 1.4</u>. Let $x \in E^{di}X$, $y \in E^{dj}X$, $\pi \subset \Sigma_k$.

(i) $\iota^* P_\pi x = x^k \in E^{dik}X$

(ii) $P_\pi 1 = 1 \in E^0(X \times B\pi)$

(iii) $P_\pi(xy) = (P_\pi x)(P_\pi y) \in E^{d(i+j)k}(X \times B\pi)$

(iv) If $i = j$ then

$$P_k(x+y) = P_k x + P_k y + \sum_{0<\ell<k} (\tau_{\ell,k-\ell})^* [(P_\ell x)(P_{k-\ell} y)]$$

(v) If E is p-local and $|\pi|$ is prime to p then $P_\pi x = \frac{1}{|\pi|} x^k \tau_\pi^* 1$.

(vi) If E is p-local then

$$P_p(x+y) = P_p x + P_p y + \frac{1}{p!}[(x+y)^p - x^p - y^p)](\tau_p^* 1).$$

(vii) If $\pi \subset \Sigma_k$ is generated by a k-cycle and $\pi' \subset \Sigma_\ell$ is generated by an ℓ-cycle then

$$(1 \times \gamma)^* P_\pi P_{\pi'} x = P_{\pi'} P_\pi x \in E^{dik\ell}(X \times B\pi \times B\pi'),$$

where $\gamma: B\pi \times B\pi' \to B\pi' \times B\pi$ switches the factors .

Proof. All parts except (vii) are immediate from 1.1. For (vii) we use the argument of [100, VIII.1.3]. If we give the set $\pi \times \pi'$ its lexicographic order we obtain a faithful action of $\Sigma_{k\ell}$ on it. Let $g \in \Sigma_{k\ell}$ be the element which switches the factors π and π'. The following diagram is readily seen to commute.

$$\begin{array}{ccccccc} \pi \times \pi' & \xrightarrow{d} & \pi \int \pi' & \xrightarrow{\iota} & \Sigma_k \int \Sigma_\ell & \xrightarrow{\beta_{k,\ell}} & \Sigma_{k\ell} \\ \downarrow \gamma & & & & & & \downarrow c_g \\ \pi' \times \pi & \xrightarrow{d} & \pi' \int \pi & \xrightarrow{\iota} & \Sigma_\ell \int \Sigma_k & \xrightarrow{\beta_{\ell,k}} & \Sigma_{k\ell} \end{array}$$

Here d is the evident diagonal and c_g is conjugation by g. By 1.1(ii) we have

$$P_{\pi'} P_\pi x = \Delta^* d^* \iota^* \beta_{k,\ell}^* \mathcal{P}_{k\ell} x = (1 \times \beta_{k,\ell} \circ \iota \circ d)^* P_{k\ell} x$$

and similarly

$$P_\pi P_{\pi'} x = (1 \times \beta_{\ell,k} \circ \iota \circ d)^* P_{k\ell} x.$$

But $(1 \times c_g)^* P_{k\ell} x = P_{k\ell} x$ since $c_g: B\Sigma_{k\ell} \to B\Sigma_{k\ell}$ is homotopic to the identity.

We conclude this section with a brief description of another kind of operation induced by H_∞^d structures, namely a multiplicative version of the transfer for finite coverings. The definition is due to May. First recall the definition of the ordinary (additive) transfer. If $p: X \to B$ is a j-fold covering then one can construct a map

$$\tilde{p}: B \to E\Sigma_j \times_{\Sigma_j} X^j$$

as in [8, p.112]. If $x \in F^i X$ is represented by $f: X \to F_i$ then $p_! x \in F^i B$ is represented by

$$B \xrightarrow{\tilde{p}} E\Sigma_j \times_{\Sigma_j} X^j \xrightarrow{1 \times f^j} E\Sigma_j \times_{\Sigma_j} (F_i)^j \longrightarrow F_i ,$$

where the last map is the Dyer-Lashof map determined by the infinite loop space structure on F_i. B Now if F is an H_∞^d ring spectrum and if $x \in F^{di} X$ is represented by $f: \Sigma(X^+) \to \Sigma^{di} F$ we define $p_\otimes x \in F^{dij} B$ to be the element represented by

$$\Sigma^\infty(B^+) \xrightarrow{\Sigma^\infty(\tilde{p}^+)} \Sigma^\infty(E\Sigma_j \times_{\Sigma_j} X^j)^+ \simeq D_j \Sigma^\infty X^+ \xrightarrow{D_j f} D_j \Sigma^{di} F \xrightarrow{\xi_{j,i}} \Sigma^{dij} F.$$

If F is merely H_∞ one can give the same definition in degree zero. Our next result records some properties of $p_\otimes$.

<u>Proposition 1.5</u> (i) $p_\otimes 1 = 1$, $p_\otimes 0 = 0$.

(ii) $p_\otimes(xy) = (p_\otimes x)(p_\times y)$

(iii) If $q:Y \to X$ is a k-fold covering then $(pq)_\otimes = p_\otimes q_\otimes$

(iv) $f^* p'_\otimes = p_\otimes g^*$ for a pullback diagram

$$\begin{array}{ccc} X & \xrightarrow{\ g\ } & X' \\ \downarrow{\scriptstyle p} & & \downarrow{\scriptstyle p'} \\ B & \xrightarrow{\ f\ } & B' \end{array}$$

(v) If Y is any space and $x \in F^{di}X$, $y \in F^{dk}Y$ then

$$(1 \times p)_\otimes(y \times x) = [(1 \times h)^* P_j y](p_\otimes x) \in F^{dj(i+k)}(Y \times B)$$

where $h:B \to B\Sigma_j$ is the classifying map of p.

<u>Proof</u>. Part (i) is trivial and parts (iii) and (iv) have the same proofs as in the additive case. For part (ii) let $f:\Sigma^\infty(X^+) \to \Sigma^{di}F$ and $g:\Sigma^\infty(X^+) \to \Sigma^{dk}F$ represent x and y. It suffices to show commutativity of the following diagram, in which Σ^∞ has been suppressed to simplify the notation.

$$\begin{array}{ccccccccc} B^+ & \xrightarrow{\tilde{p}^+} & D_j X^+ & \xrightarrow{D_j \Delta} & D_j(X^+ \wedge X^+) & \xrightarrow{D_j(f\, g)} & D_j(\Sigma^{di}F \wedge \Sigma^{dk}F) & \longrightarrow & D_j \Sigma^{d(i+k)} \\ \downarrow{\scriptstyle \Delta} & & & & \downarrow{\scriptstyle \delta} & & \downarrow{\scriptstyle \delta} & & \searrow \\ & & & & & & & & \Sigma^{dj(i+k)}F \\ & & & & & & & & \nearrow \\ B^+ \wedge B^+ & \xrightarrow{\tilde{p}^+ \wedge \tilde{p}^+} & & & D_j X^+ \wedge D_j X^+ & \xrightarrow{D_j f \wedge D_j g} & D_j \Sigma^{di}F \wedge D_j \Sigma^{dk}F & \longrightarrow & \Sigma^{dji}F \wedge \Sigma^{djk}F \end{array}$$

The pentagon commutes by I.4.3 and the remaining pieces by naturality. For part (v) it suffices by (ii) to show

$$(1 \times p)_\otimes(\pi^* y) = (1 \times h)^* P_j y$$

where $\pi:Y \times X \to Y$ is the projection. An inspection of [8, p.112] shows that the diagram

$$\begin{array}{ccc} Y^+ \wedge B^+ & \xrightarrow{\widetilde{(1 \times p)}^+} & D_j(Y^+ \wedge X^+) \\ \downarrow{\scriptstyle 1 \wedge h^+} & & \downarrow{\scriptstyle D_j \pi^+} \\ Y^+ \wedge B\Sigma_j^+ & \xrightarrow{\ \Delta^+\ } & D_j Y^+ \end{array}$$

commutes and the results follows.

Remarks 1.6.(i) Formula (v) is due to Brian Sanderson (also cf. [35, remark 6.2]). If we let $p:X \to B\Sigma_j$ be the j-fold cover associated to $E\Sigma_j \to B\Sigma_j$ and let $x = 1$ then the formula gives

$$(1 \times p)_{\otimes}(y \times 1) = P_j y,$$

so that the internal operation P_j is completely determined by the multiplicative transfer, an observation also due to Sanderson.

(ii) If $p:X \to B$ and $q:Y \to C$ are any two coverings then $p \times q$ is a covering which factors as $(p \times 1)(1 \times q)$. We can therefore compute $(p \times q)_{\otimes}(x \times y)$ in principle by using formulas (ii), (iii) and (v), but there is no simple external analog of formula (ii).

(iii) If F is H_∞^d then $\bigvee_{i \in Z} \Sigma^{di} F$ is H_∞ by II.1.3. Thus we can define a map

$$p_{\otimes} : \prod_{i \in Z} F^{di} X \longrightarrow \prod_{i \in Z} F^{di} B$$

which agrees on homogeneous elements with that already given. We leave it as an exercise for the reader to show that if x has nonzero degree then $p_{\otimes}(1 + x)$ has components $p_! x$ in degree $|x|$ and $p_\times x$ in degree $j|x|$ (cf. [35, Theorem 7.1]).

(iv) In the case $F = HZ_p$ a multiplicative version of the transfer was first defined by Evens, who called it the norm [35]. It seems likely that this agrees with $p_{\otimes}$, but we shall not give a proof. Note that in this case one always has $p_! p^* x = jx$, but it is not true that $p_{\otimes} p^* x = x^j$. For example, formula (v) gives

$$(1 \times p)_{\otimes}(1 \times p)^*(y \times 1) = (1 \times h)^* P_j y.$$

which is certainly not equal to $y^j \times 1$ in general.

2. Steenrod Operations in Ordinary Cohomology.

In this section we use the framework of §1 to construct the Steenrod operations in mod p cohomology and prove their usual properties. The construction will be similar to one given by Milgram [37, Chapter 27], except that we use stable extended powers instead of space-level ones. On the other hand, the proofs will be quite close to those of Steenrod and Epstein [100] except that we make no use of chain-level arguments.

Throughout this section and the next we write H for HZ_p, H^* for mod-p cohomology, and π for the subgroup of Σ_p generated by a p-cycle. If p is an odd prime we write m for $\frac{p-1}{2}$ as usual. For odd primes the spectrum HZ_p is H_∞^2 but not H_∞^1 (see VII.6.1), hence the power operation $\mathcal{P}_p$ can be defined in even degrees but not in odd degrees (unless one uses some form of local coefficients). The operation $\mathcal{P}_\pi$ does extend to odd degrees, as we shall now show.

Proposition 2.1. For each $i \in Z$ there is a unique map $\xi : D_\pi \Sigma^i H \to \Sigma^{pi} H$ for which the diagram

$$\begin{array}{ccc} (\Sigma^i H)^{(p)} & \xrightarrow{\ \iota\ } & D_\pi \Sigma^i H \\ & \phi \searrow \quad \swarrow \xi & \\ & \Sigma^{pi} H & \end{array}$$

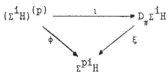

commutes, where ϕ is the iterated product map. For each $i,j \in Z$ the diagram

$$\begin{array}{ccc} D_\pi(\Sigma^i H \wedge \Sigma^j H) & \xrightarrow{\ \delta\ } & D_\pi \Sigma^i H \wedge D_\pi \Sigma^j H \\ \downarrow D_\pi \phi & & \downarrow \xi \wedge \xi \\ D_\pi(\Sigma^{i+j} H) & & \Sigma^{pi} H \wedge \Sigma^{pj} H \\ & \xi \searrow \quad \Sigma^{p(i+j)} H \quad \swarrow \phi & \end{array}$$

commutes up to the sign $(-1)^{mij}$.

The proof is the same as for I.4.5. One can in fact replace π in this result by any subgroup of the alternating group A_j, but we shall have no occasion to do so.

Using the map ξ we obtain an external operation

$$\mathcal{P}_\pi : \tilde{H}^i X \to \tilde{H}^{pi} D_\pi X$$

and an internal operation

$$P_\pi : \tilde{H}^i X \to \tilde{H}^{pi}(X \wedge B\pi^+)$$

as in §1. The uniqueness property in 2.1 implies that these operations agree with those already defined when i is even.

Since $\iota^* \mathcal{P}_\pi \Sigma 1 \in \tilde{H}^p S^p$ is the canonical generator $\Sigma^p 1$, we see that $\mathcal{P}_\pi \Sigma 1$ is an orientation for the real regular representation bundle

$$E\pi \times_\pi (\mathbf{R}^1)^p \to B\pi .$$

It follows that the element $\chi \in H^{p-1} B\pi$ defined by

$$\Sigma\chi = P_\pi \Sigma 1$$

is the Euler class of the real reduced regular representation (i.e., the sum of the nontrivial real irreducibles). In particular, χ is nonzero since each nontrivial real irreducible has nonzero Euler class.

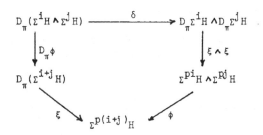

Our next result gives the basic properties of the operation P_π. Note that $\tilde{H}^*(X \wedge B\pi^+)$ is an $H^*(B\pi)$-module.

Proposition 2.2. (i) $\iota^* P_\pi x = x^p$

(ii) $P_\pi(xy) = (-1)^{m|x||y|}(P_\pi x)(P_\pi y)$

(iii) $P_\pi \Sigma x = (-1)^{m|x|}\chi(\Sigma P_\pi x)$

(iv) $P_\pi(x + y) = P_\pi x + P_\pi y$

(v) $\beta P_\pi x = 0$ if p is odd or $|x|$ is even.

Proof. Parts (i) and (ii) are immediate from 2.1 and part (iii) follows from part (ii). For part (iv) we assume first that $|x|$ is even. Then we may apply 1.4(vi) to get

$$P_p(x + y) = P_p x + P_p y + \frac{1}{p!}[(x + y)^p - x^p - y^p](\tau_p^* 1).$$

But $\tau_p^* 1 = \tau^* \iota^* 1 = p!1 = 0$ and the result follows in this case. If $|x|$ is odd this gives

$$P_\pi(\Sigma x + \Sigma y) = P_\pi \Sigma x + P_\pi \Sigma y.$$

Applying part (iii) gives the equation

$$(-1)^{m|x|}\chi(\Sigma P_\pi(x + y)) = (-1)^{m|x|}\chi(\Sigma(P_\pi x + P_\pi y))$$

and the result follows since χ is not a zero divisor in $H^*B\pi$. For part (v) we need a lemma. Let $\beta: H \to \Sigma H$ represent the Bockstein operation.

Lemma 2.3. The composite

$$D_\pi \Sigma^{2i} H \xrightarrow{\ \xi\ } \Sigma^{2pi} H \xrightarrow{\ \Sigma^{2pi}\beta\ } \Sigma^{2pi+1} H$$

factors through the transfer

$$\tau_\pi : D_\pi \Sigma^{2i} H \longrightarrow (\Sigma^{2i} H)^{(p)}.$$

The proof of 2.3 is rather technical and will be given at the end of this section. For the moment we use it to prove part (v). Let $x \in \tilde{H}^{2i} X$ be represented by $f: \Sigma^\infty X \to \Sigma^{2i} H$ and consider the following diagram, where we have suppressed Σ^∞ to simplify the notation.

$$\begin{array}{ccccccccc}
X \wedge B\pi^+ & \xrightarrow{\ \Delta\ } & D_\pi X & \xrightarrow{\ D_\pi f\ } & D_\pi \Sigma^{2i} H & \xrightarrow{\ \xi\ } & \Sigma^{2pi} H & \xrightarrow{\ \Sigma^{2pi}\beta\ } & \Sigma^{2pi+1} H \\
\downarrow{\scriptstyle 1 \tau_\pi} & & \downarrow{\scriptstyle \tau_\pi} & & \downarrow{\scriptstyle \tau_\pi} & & & \nearrow & \\
X & \xrightarrow{\ \Delta\ } & X^{(p)} & \xrightarrow{\ f^{(p)}\ } & (\Sigma^{2i} H)^{(p)} & \cdots & & &
\end{array}$$

The dotted arrows exist by 2.3 and the diagram commutes. The top row represents $\beta P_\pi x$. Thus $\beta P_\pi x$ is in the image of the transfer

$$(1\wedge\tau_\pi)^*:\tilde{H}^*X \to \tilde{H}^*(X\wedge B\pi^+).$$

But the composite of $(1\wedge\tau_\pi)^*$ with the restriction

$$(1\wedge\iota)^*:\tilde{H}^*(X\wedge B\pi^+) \to \tilde{H}^*X$$

is multiplication by p and hence vanishes. Since $(1\wedge\iota)^*$ is clearly onto we see that $(1\wedge\tau_\pi)^* = 0$ so that $\beta P_\pi x = 0$ as required. Finally, if p is odd and $x \in H^{2i-1}X$ we have

$$0 = \beta P_\pi(\Sigma x) = \beta(\chi\cdot\Sigma P_\pi x) = -\chi\cdot\Sigma(\beta P_\pi x)$$

since $\beta\chi = 0$. The result follows in this case since χ is not a zero divisor. This completes the proof of 2.2.

Now let $x \in H^qX$. If $p = 2$ we define $P^ix \in H^{q+i}X$ to be the coefficient of χ^{q-i} in $P_\pi x$. If p is odd we define $P^ix \in H^{q+2i(p-1)}X$ to be $(-1)^{mi+mq(q-1)/2}$ times the coefficient of χ^{q-2i} in $P_\pi x$. We also define an element $\omega \in H^{p-2}B\pi$ for p odd by the equation $\beta\omega = \chi$.

Proposition 2.4. (i) $P^i(x+y) = P^ix + P^iy$

(ii) $P^i(\Sigma x) = \Sigma P^ix$

(iii) $P^ix = x^p$ if $q = 2i$ and p is odd or if $q = i$ and $p = 2$. $P^ix = 0$ if $q < 2i$ and p is odd or if $q < i$ and $p = 2$.

(iv) $P^0x = x$.

(v) If $p = 2$ then $\beta P^{2i}x = P^{2i+1}x$; in particular, $\beta x = P^1x$.

(vi) If $p = 2$ then $P_\pi x = \Sigma(P^ix)\chi^{q-i}$. If p is odd then

$$P_\pi x = \Sigma(-1)^{mi+mq(q-1)/2}[(P^ix)\chi^{q-2i} + (-1)^q(\beta P^ix)\omega\chi^{q-2i-1}].$$

(vii) $P^ixy = \Sigma(P^jx)(P^{i-j}y)$.

Proof. (i), (ii) and (iii) follow from 2.2(iv), 2.2(iii) and 2.2(i) respectively. For part (iv), we observe that P^0 is a stable operation of degree 0 and hence represents an element of $H^0H \cong Z_p$. Thus P^0 is a constant multiple of the identity and the result follows since $P^01 = 1^p = 1$ by part (iii). In part (v) we can use part (ii) to reduce to the case where q is even. The result follows in that case from 2.2(v) and the relation $\beta\chi = \chi^2$. In part (vi) the $p = 2$ case is true by definition. If p is odd we can use part (ii) and 2.2(iii) to reduce to the case where q is even. We then have $P_\pi x = \iota^*P_px$. We recall from [68, Lemma 1.4] that the image of

$$\iota^*:H^*B\Sigma_p \to H^*B\pi$$

is nonzero only in dimensions of the form $2i(p-1)$ and $2i(p-1)-1$. Thus this image is generated as a ring by χ and ω and we have

$$P_\pi x = \sum (-1)^{mi+mq(q-1)/2}[(P^i x)\chi^{q-2i} + y_i \omega \chi^{q-2i-1}]$$

for some elements $y_i \in H^{q+2i(p-1)+1}X$. Now 2.2(v) imples that $y_i = (-1)^q \beta P^i x$ as required. Finally, part (vii) follows from 2.2(iv) and part (vi). This completes the proof of 2.4.

Next we shall prove the Adem relations for p odd. We use the method of proof of Bullett and MacDonald [26, §4], where the case p = 2 may be found. However, in our context the relations arise more naturally in the form given by Steiner [102]. Let U and V denote indeterminates of degree 2p-2 and define S and T by

$$S = U(1 - V^{-1}U)^{p-1}$$

$$T = V(1 - U^{-1}V)^{p-1}.$$

We shall prove that the equations

$$(1) \qquad \sum_{i,j} (P^j P^i x)U^{-j}T^{-i} = \sum_{i,j} (P^j P^i x)V^{-j}S^{-i}$$

$$(2) \qquad \sum_{i,j} (P^j \beta P^i x)U^{-j}T^{-i} = (1 - U^{-1}V) \sum_{i,j} (\beta P^j P^i x)V^{-j}S^{-i} + U^{-1}V \sum_{i,j} (P^j \beta P^i x)V^{-j}S^{-i}$$

hold for all x. The usual Adem relations can easily be obtained from these as in [102, p. 163]; the basic idea is simply to expand the right sides of (1) and (2) as power series in U and T and compare coefficients. The proof of (1) and (2), like any proof of the Adem relations, is based on the relation

$$(3) \qquad \gamma^* P_\pi P_\pi x = P_\pi P_\pi x$$

given by 1.4(vii). In order to compute $P_\pi P_\pi x$ in terms of the P^i we need to know more about the element $\chi \in H^{p-1}B\pi$. We have mentioned that χ is the Euler class of the real reduced regular representation of π, and that this representation is the sum of the nontrivial real irreducibles of π. Choose one such irreducible, and let $u \in H^2 B\pi$ denote its Euler class. Then the Euler classes of the remaining irreducibles (suitably oriented) are 2u, 3u,...,mu, and thus $\chi = \pm m! u^m$. The ambiguity in the sign arises from the question of whether the various orientations have been chosen consistently, but it turns out that we shall not need to eliminate this ambiguity. Thus we shall assume $\chi = m! u^m$ (it is in fact possible to choose the orientations so that this holds) and leave it to the reader to check that the other possibility leads to the same relations (1) and (2). We define $b \in H^1 B\pi$ by the equation $\beta b = u$, so that $\omega = m! b u^{m-1}$. Then the equation 2.4(v) may be written as follows.

$$(4) \qquad P_\pi x = \sum (-1)^{i+mq(q-1)/2}(m!)^q [P^i x + (-1)^q (\beta P^i x) b u^{-1}] u^{m(q-2i)}.$$

Since both sides of (1) and (2) are stable we may assume that q has the form 2r with r even. We define $U = -u^{2m}$, so that (4) becomes

(5) $$P_\pi x = \sum (-1)^r [P^i x + (\beta P^i x) b u^{-1}] U^{r-i}.$$

Now 2.2(ii) and 2.2(iv) give

(6) $$P_\pi P_\pi x = \sum (-1)^r [P_\pi P^i x + (-1)^m (P_\pi \beta P^i x)(P_\pi b)(P_\pi u)^{-1} (P_\pi U)^{r-i}]$$

in $H^*X \otimes H^*B\pi \otimes H^*B\pi$. We denote the copies of b and u in the second copy of $B\pi$ by c and v, and we let $V = -v^{p-1}$. Equation (4) gives the following formulas.

(7) $$P_\pi b = m! [b - ucv^{-1}] v^m$$

(8) $$P_\pi u = u^p - uv^{p-1} = u(V - U)$$

(9) $$P_\pi U = -(P_\pi u)^{p-1} = U(V - U)^{p-1} = V^{p-1} S$$

(10) $$P_\pi P^i x = \sum (-1)^r [P^j P^i x + (\beta P^j P^i x) cv^{-1}] V^{r+2im-j}$$

(11) $$P_\pi \beta P^i x = \sum (-1)^r m! [P^j \beta P^i x - ((\beta P^j \beta P^i x) cv^{-1}] V^{r+2im-j} v^m.$$

We therefore have

(12) $$P_\pi P_\pi x = (V^p S)^r \sum [P^j P^i x + (\beta P^j P^i x) cv^{-1} + (P^j \beta P_i x)(bu^{-1} - cv^{-1}) V (V - U)^{-1}$$
$$+ (\beta P^j \beta P^i x) bcu^{-1} v^{-1} V (V - U)^{-1}] V^{-j} S^{-i}.$$

Now we apply equation (3). We have $\gamma^* u = v$, $\gamma^* U = V$, and $\gamma^* S = T$. Since $V^p S = U^p T = \gamma^*(V^p S)$ we have

(13) $$P_\pi P_\pi x = \gamma^* P_\pi P_\pi x = (V^p S)^r \sum [P^j P^i x - (\beta P^j P^i x) bu^{-1}$$
$$+ (P^j \beta P^i x)(cv^{-1} - bu^{-1}) U (U - V)^{-1} - (\beta P^j \beta P^i) bcu^{-1} v^{-1} U (U-V)^{-1}] U^{-j} T^{-i}.$$

Collecting the terms in (12) and (13) which do not involve b or c gives equation (1), and the terms which involve c but not b give (2). This completes the proof.

Finally, we give the proof of Lemma 2.3. Let M be the Moore spectrum $S \cup_p e^1$ and let $i: S \to M$ be the inclusion of the bottom cell.

<u>Lemma 2.5</u>. $H^1(D_\pi M)$ has a basis $\{x,y\}$ such that $(D_\pi i)^* x = 0$, $(D_\pi i)^* y \neq 0$, and x is in the image of the transfer

$$\tau_\pi^* : H^1 M^{(p)} \to H^1 D_\pi M.$$

Proof of 2.5. We use the spectral sequence

$$H^i(\pi;H^j(M^{(p)})) \Rightarrow H^{i+j}D_\pi M$$

of I.2.4. Each of the groups $E_2^{0,1}$ and $E_2^{1,0}$ is generated by a single element. The generator of the latter group clearly survives to E_∞ and represents an element $y \in H^1D_\pi M$. Since $(i^{(p)})^*:H^0M^{(p)} \to H^0S$ is an isomorphism, so is the map induced by $D_\pi i$ on $E_2^{1,0}$. Hence $(D_\pi i)^*y \neq 0$. Now let $z \in H^1M^{(p)}$ be a generator of $H^1M \otimes H^0M \otimes \cdots \otimes H^0M$ and let $x = \tau_\pi^* z$. Clearly, x is represeneted by a generator of $E_2^{0,1}$, and $(D_\pi i)^*x = (D_\pi i)^*\tau_\pi^* z = \tau_\pi^*(i^{(p)})^*z$ which is zero since $H^1S = 0$.

Proof of 2.3. Let HZ be the spectrum representing integral cohomology. Then $H = HZ \wedge M$. Let $e:S \to HZ$ be the unit and let η be the composite

$$D_\pi M = D_\pi(S \wedge M) \xrightarrow{D_\pi(e\wedge 1)} D_\pi(HZ \wedge M) = D_\pi H \xrightarrow{\xi} H.$$

Let w be the element of $H^0D_\pi M$ represented by η. Then $(D_\pi i)^*\beta w = 0$ since β vanishes on $H^0D_\pi S = H^0B\pi$. Hence by Lemma 2.5, βw is a multiple of x and in particular it is in the image of the transfer. Thus we have a factorization

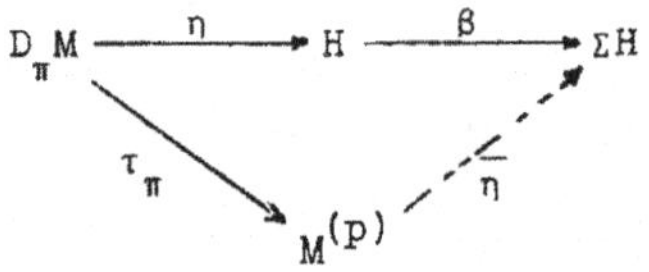

Now consider the diagram

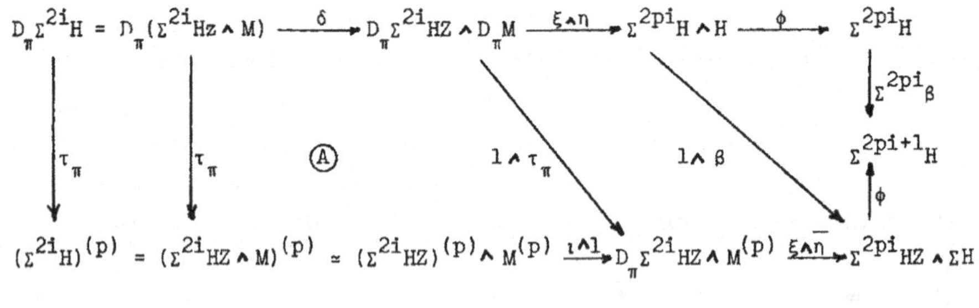

The uniqueness clause in 2.1 imples that the composite of the top row is $\xi:D_\pi\Sigma^{2i}H \to \Sigma^{2pi}H$, so it suffices to show that the diagram commutes. Part (A) commutes by VI.3.10 of the sequel, and the other parts clearly commute.

§3. Dyer-Lashof operations and the Nishida relations

An interesting feature of the treatment of Steenrod operations in §2 is that is generalizes to give the properties of Dyer-Lashof operations; thus homology operations are a special case of cohomology operations (cf. [68]). The use of stable instead of space-level extended powers is crucial for this since homology does not have a simple space-level description. We give the details in this section; IX§1 will give another approach to homology operations which generalizes to extraordinary theories. We continue to use the notations of §2, so that H denotes HZ_p.

First let M be any module spectrum over H and let Y be an arbitrary spectrum. There is a natural transformation

$$A: M^*Y \to \mathrm{Hom}(H_*Y, \pi_*M)$$

defined as follows: if $y \in M^*Y$ is represented by $f: Y \to \Sigma^i M$ then $A(y)$ is the composite

$$H_*Y = \pi_*(H \wedge Y) \xrightarrow{(1\wedge f)_*} \pi_*(H \wedge M) \longrightarrow \pi_*M,$$

which is a homomorphism raising degrees by i. Clearly A is a morphism of cohomology theories. Since it is an isomorphism for Y = S we have

Lemma 3.1. A is an isomorphism.

Now let X be a fixed H_∞ ring spectrum with structural maps θ_j (for example, X might have the form $\Sigma^\infty Z^+$ for an infinite loop space Z) and let $M = H \wedge X$. Then M is an H_∞^2 ring spectrum with structural maps

$$D_j(\Sigma^{2i}H \wedge X) \xrightarrow{\delta} D_j\Sigma^{2i}H \wedge D_jX \xrightarrow{\xi_{j,i}\wedge\theta_j} \Sigma^{2ij}H \wedge X$$

and we obtain power operations

$$\mathcal{R}_j: M^{2i}Y \to M^{2ij}D_jY$$

and $\quad R_j: M^{2i}Y \to M^{2ij}(Y \times B\Sigma_j).$

The operation $\mathcal{R}_\pi$ can be extended to odd degrees by means of the maps

$$D_\pi\Sigma^i M = D_\pi(\Sigma^i H \wedge X) \xrightarrow{\delta} D_\pi\Sigma^i H \wedge D_\pi X \xrightarrow{\xi\wedge\theta} \Sigma^{pi}H \wedge X$$

where ξ is the map given by 2.1. The unit of X gives an H_∞^2 ring map $h: H \to H \wedge X = M$ and h_* also preserves $\mathcal{P}_\pi$ in odd degrees.

Define $\bar{b}$, $\bar{u}$, $\bar{\chi}$ and $\bar{\omega}$ in $M^*B\pi$ to be the images under h_* of the elements b, u, χ and ω in $H^*B\pi$ defined in §2. Thus $\Sigma\bar{\chi} = R_\pi\Sigma 1$. Lemma 3.1 gives the following isomorphisms for any space Y.

$$M^*(Y \times B\pi) \cong (M^*Y)[[\overline{\chi}]] \quad \text{if } p = 2.$$

$$M^*(Y \times B\pi) \cong (M^*Y)[[\overline{b},\overline{u}]] \quad \text{if } p \text{ is odd}.$$

Thus we can define operations $R^i y$ for $y \in \tilde{M}^q Y$ as follows: if $p = 2$ let $R^i y$ be the coefficient of $\overline{\chi}^{q-i}$ in $R_\pi y$, and if p is odd let $R^i y$ be $(-1)^{mi+mq(q-1)/2}$ times the coefficient of $\overline{\chi}^{q-2i}$ in $R_\pi y$. Now if $Y = S^0$ there is an isomorphism $H_q X \cong \tilde{M}^{-q} S^0$ which we shall always denote by $x \longmapsto \underline{x}$. We define the Dyer-Lashof operations

$$Q^i : H_q X \to H_{q+i} X \quad \text{when } p = 2$$

$$Q^i : H_q X \to H_{q+2i(p-1)} X \quad \text{when } p \text{ is odd}$$

by the equation $\underline{Q^i x} = R^{-i}\underline{x}$. The properties of Q^i will follow from those of R_π and R^i. Our next result gives the basic facts about R_π.

<u>Proposition 3.2.</u> (i) $i^* R_\pi y = y^p$

(ii) $R_\pi(yz) = (-1)^{m|y||z|}(R_\pi y)(R_\pi z)$

(iii) $R_\pi(\Sigma y) = (-1)^{m|y|}\ \overline{\chi} \cdot \Sigma R_\pi y$

(iv) $R_\pi(y + z) = R_\pi y + R_\pi z$.

(v) $\beta R_\pi y = 0$ if p is odd or $|y|$ is even.

<u>Proof.</u> (i) and (ii) are immediate from the definitions and (iii) follows from (ii). In the proof of 2.2(v) it was observed that the transfer

$$\tau_\pi^* : H^* Y \to H^*(Y \times B\pi)$$

vanishes for all spaces Y. By 3.1 it follows that

$$\tau_\pi^* : M^* Y \to M^*(Y \times B\pi)$$

also vanishes. In particular, the map

$$\tau_p^* : M^*(pt.) \to M^*(B\Sigma_p)$$

vanishes. Part (iv) now follows by the proof of 2.2(iv). To complete the proof of part (v) it suffices to give a suitable substitute for Lemma 2.3. That lemma gives a map

$$F : (\Sigma^{2i} H)^{(p)} \to \Sigma^{2pi+1} H$$

such that $F \circ \tau_\pi$ is the composite

$$D_\pi \Sigma^{2i} H \xrightarrow{\xi} \Sigma^{2pi} H \xrightarrow{\Sigma^{2pi}\beta} \Sigma^{2pi+1} H.$$

Consider the following diagram

$$\begin{array}{ccccccc}
D_\pi(\Sigma^{2i}H\wedge X) & \xrightarrow{\delta} & D_\pi\Sigma^{2i}H\wedge D_\pi X & \xrightarrow{\xi\wedge\theta} & \Sigma^{2pi}H\wedge X & \xrightarrow{\Sigma^{2pi}\beta\wedge 1} & \Sigma^{2pi+1}H\wedge X \\
\Big\downarrow{\scriptstyle\tau_\pi} & & & \searrow{\scriptstyle\tau_\pi\wedge 1} & & \nearrow{\scriptstyle F\wedge\theta} & \\
(\Sigma^{2i}H\wedge X)^{(p)} & \cong & (\Sigma^{2i}H)^{(p)}\wedge X^{(p)} & \xrightarrow{1\wedge\iota} & (\Sigma^{2i}H)^{(p)}\wedge D_\pi X & &
\end{array}$$

The left part commutes by VI.3.10 of the sequel and the right part commutes by definition of F. Thus the top row of the diagram factors through τ_π. Using this fact in place of Lemma 2.3, the proof of 2.2(v) now goes through to prove part (v).

If we now replace P^i, χ and ω in Proposition 2.4 by R^i, $\overline{\chi}$ and $\overline{\omega}$ then every part except (iv) remains true with the same proof. If we replace U,V,S and T in the Adem relations (equations (1) and (2) of Section 2) by $\overline{U} = h_*U$, $\overline{V} = h_*V$, $\overline{S} = h_*S$ and $\overline{T} = h_*T$ then these relations remain true and have the same proof.

<u>Proposition 3.3</u>. (i) $Q^i(x+y) = Q^ix + Q^iy$

(ii) If p is odd then $Q^ix = 0$ for $2i < |x|$ and $Q^ix = x^p$ for $2i = |x|$.

If $p = 2$ then $Q^ix = 0$ for $i < |x|$ and $Q^ix = x^2$ for $i = |x|$.

(iii) $\beta Q^{2s} = Q^{2s-1}$ if $p = 2$

(iv) $Q^i(xy) = \sum_i (Q^jx)(Q^{i-j}y)$

(v) The Adem relations hold: if U and V are indeterminates of dimension $2-2p$, $S = U(1-V^{-1}U)^{p-1}$ and $T = V(1-U^{-1}V)^{p-1}$ then the equations

$$\sum_{i,j} (Q^iQ^jx)U^iT^j = \sum_{i,j} (Q^iQ^jx)V^iS^j$$

and if p is odd

$$\sum_{i,j} (Q^i\beta Q^jx)U^iT^j = (1-U^{-1}V)\sum_{i,j}(\beta Q^iQ^jx)V^iS^j$$
$$+\, U^{-1}V\sum_{i,j}(Q^i\beta Q^jx)V^iS^j$$

are valid for all x.

(vi) If X has the form $\Sigma^\infty Z^+$ for an E_∞ space Z and

$$\sigma:\tilde{H}_q\Omega Z \to H_{q+1}Z$$

is the homology suspension then $Q^i\sigma = \sigma Q^i$.

Proof. We shall prove part (vi); the remaining parts are immediate from the properties of R^i. For any space Z the retraction of Z to a point splits the cofibre sequence

$$\Sigma^\infty S^0 \longrightarrow \Sigma^\infty Z^+ \xrightarrow{\lambda} \Sigma^\infty Z$$

and gives a map

$$\nu : \Sigma^\infty Z \to \Sigma^\infty Z^+$$

Now let Z be an E_∞ space and let $X = \Sigma^\infty Z^+$, $\tilde{X} = \Sigma^\infty Z$, $W = \Sigma^\infty(\Omega Z)^+$, $\tilde{W} = \Sigma^\infty \Omega Z$. Then X and W are H_∞ ring spectra but $\tilde{X}$ and $\tilde{W}$ are not. Let ζ denote either of the composites

$$D_\pi \tilde{X} \xrightarrow{D_\pi \nu} D_\pi X \longrightarrow X \xrightarrow{\lambda} \tilde{X}$$

and

$$D_\pi \tilde{W} \xrightarrow{D_\pi \nu} D_\pi W \longrightarrow W \xrightarrow{\lambda} \tilde{W},$$

where the unmarked arrows come from the H_∞ structures on X and W. We can use the maps ζ to define operations $\tilde{R}_\pi$ in the theories represented by $H \wedge \tilde{X}$ and $H \wedge \tilde{W}$ and it is easy to see that

(1)
$$(1 \wedge \nu)_* \tilde{R}_\pi y = R_\pi (1 \wedge \nu)_* y$$

for all y. Now if $x \in \tilde{H}_q \Omega Z$ then $\underline{x} \in (H \wedge \tilde{W})^{-q} S \subset (H \wedge W)^{-q} S$, and (1) and the definition of Q^i give

(2)
$$\tilde{R}_\pi \underline{x} = \sum_i (-1)^{mi+mq(q+1)/2} \underline{Q^i x}\, \chi^{2i-q}$$

since $(1 \wedge \nu)_*$ is monic. The natural map $\varepsilon : \Sigma\Omega Z \to Z$ induces a map $\Sigma\tilde{W} \to \tilde{X}$ which will also be called ε. A fairly tedious diagram chase (given at the end of IX§7) shows that the following diagram commutes.

$$\begin{array}{ccc}
\Sigma D_\pi \tilde{W} & \xrightarrow{\Delta} & D_\pi \Sigma \tilde{W} \\
 & & \downarrow{\scriptstyle D_\pi \varepsilon} \\
\downarrow{\scriptstyle \Sigma\zeta} & & D_\pi \tilde{X} \\
 & & \downarrow{\scriptstyle \zeta} \\
\Sigma \tilde{W} & \xrightarrow{\varepsilon} & \tilde{X}
\end{array}$$

Hence the following diagram commutes, where $f : S \to \Sigma^{-q} H \tilde{W}$ represents $\underline{x}$.

$$\begin{array}{ccccccccc}
\Sigma B\pi^+ & \xrightarrow{\Sigma D_\pi f} & \Sigma D_\pi(\Sigma^{-q} H\, \tilde{W}) & \xrightarrow{\delta} & D_\pi \Sigma^{-q} H\; \Sigma D_\pi \tilde{W} & \longrightarrow & \Sigma^{-pq} H \wedge \Sigma\tilde{W} & \xrightarrow{1\wedge\varepsilon} & \Sigma^{-pq} H \wedge \tilde{X} \\
\downarrow{\scriptstyle \Delta} & & \downarrow{\scriptstyle \Delta} & & \downarrow{\scriptstyle 1\wedge\Delta} & & & \nearrow{\scriptstyle 1\wedge\zeta} & \\
D_\pi S^1 & \xrightarrow{D_\pi \Sigma f} & D_\pi(\Sigma^{-q+1} H \wedge \tilde{W}) & \longrightarrow & D_\pi \Sigma^{-q} H \wedge D_\pi \Sigma \tilde{W} & \longrightarrow & \Sigma^{-pq} H \wedge D_\pi \tilde{X} & &
\end{array}$$

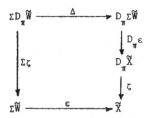

The top row of this diagram represents $(1\wedge\varepsilon)_{\pi}\Sigma\tilde{R}_{\pi}\underline{x}$ and the other composite represents $\tilde{R}_{\pi}(1\wedge\varepsilon)_{*}\Sigma\underline{x}$. Thus we have

(3) $$(1\wedge\varepsilon)_{*}\Sigma\tilde{R}_{\pi}\underline{x} = \tilde{R}_{\pi}(1\wedge\varepsilon)_{*}\Sigma\underline{x}.$$

Combining this with (1) gives

(4) $$(1\wedge\nu\varepsilon)_{*}\Sigma\tilde{R}_{\pi}\underline{x} = R_{\pi}(1\wedge\nu\varepsilon)_{*}\Sigma\underline{x} .$$

Now the definition of σ gives

(5) $$\Sigma\underline{\sigma x} = (1\wedge\nu\varepsilon)_{*}\Sigma\underline{x} .$$

Combining (5) and (2) gives

(6) $$(1\wedge\nu\varepsilon)_{*}\Sigma\tilde{R}_{\pi}\underline{x} = \sum_{i}(-1)^{mi+mq(q+1)/2}\Sigma\underline{\sigma Q^{i}x}\;\chi^{2i-q}$$

Finally, by 3.2(iii) we have

(7) $$R_{\pi}(1\wedge\nu\varepsilon)_{*}\Sigma\underline{x} = R_{\pi}\Sigma\underline{\sigma x} = (-1)^{m(q+1)}\chi(\Sigma R_{\pi}\underline{\sigma x})$$
$$= \sum_{i}(-1)^{mi+mq(q+1)/2}\Sigma\underline{Q^{i}\sigma x}\;\chi^{2i-q}.$$

The result follows from (4), (6), and (7). This completes the proof of 3.3.

We conclude this section with a proof of the Nishida relations in the form given by Steiner:

(8) $$\sum_{i,j}(\bar{P}^{i}_{*}Q^{j}x)V^{-i}S^{j} = \sum_{i,j}(Q^{i}\bar{P}^{j}_{*}x)U^{i}T^{-j}$$

and if p is odd

(9) $$\sum_{i,j}(\bar{P}^{i}_{*}\beta Q^{j}x)V^{-i}S^{j} = (1 - UV^{-1})\sum_{i,j}(\beta Q^{i}\bar{P}^{j}_{*}x)U^{i}T^{-j}$$
$$+ UV^{-1}\sum_{i,j}(Q^{i}\beta\bar{P}^{j}_{*}x)U^{i}T^{-j} ,$$

where $\bar{P}^{i}_{*}$ is the dual of the conjugate Steenrod operation $\bar{P}^{i}$ and U, V, S and T are as in 3.3(v). The usual Nishida relations can easily be obtained from these by first translating from $\bar{P}^{i}_{*}$ to P^{i}_{*} and then writing both sides as power series in U and V; see [102, p. 164]. We shall prove (8) and (9) for p odd; there is a similar proof for p = 2. The basic idea will be to show that the total Steenrod operation

$$H \longrightarrow \bigvee_{i\in Z}\Sigma^{i}H$$

is an H_{∞} ring map, and this in turn will follow easily from 1.4(vii). To make this work, however, we need a particular H_{∞} structure on $\bigvee_{i\in Z}\Sigma^{i}H$ which we now construct.

Let E^*X be the functor $H^*(X \times B\pi)$ on the category of spaces. We denote the generators of $H^1B\pi$ and $H^2B\pi$ by c and v, so that E^*X is the polynomial ring $(H^*X)[c,v]$. E^* is a multiplicative cohomology theory and hence is represented by a ring spectrum E. The projection $X \times B\pi \to X$ gives a natural transformation $H^*X \to E^*X$ which is represented by a map $g:H \to E$. Of course, E is equivalent to $\bigvee_{i\leq 0} \Sigma^i H$ with its usual ring structure and g is the inclusion of H in this wedge.

Next we define power operations in E^*. Let $\mathcal{P}_j^E$ be the composite

$$\tilde{E}^{2i}X = \tilde{H}^{2i}(X\wedge B\pi^+) \xrightarrow{j} \tilde{H}^{2ij}(D_j(X\wedge B\pi^+)) \xrightarrow{\Delta^*} \tilde{H}^{2ij}((D_jX)\wedge B\pi^+) = \tilde{E}^{2ij}D_jX.$$

It is easy to see that the $\mathcal{P}_j^E$ are consistent in the sense of Definition 1.2 and thus they determine an H_∞^2 ring structure on E by 1.3 (compare II.1.3). The operation $\mathcal{P}_\pi^E$ extends to odd degrees since $\mathcal{P}_\pi$ does, and g is an H_∞^2 ring map which also preserves $\mathcal{P}_\pi$ in odd degrees. An inspection of the definitions gives the following description of the internal operation $\mathcal{P}_\pi$.

(10) $$P_\pi^E = (1\wedge\gamma)^*P_\pi : \tilde{H}^i(X\wedge B\pi^+) \to \tilde{H}^{pi}(X\wedge B\pi^+\wedge B\pi^+)$$

Note that, with the conventions we have adopted, c and v are the generators in the second copy of $B\pi$ in this situation. As in Section 2 write b and u for the generators in the first copy of $B\pi$; thus $g_*:H^*B\pi \to E^*B\pi$ takes v to u and c to b.

Now let F^*X be the Laurent series ring $(H^*X)[[c,v,v^{-1}]] = E^*X[[v^{-1}]]$. F^* is a multiplicative cohomology theory and hence is represented by a ring spectrum F, and the inclusion $H^*X \to F^*X$ is represented by a ring map $H \to F$ which we again call g; of course F is equivalent as a ring spectrum to $\bigvee_{i\in Z} \Sigma^i H$ and g is the inclusion of H in this wedge. Now observe that the element $\mathcal{P}_j^E v \quad (H^*B\Sigma_j)[[c,v,v^{-1}]]$ is a Laurent series which is bounded above, and that by 1.1(iv) it has leading coefficient $1 \in H^0B\Sigma_j$. Hence $\mathcal{P}_j^E v$ is invertible, and it follows that we can extend the operations $\mathcal{P}_j^E$ to operations $\mathcal{P}_j^F$ in the F-cohomology of finite complexes. The $\mathcal{P}_j^F$ are consistent in the sense of 1.2 and hence give an H_∞^2 structure for F by 1.3.

Next we define the total Steenrod operation $t:H \to F$ by letting t_* be the composite

$$H^qX \xrightarrow{P_\pi} H^{pq}(X \times B\pi) = E^{pq}X \longrightarrow F^qX,$$

where the last map is multiplication by $(-1)^{mq(q-1)/2}(m!)^{-q}\, v^{-mq}$. By 2.4(vi) we have the formula

(11) $$t_*x = \sum_i [g_*P^ix + (-1)^q g_*(\beta P^ix)cv^{-1}]V^{-i},$$

where $V = -v^{p-1}$ as in Section 2. In particular, the projection of $t: \to \bigvee_{i\in Z} \Sigma^i H$ on

$\Sigma^{2k(p-1)}H$ is P^k. Either from the definition or from formula (11) we get the following equations.

(12) $$t_*c = b - cuv^{-1}$$

(13) $$t_*v = u + u^pV^{-1} = u(1 - UV^{-1})$$

(14) $$t_*V = -u^{p-1}(1 - UV^{-1})^{p-1} = U(1 - UV^{-1})^{p-1} = S.$$

t is clearly a ring map, but it turns out not to be an H_∞^2 map. However, we have

<u>Proposition 3.4</u>. Let Y be any spectrum and let $y \in H^qY$. Let $w = (1 - UV^{-1})^m$. Then

$$t_*\mathcal{P}_\pi y = w^q \mathcal{P}_\pi^F t_* y.$$

This fact will suffice for our purposes but we remark that by combining it with 7.2 below one can show that t is actually an H_∞ map. It is certainly not H_∞^2 since it does not preserve $\mathcal{P}_\pi$.

For the proof of 3.4 we need a standard lemma.

<u>Lemma 3.5</u>. For any space Y the map

$$\iota^* \oplus \Delta^*: \tilde{H}^*D_\pi Y \to \tilde{H}^*Y^{(p)} \oplus \tilde{H}^*(Y \wedge B\pi^+)$$

is monic.

For completeness we shall give a proof of 3.5 at the end of this section.

<u>Proof of 3.4</u>. Since both sides of the equation are stable (H,π,F) operations in the sense of 1.2 and 1.3 it suffices to show that they agree on finite complexes. By 3.5 it suffices to show

$$\iota^* t_* \mathcal{P}_\pi y = \iota^* w^q \mathcal{P}_\pi^F t_* y$$

and

(15) $$t_* P_\pi y = w^q P_\pi^F t_* y$$

for all y. Since $\iota^* w = 1$ and t is a ring map the first equation follows from 1.1 (iv). For the second, we first let $y = \Sigma 1$. Then

$$t_* P_\pi \Sigma 1 = t_*(\chi \cdot \Sigma 1) = (t_*\chi)\cdot(t_*\Sigma 1) = \chi w \cdot \Sigma 1$$

while

$$wP_\pi^F t_* \Sigma 1 = wP_\pi^F g_* \Sigma 1 = wg_* P_\pi \Sigma 1 = w\chi \cdot \Sigma 1.$$

Since χw is not a zero divisor, it suffices to show (15) when q is even, say

$q = 2r$. Then as elements of $(H^*Y)[[b,c,u,v,v^{-1}]]$ we have

$$t_*P_\pi y = (-1)^{mr}(m!)^{-2r}v^{-2mpr}P_\pi P_\pi y = V^{-pr}P_\pi P_\pi y$$

and

$$\begin{aligned} w^{2r}P^F_\pi t_* y &= w^{2r}(1\wedge\gamma)^*P_\pi(U^{-r}P_\pi y) \quad \text{by (10)} \\ &= w^{2r}V^{-pr}(1 - UV^{-1})^{-2mr}(1\ \ \gamma)^*P_\pi P_\pi y \\ &= V^{-pr}P_\pi P_\pi y \quad \text{by 1.4(vii),} \end{aligned}$$

and the result follows.

If we let y be the class of the identity map $H \to H$ we obtain

Corollary 3.6. The diagram

$$\begin{array}{ccc} D_\pi H & \longrightarrow & H \\ \Big\downarrow D_\pi t & & \Big\downarrow t \\ D_\pi F & \longrightarrow & F \end{array}$$

commutes, where the unmarked arrows come from the H_∞ structures of H and F.

Now let X be an H_∞ ring spectrum. Then, as we have seen, $H\wedge X$ is an H^2_∞ ring spectrum and there is an operation

$$\mathcal{R}_\pi : (H\wedge X)^q Y \to (H\wedge X)^{pq} D_\pi Y$$

for $Y \in \overline{h\mathcal{L}}$. Similarly, $F\wedge X$ is an H^2_∞ ring spectrum and we obtain an operation

$$\mathcal{R}^R_\pi : (F\wedge X)^q Y \to (F\wedge X)^{pq} D_\pi Y,$$

The unit of X induces H_∞ ring maps $h: H \to H\wedge X$ and $h' : F \to F\wedge X$.

Corollary 3.7. If Y is any spectrum and $y \in (H\wedge X)^q Y$ then the equation

$$(t\wedge 1)_* \mathcal{R}_\pi y = w^q \mathcal{R}^R_\pi (t\wedge 1)_* y$$

holds in $(F\wedge X)^{pq}Y$.

Proof. For $q = 0$ this is immediate from 3.6. If $y = \Sigma 1$ we have

$$(t\wedge 1)_* \mathcal{R}_\pi \Sigma 1 = (t\wedge 1)_* \mathcal{R}_\pi h_* \Sigma 1 = h'_* t_* \,{}_\pi \Sigma 1 = w\, \mathcal{R}^F_\pi (t\wedge 1)_* \Sigma 1$$

by 3.4. For general y let $z = \Sigma^{-q} y \in (H\wedge X)^0(\Sigma^{-q}Y)$. Then $y = (\Sigma 1)^q z$ and we have

$$(t\wedge 1)_*\mathcal{R}_\pi y = (t\wedge 1)_*\delta^*[(\mathcal{R}_\pi\Sigma 1)^q\mathcal{R}_\pi z]$$
$$= \delta^*[w^q(\mathcal{R}^F_\pi\Sigma 1)^q\mathcal{R}^F_\pi(t\wedge 1)_*z]$$
$$= w^q\mathcal{R}^F_\pi(t\wedge 1)_*y$$

as required.

Corollary 3.7 gives the following relation between the internal operations.

(16) $$(t\wedge 1)_*R_\pi y = w^qR^F_\pi(t\wedge 1)_*y$$

To prove the relations (8) and (9) one simply evaluates both sides in the special case when Y is a point. First we recall that the operation in homology induced by $P^i:H\to\Sigma^{2i(p-1)}H$ is not P^i_* but its conjugate $\overline{P}^i_*$. Since $\overline{\beta}=-\beta$ we have in particular $\beta\underline{z}=-\underline{\beta z}$. Thus (11) gives

(17) $$(t\wedge 1)_*\underline{z} = \sum_i[g_*\underline{\overline{P}^i_*z} - (-1)^q g_*\underline{\beta\overline{P}^i_*z}\ cv^{-1}]V^{-i}$$

for any $z\in H_qX$. Now let $x\in H_qX$, $y=\underline{x}$. Then we have

$$(t\wedge 1)_*R_\pi y = (t\wedge 1)_*\sum_j(-1)^{mq(q+1)/2}(m!)^{-q}[\underline{Q^jx}-(-1)^q(\underline{\beta Q^jx})bv^{-1}]v^{-mq}V^j$$
$$= (-1)^{mq(q+1)/2}(m!)^{-q}\sum_j[(t\wedge 1)_*\underline{Q^jx}-(-1)^q(t\wedge 1)_*\underline{\beta Q^jx}(t_*b)(t_*v)^{-1}](t_*v)^{-mq}(t_*V)^j$$

(18) $$= (-1)^{mq(q+1)/2}(m!)^{-q}u^{-mq}w^{-q}\sum_{i,j}[\underline{\overline{P}^i_*Q^jx} - (-1)^q\underline{\beta\overline{P}^i_*Q^jx}cv^{-1}$$
$$- (-1)^q\underline{\overline{P}^i_*\beta Q^jx}(bu^{-1}-cv^{-1})(1-UV^{-1})^{-1} + \underline{\beta\overline{P}^i_*\beta Q^jx}bcu^{-1}v^{-1}(1-UV^{-1})^{-1}]S^jV^{-i}$$

On the other hand, we have

$$w^{-q}R^F_\pi(t\wedge 1)_*y = w^{-q}R^F_\pi\sum_j[g_*\underline{\overline{P}^j_*x} - (-1)^qg_*\underline{\beta\overline{P}^j_*x}cv^{-1}]V^{-j}$$
$$= w^{-q}\sum_j[R_\pi\underline{\overline{P}^j_*x} - (-1)^q(-1)^{m(q+1)}R_\pi\underline{\beta\overline{P}^j_*x}\ (P^F_\pi c)(P^F_\pi v)^{-1}](P^F_\pi V)^{-j}$$

(19) $$= (-1)^{mq(q+1)/2}(m!)^{-q}u^{-mq}w^{-q}\sum_{i,j}[\underline{Q^i\overline{P}^j_*x} - (-1)^q\ \underline{\beta Q^i\overline{P}^j_*xbu^{-1}}$$
$$+ (-1)^qQ^i\beta\overline{P}^j_*x(bu^{-1}-cv^{-1})(1-U^{-1}V)^{-1}-\underline{\beta Q^i\beta\overline{P}^j_*}xbcu^{-1}v^{-1}(1-U^{-1}V)^{-1}]U^iT^{-j}$$

If we collect the terms in (18) and (19) not involving b or c we get (8). Collecting the terms involving b but not c gives (9).

It remains to show 3.5.

Proof of 3.5. Let p be odd; the p = 2 case is similar. We use the spectral sequence I.2.4

$$H^i(\pi;\tilde{H}^j X^{(p)}) \Longrightarrow \tilde{H}^{i+j}(D_\pi X).$$

Let $\{x_\alpha\}_{\alpha\in A}$ be an ordered basis for $\tilde{H}^*X$. Let $|\alpha|$ denote the degree of x_α. The graded group $\tilde{H}^*(X^{(p)}) \cong \tilde{H}^*(X)^{\otimes p}$ has the basis $\{x_{\alpha_1}\otimes\cdots\otimes x_{\alpha_p} \mid \alpha_1,\ldots,\alpha_p \in A\}$ and the E_2-term has a basis consisting of representatives for the elements $\{b^\varepsilon u^i \mathcal{P}_\pi x_\alpha \mid \alpha \in A,\ \varepsilon = 0 \text{ or } 1,\ i \geq 0\}$ and $\{\tau_\pi(x_{\alpha_1}\otimes\cdots\otimes x_{\alpha_p}) \mid \alpha_1 = \min \alpha_i \neq \max \alpha_i\}$ (in particular, the spectral sequence collapses, as we also know from I.2.3). Hence these elements form a basis for $\tilde{H}^*(D_\pi X)$. Let $z \in \tilde{H}^*D_\pi X$ be a nonzero element with $\iota^*z = \Delta^*z = 0$. Since $\iota^*z = 0$, z is a finite sum of the form

$$\sum_{\alpha,\varepsilon,i} \lambda_{\alpha,\varepsilon,i} b^\varepsilon u^i \mathcal{P}_\pi x_\alpha .$$

Since $\Delta^*z = 0$, we have

$$(20)\quad 0 = \sum_{\alpha,\varepsilon,i} \lambda_{\alpha,\varepsilon,i} b^\varepsilon u^i P_\pi x_\alpha$$

$$= \sum (-1)^{j+m|\alpha|(|\alpha|-1)/2}(m!)^{|\alpha|}\lambda_{\alpha,\varepsilon,i} u^{i+m|\alpha|-2j} m_b{}^\varepsilon[P^j x_\alpha + (-1)^{|\alpha|}(\beta P^j x_\alpha)bu^{-1}]$$

by equation (4) of section 2. Now let K be $\max\{i+m|\alpha| \mid \lambda_{\alpha,\varepsilon,i} \neq 0\}$ and let S be the set of triples (α,ε,i) with $\lambda_{\alpha,\varepsilon,i} \neq 0$ and $i+m|\alpha| = K$. Then the coefficient of u^k in line (20) is

$$\sum_{(\alpha,\varepsilon,i)\in S} (-1)^{m|\alpha|(|\alpha|-1)/2}(m!)^{|\alpha|}\lambda_{\alpha,\varepsilon,i} b^\varepsilon x_\alpha$$

since all other terms in line (20) involve smaller powers of u. But this is a contradiction since the x_α are linearly independent.

§4. Atiyah's power operations in K-theory

In this section we show that the power operations in KU and KO defined by Atiyah [17] give H_∞^d structures for these spectra which agree with those constructed in VII §7. We shall work with complex K-theory, but everything is similar for KO.

We begin by recalling the definition of Atiyah's operations. Let G be a finite group. If Y is a G-space let $\mathrm{Vect}_G Y$ be the set of isomorphism classes of equivariant vector bundles over Y; we write Vect Y for the case where G is the trivial group. If Y is a free G-space there is a natural bijection

$$\mathrm{Vect}_G Y \cong \mathrm{Vect}(Y/G)$$

(see [18, 1.6.1]). If Y is any G-space we write Λ for the composite

$$\mathrm{Vect}_G Y \to \mathrm{Vect}_G(EG \times Y) \to \mathrm{Vect}(EG \times_G Y),$$

where the first map is induced by the projection $EG \times Y \to Y$. The map Λ is additive and hence if Y is a finite G-complex we obtain a map

$$K_G Y \to K(EG \times_G Y)$$

which will also be denoted by Λ. Now if X is a finite nonequivariant complex and we let Σ_j act on X^j by permuting the factors then the j-fold tensor power gives a map

$$\bar{\mathcal{P}}_j : \mathrm{Vect}\ X \to \mathrm{Vect}_{\Sigma_j} X^j \to K_{\Sigma_j} X^j$$

which however is not additive. In order to extend it to virtual bundles and to the relative case we must use the "difference construction" [94, Proposition 3.1]. Let (Y,B) be a G-pair and consider the set of complexes

$$0 \leftarrow E_0 \xleftarrow{d_1} E_1 \leftarrow \cdots \xleftarrow{d_n} E_n \leftarrow 0$$

of G-vector bundles E_i over Y which are acyclic over B. We write $\mathcal{D}_G(Y,B)$ for the set of isomorphism classes of such complexes. Two elements E_* and E'_* of $\mathcal{D}_G(Y,B)$ are <u>homotopic</u>, denoted $E_* \simeq E'_*$, if there is an element H_* $\mathcal{D}_G(Y \times I, B \times I)$ (with G acting trivially on I) which restricts to E_* and E'_* at the two ends. We say that E_* and E'_* are <u>equivalent</u>, written $E_* \sim E'_*$, if there are complexes F_* and F'_* which are acyclic on Y such that

$$E_* \oplus F_* \simeq E'_* \oplus F'_*.$$

It is shown in [94, appendix] that there is a natural epimorphism

$$\Gamma : \mathcal{D}_G(Y,B) \to K_G(Y,B)$$

which induces a bijection from the equivalence classes in $\mathcal{D}_G(Y,B)$ to $K_G(Y,B)$. If B is empty Γ is easy to describe: it takes E_* to $\sum_i (-1)^i E_i$. Γ is additive and multiplicative if we define addition and multiplication in $\mathcal{D}_G$ to be the direct sum and tensor product of complexes. Now if (X,A) is any pair of finite CW complexes the j-fold tensor product of complexes give a map

$$\mathcal{D}(X,A) \to \mathcal{D}_{\Sigma_j}((X,A)^j).$$

If E_* and E'_* in $\mathcal{D}(X,A)$ are homotopic by a homotopy H_* then the restriction of $H_*^{\otimes j}$ along the diagonal map

$$(X,A)^j \times I \to (X,A)^j \times I^j$$

gives a homotopy between $E_*^{\otimes j}$ and $(E_*^!)^{\otimes j}$. If F_* is acyclic on X then the inclusion $(E_*)^{\otimes j} \to (E_* \oplus F_*)^{\otimes j}$ is Σ_j-equivariantly split and is a homology equivalence by the Künneth theorem, so that $E_*^{\otimes j} \sim (E_* \oplus F_*)^{\otimes j}$. It follows that the j-fold tensor product preserves equivalence and we can pass to equivalence classes to obtain a map

$$\overline{\mathcal{P}}_j : K(X,A) \to K_G((X,A)^j).$$

Letting A be the basepoint * of X we write $\mathcal{P}_j$ for the composite

$$\tilde{K}X = K(X,*) \longrightarrow K_{\Sigma_j}((X,*)^j) \xrightarrow{\Lambda} K(E\Sigma_j \times_{\Sigma_j} (X,*)^j) = \tilde{K}D_jX.$$

We can extend $\mathcal{P}_j$ to all even dimensions by letting it take the Bott element $b \in \tilde{K}^{-2}(S^0)$ to b^j. It is easy to see that the $\mathcal{P}_j$ are consistent in the sense of 1.3, so by 1.2 and 1.3 we have

Theorem 4.1. KU (resp. KO) has a unique H_∞^2 (resp. H_∞^8) ring structure for which the power operations are those defined by Atiyah.

We shall see in Section 6 that the H_∞^8 structure on KO extends to an H_∞^4 structure. Our next result answers an obvious question.

Proposition 4.2. The structures on KO and KU given by 4.1 are the same as those given by VII.7.2.

For the proof we need a lemma.

Lemma 4.3. Let X be a based space and let $\lambda : X^+ \to X$ be the based map which is the identity on X. Then

$$(D_j\lambda)^* : \tilde{F}^* D_j X \to \tilde{F}^*(D_j(X^+))$$

is a split monomorphism for any theory F.

Proof of 4.3. If $\nu : \Sigma^\infty X \to \Sigma^\infty X^+$ is the map given in the proof of 3.3 then $(D_j\nu)^*(D_j\lambda)^* = (D_j(\Sigma^\infty\lambda \circ \nu))^* = 1$.

Proof of 4.2. Let $\mathcal{P}_j$ be Atiyah's power operation and let $\mathcal{P}'_j$ be that given by VII.7.2. By VII.7.7 we have

$$\mathcal{P}'_j(\Sigma^2 b) = (\mathcal{P}'_j \Sigma^2 1) \cdot b^j$$

while by 1.1(iii) we have

$$\mathcal{P}'_j(\Sigma^2 b) = (\mathcal{P}'_j\Sigma^2 1)\cdot \mathcal{P}'_j b.$$

Since $\mathcal{P}'_j\Sigma^2 1$ is an orientation for the Thom complex $D_j S^2$ this implies $\mathcal{P}'_j b = b^j = \mathcal{P}_j b$. It therefore suffices by 1.3 to show that $\mathcal{P}_j$ and $\mathcal{P}'_j$ are equal on $\tilde{K}X$ for any finite complex X, and by 4.3 it suffices to show that they agree on $\tilde{K}(X^+) = KX$. They do agree on Vect X by [71, VIII.1.2]. But any element x of KX can be written in the form V-W with V,W ε Vect X, and we have

$$\mathcal{P}_j V = \mathcal{P}_j(x + W) = \mathcal{P}_j x + \mathcal{P}_j W + \sum_{i=1}^{j-1} \tau_{i,j-i}[(\mathcal{P}_i x)(\mathcal{P}_{j-i} W)]$$

by 1.1(vi), and similarly for $\mathcal{P}'_j$. Hence

$$\mathcal{P}_j x = \mathcal{P}_j V - \mathcal{P}_j W - \sum_{i=1}^{j-1} \tau_{i,j-i}[(\mathcal{P}_i x)(\mathcal{P}_{j-i} W)]$$

and similarly for $\mathcal{P}'_j$. We therefore have $\mathcal{P}_j x = \mathcal{P}'_j x$ by induction on j.

By analogy with Section 2 we now ask what operations in K-theory can be obtained from the internal power operation

$$P_\pi : KX \to K(X \times B\pi)$$

The structure of $K(B\pi)$ has been determined by Atiyah [16]: $\tilde{K}(B\pi)$ is a $\hat{Z}_p$-module and the composite

$$IR(\pi) \otimes \hat{Z}_p \xrightarrow{\Lambda \otimes 1} \tilde{K}(B\pi) \otimes \hat{Z}_p \longrightarrow \tilde{K}(B\pi)$$

is an isomorphism, where $IR(\pi)$ is the augmentation ideal. If ρ is the automorphism group of π then the invariant subgroup $\tilde{K}(B\pi)^\rho$ is generated by $\Lambda(N-p)$, where N is the regular representation of π. Atiyah also shows that $K^1 B\pi = 0$. In particular, $K^* B\pi$ is flat over $K^*(pt)$ and we obtain a Künneth isomorphism

$$KX \otimes K(B\pi) \cong K(X \times B\pi)$$

for finite complexes X. Since P_π is the restriction of P_p we see that P_π actually lands in the invariant subring $KX \otimes K(B\pi)^\rho$. We can therefore define operations

$$\varphi^p : KX \to KX$$

and

$$\theta^p : KX \to KX \times \hat{Z}_p$$

by the equation

(1) $$P_\pi x = \varphi^p x \otimes 1 + \theta^p x \otimes \Lambda(N - P).$$

By 1.4(i) we have

(2) $\varphi^p x = x^p.$

Atiyah proves the relation

(3) $p\theta^p x = x^p - \psi^p x$

in [17]. Since the representation N of π is induced from the trivial representation of the trivial group we have $\Lambda(N) = \tau_\pi 1$ and therefore (1), (2) and (3) give

(4) $P_\pi x = \psi^p x \otimes 1 + \theta^p x \otimes \tau_\pi 1,$

an equation which will be used in §7.

We can in fact lift θ^p to KX by using the equivariant internal operation $\overline{P}_\pi$. This is the composite

$$KX \xrightarrow{\bar{\mathcal{P}}_\pi} K_\pi(X^p) \xrightarrow{\Delta^*} K_\pi X,$$

where Δ is the diagonal map from X with its trivial π-action to X^p with its permutation action. Clearly $P_\pi = \Lambda \circ \overline{P}_\pi$. Since π acts trivially on X, we have $K_\pi X \cong KX \otimes R\pi$. The ρ-invariant subring of $R\pi$ is generated by 1 and N-p, so we may define $\theta^p x$ as an element of KX by the equation

$$\overline{P}_\pi x = x^p \otimes 1 + \theta^p x \otimes (N - p).$$

The operation $\overline{P}_\pi$ satisfies the obvious analog of 1.4 and one can use its properties to obtain additivity and multiplicity formulas for θ^p and ψ^p (using equation (3) as the definition of ψ^p). One can also obtain the G-equivariant Adams operations in this way by starting with a G-complex X and constructing operations

$$\bar{\mathcal{P}}_j : K_G X \to K_{\Sigma_j \int G} X^j$$

exactly as before. The reader is referred to [34] for details.

§5. tom Dieck's operations in cobordism

In [31], tom Dieck constructed "Steenrod operations" (power operations in our terminology) for the cobordism spectra associated to the classical groups. In this section we use these operations to give H_∞^d structures for these spectra. A wider class of cobordism spectra will be investigated by Lewis in the sequel, and he will show that they have not just H_∞ but E_∞ structures. His results do not quite include those of this section, however, since his methods do not give the "d-structure" (i.e., the Σ_j-orientations) for the classical spectra.

Throughout this section we write G for any of the classical groups O, SO, $Spin^c$, U, SU, Sp or Spin. Let d = 1,2,2,2,4,4,4 respectively. We depart somewhat from standard notation (in this section only) by writing $G(i)$ for the group which acts on $\mathbf{R}^{di}$. Let p_i be the universal $G(i)$-vector bundle over $BG(i)$, let $S(p_i)$ be its fibrewise one-point compactification, and let $T(p_i)$ be the Thom complex obtained by collapsing the points at ∞. We shall always identify principal $G(i)$-bundles with free $G(i)$-spaces, so that the principal bundle associated to p_i is $EG(i)$. If q is any $G(i)$-vector bundle with principle bundle Q, there is a bundle map $F:q \to p_i$ and induced maps $S(F):S(q) \to S(p_i)$ and $T(f):T(q) \to T(p_i)$. If F' is another such map we shall need to know that $T(F')$ is homotopic to $T(F)$ (of course this is well-known for the maps of base spaces induced by F and F'). Now F has the form $\tilde{F} \times_{G(i)} \mathbf{R}^{di}$ for some $G(i)$-map $\tilde{F}:Q \to EG(i)$ and $S(F)$ is equal to $\tilde{F} \times_{G(i)} S^{di}$, and similarly for F' and $S(F')$. It is shown in [32] that there is at most one $G(i)$-equivariant homotopy class of $G(i)$-maps from any $G(i)$-space into $EG(i)$, so it follows that $S(F)$ is homotopic to $S(F')$ by a homotopy preserving the base points in each fibre, and hence $T(F) \simeq T(F')$ as required.

Now we define the Thom prespectrum TG by letting $(TG)_{di} = T(p_i)$ with

$$\sigma:\Sigma^d T(p_i) \to T(p_{i+1})$$

induced by any bundle map from $p_i \oplus \mathbf{R}^d$ to p_{i+1}. We wish to show that TG is an H_∞^d ring prespectrum. For this we need some bundle theoretic observations.

Let p be a $G(i)$-vector bundle over X with associated principal bundle P. Then $E\Sigma_j \times_{\Sigma_j} p^j$ is a vector bundle over $E\Sigma_j \times_{\Sigma_j} X^j$; we wish to give it a canonical $G(ij)$-bundle structure. Let $H = G(i)^j$. Then p^j is an H bundle over X^j with principal bundle P^j, and Σ_j acts on everything on the left. However, its action on P^j does not commute with the right H-action (P^j is not a "Σ_j-equivariant principal H-bundle"). Instead we have $\sigma(ph) = (\sigma p)(\sigma h)$ for $\sigma \in \Sigma_j$, $p \in P^j$, $h \in H$. Now let $Q = P^j \times_h G(ij)$. This is a principal $G(ij)$-bundle over X^j with associated vector bundle p^j. Because of our choice of d the permutation action of Σ_j on $(\mathbf{R}^{di})^j$ lifts to a homomorphism $\Sigma_j \to G(ij)$ denoted $\sigma \mapsto \bar{\sigma}$, and we have $\sigma(h) = \bar{\sigma}h\bar{\sigma}^{-1}$ for all $h \in H$. We define a left action of Σ_j on Q by $\sigma(p,g) = (\sigma p, \overline{\sigma g})$; it is easy to check that this action is well-defined and that it commutes with the right action of $G(ij)$. Thus Q is a Σ_j-equivariant principal $G(ij)$-bundle and hence so is its pullback $E\Sigma_j \times Q$ to $E\Sigma_j \times_{\Sigma_j} X^j$. Since Σ_j acts freely on $E\Sigma_j \times Q$ and commutes with $G(ij)$ we can divide out by its action to get a principal $G(ij)$-bundle $E\Sigma_j \times_{\Sigma_j} Q$ over $E\Sigma_j \times_{\Sigma_j} X^j$. The reader can check that the associated vector bundle is $E\Sigma_j \times_{\Sigma_j} p^j$.

Since $T(E\Sigma_j \times_{\Sigma_j} p^j)$ is naturally homeomorphic to $D_j T(p)$ we obtain maps

$$\zeta_{j,i}:D_j(TG)_{di} \cong T(E\Sigma_j \times_{\Sigma_j} p_i^j) \longrightarrow T(p_{ij}) = (TG)_{dij}$$

for all $i,j \geq 0$. The diagrams of Definition VII.5.1 commute since in each case the two composites are induced by bundle maps into a universal bundle. Thus we have shown

<u>Proposition 5.1.</u> The maps $\zeta_{j,i}$ are an H_∞^d structure for TG.

Now define MG = Z(TG). Every G(i)-vector bundle q has a canonical Thom class in this theory represented by the map

$$T(q) \longrightarrow T(p_i) \xrightarrow{\kappa} (MG)_{di}$$

At this point we need some $\lim^1$ information.

<u>Lemma 5.2.</u> All of the pairs (TG,MG'), (TG,KU), (TG,KO), (TG,ku) and (TG,kO) are $\lim^1$-free.

<u>Proof.</u> First consider (TG,MG'). The pair (TU,MU) is clearly $\lim^1$-free since the spectral sequence $E_r(TU_{2i};MU)$ collapses for dimensional reasons. For each other choice of G and G' there are maps $f:MU \to MG'$ and $g:TG \to TU$ satisfying the hypotheses of VII.4.4, hence each pair (TG,MG') is $\lim^1$-free. A similar argument gives the remaining cases.

<u>Corollary 5.3.</u> MG is an H_∞^d ring spectrum.

On the other hand, it was shown in [71,IV§2] that MG has an E_∞ ring structure. Such structures always determine H_∞ structures, as mentioned in I§4; see [Equiv, VII§2] for the details. Let $\xi_j^E:D_j MG \to MG$ be the structural maps obtained in this way and let ξ_j^H be those obtained from 5.1 and 5.3. As one would expect, the two structures agree:

<u>Proposition 5.4.</u> For each j, $\xi_j^E \simeq \xi_j^H$.

<u>Proof.</u> We use the notations and Definitions of VII§8. Fix i and let $a = a_i$. It suffices to show that the elements z_i^H and z_i^E in cobordism represented by the composites

$$T(\eta_i) \wedge_{\Sigma_j} T(p_i)^{(j)} \xrightarrow{\kappa} (D_j MG)_a \xrightarrow{(\xi_j^E)_a} (MG)_a$$

and

$$T(\eta_i) \wedge_{\Sigma_j} T(p_i)^{(j)} \xrightarrow{\kappa} (D_j MG)_a \xrightarrow{(\xi_j^E)_a} (MG)_a$$

are equal. An inspection of the proofs of [71, IV.2.2] and [Equiv. VII.2.4] shows that the second composite is induced by a bundle map from $\eta_i \oplus (p_i)^j$ into the universal bundle p_a, hence z_i^E is the canonical Thom class in $MG^a(T(\eta_i) \wedge_{\Sigma_j} T(p_i)^{(j)})$. On the other hand by Proposition VII.8.1 there is a relative Thom isomorphism

$$\Psi:(MG)^a(T(n_i)\wedge_{\Sigma_j}(T(p_i)^{(j)}) \longrightarrow (MG)^{a+dij}(\Sigma^a T(E\Sigma_j \times_{\Sigma_j}(p_i)^j))$$

which takes z_i^H to the canonical Thom class in the target group. Since the canonical Thom class of a Whitney sum is the product of the Thom classes, the relative Thom isomorphism Ψ takes the Thom class of $T(n_i)\wedge_{\Sigma_j}(T(p_i)^{(j)})$ to that of $\Sigma^a T(E\Sigma_j \times_{\Sigma_j}(p_i)^j)$. Thus $\Psi z_i^H = \Psi z_i^E$ and the result follows.

We conclude this section with a discussion of cobordism operations related to P_π. The situation in unoriented cobordism is quite simple: there is a Künneth isomorphism

$$MO^*(X \times BZ_2) \cong (MO^*X)[[\chi]]$$

where χ is the MO^* Euler class of the Hopf bundle, and we can define operations

$$R^i : MO^q X \to MO^{q+i} X$$

for $i \in Z$ by the equation

$$P_2 x = \sum_i (R^i x)\chi^{q-i}.$$

One can prove various properties of the R^i exactly as in §2 (see [31, §15]).

To deal with the case of complex cobordism we need some formal-groups notation. Let $F(x,y)$ be the formal group of MU and let $[n](x)$ be the power series defined inductively for $n \geq 0$ by $[1](x) = x$ and $[n+1](x) = F([n](x),x)$. There is a Künneth theorem due to Landweber [49]:

$$MU^*(X \times B\pi) \cong (MU^*X)[[u]]/[p](u),$$

where u is the Euler class of a nontrivial irreducible <u>complex</u> representation of π. The power series $[p](u)$ has leading term pu but is not divisible by p, so that in particular $MU^*B\pi$ is torsion free. We cannot continue as in the unoriented case since the power series $[p](u)$ and the ring $MU^*B\pi$ admit no simple descriptions. There is however a relation between P_π and the Landweber-Novikov operations s_α which is due to Quillen and was used by him to give a proof of the structure theorem for π_*MU. Let $a_j(x)$ for $j \geq 1$ be the coefficient of y^j in the power series $\prod_{i=1}^{p-1} F([i](x),y)$. For a multi-index $\alpha = (\alpha_1,\dots,\alpha_k)$ let $a(x)^\alpha = a_1(x)^{\alpha_1}\cdots a_k(x)^{\alpha_k}$. Define $\chi \in MU^{2p-2}B\pi$ by the equation $\chi\cdot\Sigma^2 1 = P_\pi \Sigma^2 1$; thus χ is the Euler class of the complex reduced regular representation.

Proposition 5.5. For any finite complex X there is an integer $m \geq 0$ such that the equation

(1) $$(P_\pi x)\chi^{m-q} = \sum_{|\alpha| \leq m} (S_\alpha x)a(u)^\alpha \chi^{m-|\alpha|}$$

holds for all $x \in MU^{2q}X$.

For the proof see [93] or [11]. There is a similar relation between P_π and s_α in the unoriented case. Since the right side of equation (1) is additive in x we have

Corollary 5.6. $(P_\pi)(x+y) - P_\pi x - P_\pi y)\cdot\chi^m = 0$ for large m.

§6. The Atiyah-Bott-Shapiro orientation.

It is well-known that the KU and KO orientations constructed by Atiyah, Bott and Shapiro in [19] give rise to ring maps

$$\phi^U : MSpin^c \to KU$$

and $$\phi^O : MSpin \to KO$$

In this section we shall prove

Theorem 6.1. ϕ^U is an H_∞^2 ring map and ϕ^O is an H_∞^8 ring map.

Remark 6.2. MSpin actually has an H_∞^4 structure, as shown in §5. By combining 6.1 with VII.6.2 we see that the H_∞^8 structures for KO and kO constructed in §4 and in VII§7 extend to H_∞^4 structures.

We shall give the proof of 6.1 only for ϕ^O, which will henceforth be denoted by ϕ; the remaining case is similar. If p is a Spin(8i)-vector bundle we denote its Atiyah-Bott-Shapiro orientation in KO(T(p)) by $\mu(p)$.

First we translate 6.1 to a bundle-theoretic statement. As usual, let p_{8i} be the universal Spin(8i)-vector bundle. If $X \subset BSpin(8i)$ is any finite complex, we obtain an orientation class

$$\mu(p_{8i}|X) \quad \widetilde{KO}(T(p_{8i}|X)).$$

These classes are consistent as X varies, hence by 5.2 and VII.4.2 they determine a unique class in $\widetilde{KO}(TSpin_{8i})$ which is represented by a map

$$\mu_i : TSpin_{8i} \to BO \times Z.$$

The sequence $\{\mu_i\}$ is a map of prespectra, and ϕ is defined to be $Z\{\mu_i\}$ (see VII§1). The multiplicative property [19, 11.1 and 11.3] of the Atiyah-Bott-Shapiro orientation implies at once that $\{\mu_i\}$ is a ring map, and hence so is ϕ by 5.2 and VII.2.3. Similarly, Theorem 6.1 is a consequence of the following property of μ.

Proposition 6.3. If p is any Spin(8i)-vector bundle then

$$\mu(E\Sigma_j \times_{\Sigma_j} p^j) = \mathcal{P}_i\mu(p),$$

where $\mathcal{P}_j$ is the power operation defined in §4.

In the terminology of §1, Proposition 6.3 says that $\mathcal{P}_j$ satisfies tom Dieck's axiom P4. tom Dieck gives a simple proof of the analogous statement for the KU-orientation of complex bundles in [31, §12].

For the proof of 6.3 we need to recall several technical facts from [19]. The first is the "shrinking" construction in $\mathcal{D}(D,Y)$. Let

$$E_*: \qquad 0 \leftarrow E_0 \xleftarrow{d_1} E_1 \leftarrow \cdots \xleftarrow{d_n} E_n \leftarrow 0$$

be a complex of real vector bundles over X which is acyclic over Y. Choose Euclidean metrics in each E_i and let $\delta_i : E_{i-1} \to E_i$ be the adjoint of d_i with respect to the chosen metrics. Let

$$s(E_*): \qquad 0 \leftarrow s(E)_0 \xleftarrow{D} s(E)_1 \leftarrow 0$$

be the complex with $s(E)_0 = \bigoplus_{i \text{ even}} E_i$, $s(E)_1 = \bigoplus_{i \text{ odd}} E_i$, and differential

$$D(e_1, e_3, \ldots) = (d_1e_1, \delta_2e_1 + d_3e_3, \delta_4e_3 + d_5e_5, \ldots)$$

Then s(E) is in $\mathcal{D}(X,Y)$ and it defines the same element in KO(X,Y) that E does (see [19, p.22]). The same construction works G-equivariantly provided that the chosen Euclidean metrics are G-invariant.

Next we need the Clifford algebra C_i. By definition, C_i is the quotient of the tensor algebra $T(\mathbb{R}^i)$ by the ideal generated by the set $\{x \otimes x - \|x\|^2 \cdot 1 \mid x \in \mathbb{R}^i\}$. The grading on $T(\mathbb{R}^i)$ gives C_i a Z_2-grading by even and odd degrees and we will write $\boxtimes$ for the Z_2-graded tensor product of two Z_2-graded objects. By a <u>module</u> M over C_i we mean a Z_2-graded real vector space with a map

$$C_i \boxtimes M \to M$$

satisfying the usual properties. Equivalently, such a structure is given by two maps

$$\mathbb{R}^i \otimes M^0 \to M^1$$

and $$\mathbb{R}^i \otimes M^1 \to M^0,$$

each denoted by $x \otimes m \longmapsto xm$, such that

(1) $$x(xm) = -\|x\|^2 m$$

for all x,m. In particular, the latter description shows that if M is a C_i-module and N is a C_j-module then $M \boxtimes N$ is a C_{i+j}-module with

$$(x \oplus y)(m \otimes n) = xm \otimes n + (-1)^{|m|} x \otimes yn$$

for all $x \in \mathbb{R}^i$, $y \in \mathbb{R}^j$, $m \in M$, $n \in N$. If M is any module over C_i we can define a complex

$$E(M): \qquad 0 \longleftarrow E_0(M) \xleftarrow{d} E_1(M) \longleftarrow 0$$

of real vector bundles over $\mathbb{R}^i$ by letting $E_0(M) = \mathbb{R}^i \times M^0$, $E_1(M) = \mathbb{R}^i \times M^1$, and $d(x,m) = (x,xm)$. Equation (1) shows that this is acyclic except at 0, and in particular it defines an element of $KO(D^i,S^{i-1})$.

We can now define two complexes over $(\mathbb{R}^i)^j$, namely $E(M^{\boxtimes j})$ and the external tensor product $E(M)^{\otimes j}$. The first has length 2 and the second has length j+1. We need to be able to compare them.

Lemma 6.4. The inner product in $E(M)^{\otimes j}$ can be chosen so that $s(E(M)^{\otimes j})$ is isomorphic to $E(M^{\boxtimes j})$.

Proof. It is shown in [19,p. 25] that one can choose inner products in M^0 and M^1 so that the adjoint of $x:M^1 \to M^0$ is $-x:M^0 \to M^1$ for each $x \in \mathbb{R}^i$. We define an inner product in $M^{\otimes j}$ by

$$\langle m_1 \otimes \cdots \otimes m_j, m_1' \otimes \cdots \otimes m_j' \rangle = \langle m_1, m_1' \rangle \cdots \langle m_j, m_j' \rangle$$

with the understanding that $\langle m,m' \rangle = 0$ if $|m| \neq |m'|$. Then $s(E(M)^{\times j})$ and $E(M^{\boxtimes j})$ clearly involve the same two bundles, but they have different differentials, say d and d'. The definition of the shrinking construction gives

$$d(x,m_1 \otimes \cdots \otimes m_j) =$$
$$\sum_{k=1}^{j} (-1)^{|m_1|+\cdots+|m_i|-1} (x,m_1 \otimes \cdots \otimes m_{i-1} \otimes x_i m_i \otimes m_{i-1} \otimes \cdots \otimes m_j)$$

if $x = x_1 \oplus \cdots \oplus x_j \in (\mathbb{R}^i)^j$, while the definition of $M^{\boxtimes j}$ as a C_{ij}-module gives

$d'(x,m_1 \otimes \cdots \otimes m_j) =$

$$\sum_{k=1}^{j} (-1)^{|m_1|+\cdots+|m_{i-1}|} (x, m_1 \otimes \cdots \otimes m_{i-1} \otimes x_i m_i \otimes m_{i+1} \otimes \cdots \otimes m_j).$$

The required isomorphism is given by taking $(x,m_1 \otimes \cdots \otimes m_j)$ to itself if $|m_1| + \cdots + |m_j|$ is congruent to 0 or 1 mod 4 and to its negative in the remaining cases.

Next we recall that Spin(i) is a subgroup of the group of units of C_i (in fact this is the definition of Spin(i) in [19, p.8]) and that the resulting conjugation action on $R^i \subset C_i$ agrees with its usual action on R^i. We can therefore define an action of Spin(i) on E(M) through automorphisms by $g(x,m) = (gxg^{-1},gm)$. Now if P is a principal Spin(i)-bundle over X with associated vector bundle $p:V \to X$ we can define a complex E(M,P) over $V = P \times_{Spin(i)} R^i$ by

$$E(M,P) = P \times_{Spin(i)} E(M).$$

This complex defines an element of $\mathcal{D}(BV,SV)$ and hence of $\widetilde{KO}(T(p))$. If P is a G-equivariant principal bundle for some G (i.e., G acts from the left on P and commutes with the right action of Spin(i)) then E(M,P) has a left G-action and defines an element of $\widetilde{KO}_G(T(p))$. If G acts freely on P we can divide out by its action , and it is easy to see that the quotient complex E(M,P)/G is just E(M,P/G).

Atiyah, Bott and Shapiro specify a module λ over C_8 for which $E(\lambda)$ represents the Bott element in $\widetilde{KO}(S^8)$ (see [19, p.15]), and if P is a principal Spin(8i)-bundle they define $\mu(p) \in \widetilde{KO}(T(p))$ to be the element represented by $E(\lambda^{\boxtimes i},P)$.

From now on we fix i, P and p and denote $\lambda^{\boxtimes i}$ by M. Let $q = p^j$ with its permutation action by Σ_j and let Q be the associated Σ_j-equivariant Spin(8ij)-bundle as defined in Section 5. To prove 6.3 it suffices to show that $E(M^{\boxtimes j},Q)$ and the external tensor product $E(M,P)^{\otimes j}$ define the same element of $\widetilde{KO}_{\Sigma_j}(T(q))$. We can describe these complexes more simply: the first is

$$P^j \times_{Spin(8i)^j} E(M^{\boxtimes j})$$

and the second is

$$P^j \times_{Spin(8i)^j} (E(M)^{\otimes j});$$

in each case Σ_j acts through permutations of both factors. Now it is shown in [19, p. 25] that the inner products on M^0 and M^1 used in the proof of Lemma 6.4 can be chosen to be invariant under Spin(8i), hence the inner product on $E(M)^{\otimes j}$ used in the proof of that lemma is invariant under both $(Spin(8i))^j$ and Σ_j, and so is the isomorphism $s(E(M)^{\otimes j}) \cong E(M^{\boxtimes j})$. It follows that $s(E(M,P)^{\otimes j})$ is isomorphic to $E(M^{\boxtimes j},Q)$ as required.

§7. p-local H_∞ ring maps.

In this section we make some general observations about p-local H_∞ ring maps and apply them to show that the Adams operations are H_∞ ring maps and that the Adams summand of $KU_{(p)}$ is an H_∞^2 ring spectrum. We also obtain a sufficient condition for BP to be an H_∞^2 ring spectrum.

Throughout this section we let p be a fixed prime and let $\pi \subset \Sigma_p$ be generated by a p-cycle.

Lemma 7.1. Let F be a p-local spectrum and let Y be any spectrum. The map

$$\beta^*: F^*(D_{jp}Y) \to F^*(D_j D_\pi Y)$$

is split monic, and if j is prime to p the map

$$\alpha^*: F^*D_j Y \to F^*(Y \wedge D_{j-1}Y)$$

is split monic.

Proof. The subgroup $\Sigma_j \int \pi$ of Σ_{jp} has index prime to p, and hence the composite

$$H^*(\Sigma_{jp};M) \longrightarrow H^*(\Sigma_j \int \Sigma_p;M) \xrightarrow{\tau} H^*(\Sigma_{jp};M)$$

is an isomorphism for any p-local Σ_{jp}-module M. Thus

$$F^*D_{jp}Y \xrightarrow{\iota^*} F^*D_{\Sigma_j \int \pi} Y$$

is split monic by I.2.4. The result for β^* follows since β factors as

$$D_j D_\pi Y \simeq D_{\Sigma_j \int \pi} Y \xrightarrow{\iota} D_{jp} Y$$

and the result for α^* is similar.

As an application, we have

Proposition 7.2. Let E and F be H_∞^d ring spectra with power operations $\mathcal{P}_j$ and $\mathcal{P}'_j$. Suppose that F is p-local. Let $f: E \to F$ be a ring map such that the equation

$$(1) \qquad f_* \circ \mathcal{P}_p = \mathcal{P}'_p \circ f_*$$

holds on $E^{di}Y$ for all $i \in Z$ and all spectra Y. Then f is an H_∞^d ring map.

Proof. We shall show that $f_* \circ \mathcal{P}_j = \mathcal{P}'_j \circ f_*$ for all j by induction on j. This is trivial for j = 1 since $\mathcal{P}_1$ is the identity. Suppose it is true for all k < j. If j is prime to p we have $\alpha^* f_* \mathcal{P}_j y = (f_* y)(f_* \mathcal{P}_{j-1} y)$ and $\alpha^* \mathcal{P}'_j f_* y = (f_* y)(\mathcal{P}'_{j-1} f_* y)$. If j has the form kp we have $\beta^* f_* \mathcal{P}_j y = f_* \mathcal{P}_k \mathcal{P}_\pi x$ and $\beta^* \mathcal{P}'_j f_* x = \mathcal{P}'_k \mathcal{P}'_\pi f_* x$. In either case the result follows from 7.1 and the inductive hypothesis.

Under the usual $\lim^1$ hypotheses, it suffices to check equation (1) for spaces of for finite CW complexes. However, for actual calcualtions it is much easier to deal with the internal operation P_π than with $\mathcal{P}_\pi$. Our next result allows us to reduce to this case when we are dealing with spectra like KU or MU.

Proposition 7.3. Let F be a p-local spectrum such that $\pi_* F$ is free over $Z_{(p)}$ in even dimensions and zero in odd dimensions. Let X be a space such that $H_*(X;Z)$ is free abelian in even dimensions and zero in odd dimensions. Suppose that X and F have finite type. Then the map

$$\iota^* \oplus \Delta^* : \tilde{F}^* D_\pi X \to \tilde{F}^* X^{(p)} \oplus \tilde{F}^*(X \wedge B\pi^+)$$

is monic.

Proof. First let $F = HZ_{(p)}$. The Bockstein on $\tilde{H}^*(D_\pi X; Z_p)$ is given by II.5.5 and it follows that $E_2 = E_\infty$ in the Bockstein spectral sequence. Thus $\tilde{H}^*(D_\pi X; Z_{(p)})$ is a direct sum of copies of $Z_{(p)}$ and Z_p, so it suffices to show that the maps $(\iota^* \oplus \Delta^*) \otimes Q$ and $(\iota^* \oplus \Delta^*) \otimes Z_p$ are monic. For the first we observe that $\iota^* \otimes Q$ is a split injection by a simple transfer argument. For the second we use 3.5 and the universal coefficient theorem. This completes the proof for $F = HZ_{(p)}$. For the general case, we observe that $\iota^* \oplus \Delta^*$ induces a monomorphism on E_2 of the Atiyah-Hirzebruch spectral sequence and that the spectral sequences for $X^{(p)}$ and $X \wedge B\pi^+$ collapse for dimensional reasons.

Our first application is to the Adams operation

$$\psi^k : KU_{(p)} \to KU_{(p)}$$

with k prime to p. This is well-known to be a ring map.

Theorem 7.4. If Y is any spectrum and $y \in KU^{2n} Y$ then $\psi^k \mathcal{P}_j y = k^{-jn} \mathcal{P}_j (k^n \psi^k y)$. In particular, ψ^k is an H_∞ ring map but not an H_∞^2 ring map.

Proof. Let $\mathcal{P}'_j y = k^{-jn} \mathcal{P}_j k^n y$ for y $K^{2n} Y$. We must show $\psi^k \mathcal{P}_j = \mathcal{P}'_j \psi^k$. The $\mathcal{P}'_j$ are consistent in the sense of 1.2 and thus define another H_∞^2 structure on $KU_{(p)}$ (which agrees with the standard H_∞ structure but has different Σ_j-orientations). By 7.2 it suffices to show $\psi^k \mathcal{P}_p = \mathcal{P}'_p \psi^k$, and by 1.3 it suffices to show this for finite complexes. Since ψ^k is a ring map we clearly have $\iota^* \psi^k \mathcal{P}_p = \iota^* \mathcal{P}'_p \psi^k$, so by 7.3 it suffices to show

(2) $$\psi^k P_\pi x = P'_\pi \psi^k x$$

for all $x \in K^{2n}X$ whenever X is a finite complex. If x is the Bott element b then $\psi^k b = kb$ and $P_\pi b = b^p$ so (2) is satisfied in this case. Thus we may assume $n = 0$. Since ψ^k is a stable map it commutes with the transfer, and thus (2) will follow from equation (4) of section 4 once we show that ψ^k commutes with θ^p. It suffices to show this for the universal case $BU \times Z$, and since $K(BU \times Z)$ is torsion free it suffices to show that ψ^k commutes with $p\theta_p$. But this is immediate from equation (3) of Section 4.

Next we recall the Adams idempotents

$$E_a : KU_{(p)} \to KU_{(p)}, \quad a \in Z_{p-1}$$

defined in [5, Lecture 4]. These idempotents split off pieces of $KU_{(p)}$ which we shall denote by $L_0, \dots, L_{p-2}$. Thus the idempotent E_a factors into a projection map and an inclusion map:

$$KU_{(p)} \xrightarrow{r_a} L_a \xrightarrow{s_a} KU_{(p)}$$

with $r_a s_a = 1$. Since $\sum_{a \in Z_{p-1}} E_a = 1$ we have $KU_{(p)} \simeq L_0 \vee \dots \vee L_{p-2}$. The E_a satisfy the formulas $E_0 1 = 1$,

(3) $$E_a b^n = \begin{cases} 0 & \text{if } n \not\equiv a \bmod p-1 \\ b^n & \text{otherwise} \end{cases}$$

and

(4) $$E_a(xy) = \sum (E_{a'}x)(E_{a-a'}y).$$

In particular, the image of E_0 is a subring of K^*X and hence L_0 has a unique structure for which s_0 is a ring map. On the other hand, (3) implies that the kernel of E_0 is not an ideal and hence there is no ring structure on L_0 for which r_0 is a ring map.

Proposition 7.5. L_0 has a unique H_∞^2 ring structure for which s_0 is an H_∞^2 ring map.

Proof. We must show that $\mathcal{P}_j$ takes the image of E_0 to itself, i.e., that the equation

(5) $$E_0 \mathcal{P}_j E_0 y = \mathcal{P}_j E_0 y$$

holds on $K^{2n}Y$ for every $n \in Z$ and every spectrum Y.

Let ch be the Chern character and let X be a finite complex. We have $ch(\psi^p E_a x) = ch(E_a \psi^p x)$ for all $a \in Z_{p-1}$ and all $x \in KX$ by [5, p.84-85] and [1, 5.1(vi)]. Hence $\psi^p E_a = E_a \psi^p$ by [5, Lemma 4 of lecture 4]. As in the proof of 7.4 it follows that $E_a \theta^p = \theta^p E_a$ and that $E_a P_\pi x = P_\pi E_a x$ for all $x \in KX$. Now let $n \in Z$ and let a be the class of n in Z_{p-1}. Then we have

$$\begin{aligned} E_0 P_\pi E_0(b^n x) &= E_0 P_\pi (b^n E_{-a} x) = E_0(b^{pn} P_\pi E_{-a} x) \\ &= b^{pn} E_{-a} P_\pi E_{-a} x = b^{pn} P_\pi E_{-a} x \\ &= P_\pi(b^n E_{-a} x) = P_\pi E_0(b^n x) \end{aligned}$$

for all $x \in KX$. As in the proof of 7.4 it follows that (5) holds on the space level with $\mathcal{P}_j$ replaced by $\mathcal{P}_\pi$. Since both sides of (5) are stable in the sense of 1.2 and 1.3, it follows that (5) holds on the spectrum level with $\mathcal{P}_j$ replaced by $\mathcal{P}_\pi$. The rest of the proof is an induction on j just like that in the proof of 7.2. We give the inductive step when j has the form kp:

$$\begin{aligned} \beta^* E_0 \mathcal{P}_j E_0 y &= E_0 \beta^* \mathcal{P}_j E_0 y = E_0 \mathcal{P}_k \mathcal{P}_\pi E_0 y \\ &= E_0 \mathcal{P}_k (E_0 \mathcal{P}_\pi E_0 y) = (\mathcal{P}_k E_0) \mathcal{P}_\pi E_0 y \quad \text{by inductive hypothesis} \\ &= \mathcal{P}_k \mathcal{P}_\pi E_0 y = \beta^* \mathcal{P}_j E_0 y \,, \end{aligned}$$

so that (5) holds in this case by 7.1. The remaining case is similar.

It would obviously be desirable to have an analog of 7.5 for BP. In this case the Quillen idempotent ε factors into a projection and an inclusion

$$MU_{(p)} \xrightarrow{r} BP \xrightarrow{s} MU_{(p)}$$

which are both ring maps. We could therefore attempt to factor the operations $\mathcal{P}_j$ either through the inclusion (as in the proof of 7.5) or through the projection (or both). The proof of 7.5 shows that the $\mathcal{P}_j$ factor through s_* if and only if the following equation holds for all finite complexes X and all $x \in MU^{2i}X$.

(6) $$\varepsilon P_\pi \varepsilon x = P_\pi \varepsilon x.$$

Similarly, the $\mathcal{P}_j$ factor through r_* if and only if the equation

(7) $$\varepsilon P_\pi \varepsilon x = \varepsilon P_\pi x$$

holds. In either case the resulting structural maps on BP would be the composites

$$\xi_j^! : D_j BP \xrightarrow{D_j s} D_j MU \xrightarrow{\xi_j} MU \xrightarrow{r} BP.$$

The point is that, while these maps ξ_j' clearly satisfy the first and third diagrams of Definition I.4.3, the diagram involving β is much harder to verify and equations (6) and (7) give two sufficient conditions for it to commute. We conclude this section by giving some weaker sufficient conditions.

Lemma 7.6. Equation (6) or (7) holds in general if it does when x is the Euler class $v \in MU^2CP^\infty$ of the Hopf bundle over CP^∞.

Proof. Suppose $\varepsilon P_\pi \varepsilon v = \varepsilon P_\pi v$. Since ε is a ring map we have $\varepsilon \mathcal{P}_\pi \varepsilon v = \varepsilon \mathcal{P}_\pi v$ by 7.3 (with $X = CP^\infty$). Now $\varepsilon \mathcal{P}_\pi \varepsilon$ and $\varepsilon \mathcal{P}_\pi$ both satisfy tom Dieck's axioms P1, P2, and P3, so Theorem 11.2 of [31] implies that they are equal, hence $\varepsilon P_\pi \varepsilon = \varepsilon P_\pi$ for all spaces as required. The other case is similar.

Next we need some notation. Let $f(x) = \frac{[p](x)}{x} \in MU^*[[x]]$ where $[p](x)$ is the power series defined at the end of Section 5. Let $[p]'(x) \in BP^*[[x]]$ be $r_*[p](x)$ and let $f'(x) = r_*f(x)$. Let $u' \in BP^*B\pi$ be r_*u, so that u' is the BP-Euler class of a nontrivial complex irreducible representation of π. Landweber's Künneth theorem for $MU^*(X \times B\pi)$ given in Section 5 implies

$$BP^*(X \times B\pi) \cong (BP^*X)[[u']]/[p'](u')$$

Lemma 7.7. Equation (7) holds for all X if and only if equation

(8) $$r_*P_\pi \varepsilon [CP^n] = r_*P_\pi [CP^n] \mod f'(u')$$

holds in $BP^*B\pi$ for all $n \geq 0$.

Proof. Assume that (8) holds. We shall show that $r_*P_\pi \varepsilon v = r_*P_\pi v$, where v is as in 7.6. Let M^*X denote the even-dimensional part of $MU^*_{(p)}X$ and let P be the composite

$$M^*X \xrightarrow{P_\pi} M^*B\pi \cong (M^*X)[[u]]/[p](u) \longrightarrow (M^*X)[[u]]/f(u).$$

If M^*X has no p-torsion then, since $f(x)$ has constant term p, u is not a zero-divisor in $M^*(X)[[u]]/f(u)$. The element χ of Corollary 5.6 has leading term $(p-1)!u^{p-1}$, hence χ is also not a zero divisor. Thus 5.6 implies that P is additive for such X. It is also multiplicative by 1.4(iii). In particular we have a ring homomorphism

$$P: M^*(pt) \to M^*(pt))[[u]]/f(u).$$

Since the elements $[CP^n]$ generate $M^*(pt) \otimes Q$ as a ring and since $MU^*(B\pi)$ is torsion free, equation (8) implies

(9) $$r_*P_\pi\varepsilon x = r_*P_\pi x \mod f'(u')$$

for all $x \in MU^*(pt)$.

Now let $\varepsilon v = \sum_{i=1}^{\infty} b_i v^i$. Since ε is an idempotent we have $b_1 = 1$ and $\varepsilon b_i = 0$ for $i \geq 2$. Hence (9) gives

$$r_*P_\pi b_i = 0 \mod f'(u')$$

for all $i \geq 2$. Now the ring homomorphism

$$P:M^*(CP^\infty) \to M^*(CP^\infty \times B\pi) \cong M^*[[v,u]]/f(u)$$

is continuous with respect to the usual filtrations by [31, Theorem 5.1] and hence we have

$$r_*P_\pi\varepsilon v \equiv r_*P_\pi \sum_{i=1}^{\infty} b_i v \equiv \sum_{i=1}^{\infty} (r_*P_\pi b_i)(r_*P_\pi v)^i \equiv r_*P_\pi v \mod f'(u').$$

Finally, we observe that the map

$$BP^*(CP^\infty \times B\pi) \cong BP^*[[v',u']]/[p]'(u') \to BP^*[[v',u']]/u' \oplus BP^*[[v',u']]/f'(u')$$

is monic since u' and $f'(u')$ are relatively prime. We have shown that $r_*(P_\pi\varepsilon v - P_\pi v)$ goes to zero in the second summand, so we need only show that it goes to zero in the first. But the map

$$BP^*(CP^\infty \times B\pi) \to BP^*[[v',u']]/u' \cong BP^*[[v']]$$

can be identified with the restriction

$$(1 \times \iota)^*:BP^*(CP^\infty \times B\pi) \to BP^*CP^\infty$$

and the result follows since

$$(1 \times \iota)^*r_*(P_\pi\varepsilon v - P_\pi v) = r_*((\varepsilon v)^p - v^p) = (r_*v)^p - (r_*v)^p = 0.$$

We can now use Quillen's formula 5.5 to give a very explicit equation which is equivalent to (7).

Corollary 7.8. Equation (7) holds for all X if and only if the element

$$\sum_{|\alpha| \leq n} (c_\alpha, b^{-n-1}) r_*[CP^{n-|\alpha|}] r_*(a(u)^\alpha)(r_*\chi)^{n-|\alpha|}$$

of $BP^*B\pi$ is zero for each n not of the form p^k-1. Here the (c_α, b^{-n-1}) are certain numerical coefficients defined in [6, Theorem 4.1 of part I].

Proof. This is immediate from 5.5, 7.7, and [6, Theorems I.4.1 and II.15.2].

There is no obvious reason for the elements specified in 7.8 to be zero. If they were zero, it would be evidence of a rather deep connection between P_π and ε. The author's opinion is that there is no such deep connection and that neither equation (7) nor equation (6) holds in general.

CHAPTER IX

THE MOD p K-THEORY OF QX

by J. E. McClure

In this chapter we use the theory of H_∞ ring spectra to construct and analyze Dyer-Lashof operations in the complex K-theory of infinite loop spaces analogous to the usual Dyer-Lashof operations in ordinary homology. As an application we compute $K_*(QX;Z_p)$ in terms of the K-theory Bockstein spectral sequence of X.

Dyer-Lashof operations in K-theory were first considered by Hodgkin, whose calculation of $K_*(QS^0;Z_p)$ [41] led him to conjecture the existence of a single operation analogous to the sequence of operations in ordinary homology. He constructed such an operation, denoted by Q, for odd primes [42]; a similar construction for p = 2 was given independently by Snaith, who later refined Hodgkin's construction for odd primes and analyzed the properties of Q. The construction of Hodgkin and Snaith was based on the E^∞ term of a certain spectral sequence (namely the spectral sequence of I.2.4) and therefore had indeterminacy, and Hodgkin showed that in fact any useful operation in the mod p K-homology of infinite loop spaces must have indeterminacy. He also observed that the Dyer-Lashof method for calculating $H_*(QX;Z_p)$ by use of the Serre spectral sequence completely failed to generalize to K-theory. The indeterminacy was a considerable inconvenience, but the operation was still found to have applications, notably in the calculation of $K_*(QRP^n;Z_2)$ given by Miller and Snaith [84]. This result, which was proved by using the Eilenberg-Moore spectral sequence starting from Hodgkin's calculation of $K_*(QS^0;Z_p)$, was the first indication that $K_*(QX;Z_p)$ might be tractable in the presence of torsion in X. The main technical difficulty in the proof was in determining exactly how many times Q could be iterated on a given element, since Q could be defined only on the kernel of the Bockstein β. (Incidentally, a joint paper of Snaith and the present author showed that the odd-primary construction of Q contained an error and that in this case as well Q could only be defined on the kernel of β.) The answer for RP^n was that Q could be iterated on an element exactly as many times as the element survived in the Bockstein spectral sequence. Unfortunately, the methods used in this case did not extend to spaces more complicated than RP^n.

In view of these facts, it is rather surprising that there is in fact a theory of primary Dyer-Lashof operations in K-theory for which practically every statement about ordinary Dyer-Lashof operations, including the calculation of $H_*(QX;Z_p)$, has a precise analog. We shall remove the indeterminacy of Q by constructing it as an operation from mod p^2 to mod p K-theory, and more generally from mod p^{r+1} to mod p^r K-theory. It follows that Q can be iterated on any element precisely as often as the element survives in the Bockstein spectral sequence. There are also operations $\mathcal{Q}$ and R taking mod p^r to mod p^{r+1} K-theory in even and odd dimensions respectively ($\mathcal{Q}$ is the K-theory analog of the Pontrjagin p-th power [57, 28], while R has no analog in ordinary homology). These will play a key role in determining the properties of the Q-operation and in our calculation of $K_*(QX;Z_p)$. They also give indecomposable generators in the K-theory Bockstein spectral sequence for QX.[1] The operations Q, $\mathcal{Q}$ and R form a complete set of Dyer-Lashof operations in the sense that they exhaust the possibilities in a certain universal case; see Section 8. The key to defining primary operations in higher torsion is the machinery of stable extended powers, which gives a very satisfactory replacement for the chain-level machinery in ordinary homology; more precisely, it allows questions about the operations to be reduced to a universal case in the same way that chain-level arguments allow reduction to $B\Sigma_p$. In applying this machinery to K-theory we make essential use of the fact that periodic K-theory is an H_∞ ring spectrum, as shown in VII §7 and VIII §4, and the fact that the Adams operations are p-local H_∞ maps as shown in VIII §7.

This chapter is largely self-contained, and in particular it does not depend logically on the earlier work of Hodgkin, Snaith, Miller and the author. The organization is as follows. In section 1 we give a very general definition of Dyer-Lashof operations in E-homology for an H_∞ ring spectrum E. When E is HZ_p we recover the ordinary Dyer-Lashof operations. In section 2 we use some of the properties developed in section 1 to give a new way of computing $H_*(QX;Z_p)$ for connected X without use of the Serre spectral sequence, the Kudo transgression theorem, or even the equivalence $\Omega Q\Sigma X \simeq QX$; instead the basic ingredients are the approximation theorem and the transfer. In section 3 we give the properties of Q, $\mathcal{Q}$ and R and the statement of our calculation of $K_*(QX;Z_p)$; up to isomorphism the result depends only on the K-theory Bockstein spectral sequence of X, but for functoriality we need a more precise description. Section 4 contains the calculation of $K_*(QX;Z_p)$, which is modeled on that in section 2. Sections 5 through 8 give the construction and properties of Q, $\mathcal{Q}$, and R. In section 5 we lay the groundwork by giving very precise descriptions of the groups $K_*(D_pS^n;Z_{p^r})$. Section 6 gives enough information about Q to calculate $K_*(D_pX;Z_p)$, a result needed in section 4. The argument differs from that in [77] in three ways: it is shorter (but less elementary), it gives a more precise result, and it applies to the case p = 2. Sections 7 and 8 complete

---

*It was asserted in the original version of this work ([76, Theorem 5]) that certain composites of Q and R gave indecomposable generators in $K_*(QX;Z_p)$. Doug Ravenel has since pointed out to the author that this is incorrect: his argument is given in Remark (ii) following Theorem 3.6 below. The corrected versions of [76, Theorems 5 and 6] are also given in Section 3. (The mistake in the original version was in the proof of Lemma 4.7 for $M = \Sigma M_r$, where it was asserted that the $r > 1$ and $r = 1$ cases are similar. They are not.)

the construction of Q, $\mathcal{Q}$, and R. In section 9 we prove a purely algebraic fact needed in section 4; this fact is considerably more difficult than its analog in homology because of the nonadditivity of the operations.

I would like to thank Vic Snaith for introducing me to this subject and for the many insights I have gotten from his book and his papers with Haynes Miller. I would also like to thank Doug Ravenel for pointing out the mistake mentioned above. I owe Gaunce Lewis many commutative diagrams, as well as the first version of Definition 1.7. Finally, I would like to thank Peter May for encouragement and for his careful reading of the manuscript.

1. Generalized Homology Operations

Let E be a fixed H_∞ ring spectrum. In this section we shall construct generalized Dyer-Lashof operations in the E-homology of H_∞ ring spectra X. When E is HZ_p these are (up to reindexing) the ordinary Dyer-Lashof operations defined by Steinberger in chapter III, and for E = S they are Bruner's homotopy operations. When E is the spectrum K representing integral K-theory we obtain the operations referred to in the introduction which will be studied in detail in sections 3-9.

For simplicity, we shall begin by defining operations in E_*X, although ultimately (for the application to K-theory) we must introduce torsion coefficients. Fix a prime p. For each $n \in Z$ the operations defined on E_nX will be indexed by $E_*(D_pS^n)$, i.e., for each $e \in E_m(D_pS^n)$ we shall define a natural operation

$$Q_e : E_nX \to E_mX$$

in the E-homology of H_∞ ring spectra called the <u>internal Dyer-Lashof operation determined by</u> e. As usual, Q_e will be the composite of the structural map

$$(\xi_p)_* : E_mD_pX \to E_m X$$

with an <u>external</u> operation

$$Q_e : E_nX \to E_mD_pX$$

which is defined for arbitrary spectra X and is natural for arbitrary maps $X \to Y$. Throughout this chapter we shall use the same symbol for corresponding internal and external Dyer-Lashof operations, with the context indicating which is intended. In this section we shall be concerned only with the external operations, and thus X and Y will always denote arbitrary spectra.

In order to motivate the definition of the external operation Q_e we give it in stages. Fix $m,n \in Z$ and $e \in E_mD_pS^n$. Let $u \in E_0S$ denote the unit element. We define Q_e first on the element $\Sigma^nu \in E_nS^n$ by $Q_e(\Sigma^nu) = e$. If $x \in E_nX$ happens to be spherical, then there is a map $g:S^n \to X$ with $g_*(\Sigma^n u) = x$, and naturality requires us to define $Q_e x = (D_p g)_* e$. Now any element $x \in E_n X$ is represented by a map $f:S^n \to E \wedge X$, and to complete the definition of Q_e it suffices to give an analog for general x of the homomorphism $(D_p g)_*$ which exists when x is spherical. It is useful to do this in a somewhat more general context, so let Y be any spectrum and let $f:Y \to E \wedge X$ be any map. First we define f_{**} to be the composite

$$E_*Y = \pi_*(E \wedge Y) \xrightarrow{(1 \wedge f)_*} \pi_*(E \wedge E \wedge X) \xrightarrow{(\phi \wedge 1)_*} \pi_*(E \wedge X) = E_*X,$$

where ϕ is the product on E. Note that $f_{**}\Sigma^n u = x$ if $f:S^n \to E \wedge X$ represents x. Next define $\overline{D}_\pi f$ for any $\pi \subset \Sigma_j$ to be the composite

$$D_\pi Y \xrightarrow{D_\pi f} D_\pi(E \wedge X) \xrightarrow{\delta} D_\pi E \wedge D_\pi X \xrightarrow{\xi \wedge 1} E \wedge D_\pi X,$$

where ξ comes from the H_∞ structure of E. Combining these definitions we obtain a map

$$(\overline{D}_\pi f)_{**}:E_*D_\pi Y \longrightarrow E_*D_\pi X .$$

Definition 1.1. If x $E_n X$ is represented by $f:S^n \to E \wedge X$ and e is an element of $E_m D_p S^n$ then

$$Q_e x = (\overline{D}_p f)_{**}(e) \in E_m D_p X.$$

Of course, this agrees with the definition given earlier when x is spherical, and in particular when E = S we recover the external version of Bruner's operation. Next let $E = HZ_p$. The standard external operation (as defined by Steinberger) is denoted $e_i \otimes x^p$, where e_i is the generator of $H_i(\Sigma_p;Z_p(n))$ defined in [68,section 1] (recall that $Z_p(n)$ is Z_p with Σ_p acting trivially if n is even and via the sign representation if n is odd). Now it is easy to see that the map

$$\Phi:H_i(\Sigma_p;Z_p(n)) \longrightarrow H_{i+2pn}(D_p S^n;Z_p)$$

given by $e_i \mapsto e_i \otimes (\Sigma^n u)^p$ is an isomorphism, and we have

Proposition 1.2. If $e = \Phi(e_i)$ then $Q_e x = e_i \otimes x^p$ for all x.

The proof of 1.2 will be given later in this section.

It is possible to put Definition 1.1 in a more categorical context. Let $\mathcal{C}_E$ be the category in which objects are spectra and the morphisms from X to Y are the stable maps from X to $E \wedge Y$. The composite in $\mathcal{C}_E$ of $f:X \to E \wedge Y$ and $g:Y \to E \wedge Z$ is the following composite of stable maps

$$X \xrightarrow{f} E\wedge Y \xrightarrow{1\wedge g} E\wedge E\wedge Z \xrightarrow{\phi\wedge 1} E\wedge Z .$$

The construction $\overline{D}_\pi$ on morphisms, combined with D_π on objects, gives a functor $\overline{D}_\pi : \mathcal{C}_E \to \mathcal{C}_E$, and we can also define a smash product $\overline{\wedge}$ on E by letting $f_1 \overline{\wedge} f_2$ be the composite

$$X_1 \wedge X_2 \xrightarrow{f_1\wedge f_2} E\wedge X_1 \wedge E\wedge X_2 \simeq E\wedge E\wedge X_1 \wedge X_2 \longrightarrow E\wedge X_1 \wedge X_2 .$$

Finally, E homology is a functor on $\mathcal{C}_E$ which takes f to f_{**}, and the following lemma shows that both Q_e and the external product in E-homology are natural transformations.

Lemma 1.3. (i) $(\overline{D}_p f)_{**} Q_e y = Q_e f_{**} y$ for any $y \in E_* Y$ and any $f: Y \to E\wedge X$.

(ii) $(f_{1**} y_1) \otimes (f_{2**} y_2) = (f_1 \overline{\wedge} f_1)_{**} (y_1 \otimes y_2)$.

As one would expect, the maps ι, α, β and δ of I§1 also give natural transformations.

Lemma 1.4. (i) $\iota_* (\overline{D}_\pi f)_{**} = (\overline{D}_\rho f)_{**} \iota_*$ if $\pi \subset \rho$.

(ii) $\alpha_* (\overline{D}_\pi f \ \overline{\wedge}\, \overline{D}_\rho f)_{**} = (\overline{D}_{\pi\times\rho} f)_{**} \alpha_*$.

(iii) $\beta_* (\overline{D}_\pi \overline{D}_\rho f)_{**} = (\overline{D}_{\pi \int \rho} f)_{**} \beta_*$.

(iv) $\delta_* (\overline{D}_\pi (f_1 \overline{\wedge} f_2))_{**} = (\overline{D}_\pi f_1 \overline{\wedge}\, \overline{D}_\pi f_2)_{**} \delta_*$.

We shall need two further transformations, namely the "diagonal" $\Delta : \Sigma D_\pi X \to D_\pi \Sigma X$ and the transfer $\tau : D_\rho X \to D_\pi X$. The first of these was constructed in II§3. The transfer was defined in II§1 for certain special cases, and will be defined in IV§3 of the sequel whenever $\pi \subset \rho$.

Lemma 1.5. (i) $(\overline{D}_\pi \Sigma f)_{**} \Delta_* = \Delta_* (\Sigma \overline{D}_\pi f)_{**}$.

(ii) $\tau_* (\overline{D}_\rho f)_{**} = (\overline{D}_\pi f)_{**} \tau_*$.

The proofs of 1.3, 1.4 and 1.5 are routine diagram chases (using [Equi.,VI.3.9] for 1.4(ii) and (iii) and [Equi.,IV.§3] for 1.5(ii)).

Next we would like to define Dyer-Lashof operations in E-homology with torsion coefficients. We shall always abbreviate $E_*(X; Z_{p^r})$ by $E_*(X;r)$. If M_r denotes the Moore spectrum $S^{-1} \cup_{p^r} S^0$ and E_r denotes $E\wedge \Sigma M_r$ then by definition we have $E_n(X;r) = \pi_n(E_r \wedge X)$. Thus if E_r is an H_∞ ring spectrum (for example, if E is ordinary integral homology) we can apply Definition 1.1 directly to E_r. However, it is a melancholy fact that in general E_r is not an H_∞ ring spectrum, as shown by the following, which will be proved at the end of section 7.

Proposition 1.6. K_r is not an H_∞ ring spectrum for any r.

Thus we must generalize 1.1. First of all, if $f:Y \to E \wedge X$ is any map we define f_{**} to be the composite

$$E_*(Y;r) = \pi_*(E_r \wedge Y) \xrightarrow{(1\wedge f)_*} \pi_*(E_r \wedge E \wedge X) \longrightarrow \pi_*(E_r \wedge X) = E_*(X;r).$$

Next observe that the Spanier-Whitehead dual of ΣM_r is M_r, so that there is a natural isomorphism

$$E_n(X;r) \cong [\Sigma^n M_r, E \wedge X].$$

In particular, any $x \in E_n(X;r)$ is represented by a map $f:\Sigma^n M_r \to E \wedge X$ and there results a homomorphism

$$(\overline{D}_p f)_{**}: E_*(D_p \Sigma^n M_r; s) \to E_*(D_p X; s)$$

for any $s \geq 1$. Note that $f_{**}\Sigma^n u_r = x$, where u_r is the composite $M_r = S \wedge M_r \xrightarrow{u\wedge 1} E \wedge M_r$. We shall call u_r the fundamental class of M_r.

Definition 1.7. Let $e \in E_m(D_p \Sigma^n M_r; s)$. Then

$$Q_e: E_n(X;r) \to E_m(D_p X; s)$$

is defined by $Q_e x = (\overline{D}_p f)_{**}(e)$, where $f:\Sigma^n M_r \to E \wedge X$ is a map representing x.

Lemmas 1.3, 1.4, and 1.5 remain valid in this generality.

When E is integral homology and r = s = 1 Definition 1.7 provides another way of constructing ordinary Dyer-Lashof operations, which are of course the same as those given by Definition 1.1. However, even in this case 1.7 has certain technical advantages; for example, it gives the relation between the Bockstein and the Dyer-Lashof operations, and by allowing r and s to be greater than 1 one obtains the Pontryagin p-th powers.

We conclude with the proof of 1.2. We write E for HZ_p. The result holds by definition when $x = \Sigma^n u \in E_n S^n$, so it suffices to show that

$$(\overline{D}_p f)_{**}(e_i \otimes y^p) = e_i \otimes (f_{**}y)^p$$

for all $f:Y \to E \wedge X$. We shall do this by a direct comparison with the mod p chain level. If A_* is any chain complex over Z_p we write $D_p A_*$ for $W \otimes_{\Sigma_p} (A_*)^{\otimes p}$, where W is a fixed resolution of Z_p by free $Z_p[\Sigma_p]$-modules. We let C_* denote the mod p cellular chains functor on CW-spectra, and we have a natural equivalence $D_pC_* \simeq C_*D_p$ by I.2.1. If Γ_* denotes the trivial chain complex with Z_p in dimension zero then there is a natural equivalence between E^0X and the chain-homotopy classes of degree zero maps from C_*X to Γ_*. In particular, we obtain chain maps $\theta: C_*E \to \Gamma_*$ and $\theta': D_pC_*E \to \Gamma_*$ representing the identity $E \to E$ and the structural map $D_pE \to E$. If ε denotes the composite $D_p\Gamma_* = W/\Sigma_p \to \Gamma_*$ (in which the second map is the augmentation) then $\varepsilon \circ D_p\theta$ is a chain map which, like θ', represents an element of $E^0(D_pE)$ extending the product map $E^{(p)} \to E$. But the proof of I.3.6 shows that there is only one such element, hence we have we have $\varepsilon \circ D_p\theta \simeq \theta'$. Next, observe that f_{**} is equal to the composite

$$E_*Y \longrightarrow E_*(E \wedge X) \longrightarrow E_*X,$$

where the second map is the slant product with the identity class in E^0E. Hence f_{**} is represented on the chain level by the composite

$$h: C_*Y \longrightarrow C_*(E \wedge X) \simeq C_*E \otimes C_*X \xrightarrow{\theta \otimes 1} \Gamma_* \otimes C_*X \simeq C_*X.$$

Since h is a chain map we have

$$(D_ph)_*(e_i \otimes y^p) = e_i \otimes (h_*y)^p = e_i \otimes (f_{**}\, y)^p,$$

so it suffices to show $(\overline{D}_pf)_{**} = (D_ph)_*$. Now $(\overline{D}_pf)_{**}$ is equal to the composite

$$E_*D_pY \longrightarrow E_*(D_p(E \wedge X)) \xrightarrow{\delta_*} E_*(D_pE \wedge D_pX) \longrightarrow E_*D_pX\ ,$$

where the last map is the slant product with the structural map in E^0D_pE. Hence $(\overline{D}_pf)_{**}$ is represented on the chain level by the composite H around the outside of the following diagram

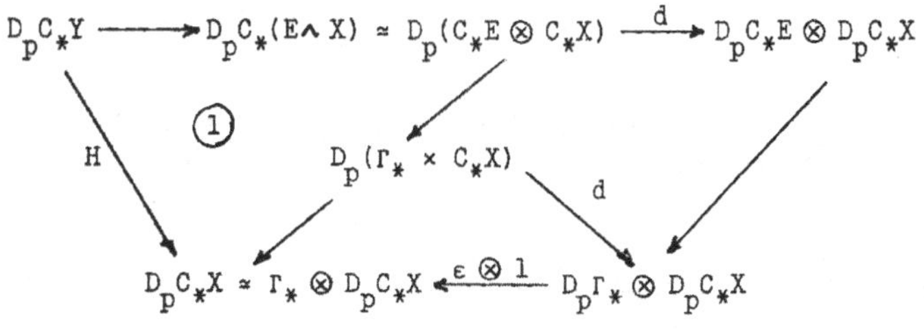

Here d is the evident diagonal transformation and the diagram clearly commutes. Inspection of the piece marked (1) shows that $H \simeq D_ph$ as required.

2. The Homology of CX

Our main aim in this chapter is the computation of $K_*(CX;1)$. In this section we illustrate the basic method in a simpler and more familiar situation, namely the computation of the ordinary mod p homology of CX. (All homology in this section is to be taken with mod p coefficients for an odd prime p; the p = 2 case is similar.) This result is of course well-known, but in fact our method gives some additional generality, since both the construction CX and our computation of H_*CX generalize to the situation where X is a (unital) spectrum, while the usual method of computation does not.

We begin by listing the relevant properties of this spectrum-level construction (which is due to Steinberger); a complete treatment will be given in [Equi., chapter VII]. By a unital spectrum we simply mean a spectrum X with an assigned map $S \to X$ called the unit. For any unital spectrum X one can construct an E_∞ ring spectrum CX, and this construction is functorial for unit-preserving maps. In particular, X might be $\Sigma^\infty Y^+$ for some based space Y, and there is then an equivalence $CX \simeq \Sigma^\infty(CY)^+$ relating the space-level and spectrum-level constructions. There is a natural filtration F_kCX of CX and natural equivalences $F_1CX \simeq X$ and

$$F_kCX/F_{k-1}CX \simeq D_k(X/S).$$

Finally, there are natural maps $F_jCX \wedge F_kCX \to F_{j+k}CX$ and $D_jF_kCX \to F_{jk}CX$ for which the following diagrams commute.

$$\begin{array}{ccc} CX \wedge CX & \longrightarrow & CX \\ \uparrow & & \uparrow \\ F_jCX \wedge F_kCX & \longrightarrow & F_{j+k}CX \\ \downarrow & & \downarrow \\ D_j(X/S) \wedge D_k(X/S) & \xrightarrow{\alpha} & D_{j+k}(X/S) \end{array} \qquad \begin{array}{ccc} D_jCX & \longrightarrow & CX \\ \uparrow & & \uparrow \\ D_jF_kCX & \longrightarrow & F_{jk}CX \\ \downarrow & & \downarrow \\ D_jD_k(X/S) & \xrightarrow{\beta} & D_{jk}(X/S) \end{array}$$

Now let X be a unital spectrum and assume the element $\eta \in H_0X$ induced by the unit map is nonzero. We can then choose a set $A \subset H_*X$ such that $A \cup \{\eta\}$ is a basis for H_*X. Let CA be the free commutative algebra generated by the set

$$\{Q^Ix \mid x \in A,\ I \text{ is admissible and } e(I) + b(I) > |x|\}$$

(here $|x|$ denotes the degree of x; see [28, I.2] for the definitions of admissibility, e(I) and b(I)). The elements of this set, which will be called the standard indecomposables for CA, are to be regarded simply as indeterminates since the Q^I do not act on H_*X. The basis for CA consisting of products of standard indecomposables will be called the standard basis for CA. Using the inclusion $X \to CX$ and the fact that CX is an E_∞ ring spectrum we obtain a ring map

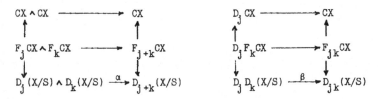

$$\lambda : CA \to H_*CX$$

and we shall show

Theorem 2.1. λ is an isomorphism.

We shall derive this theorem from an analogous fact about extended powers. Let Y be any spectrum and let A be a basis for H_*Y. CA is defined as before, and we make it a filtered ring by giving $Q^I x$ filtration $p^{\ell(I)}$. Let $D_kA = F_kCA/F_{k-1}CA$ for $k \geq 1$; this has a standard basis consisting of the standard basis elements in $F_kCA - F_{k-1}CA$. There is an additive map

$$\lambda_k : D_kA \to H_*D_kY$$

defined as follows. If all Dyer-Lashof operations and products are interpreted externally then a standard basis element of D_kA represents an element of $H_*((D_p)^{j_1}Y \wedge \cdots \wedge (D_p)^{j_s}Y)$ with $p^{j_1} + \cdots + p^{j_s} = k$; here $(D_p)^j$ denotes the j-th iterate of D_p. Applying the natural maps α_* and β_* gives an element H_*D_kY which by definition is the value of λ_k for the original basis element. We then have

Theorem 2.2. λ_k is an isomorphism for all $k \geq 1$.

Assuming 2.2 for the moment, we give the proof of 2.1. Let X be a unital spectrum and let $A \cup \{\eta\}$ be a basis for H_*X. Let $Y = X/S$. Then A projects to a basis for H_*Y which we also denote by A. For each $k \geq 1$ the map $\lambda | F_kCA$ lifts to a map $\lambda^{(k)} : F_kCA \to H_*F_kCX$ and the following diagram commutes.

$$\begin{array}{ccccccccc}
0 & \longrightarrow & F_{k-1}CA & \longrightarrow & F_kCA & \longrightarrow & D_kA & \longrightarrow & 0 \\
 & & \downarrow{\scriptstyle \lambda^{(k-1)}} & & \downarrow{\scriptstyle \lambda^{(k)}} & & \downarrow{\scriptstyle \lambda_k} & & \\
 & & H_*F_{k-1}CX & \longrightarrow & H_*F_kCX & \xrightarrow{\gamma} & H_*D_kY & &
\end{array}$$

Since λ_k is an isomorphism, the map γ is onto and hence the bottom row is short exact. It now follows by induction and the five lemma that $\lambda^{(k)}$ is an isomorphism for all k, and 2.1 follows by passage to colimits.

We begin the proof of 2.2 with a special case

Lemma 2.3. λ_p is an isomorphism for all Y.

The proof of the lemma is a standard chain-level calculation which will not be given here (see [68, section 1]). It is interesting to note, however, that one can prove 2.3 without any reference to the chain-level using the methods of section 6 below.

Next we use the machinery of section 1 to reduce to the case where Y is a wedge of spheres. For each $x \in A$ choose a map $f_x : S^{|x|} \to H \wedge Y$ representing x. Let $Z = \bigvee S^{|x|}$ and let $f : Z \to H \wedge Y$ be the wedge of the f_x. Then $f_{**} : H_*Z \to H_*Y$ is an isomorphism. We claim that 2.2 will hold for Y if it holds for Z (where H_*Z is given the basis B consisting of the fundamental classes of the $S^{|x|}$). To see this, consider the following diagram

$$\begin{array}{ccc} D_k B & \xrightarrow{D_k(f_{**})} & D_k A \\ \downarrow{\scriptstyle \lambda_k} & & \downarrow{\scriptstyle \lambda_k} \\ H_* D_k Z & \xrightarrow{(\overline{D}_k f)_{**}} & H_* D_k Y \end{array}$$

The map $D_k(f_{**})$ is induced by f_{**}, which clearly takes B to A. Thus $D_k(f_{**})$ is an isomorphism. The diagram commutes by 1.3 and 1.4(ii) and (iii). The claim now follows from

<u>Lemma 2.4.</u> Let $h : W \to H \wedge X$ be any map. If h_{**} is an isomorphism, so is $(\overline{D}_k h)_{**}$ for all k.

<u>Proof</u>. The proof is by induction on k. First suppose that $k = jp$. Since the case $k = p$ of 2.4 follows from 2.3 we may assume $j > 1$. Let $\pi = \Sigma_j \int \Sigma_p$ and consider the following diagram

$$\begin{array}{ccccccc} H_*D_kW & \xrightarrow{\tau_*} & H_*D_\pi W & \xleftarrow{\beta_*} & H_*D_jD_pW & \xrightarrow{\beta_{jp*}} & H_*D_kW \\ \downarrow{\scriptstyle (\overline{D}_k h)_{**}} & & \downarrow{\scriptstyle (\overline{D}_\pi h)_{**}} & & \downarrow{\scriptstyle (\overline{D}_j\overline{D}_p h)_{**}} & & \downarrow{\scriptstyle (\overline{D}_k h)_{**}} \\ H_*D_kX & \xrightarrow{\tau_*} & H_*D_\pi X & \xleftarrow{\beta_*} & H_*D_jD_pX & \xrightarrow{\beta_{jp*}} & H_*D_kW \end{array}$$

The diagram commutes by 1.4(i) and (iii) and 1.5(ii). The map β_* is an isomorphism. The map $(\overline{D}_p h)_{**}$ is an isomorphism by the case $k = p$, hence so is $(\overline{D}_j\overline{D}_p h)_{**}$ by inductive hypothesis. Our assumption on k implies that τ_* is monic and β_{jp*} is onto, hence $(\overline{D}_k h)_{**}$ is monic by inspection of the first square and onto by inspection of the third. The proof is the same when k is prime to p, except that we let π be $\Sigma_{k-1} \times \Sigma_1$.

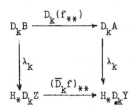

Next we reduce to the case of a single sphere. To simplify the notation we assume that Z is a wedge of two spheres $S^m \vee S^n$; the argument is the same in the general case. Let B_1 and B_2 be the bases for H_*S^m and H_*S^n consisting of the fundamental classes, so that $B = B_1 \cup B_2$. There is an evident map $CB_1 \otimes CB_2 \to CB$ and passing to the associated graded gives a map

$$\varphi\colon \sum_{i=0}^{k} (D_i B_1 \otimes D_{k-i} B_2) \longrightarrow D_k B.$$

Recall the equivalence

$$\bigvee_{i=0}^{k} (D_i S^m \wedge D_{k-1} S^n) \simeq D_k(S^m \vee S^n) = D_k Z$$

constructed in II§1.

<u>Lemma 2.5</u>. φ is an isomorphism, and the diagram

$$\begin{array}{ccc} \sum_{i=0}^{k} (D_i B_1 \otimes D_{k-i} B_2) & \xrightarrow{\ \varphi\ } & D_k B \\ \Big\downarrow \sum (\lambda_i \otimes \lambda_{k-i}) & & \Big\downarrow \lambda_k \\ \sum_{i=0}^{k} (H_* D_i S^m \otimes H_* D_{k-i} S^n) & \longrightarrow & H_* D_k Z \end{array}$$

commutes.

<u>Proof</u>. φ is an isomorphism since it takes the standard basis on the left to that on the right. The commutativity of the diagram is immediate from the definitions.

By Lemma 2.5 we see that 2.2 will hold for Z once we have shown the following. Let $x \in H_n S^n$ be the fundamental class.

<u>Lemma 2.6</u>. $\lambda_k\colon D_k\{x\} \to H_* D_k S^n$ is an isomorphism for all $k \geq 1$ and all integers n.

<u>Proof</u>. By induction on k. First assume that $k = jp$ for some $j > 1$. For the proof in this case we use the following diagram, which will be denoted by (*).

(*)

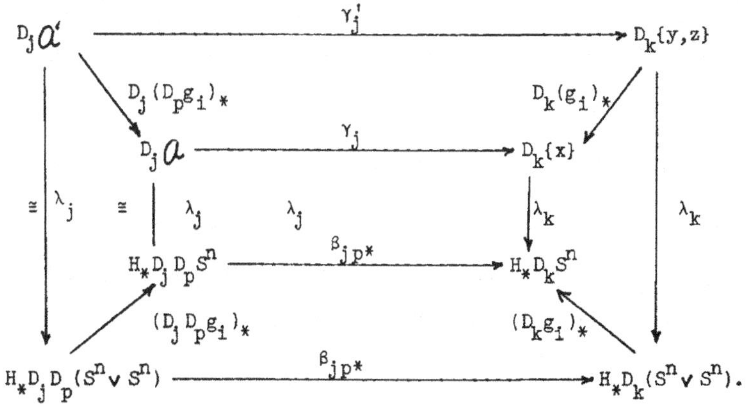

Here $y,z \in H_n(S^n \vee S^n)$ are the fundamental classes of the first and second summands. The set $\mathcal{A} \subset H_*D_pS^r$ is $\{\beta^\varepsilon Q^s x \mid 2s-\varepsilon \geq n\}$. (The reader is warned as this point to distinguish carefully between the Bockstein β and the natural map β of section I.1. This is made easier by the fact that we never use the latter map per se, only the homomorphism β_* induced by it.) The set $\mathcal{A}' \subset H_*D_p(S^n \vee S^n)$ is $\{\beta^\varepsilon Q^s y, \beta^\varepsilon Q^s z \mid 2s-\varepsilon \geq n\}$ if n is odd and is the union of this set with $\{y^i z^{p-i} \mid 1 \leq i \leq p-1\}$ when n is even. Lemma 2.3 implies that $\mathcal{A}$ and $\mathcal{A}'$ are bases, and hence the maps λ_j are isomorphisms by inductive hypothesis. The maps $g_i : S^n \vee S^n \to S^n$ are defined for i = 0,1 and 2 by $g_0 = 1 \vee 1$, $g_1 = 1 \vee *$ and $g_2 = * \vee 1$, where 1 and * denote the identity map and the trivial map of S^n. To complete the construction of the diagram we require

Lemma 2.7. There exist maps γ_j and γ_j', independent of i, such that diagram (*) commutes for i = 0,1 and 2.

The proof of 2.7 is given at the end of this section; all that is involved is to "simplify" expressions in $D_j\mathcal{A}'$ and $D_j\mathcal{A}$ using the Adem relations and the Cartan formula in a sufficiently systematic way.

Now consider the inner square of diagram (*). By assumption on k we see that $\beta_{jp*} \circ \tau_*$ is an isomorphism, hence λ_k is onto. Let $\theta : D_k\{x\} \to D_k\{x\}$ be the composite $\gamma_j \circ \lambda_j^{-1} \circ \tau_* \circ \lambda_k$. Clearly λ_k will be monic if θ is. In fact we shall show that θ is an isomorphism. We claim first of all that θ takes the subspace $\mathcal{D} \subset D_k\{x\}$ generated by the decomposable standard basis elements isomorphically into itself. To see this we use the outer square of diagram (*). Let $\theta' : D_k\{y,z\} \to D_k\{y,z\}$ be the composite $\gamma_j' \circ \lambda_j^{-1} \circ \tau_* \circ \lambda_k$. Let $\mathcal{D}' \subset D_k\{y,z\}$ be the image of $\sum_{i=1}^{k-1} (D_i\{y\} \otimes D_{k-i}\{z\})$ under the map φ of Lemma 2.5. Then $\mathcal{D}'$ is the kernel of the map

$$D_k(g_1)_* \oplus D_k(g_2)_* : D_k\{y,z\} \longrightarrow D_k\{x\} \oplus D_k\{x\}$$

and hence θ' takes $\mathcal{D}'$ into itself. But $D_k(g_0)_*(\mathcal{D}') = \mathcal{D}$ and $D_k(g_0)_* \circ \theta' = \theta \circ D_k(g_0)_*$, hence θ takes $\mathcal{D}$ into itself and we have the commutative diagram

$$\begin{array}{ccccc}
\mathcal{D}' & \xrightarrow{D_k(g_0)_*} & \mathcal{D} & \longrightarrow & 0 \\
\downarrow{\scriptstyle \theta'} & & \downarrow & & \\
\mathcal{D}' & \xrightarrow{D_k(g_0)_*} & \mathcal{D} & \longrightarrow & 0 .
\end{array}$$

Since both $\mathcal{D}$ and $\mathcal{D}'$ have finite type $\theta : \mathcal{D} \to \mathcal{D}$ will be an isomorphism if $\theta' : \mathcal{D}' \to \mathcal{D}'$ is monic. But λ_k is monic on $\mathcal{D}'$ by 2.5 and the inductive hypothesis, hence θ' is also monic on $\mathcal{D}'$ since $\lambda_k \circ \theta' = (\beta_{jp*} \circ \tau_*) \circ \lambda_k$.

Now let $\mathcal{J} = D_k\{x\}/\mathcal{B}$. This has the basis $\{Q^I x \mid I$ admissible, $p^{\ell(I)} = k$, $e(I) + b(I) > n\}$. We wish to show that the map $\overline{\theta}\colon \mathcal{J} \to \mathcal{J}$ induced by θ is an isomorphism. The basic idea is to use the homology suspension, or rather its external analog which is the map $\Delta_*\Sigma\colon H_i D_p S^n \to H_{i+1} D_p S^{n+1}$, to detect elements of $\mathcal{J}$. Let $\tilde{x} \in H_{n+1} S^{n+1}$ be the fundamental class. We define $\Gamma\colon \mathcal{J} \to D_k\{\tilde{x}\}$ by $\Gamma(Q^I x) = Q^I\tilde{x}$, where we interpret $Q^I\tilde{x}$ as zero if $e(I) < n+1$ and as a p-th power in the usual way if $e(I) = n+1$ and $b(I) = 0$. The key fact is the following, which will be proved at the end of this section.

<u>Lemma 2.8.</u> The diagram

$$\begin{array}{ccc} \mathcal{J} & \xrightarrow{\ \overline{\theta}\ } & \mathcal{J} \\ \Big\downarrow \Gamma & & \Big\downarrow \Gamma \\ D_k\{\tilde{x}\} & \xrightarrow{\ \theta\ } & D_k\{\tilde{x}\} \end{array}$$

commutes.

We also need the fact that the evident action of the Bockstein on $\mathcal{J}$ commutes with θ; this will be clear from the proof of 2.7.

Now let $\mathcal{J}_n$ be the subspace of $\mathcal{J}$ spanned by the set $\{Q^I x \mid I$ admissible, $p^{\ell(I)} = k$, $e(I) + b(I) \le n+m\}$. We shall show first that $\overline{\theta}$ is monic on $\mathcal{J}_1$. Let $\mathcal{J}_1'$ be the subspace of $\mathcal{J}_1$ spanned by the set $\{Q^I x \mid I$ admissible, $p^{\ell(I)} = k$, $e(I) = n+1$, $b(I) = 0\}$. Then $\mathcal{J}_1 = \mathcal{J}_1' \oplus \beta\mathcal{J}_1'$. From the definition of Γ we see that $\beta\mathcal{J}_1'$ is the kernel of Γ, that Γ is monic on $\mathcal{J}_1'$ and that $\Gamma(\mathcal{J}_1') = \Gamma(\mathcal{J}) \cap \mathcal{B}$. Let w be a nonzero element of $\mathcal{J}_1'$. We claim that $\overline{\theta}w$ lies in $\mathcal{J}_1$, so that it can be written uniquely in the form $w' + \beta w''$ with $w', w''\ \mathcal{J}_1'$, and furthermore we claim that $w' \neq 0$. To see this note that Γw is a nonzero decomposable, hence $\theta\Gamma w$ is also a nonzero decomposable, hence $\Gamma\overline{\theta}w = \theta\Gamma w$ is a nonzero element of $\Gamma(\mathcal{J}) \cap \mathcal{B} = \Gamma(\mathcal{J}_1')$. Thus there is a nonzero element w' of $\mathcal{J}_1'$ with $\Gamma w' = \Gamma\overline{\theta}w$, so that $\overline{\theta}w - w'$ is in $\ker\Gamma = \beta\mathcal{J}_1'$ as required. Now let w_1, w_2 be any elements of $\mathcal{J}_1'$ with $\overline{\theta}w_1 = w_1' + \beta w_1''$ and $\overline{\theta}w_2 = w_2' + \beta w_2''$. Suppose that $v = w_1 + \beta w_2$ is the kernel of $\overline{\theta}$. Then $0 = \overline{\theta}v = w_1' + \beta w_1'' + \beta w_2'$, hence $w_1' = 0$ and $w_1'' + w_2' = 0$. But $w_1' = 0$ implies $w_1 = 0$, hence $w_1'' = 0$. Thus $w_2' = 0$, whence $w_2 = 0$ and $v = 0$, showing that $\overline{\theta}$ is monic on $\mathcal{J}_1$.

Next we claim that $\overline{\theta}$ is monic on $\mathcal{J}_m$ for all $m \ge 1$. Let $w \in \mathcal{J}_m$ with $\overline{\theta}w = 0$. Let $\tilde{\mathcal{J}} = D_k\{\tilde{x}\}/\mathcal{B}$ and let $\overline{\Gamma}$ be the composite $\mathcal{J} \to D_k\{\tilde{x}\} \to \tilde{\mathcal{J}}$. Then $\overline{\Gamma}w$ is in the subspace $\tilde{\mathcal{J}}_{m-1}$ generated by $Q^I\tilde{x}$ with I admissible, $p^{\ell(I)} = k$ and $e(I) + b(I) - (n+1) \le m-1$. Since $\overline{\theta}\,\overline{\Gamma}w = \overline{\Gamma}\,\overline{\theta}w = 0$ and since (by induction on m) θ is monic on $\tilde{\mathcal{J}}_{m-1}$ we see that $\overline{\Gamma}w = 0$. Now the kernel of $\overline{\Gamma}$ is precisely $\mathcal{J}_1$, and we have shown already that $\overline{\theta}$ is monic on $\mathcal{J}_1$, hence $w = 0$ as required. Thus $\overline{\theta}\colon \mathcal{J} \to \mathcal{J}$

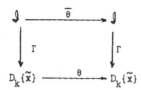

is monic, and since has finite type Ψ is an isomorphism. This completes the proof of 2.6 for the case $k = jp$.

Now suppose k is prime to p and consider the following diagram

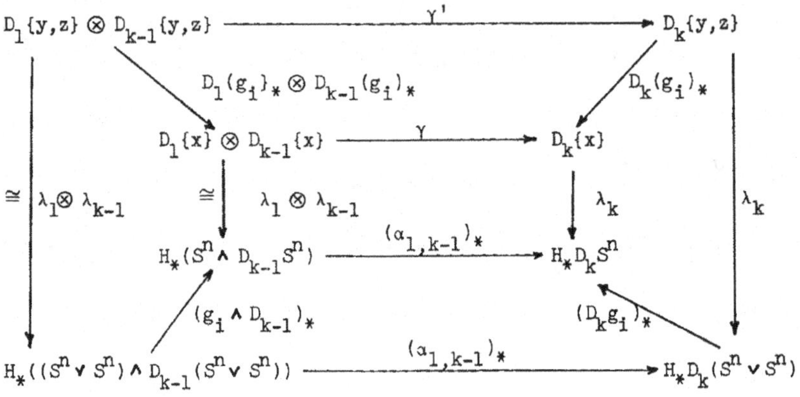

Here γ and γ' are obtained from the products in $C\{x\}$ and $C\{y,z\}$ by passage to the associated graded. The diagram clearly commutes. The analysis of this diagram proceeds as before, except that in this case the map $D_k(g_0)_*$ takes the kernel of $D_k(g_1)_* \oplus D_k(g_2)_*$ onto all of $D_k\{x\}$, so that we can conclude at once that λ_k is an isomorphism without having to consider indecomposables.

This completes the proof of 2.6, and thereby of 2.2, except that we must still verify 2.7 and 2.8. For these we need certain properties of the external Q^s. First of all these operations are additive, and $Q^s x = \iota_*(x^{(p)})$ if $2s = |x|$. The external Cartan formula is

$$\delta_* Q^s(x \otimes y) = \sum_{i=0}^{s} Q^i x \otimes Q^{s-i} y.$$

The external Adem relations are obtained by prefixing β_{pp*} to both sides of the standard Adem relations. All of these relations can be obtained directly from the definitions of section 1, without any use of internal operations (compare sections 7 and 8 below). They can also be derived from the corresponding properties for internal operations by means of the equivalence

$$C(X \vee S^0) \simeq \bigvee_{k>0} D_k X$$

proved in [Equi., VII§5].

<u>Proof of 2.7.</u> Every standard indecomposable in $C\mathcal{A}$ has the form $Q^I(\beta^\varepsilon Q^s x)$. We can formally simplify such an expression by means of the Adem relations into a sum of admissible sequences acting on x (for definiteness we assume that at each step the

Adem relations are applied at a position in the sequence as far to the right as possible). The result is an element of $C\{x\}$, where we agree to interpret all sequences with excess less than $|x|$ as zero, and we extend multiplcatively to get a map $F_j C\mathcal{Q} \to F_k C\{x\}$. The map γ_j is obtained by passage to quotients. The map γ_j' is obtained in the same way except that we use the Cartan formula to simplify expressions of the form $Q^I(y^i z^{p-i})$ with $0 < i < p$. The inner and outer squares of diagram (*) commute as a consequence of the external Cartan formula and Adem relations, and the upper trapezoid clearly commutes when i is 1 or 2. When i is zero the element $y^i z^{p-i}$ of $\mathcal{Q}'$ goes to $Q^{n/2}x$, and so it is necessary to check that the result of simplifying $Q^I Q^{n/2}x$ with the Adem relations is the same as using the Cartan formula on $Q^I x^p$; the result in each case is zero unless all entries of I are divisible by p, in which case it is $(Q^{I/p}x)^p$.

Finally, we give the proof of 2.8. We need two facts about $\Delta_*: H_*(\Sigma D_k X) \to H_*(D_k \Sigma X)$, namely that $\Delta_* \Sigma Q^s x = Q^s \Sigma x$ if $k = p$ and that $\Delta_* \Sigma(\alpha_{i,k-i})_*(x \otimes y)$ is zero for $0 < i < k$. The first of these, which is the external version of the stability of Q^s, was proved in II.5.6. For the second, which is the external analog of the fact that the homology suspension annihilates decomposables, we use the third diagram of II.3.1 with $X = S^1$, noting that the diagonal $\Delta: S^1 \to S^1 \wedge S^1$ is nullhomotopic. Now 2.8 is immediate from the commutativity of the following diagram.

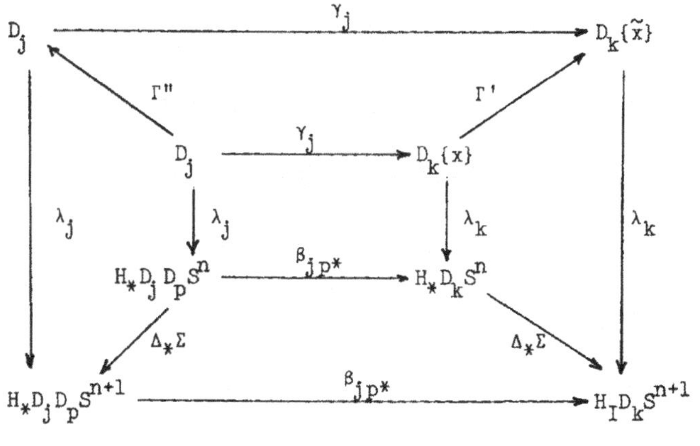

Here γ_j is the map constructed in the proof of 2.7 and Γ' is the composite $D_k\{x\} \longrightarrow \mathcal{J} \xrightarrow{\Gamma} D_k\{\tilde{x}\}$. We define Γ'' to take decomposables to zero and $Q^I(\beta^\varepsilon Q^s x)$ to $Q^I(\beta^\varepsilon Q^s \tilde{x})$. Commutativity of the left and right trapezoids follow from the two formulas given above. Commutativity of the upper trapezoid is obvious except on elements of the form $Q^I(\beta^\varepsilon Q^s x)$ with $e(I) = n+1 + 2s(p-1) - \varepsilon$ and $b(I) = 0$, and it follows in this case from a simple calculation.

3. Dyer-Lashof Operations in K-Theory

In this section we give our main results about K-theory Dyer-Lashof operations. We begin by fixing notations. We shall work in the stable category, so that X will always denote a spectrum. Homology operations are to be interpreted as internal rather than external. We use Z_2-graded K-theory, with $|x|$ denoting the mod 2 degree of x. There are evident natural maps

$$\pi : K_\alpha(X;r) \longrightarrow K_\alpha(X;r-1) \quad \text{if } r \geq 2$$

$$p_*^s : K_\alpha(X;r) \longrightarrow K_\alpha(X;r+s) \quad \text{if } s \geq 1$$

$$\beta_r : K_\alpha(X;r) \longrightarrow K_{\alpha+1}(X;r)$$

$$\Sigma : K_\alpha(X;r) \longrightarrow K_{\alpha+1}(\Sigma X;r) .$$

(Recall that ΣX means $S^1 \wedge X$ in this chapter, not $X \wedge S^1$ as in chapters I-VII.) β_1 will usually be written simply as β. We write π^s for the s-th iterate of π. It will often be convenient to denote the identity map either by π^0 or p_*^0. We write π^∞ for the reduction map $K_\alpha(X;Z) \to K_\alpha(X;r)$. Our first two results give some useful elementary facts about mod p^r K-theory; the proofs may be found in [13] (except for 3.2(iii), which is Lemma 6.4 of [63], and 3.2(iv), which will be proved in section 7).

Proposition 3.1. (i) $K_*(X;r)$ is a Z_{p^r}-module.

(ii) If $s \geq 1$ then $\pi^s \beta_{r+s} p_*^s = \beta_r$.

(iii) πp_* and $p_* \pi$ are multiplication by p.

(iv) $\beta_r \beta_r = 0$.

Proposition 3.2. For each $r \geq 1$ there is an external product

$$K_\alpha(X;r) \otimes K_{\alpha'}(Y;r) \to K_{\alpha+\alpha'}(X \wedge Y;r),$$

denoted by $x \otimes y$, which has the following properties.

(i) $\otimes$ is natural, bilinear and associative.

(ii) If u $K_0 S$ is the unit then $x \otimes \pi^\infty u = \pi^\infty u \otimes x = x$.

(iii) $\pi(x \otimes y) = \pi x \otimes \pi y$ and $\pi^\infty(x \otimes y) = \pi^\infty x \otimes \pi^\infty y$.

(iv) $p_*(x \otimes \pi y) = (p_* x) \otimes y$.

(v) $\beta_r(x \otimes y) = \beta_r x \otimes y + (-1)^{|x|} x \otimes \beta_r y$.

(vi) $\Sigma(x \otimes y) = \Sigma x \otimes y = (-1)^{|x|} x \otimes \Sigma y$.

If p is odd then the following also holds, where $T:X\wedge Y \to Y\wedge X$ switches the factors.

(vii) $T_*(x\otimes y) = (-1)^{|y||x|} y\otimes x$

If p = 2 there are two external products for each r satisfying (i), (ii), (v) and (vi). If these are denoted by $\otimes$ and $\otimes'$ the relation

(viii) $x\otimes y = x\otimes' y + 2^{r-1}\beta_r x\otimes \beta_r y$

holds. Relations (iii) and (iv) hold when either mod 2^r product is paired with either mod 2^{r-1} product. If $r\geq 2$ then (vii) holds for both $\otimes$ and $\otimes'$, while if r = 1 then the following holds.

(vii)' $T_*(x\otimes y) = y\otimes' x = y\otimes x + \beta y\otimes \beta x.$

We shall actually give a canonical choice of mod 2^r multiplications in Remark 3.4(iv) below. When X is a ring spectrum we obtain an internal product denoted xy. We write $\eta\in K_0(X;r)$ for the unit in this case, reserving the letter u for the unit of K_0S.

Our next result gives the properties of our first operation, which is denoted by Q. In order to relate Q to the K-homology suspension we must restrict to the space level, and we fix notations for dealing with this case. If Y is any space we write $K_*(Y;r)$ for $K_*(\Sigma^\infty Y^+;r)$ and, if Y is based, we write $\tilde{K}_r(Y;r)$ for $K_*(\Sigma^\infty Y;r)$. The homology suspension σ is the composite

$$\tilde{K}_\alpha(\Omega Y;r) \xrightarrow{\Sigma} \tilde{K}_{\alpha+1}(\Sigma\Omega Y;r) \longrightarrow \tilde{K}_{\alpha+1}(Y;r) \subset K_{\alpha+1}(Y;r).$$

If Y is an H_∞ space then ΩY is also an H_∞ space and $\Sigma^\infty Y^+$ is an H_∞ ring spectrum; see I.3.7 and I.3.8.

<u>Theorem 3.3</u>. Let X be an H_∞ ring spectrum. For each $r\geq 2$ and $\alpha\in Z_2$ there is an operation

$$Q:K_\alpha(X;r) \to K_\alpha(X;r-1)$$

with the following properties, where x,y $K_*(X;r)$.

(i) Q is natural for H_∞ maps of X.

(ii) $Q\eta = 0$.

(iii) $Q\pi x = \pi Qx$ if $r\geq 3$.

$$\text{(iv)}\quad Qp_*x = \begin{cases} x^p & \text{if } |x| = 0 \text{ and } r = 1 \\ p_*Qx - (p^{p-1} - 1)x^p & \text{if } |x| = 0 \text{ and } r \geq 2 \\ 0 & \text{if } |x| = 1 \text{ and } r = 1 \\ p_*Qx & \text{if } |x| = 1 \text{ and } r \geq 2 \end{cases}$$

$$\text{(v)}\quad \beta_{r-1}Qx = \begin{cases} Q\beta_r x - p\pi(x^{p-1}\beta_r x) & \text{if } |x| = 0 \\ (\pi\beta_r x)^p + pQ\beta_r x & \text{if } |x| = 1. \end{cases}$$

$$\text{(vi)}\quad Q(x+y) = \begin{cases} Qx + Qy - \pi[\sum_{i=1}^{p-1} \frac{1}{p} \binom{p}{i} x^i y^{p-i}] & \text{if } p \text{ is odd and } |x| = |y| = 0 \\ Qx + Qy - \pi(xy) + 2^{r-2}(\pi\beta_r x)(\pi\beta_r y) & \text{if } p = 2 \text{ and } |x| = |y| = 0 \\ Qx + Qy & \text{if } |x| = |y| = 1. \end{cases}$$

$$Q(kx) = kQx - \frac{1}{p}(k^p - k)(\pi x)^p \quad \text{if } k \quad Z, \ |x| = 0.$$

(vii) Let $|x| = |y| = 0$. Then

$$Q(xy) = \begin{cases} Qx\cdot\pi(y^p) + (x^p)\cdot Qy + p(Qx)(Qy) & \text{if } p \text{ is odd} \\ Qx\cdot\pi(y^2) + \pi(x^2)\cdot Qy + 2(Qx)(Qy) + 2^{r-2}\pi(x\beta_r x)\pi(y\beta_r y) & \\ \qquad + 2^{2r-4}(Q\beta_r x)(Q\beta_r y) & \text{if } p = 2. \end{cases}$$

Let $|x| = 1$, $|y| = 0$. Then

$$Q(xy) = \begin{cases} Qx\cdot\pi(y^p) + p(Qx)(Qy) & \text{if } p \text{ is odd} \\ Qx\cdot\pi(y^2) + 2(Qx)(Qy) + 2^{2r-4}(\pi\beta_r x)^2(Q\beta_r y) & \text{if } p = 2. \end{cases}$$

Let $|x| = |y| = 1$. Then

$$Q(xy) = \begin{cases} (Qx)(Qy) & \text{if } p \text{ is odd} \\ (Qx)(Qy) + 2^{r-2}\pi(x\beta_r x)\pi(y\beta_r y) + 2^{2r-4}(\pi\beta_r x)^2(Q\beta_r y) & \\ \qquad + 2^{2r-4}(Q\beta_r x)(\pi\beta_r y)^2 & \text{if } p = 2. \end{cases}$$

(viii) If Y is an H_∞ space and $x \in \tilde{K}_\alpha(\Omega Y;r)$ then $Qx \in \tilde{K}_\alpha(\Omega Y;r-1)$ and

$$\sigma Qx = \begin{cases} Q\sigma x & \text{if } |x| = 0 \\ (\pi\sigma x)^p + pQ\sigma x & \text{if } |x| = 1. \end{cases}$$

(ix) If k is prime to p then $\psi^k Qx = Q\psi^k x$, where ψ^k is the k-th Adams operation.

(x) If p = 2 and $|x| = 1$ then

$$x^2 = \begin{cases} Q\beta_2 2_* x & \text{if } r = 1 \\ 2^{r-2}\beta_r 2_* Qx & \text{if } r \geq 2. \end{cases}$$

In particular $(\pi^{r-1}x)^2 \in K_0(X;1)$ is zero if $r \geq 3$ and is equal to $(\pi\beta_2 x)^2$ if r = 2.

Remarks 3.4. (i) There are no analogs for the Adem relations.

(ii) We shall write $Q^s: K_\alpha(X;r) \to K_\alpha(X;r-s)$ for the s-th iterate of Q when r > s (and similarly for the operations R and $\mathcal{Q}$ to be introduced later).

(iii) If $x \in K_*(X;1)$ has $\beta x = 0$ then x lifts to $y \in K_*(X;2)$. Thus one can define a secondary operation $\overline{Q}$ on the kernel of β by $\overline{Q}x = Qy$. The element y is well-defined modulo the image of p_* and thus 3.3(iv) shows that $\overline{Q}x$ is well-defined modulo p-th powers if $|x| = 0$ and has no indeterminacy if $|x| = 1$. This is essentially the operation defined by Hodgkin and Snaith [42,99] (although their construction is incorrect when p is odd, as shown in [77]).

(iv) When p = 2, parts (vi) and (vii) are corrected versions of the corresponding formulas in [76]. Note that $2^{2r-4} = 0 \bmod 2^{r-1}$ unless r = 2. The formula for Q(xy) with $|x| = |y| = 1$ and p = 2 implicitly assumes that the mod 2^r multiplications for $r \geq 2$ have been suitably chosen, since the evaluation of $Q(xy + 2^{r-1}(\beta_r x)(\beta_r y))$ by means of 3.3(vi) and (vii) gives a different formula. Thus we may (inductively) fix a canonical choice of mod 2^r multiplications by choosing the mod 2 multiplication arbitrarily and requiring the formula to hold as stated for $r \geq 2$. From now on we shall always use this choice of multiplications.

Our next result shows that, in contrast to ordinary homology, $K_*(X;1)$ will in general have nilpotent elements.

Corollary 3.5. If X is an H_∞ ring spectrum and x $K_1(X;r)$ then $(\pi^{r-1}\beta_r x)^{p^r} = 0$ in $K_0(X;1)$.

Proof of 3.5. (By induction on r). If r = 1 then

$$(\beta x)^p = (\pi\beta_2 p_* x)^p = \beta Q p_* x = 0$$

by 3.1(ii), 3.3(v) and 3.3(iv). If $r \geq 2$ then

$$(\pi^{r-1}\beta_r x)^{p^r} = [(\pi^{r-1}\beta_r x)^p]^{p^{r-1}} = (\pi^{r-2}\beta_{r-1}Qx)^{p^{r-1}} = 0$$

by 3.3(v) and the inductive hypothesis.

It turns out that iterated Q-operations on r-th Bocksteins are also nilpotent. In order to see this we must make use of the operation R described in our next theorem.

Theorem 3.6. Let X be an H_∞ ring spectrum. For each $r \geq 1$ there is an operation

$$R:K_1(X;r) \to K_1(X;r+1)$$

with the following properties, where $x,y \in K_1(X;r)$.

(i) R is natural for H_∞ maps of X

(ii) $\pi Rx = Qp_*x - x(\beta_r x)^{p-1}$, and if $r \geq 2$ then $R\pi x = Qp_*x - p^{p-1}x(\beta_r x)^{p-1}$

(iii) $p_*Rx = Rp_*x$

(iv) $\beta_{r+1}Rx = Q\beta_{r+2}p_*^2 x$

(v) $R(x+y) = Rx + Ry - \sum_{i=1}^{p-1} [\frac{1}{p}\binom{p}{i}(p_*x)(\beta_{r+1}p_*x)^{i-1}(\beta_{r+1}p_*y)^{p-i}$

$$+ \binom{p-1}{i}\beta_{r+1}p_*(xy)(\beta_{r+1}p_*x)^{i-1}(\beta_{r+1}p_*y)^{p-i-1}]$$

(vi) If Y is an H_∞ space and $x \in \tilde{K}_1(Y;r)$ then

$$\sigma Rx = \begin{cases} p_*[(\sigma x)^p] & \text{if } r = 1 \\ p_*[(\sigma x)^p] + p_*^2 Q\sigma x & \text{if } r \geq 2. \end{cases}$$

(vii) If k is prime to p then $\psi^k Rx = R\psi^k x$.

(viii) If $r \geq 2$ then $QRx = RQx$. If $r = 1$ then $QRx = 0$.

Remarks (i) Let $x \in K_1(X;r)$ and let $s \geq 1$. By 3.3(v) we have $(\pi^{r+s-1}\beta_{r+s}R^s x)^{p^r} = \pi^{s-1}\beta_s Q^r R^s x$. But $Q^r R^s x = R^{s-1}QR(Q^{r-1}x) = 0$ by 3.6(viii). We therefore have the following nilpotency relation.

$$(\pi^{r+s-1}\beta_{r+s}R^s x)^{p^r} = 0.$$

Note that this is a smaller exponent than would be given by 3.5. In terms of the Q-operation this relation may be written $(\pi^{r-s-1}Q^s\beta_r x)^{p^r} = 0$ for $s < r$ and $(Q^s\beta_{s+1}p_*^{s-r+1}x)^{p^r} = 0$ for $s \geq r$.

(ii) The second statement of 3.6(viii) was not in the original version of this work (cf. [76, Theorem 3(iv)]). The decomposability of QRx when r = 1 (which actually implies its vanishing, as we shall see in Section 8) had been asserted by Snaith when p = 2 ([99, Proposition 5.2(ii)]), but was not included in [76] because the author erroneously thought he could prove QRx to be indecomposable in $K_1(QX;1)$ whenever $x \in K_1(X;1)$ had nonzero Bockstein (cf. [76, Theorem 4]). This point was recently settled by Doug Ravenel, who observed that if one starts with the description of $K_*(Q(S^1 \cup_p e^2);1)$ given in [76, Theorem 4] and applies the Rothenberg-Steenrod spectral sequence (which collapses) then one can see that the only indecomposable in $K_1(Q(S^2 \cup_p e^3);1)$ is the generator of $K_1(S^2 \cup_p e^3;1)$, and in particular QR of this generator is decomposable. This contradicts part of [76, Theorem 4] and a corrected version of that result will be given later in this section. We shall give a completely different argument in Section 8 to show that QRx is decomposable, and in fact vanishes, for all $x \in K_1(X;1)$.

We next introduce an operation $\mathcal{Q}$ which is the K-theoretic analog of the Pontrjagin p-th power [57, 28]. This operation is a necessary tool in our calculation of $K_*(QX;1)$ and will also be used to give generators for the higher terms of the Bockstein spectral sequence.

<u>Theorem 3.7.</u> Let X be an H_∞ ring spectrum. For each $r \geq 1$ there is an operation

$$\mathcal{Q}: K_0(X;r) \to K_0(X;r+1)$$

with the following properties, where $x,y \in K_*(X;r)$.

(i) $\mathcal{Q}$ is natural for H_∞ maps of X.

(ii) $\pi\mathcal{Q}x = x^p$, and if $r \geq 2$ then $\pi x = x^p$.

(iii) $\mathcal{Q}p_*x = p^{p-1}p_*\mathcal{Q}x$.

(iv) $\pi\beta_{r+1}\mathcal{Q}x = x^{p-1}\beta_r x$

(v) $$\mathcal{Q}(x+y) = \begin{cases} \mathcal{Q}x + \mathcal{Q}y + \sum_{i=1}^{p-1} \frac{1}{p}\binom{p}{i} p_*(x^i y^{p-i}) & \text{if } p \text{ is odd or } r \geq 2 \\ \mathcal{Q}x + \mathcal{Q}y + 2_*(xy) + (\beta_2 2_* x)(\beta_2 2_* y) & \text{if } p = 2 \text{ and } r = 1. \end{cases}$$

(vi) Let $|x| = |y| = 0$. Then $\mathcal{Q}(xy) = (\mathcal{Q}x)(\mathcal{Q}y)$ if p is odd, while if p = 2 there is a constant $\varepsilon_r \in Z_2$, independent of x and y, with

$$\mathcal{Q}(xy) = \begin{cases} (\mathcal{Q}x)(\mathcal{Q}y) + (1+2\varepsilon_1)(\beta_2\mathcal{Q}x)(\beta_2\mathcal{Q}y) & \text{if } r = 1 \\ (\mathcal{Q}x)(\mathcal{Q}y) + 2^r\varepsilon_r(\beta_{r+1}\mathcal{Q}x)(\beta_{r+1}\mathcal{Q}y) & \text{if } r \geq 2. \end{cases}$$

Let $|x| = 1$, $|y| = 0$. Then

$$R(xy) = \begin{cases} (Rx)(\mathcal{Q}y) & \text{if } p \text{ is odd and } r = 1 \\ (Rx)(\mathcal{Q}y) + p_*^2[(Qx)(Qy)] & \text{if } p \text{ is odd and } r \geq 2 \\ (Rx)(\mathcal{Q}y) - (1+2\varepsilon_1)(\beta_2 Rx)(\beta_2\mathcal{Q}y) & \text{if } p = 2 \text{ and } r = 1 \\ (Rx)(\mathcal{Q}y) + 4_*[(Qx)(Qy)] + 2^{r-2}(\beta_{r+1}4_*Qx)(\beta_{r+1}\mathcal{Q}y) & \\ \qquad + 2^r\varepsilon_r(\beta_{r+1}Rx)(\beta_{r+1}\mathcal{Q}y) & \text{if } p = 2 \text{ and } r \geq 2, \end{cases}$$

and $R(yx) = (\mathcal{Q}y)(Rx) + (1+2\varepsilon_1)(\beta_2\mathcal{Q}y)(\beta_2 Rx)$ if $p = 2$ and $r = 1$. Let $|x| = |y| = 1$. Then there is a constant $\varepsilon_r' \in Z_p$, independent of x and y, with

$$\mathcal{Q}(xy) = \begin{cases} p^r\varepsilon_r'(Rx)(Ry) & \text{if } p \text{ is odd} \\ (1+2\varepsilon_1')(Rx)(Ry) - (1+2\varepsilon_1+2\varepsilon_1')(\beta_2 Rx)(\beta_2 Ry) & \text{if } p = 2 \text{ and } r = 1 \\ 2^r\varepsilon_r'(Rx)(Ry) + 2^{r-2}(Rx)(4_*Qy) + 2^{r-2}(4_*Qx)(Ry) & \\ \qquad + 2^{2r-4}(\beta_{r+1}4_*Qx)(\beta_{r+1}4_*Qy) & \text{if } p = 2 \text{ and } r \geq 2. \end{cases}$$

(vii) Let Y be an H_∞ space and let $x \in \tilde{K}_0(Y;r)$. If $p = 2$ then $\sigma\mathcal{Q}x = 2^r R(\sigma x)$, while if p is odd there is a constant ε_r'', independent of x, with $\sigma\mathcal{Q}x = p^r\varepsilon_r'' R(\sigma x)$.

(viii) If k is prime to p then $\psi^k\mathcal{Q}x = \mathcal{Q}\psi^k x$.

(ix) $$Q\mathcal{Q}x = \begin{cases} 0 & \text{if } r = 1 \\ \sum_{i=1}^{p} \binom{p}{i} p^{i-2} x^{p^2-ip} p_*[(Qx)^i] & \text{if } r \geq 2. \end{cases}$$

The undetermined constants ε_r in part (vi) depend on the choice of multiplications; they can be made equal to zero for a suitable choice but it is not clear what their values are for our canonical choice. It is quite possible that the ε_r, ε'_r and ε''_r are all zero.

Next we shall use the operations Q and R to describe $K_*(CX;1)$ for an arbitrary unital spectrum X. If Y is a based space then the homology equivalence of [28, Theorem I.5.10] is also a K-theory equivalence (by the Atiyah-Hirzebruch spectral sequence), hence

$$K_*(QY;1) \cong (\pi_0 Y)^{-1} K_*(CY;1) = (\pi_0 Y)^{-1} K_*(C\Sigma^\infty(Y^+);1)$$

so that our calculation will also give $K_*(QY;1)$.

First recall the K-theory Bockstein spectral sequence $E^r_* X$ (abbreviated BSS) from [13, section 11]. X was assumed to be a finite complex in [13] but we wish to work in greater generality. The finiteness assumption is necessary for those results which deal with the E^∞ term, since in general there is no useful relation between $E^\infty_* X$ and $K_* X$ (for example, $E^\infty_* RP^\infty$ is concentrated in dimension zero, while $K_* RP^\infty$ is concentrated in dimension one). On the other hand, the results of [13] which deal with E^r for r finite remain valid for arbitrary spectra X. In particular, any (r-1)-cycle x can be lifted to an element $y \in K_*(X;r)$ and we have $d_r x = \pi^{r-1}\beta_r y$. The element y has order p^r if and only if x is nonzero in E^r. If we write $K_*(X;\infty)$ for the inverse limit of the $K_*(X;r)$ then an infinite cycle always lifts to $K_*(X;\infty)$; we shall frequently use this notation. Our next definition gives the kind of data necessary for the description of $K_*(CX;1)$.

Definition 3.8. Let $1 \le n \le \infty$. A set $A = \bigcup_{1 \le r \le n} A_r$ with $A_r \subset K_*(X;r)$ is called a subbasis of height n for X if for each $s \le n$ the set

$$\{\pi^{r-1}x \mid x \in A_r,\ s \le r \le n\} \cup \{\pi^{r-1}\beta_r x \mid x \in A_r,\ s \le r < n\}$$

projects to a basis for $E^s_* X$.

If the height of a subbasis is not specified, it will always be assumed to be infinite. Subbases with finite height will occur only in sections 7 and 8. It is not hard to see that any spectrum has a subbasis of any given height. The term subbasis is motivated by our next result, which is an easy consequence of the results of [13,§11]. Recall that a subset S of an abelian group G is a basis for G if G is the direct sum of the cyclic subgroups generated by the elements of S.

Proposition 3.9. If $A = \bigcup_{1 \le r \le n} A_r$ is a subbasis of height n for X and if $s \le n$ (with $s < \infty$ if $n = \infty$) then the set

$$\{\pi^{r-s}x \mid x \in A_r,\ s \leq r \leq n\} \cup \{\pi^{r-s}\beta_r x \mid x \in A_r,\ s \leq r < n\}$$

$$\cup \{p_*^{s-r}x \mid x \in A_r,\ r < s\} \cup \{\beta_s p_*^{s-r}x \mid x \in A_r,\ r < s\}$$

is a basis for $K_*(X;s)$. The elements of the form $p_*^{s-r}x$ and $\beta_s p_*^{s-r}x$ have order p^r and the remaining basis elements have order p^s.

Now let X be a unital spectrum. Let $\eta \in K_0(X;\infty)$ be the unit and suppose that $\pi^\infty\eta$ is nonzero in $K_0(X;1)$. Then we may choose a set $A = \bigcup_{1\leq r\leq\infty} A_r$ such that $A \cup \{\eta\}$ is a subbasis for X. We write $A_{r,0}$ and $A_{r,1}$ for the zero- and one-dimensional subsets of A_r. Let p be odd, and let CA be the quotient of the free commutative algebra generated by the three sets

$$\{\pi^{r-s-1}Q^s x \mid x \in A_r,\ 0 \leq s < r \leq \infty\}$$

$$\{\pi^{r-s-1}\beta_{r-s}Q^s x \mid x \in A_{r,0},\ 0 \leq s < r < \infty\}$$

and

$$\{\pi^{r+s-1}\beta_{r+s}R^s x \mid x \in A_{r,1},\ r < \infty,\ 0 \leq s < \infty\}$$

by the ideal generated by the set

$$\{(\pi^{r+s-1}\beta_{r+s}R^s x)^{p^r} \mid x \in A_{r,1},\ r < \infty,\ 0 \leq s < \infty\}.$$

The elements of the first three sets will be called the <u>standard indecomposables</u> of CA. Here symbols like $\pi^{r-s-1}Q^s x$ are simply indeterminates, since the Dyer-Lashof operations are not defined on $K_*(X;r)$. However, by means of the inclusion $X \to CX$ we may interpret these symbols as elements of $K_*(CX;1)$. Thus we obtain a ring map

$$\lambda: CA \to K_*(CX;1).$$

Our main theorem is

<u>Theorem 3.10.</u> λ is an isomorphism.

We could have defined CA in terms of the Q-operation alone, without using R, since the third generating set is equal to

$$\{\pi^{r-s-1}Q^s\beta_r x \mid x \in A_{r,1},\ r < \infty,\ 0 \leq s \leq r\} \cup \{Q^s\beta_{s+1}p_*^{s-r+1}x \mid x \in A_{r,1},\ r < \infty,\ s > r\}$$

The definition we have given is more convenient for our purposes, however, since it allows us to treat the cases $s \leq r$ and $s > r$ in a unified way.

Theorem 3.10 also holds for p = 2, but the definition of CA in this case is more complicated since mod 2 K-theory is not commutative. Recall from 3.2(vii)' that the commutator of two elements is the product of their Bocksteins. To build this into the definition of CA we define the modified tensor product $C_1 \tilde{\otimes} C_1$ of two Z_2-graded differential algbebras over Z_2 to be their Z_2-graded tensor product with multiplication given by

$$(x \otimes y)(x' \otimes y') = xx' \otimes yy' + x(dx') \otimes (dy)y'.$$

We can define the modified tensor product of finitely many C_i similarly and of infinitely many C_i by passage to direct limits. Now for each $x \in A_{r,0}$ we define C_x to be the free strictly commutative algebra generated by $\{\pi^{r-s-1}Q^s x \mid 0 \leq s \leq r\}$ and if $r < \infty$, $\{\pi^{r-s-1}\beta_{r-s}Q^s x \mid 0 \leq s < r\}$. Give this the differential which takes $Q^{r-1}x$ to $\beta Q^{r-1}x$ and all other generators to zero. For each $x \in A_{r,1}$ we define C_x to be the commutative algebra generated by the sets $\{\pi^{r-s-1}Q^s x \mid 0 \leq s < r\}$ and, if $r < \infty$, $\{\pi^{r+s-1}\beta_{r+s}R^s x \mid 0 \leq s < r\}$, with the relations

(i) $(\pi^{r+s-1}\beta_{r+s}R^s x)^{2^r} = 0$

and

$$\text{(ii)}\quad (\pi^{r-s-1}Q^s x)^2 = \begin{cases} 0 & \text{if } 0 \leq s < r-2 \\ (\pi^{r-1}\beta_r x)^{2^{r-1}} & \text{if } s = r-2 \\ (\pi^r \beta_{r+1} Rx)^{2^{r-1}} & \text{if } s = r-1. \end{cases}$$

(Relation (ii) is motivated by 3.3(x)). Give C_x the differential which takes $Q^{r-1}x$ to $(\pi^{r-1}\beta_r x)^{2^{r-1}}$ and all other generators to zero. Finally, we define CA to be the modified tensor product $\tilde{\bigotimes}_{x \in A} C_x$. There is an evident ring map $\lambda: CA \to K_*(CX;1)$ and with these definitions Theorem 3.10 and its proof are valid.

Remarks 3.11. (i) When $X = S^0$, or when $p = 2$ and X is a sphere or a real projective space, we recover the calculations of Hodgkin [41] and Miller and Snaith [83,84].

(ii) We can describe the additive structure of CA more explicitly as follows. When $p = 2$ we define the standard indecomposables of CA to be the same three sets as in the odd-primary case. If we give these some fixed total ordering then CA has an additive basis consisting of all ordered products of standard indecomposables in which each of the odd-dimensional indecomposables occurs no more than once and each $\pi^{r+s-1}\beta_{r+s}R^s x$ occurs less than 2^r times. This basis will be called the standard basis for CA. We define the standard basis in the same way when p is odd.

Next we discuss the functoriality of the description given by 3.10. If X and X' are unital spectra with subbases $A \cup \{\eta\}$ and $A' \cup \{\eta\}$ then a unit-preserving map $f: X \to X'$ will be called based if $f_* A_r \subset A'_r \cup \{0\}$ for all $r \geq 1$. Such a map clearly induces a map $f_*: CA \to CA'$, and we have $\lambda \circ f_* = (Cf)_* \circ \lambda$. If f is not based, it is still possible in principle to determine $(Cf)_*$ on $K_*(CX;1)$ by using 3.3, 3.6 and 3.9 (although in practice the formulas may become complicated). For example, if $f:S^2 \to S^2$ is the degree p map and $x \in K_0(S^2;2)$ is the generator then

$$(Cf)_*Qx = Q(f_*x) = Q(px) = \pi(x^p) \neq 0$$

in $K_0(CS^2;1)$. Since $f_*:K_*(S^2;1) \to K_*(S^2;1)$ is zero this gives another proof of Hodgkin's result that $K_*(CX;1)$ cannot be an algebraic functor of $K_*(X;1)$. A similar calculation for the degree p^r map shows that $K_*(CX;1)$ is not a functor of $K_*(X;r)$ for any $r < \infty$. Finally, the projection $S^1 \cup_p e^2 \to S^2$ onto the top cell induces the zero map in integral K-homology but is nonzero on $K_*(C(S^1 \cup_p e^2);1)$ so that $K_*(CX;1)$ is not a functor of $K_*(X;Z)$. Thus it seems that the use of subbases cannot be avoided.

We conclude this section by determining the BSS for CX.

Theorem 3.12. For $1 \leq m < \infty$, E^m_* CX is additively isomorphic to the quotient of the free strictly commutative algebra generated by the six sets

$$\{\pi^{r-s-1}Q^s x \mid x \in A_r,\ m \leq r-s,\ 0 \leq s < r\}$$

$$\{\pi^{r-s-1}\beta_{r-s}Q^s x \mid x \in A_{r,0},\ m \leq r-s < \infty,\ 0 \leq s < r\}$$

$$\{\pi^{m-1}\mathfrak{Q}^{m-r+s}Q^s x \mid x \in A_{r,0},\ 1 \leq r-s < m\}$$

$$\{\pi^{m-1}\beta_m\mathfrak{Q}^{m-r+s}Q^s x \mid x \in A_{r,0},\ 1 \leq r-s < m\}$$

$$\{\pi^{m-1}R^{m-r+s}Q^s x \mid x \in A_{r,1},\ 1 \leq r-s < m\}$$

and $$\{\pi^{r+s-1}\beta_{r+s}R^s x \mid x \in A_{r,1},\ m \leq r+s < \infty\}$$

by the ideal generated by the set

$$\{(\pi^{r+s-1}\beta_{r+s}R^s x)^{p^t} \mid x \in A_{r,1},\ m \leq r+s < \infty,\ t = \min(r,r+s+1-m)\}.$$

If p is odd or $m \geq 3$ the isomorphism is multiplicative.

The proof of 3.12 is the usual counting argument, and is left to the reader. In order to determine the differential in E^m_*CX one needs the formula

$$\pi^{r-s+t-1}\beta_{r-s+t}R^tQ^s x = (\pi^{r+t-1}\beta_{r+t}R^t x)^{p^s}$$

for $x \in A_{r,1}$, $0 \leq s < r < \infty$, $t \geq 0$; this is is a consequence of 3.3(viii) and 3.3(v).

4. Calculation of $K_*(CX;Z_p)$

In this section we give the proof of Theorem 3.10, except for two lemmas which will be dealt with in Sections 6 and 9. The argument is very similar to that given in Section 2 for ordinary homology, and in several places we shall simply refer to that section.

First we reformulate 3.10 as a result about extended powers. Let Y be any spectrum and let A be a subbasis for Y. We define CA with its standard indecomposables and standard basis as in Section 3. We make CA a filtered ring by giving elements of A filtration 1 and requiring Q and R to multiply filtration by p. Let $D_kA = F_kCA/F_{k-1}CA$ for $k \geq 1$; this has a standard basis consisting of the standard basis elements in $F_kCA - F_{k-1}CA$. There is an additive map

$$\lambda_k : D_kA \to K_*(D_kY;1)$$

defined as in Section 2 by interpreting Q,R and the multiplication externally and then applying α_* and β_*. We shall prove

Theorem 4.1. λ_k is an isomorphism for all $k \geq 1$.

Remark 4.2. Using 4.1 and the external versions of 3.3(v), 3.6(iv) and 3.7(iv) (which will be proved in sections 7 and 8) one can determine the BSS for D_kY as follows. If $m \geq 1$ let C^mA denote the algebra whose generators and relations are given in 3.12. We make C^mA a filtered ring by giving elements of A filtration 1 and requiring R, Q and $\mathcal{Q}$ to multiply filtration by p. If D_k^mA is the k-th subquotient of C^mA there is an isomorphism $D_k^mA \to E_*^mD_kX$. The proof is similar to that for 3.12 and is left to the reader.

The derivation of 3.10 from 4.1 is the same as that given for 2.1 in section 2. We therefore turn to the proof of 4.1. We need the following special case, which will be proved in section 6.

Lemma 4.3. λ_p is an isomorphism for all Y.

We shall reduce the proof of 4.1 to the case where Y is a wedge of Moore spectra. First we need some notation. As in section 1 we write M_r for $S^{-1} \cup_{p^r} e^0$. The set $\{u_r\}$ is a subbasis for M_r. We write M_∞ for the colimit of the M_r with respect to the maps $M_r \to M_{r+1}$ having degree p on the bottom cell. Then $K_1(M_\infty;r) = 0$ for all r and $K_0(M_\infty;r)$ is a copy of Z_{p^r} generated by the image of u_r. Let $u_\infty \in K_0(M_\infty;\infty)$ be the element which projects to the image of u_r for all r. Then $\{u_\infty\}$ is a subbasis for M_∞.

For each $x \in A_r$ we can choose a map $f_x : \Sigma^{|x|}M_r \to K \wedge Y$ representing x. (If $r = \infty$ we let f_x be any map which restricts on each $\Sigma^{|x|}M_r$ to a representaive for the mod p^r reduction of x.) Let $Z = \bigvee_{1 \leq r \leq \infty} \bigvee_{x \in A_r} \Sigma^{|x|}M_r$ and let $f : Z \to K \wedge Y$ be the wedge of the f_x. We give Z the subbasis B consisting of the fundamental classes of the

$\Sigma^{|x|}M_r$. Then $f_{**}:K_*(Z;r) \to K_*(Y;r)$ gives a one-to-one correspondence between B_r and A_r, and in particular it is an isomorphism for all r. Now consider the diagram

$$\begin{array}{ccc} D_kB & \xrightarrow{D_k(f_{**})} & D_kA \\ \downarrow{\lambda_k} & & \downarrow{\lambda_k} \\ K_*(D_kZ;1) & \xrightarrow{(\bar{D}_kf)_{**}} & K_*(D_kY;1) \end{array},$$

which commutes by 1.3 and 1.4(ii) and (iii). If 4.1 holds for Z, its validity for Y will be immediate from the diagram and the following lemma.

Lemma 4.4. Let $h:W \to K \wedge X$ be any map. If $h_{**}:K_*(W;1) \to K_*(X;1)$ is an isomorphism, then

(i) $f_{**}:K_*(W;r) \to K_*(X;r)$ is an isomorphism for all r, and

(ii) $(\bar{D}_kf)_{**}:K_*(D_kW;1) \to K_*(D_kX;1)$ is an isomorphism for all k.

Proof. (i) By induction on r. Suppose the result is true for some $r \geq 1$ and consider the short exact sequence

$$0 \longrightarrow Z_p \longrightarrow Z_{p^{r+1}} \longrightarrow Z_{p^r} \longrightarrow 0 .$$

This gives rise to the following commutative diagram with exact rows.

$$\begin{array}{ccccccccc} K_{\alpha+1}(W;r) & \longrightarrow & K_\alpha(W;1) & \longrightarrow & K_\alpha(W;r+1) & \longrightarrow & K_\alpha(W;r) & \longrightarrow & K_{\alpha-1}(W;1) \\ \downarrow{f_{**}} & & \downarrow{f_{**}} & & \downarrow{f_{**}} & & \downarrow{f_{**}} & & \downarrow{f_{**}} \\ K_{\alpha+1}(X;r) & \longrightarrow & K_\alpha(X;1) & \longrightarrow & K_\alpha(X;r+1) & \longrightarrow & K_\alpha(X;r) & \longrightarrow & K_{\alpha-1}(X;1) \end{array}$$

Part (i) follows by the five lemma. The proof of part (ii) is now completely parallel to that of Lemma 2.4.

Next we reduce to the case of a single Moore spectrum. We assume for simplicity that Z is a wedge of two Moore spectra $\Sigma^m M_r \vee \Sigma^n M_s$; the argument is the same in the general case. Let B_1 and B_2 be the subbases $\{\Sigma^m u_r\}$ and $\{\Sigma^n u_s\}$, so that $B = B_1 \cup B_2$. There is an evident map $CB_1 \otimes CB_2 \to CB$ which on passage to the associated graded gives a map

$$\varphi: \sum_{i=0}^{k} (D_iB_1 \otimes D_{k-i}B_2) \to D_kB.$$

Lemma 4.5. φ is an isomorphism, and the diagram

$$\begin{array}{ccc}
\sum_{i=0}^{k} (D_i B_1 \otimes D_{k-i} B_2) & \xrightarrow{\ \varphi\ } & D_k B \\
\Big\downarrow {\scriptstyle \Sigma(\lambda_i \otimes \lambda_{k-i})} & & \Big\downarrow {\scriptstyle \lambda_k} \\
\sum_{i=0}^{k} (K_*(D_i \Sigma^m M_r;1) \otimes K_*(D_{p-i} \Sigma^n M_s;1)) & \xrightarrow{\ \cong\ } & K_*(D_k Z;1)
\end{array}$$

commutes

The proof is the same as for 2.5. The lemma implies that 4.1 will hold for Z once we have shown the following. We write x for $\Sigma^n u_r \in K(\Sigma^n M_r;r)$.

Lemma 4.6. $\lambda_k : D_k\{x\} \to K_*(D_k \Sigma^n M_r;1)$ is an isomorphism for all $k \geq 1$ and all n.

Proof. By induction on k. First let $k = jp$ with $j > 1$. We need the commutativity of the following diagram for i = 0,1 and 2.

(*)

$$\begin{array}{ccccccc}
D_j \mathfrak{A}' & & & \xrightarrow{\ \gamma_j'\ } & & & D_k\{y,z\} \\
 & \searrow^{F_i} & & & & \swarrow^{D_k(g_i)_*} & \\
 & & D_j \mathfrak{A} & \xrightarrow{\ \gamma_j\ } & D_k\{x\} & & \\
\cong \Big\downarrow \lambda_j & & \cong \Big\downarrow \lambda_j & & \Big\downarrow \lambda_k & & \Big\downarrow \lambda_k \\
 & & K_*(D_j D_p M;1) & \xrightarrow{\ \beta_{jp*}\ } & K_*(D_k M;1) & & \\
 & \nearrow^{(D_j D_p g_i)_*} & & & & \nwarrow^{(D_k g_i)_*} & \\
K_*(D_j D_p(M \vee M);1) & & & \xrightarrow{\ \beta_{jp*}\ } & & & K_*(D_k(M \vee M);1)
\end{array}$$

Here M denotes $\Sigma^n M_r$ and $y,z \in K_*(M\ \ M;r)$ are the fundamental classes of the first and second summands. The sets $\mathfrak{A}$ and $\mathfrak{A}'$ are subbases for $D_p M$ and $D_p(M \vee M)$ which will be specified later. The maps $g_i : M \vee M \to M$ are defined by $g_0 = 1 \vee 1$, $g_1 = 1 \vee *$, and $g_2 = * \vee 1$, and the F_i are determined uniquely by the requirement that the left-hand trapezoid commute. To complete the diagram we need

Lemma 4.7. There exist $\mathfrak{A}$, $\mathfrak{A}'$, γ_j and γ_j' independent of i such that diagram (*) commutes for i = 0,1 and 2.

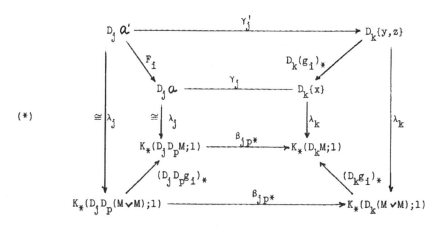

The proof will be given in Section 9. Like the proof of 2.7, it consists of systematic simplifications of the elements of $D_j\mathcal{A}$ and $D_j\mathcal{A}'$. The details are much more complicated, however, because of the nonadditivity of the operations.

Now consider the inner square of the diagram. Since $\beta_{jp*} \circ \tau_*$ is an isomorphism, we see that λ_k is onto. Letting $\theta = \gamma_j \circ \lambda_j^{-1} \circ \tau_* \circ \lambda_k$, we see as in section 2 that θ induces an isomorphism of the subspace $\mathcal{D}$ of $D_k\{x\}$ spanned by the decomposable standard basis elements. In particular, λ_k is monic on $\mathcal{D}$.

The remainder of the proof differs from that in Section 2, and is in fact considerably simpler since there are only a few indecomposables. It suffices to show the following.

Lemma 4.8. Let $w \in \mathcal{D}$. If $n = 1$ then

(i) $\lambda_k(\pi^{r-s-1}Q^s x - w) \neq 0$, where $k = p^s$, $2 \le s < r \le \infty$

(ii) $\lambda_k(\pi^{r+s-1}\beta_{r+s}R^s x - w) \neq 0$, where $k = p^s$, $r < \infty$, $2 \le s < \infty$.

If $n = 0$ then

(iii) $\lambda_k(\pi^{r-s-1}Q^s x - w) \neq 0$, where $k = p^s$, $2 \le s < r < \infty$

(iv) $\lambda_k(\pi^{r-s-1}\beta_{r-s}Q^s x - w) \neq 0$, where $k = p^s$, $2 \le s < r < \infty$.

Proof. We need two facts about the map $\Delta_*: K_*(\Sigma D_k X;r) \to K_*(D_k\Sigma X;r)$, namely that $\Delta_*\Sigma(\alpha_{i,k-1})_*(x \otimes y) = 0$ for $0 < i < k$ and that, when $k = p$,

$$\Delta_*\Sigma Qx = \begin{cases} Q(\Sigma x) & \text{if } |x| = 0 \\ \pi\iota_*(\Sigma x)^{(p)} + pQ\Sigma x & \text{if } |x| = 1 . \end{cases}$$

The first fact is shown as in the proof of 2.8, while the second, which is the external version of 3.3(viii), will be shown in section 7.

Now consider part (i). We have $\Delta_*\Sigma w = 0$ and

$$\Delta_*\Sigma\pi^{r-s-1}Q^s x = \pi^{r-1}\iota_*(\Sigma x)^{p^s} .$$

But $\pi^{r-1}\iota_*(\Sigma x)^{p^s}$ is nonzero since λ_k is monic on decomposables.

Combining part (i) with the fact that λ_k is onto and is monic on decomposables, we see that

$$\lambda_k: D_k\{x\} \to K_*(D_k\Sigma M_r;1)$$

is an isomorphism in degree 1 and is onto in degree zero. It is monic in degree 0 if and only if part (ii) holds. But if not then $K_0(D_k\Sigma M_r;1)$ and $K_1(D_k\Sigma M_r;1)$ would have different dimensions as vector spaces, and therefore the Bockstein spectral sequence $E^m_*(D_k\Sigma M_r)$ would be nonzero for all m. But the transfer embeds $E^m_*D_k\Sigma M_r$ in $E^m_*D_jD_p\Sigma M_r$, and the latter is zero for $p^{m-r-1} > j$ by Remark 4.2 and the inductive hypothesis of 4.6.

Finally, part (iii) follows from (i) and the equation

$$\Delta_*\Sigma\pi^{r-s-1}Q^s x = \pi^{r-s-1}Q^s\Sigma x,$$

while (iv) follows from (iii) using the argument given for (ii).

This completes the proof of 4.6 for the case k = jp. The remaining case, when k is prime to p, is handled exactly as in Section 2.

5. Calculation of $\tilde{K}_*(D_pS^n;Z_{p^r})$

In order to construct and analyze the Q-operation we shall need a precise description of $K_*(D_p\Sigma^n M_r;r-1)$. In this section we give some facts about $K_*(D_pS^n;r)$ which will be used in Sections 6 and 7 to obtain such a description. We work with K-theory on spaces in this section.

If X is a space there is a relative Thom isomorphism

$$\Phi:\tilde{K}_*(D_pX;r) \xrightarrow{\cong} \tilde{K}_*(D_p\Sigma^2X;r)$$

corresponding to the bundle

$$E\Sigma_p \times_{\Sigma_p} (X^{(p)} \times R^{2p}) \to E\Sigma_p \times_{\Sigma_p} X^{(p)}$$

and the inclusion

$$E\Sigma_p \times_{\Sigma_p} (*) \to E\Sigma_p \times_{\Sigma_p} X^{(p)}.$$

As we have seen in VII§3 and VII§8, this isomorphism can in fact be defined for an arbitrary spectrum X. In calculating $\tilde{K}_*(D_pS^n;r)$ we may therefore assume n = 0 or n = 1; in the former case we have $D_pS^0 = B\Sigma_{p_+}$.

<u>Lemma 5.1</u>. $K_\alpha(B\Sigma_p;1)$ is zero if $\alpha = 1$ and $Z_p \oplus Z_p$ if $\alpha = 0$. $\tilde{K}_\alpha(D_pS^1;1)$ is zero if $\alpha = 0$ and Z_p if $\alpha = 1$.

<u>Proof</u>. We use the Atiyah-Hirzebruch spectral sequence for mod p K-homology. By [40, III.1.2] the differentials d_i vanish for $i < 2p-1$ and d_{2p-1} is $\beta P^1_* - P^1_*\beta$ (here P^1 denotes Sq^2 if p = 2). For spaces of the form D_pX, a basis for the E^2-term consisting of external Dyer-Lashof operations is given in [68, 1.3 and 1.4]. The differential d_{2p-1} can be evaluated using the external form of the Nishida relations

[68, 9.4]; the explicit result is that $d_{2p-1}(e_i \otimes y^p)$ is a nonzero multiple of

$$(\beta e_{i+2-2p}) \otimes y^p - e_{i+1-p} \otimes (\beta y)^p$$

for any $y \in H_*(X;1)$. Letting $X = S^0$ or S^1 we see that E^{2p} is generated by $e_0 \otimes u^p$ and $e_{2p-2} \otimes u^p$ in the former case and by $e_{p-1} \otimes (\Sigma u)^p$ in the latter. Then $E^{2p} = E^\infty$ for dimensional reasons and the result follows.

Using 5.1 and the K-theory BSS we conclude that $K_*(B\Sigma_p;r)$ is free over Z_{p^r} on two generators in dimension zero and that $\tilde{K}_*(D_pS^1;r)$ is free over Z_{p^r} on one generator in dimension one. We wish to give explicit bases. It is convenient to work in K-cohomology, as we may by the following.

Lemma 5.2. The natural map

$$\tilde{K}^*(D_pS^n;r) \to \mathrm{Hom}(\tilde{K}_*(D_pS^n;r), Z_{p^r})$$

is an isomorphism for all $r < \infty$.

Proof. When $r = 1$ a cell-by-cell induction and passage to limits gives the results for an arbitrary space; in particular it holds for D_pS^n. The result for general r follows from the BSS.

Next we give a basis for $K^0(B\Sigma_p;r)$. We write 1 for the unit in this group and $1_{(e)}$ for the unit of $K^0(pt.;r)$. Let τ be the transfer $\Sigma^\infty(B\Sigma_p^+) \to \Sigma^\infty(Be^+) = S$.

Proposition 5.3. $K^*(B\Sigma_p;r)$ is freely generated over Z_{p^r} by 1 and $\tau^*1_{(e)}$.

Proof. Let $\pi = Z_p$ and denote the inclusion $\pi \subset \Sigma_p$ by ι. Then $K^1(B\pi;r) = 0$ and the natural map

$$R\pi \otimes Z_{p^r} \to K^0(B\pi;r)$$

is an isomorphism. If ρ is the group of automorphisms of π then a standard transfer argument shows that the restriction

$$\iota^*: K^*(B\Sigma_p;r) \to K^*(B\pi;r)$$

is a monomorphism whose image is contained in the invariant subring $K^*(B\pi;r)^\rho$. Now ι^*1 is the unit 1_π of $K^0(B\pi;r)$, while the double coset formula gives $\iota^*\tau^*1_{(e)} = (p-1)!(\tau')^*1_{(e)}$, where τ' is the transfer $\Sigma^\infty(B\pi^+) \to S$. Since 1_π and $\tau'1_{(e)}$ form a basis for $K^*(B\pi;r)^\rho$ the result follows.

In order to give a specific generator for $\tilde{K}^*(D_pS^1;r)$ we consider the map

$$\Delta^*:\tilde{K}^*(D_pS^{n+1};r) \to \tilde{K}^*(\Sigma D_pS^n;r).$$

Lemma 5.4. The composite

$$\tilde{K}^0(D_pS^2;r) \xrightarrow{\Delta^*} \tilde{K}^0(\Sigma D_pS^1;r) \xrightarrow{(\Sigma\Delta)^*} \tilde{K}^0(\Sigma^2 D_pS^0;r) \cong K^0(B\Sigma_p;r)$$

takes $\Phi(1)$ to $\frac{1}{(p-1)!}(p! - \tau^*1_{(e)})$ and $\Phi(\tau^*1_{(e)})$ to zero.

As an immediate consequence we have

Corollary 5.5. $\Sigma\Delta^*\Phi(1)$ generates $\tilde{K}^*(D_pS^1;r)$.

Before proving 5.4 we give the desired bases for $K_*(B\Sigma_p;r)$ and $\tilde{K}_*(D_pS^1;r)$.

Definition 5.6. The canonical basis for $K_*(B\Sigma_p;r)$ is the dual of the basis $\{1, \frac{1}{(p-1)!}(p! - \tau^*1_{(e)})\}$. The canonical basis for $\tilde{K}_*(D_pS^1;r)$ is the dual of $\{\Sigma\Delta^*\Phi(1)\}$.

Note that the unit η in $K_0(B\Sigma_p;r)$ is the first element of the canonical basis for this group. We shall always write v for the remaining element and v' for the basis element in $K_1(D_pS^1;r)$.

Proof of 5.4. Consider the subset of $E\Sigma_p \times_{\Sigma_p} (\mathbb{R}^2)^p$ consisting of points for which the sum of the $\mathbb{R}^2$-coordinates is zero. The projection to $B\Sigma_p$ makes this subset the total space of a bundle ξ over $B\Sigma_p$. Now D_pS^2 is homeomorphic to the second suspension of the Thom complex $T\xi$ of ξ, and under this homeomorphism the map $\Delta \circ \Sigma\Delta:\Sigma^2 D_pS^0 \to D_pS^2$ is the second suspension of the inclusion $B\Sigma_p^+ \subset T\xi$, while $\Phi(1)$ agrees with the Aityah-Bott-Shapiro orientation for ξ. Thus it suffices to show that the Euler class of ξ is $\frac{1}{(p-1)!}(p! - \tau^*1_{(e)})$. If $\pi = Z_p$ and $\iota: \pi \subset \Sigma_p$ is the inclusion it suffices to show that the pullback $(B\iota)^*\xi$ has Euler class $p - (\tau')^*1_{(e)}$ in $K^0(B\pi) \cong R\pi \otimes Z_p^\wedge$, where τ' is the transfer $\Sigma^\infty(B\pi^+) \to S$. Let $x \in R\pi$ be any nontrivial irreducible. Then $(B\iota)^*\xi$ is the sum of the bundles over $B\pi$ induced by $x, x^2, \ldots, x^{p-1}$. These bundles have Euler classes $1-x, \ldots, 1-x^{p-1}$, hence $(B\iota)^*\xi$ has Euler class $(1-x)\cdots(1-x^{p-1})$. Evaluation of characters shows that

$$(1-x)\cdots(1-x^{p-1}) = p - (1 + x + \cdots + x^{p-1})$$

and the result follows.

Next we collect some information about the elements η, v and v' for use in section 7.

Proposition 5.7. (i) $\pi:\tilde{K}_*(D_pS^n;r) \to \tilde{K}_*(D_pS^n;r-1)$ takes v to v and v' to v'.

(ii) $\Delta_*:\tilde{K}_1(\Sigma(B\Sigma_p^+);r) \to \tilde{K}_1(D_pS^1;r)$ takes $\Sigma\eta$ to zero and Σv to v'.

(iii) $\Delta_*:\tilde{K}_0(\Sigma D_pS^1;r) \to \tilde{K}_0(D_pS^2;r)$ takes $\Sigma v'$ to $\Phi(\eta + pv)$.

(iv) $\tau_*:\tilde{K}_*(D_pS^n;r) \to \tilde{K}_*((S^n)^{(p)};r)$ takes η to $p!u$ and v to $-(p-1)!u$ when $n = 0$ and takes v' to zero when $n = 1$.

(v) $\delta_*:K_0(B\Sigma_p;r) \to K_0(B\Sigma_p \times B\Sigma_p;r)$ takes η to $\eta \otimes \eta$ and v to $v \otimes \eta + \eta \otimes v + p(v \otimes v)$.

(vi) $\delta_*:\tilde{K}_1(D_pS^1;r) \to \tilde{K}_1(D_pS^1 \wedge B\Sigma_p^+;r)$ takes v' to $v' \otimes \eta + p(v' \otimes v)$.

(vii) $\delta_*:\tilde{K}_0(D_pS^2;r) \to \tilde{K}_0(D_pS^1 \wedge D_pS^1;r)$ takes $\Phi(\eta)$ to zero and $\Phi(v)$ to $v' \otimes v'$.

For the proof we need a preliminary result.

Lemma 5.8. (i) If X is a spectrum with $E^1 = E^r$ in the K-theory BSS and if Y is any spectrum then the external product map

$$K_*(X;r) \otimes K_*(Y;r) \to K_*(X \wedge Y;r)$$

is an isomorphism, where the tensor product is taken in the Z_2-graded sense.

(ii) If in addition $K_*(X;1)$ and $K_*(Y;1)$ are finitely generated then the external product map

$$K^*(X;r) \otimes K^*(Y;r) \to K^*(X \wedge Y;r)$$

is an isomorphism.

Proof When $r = 1$ the first statement is well-known (see [13, Theorem 6.2], for example). It follows that the external product induces an isomorphism of K-theory Bockstein spectral sequences. Hence if B is a basis for $K_*(X;r)$ and A is a subbasis of height r for Y then the set $\{\pi^{r-s}x \otimes y \mid x \in B,\ y \in A_s\}$ is a subbasis of height r for $X \wedge Y$ and part (i) follows. The case $r = 1$ of part (ii) follows from part (i) by duality, and the general case follows from it as in part (i).

Next we turn to the proof of 5.7, which will conclude this section. In each case it suffices by 5.8 to show the dual. Then (i) is immediate and (ii) and (iii) follow from 5.4. The first and second statements of part (iv) are trivial, as is the third when $p = 2$. When p is odd we observe that τ_*v' must be invariant under the Σ_p action on $\tilde{K}_*((S^1)^{(p)};r)$. Clearly zero is the only invariant element.

For part (v) we observe that $\tau^*1_{(e)} \otimes \tau^*1_{(e)}$ is $\tau^*(\iota^*\tau^*1_{(e)})$ by Frobenius reciprocity. Now $\iota^*\tau^*1_{(e)} = p!1_{(e)}$, and thus

$$\left[\frac{1}{(p-1)!}(p! - \tau^*1_{(e)})\right]^2 = \frac{p}{(p-1)!}(p! - \tau^*1_{(e)})$$

in $K^0(B\Sigma_p;r)$; the result follows by duality.

For part (vi), consider the composite

$$\tilde{K}_1(\Sigma(B\Sigma_p^+);r) \xrightarrow{\Delta_*} \tilde{K}_1(D_pS^1;r) \xrightarrow{\delta_*} \tilde{K}_1(D_pS^1 \wedge B\Sigma_p^+;r) .$$

We have $\Delta_*\Sigma v = v'$, and

$$\begin{aligned}\delta_*\Delta_*\Sigma v &= (\Delta \wedge 1)_*\Sigma\delta_* v\\ &= (\Delta_*\Sigma v) \otimes \eta + (\Delta_*\Sigma\eta) \otimes v + p(\Delta_*\Sigma v) \otimes v\\ &= v' \otimes \eta + p(v' \otimes v).\end{aligned}$$

For part (vii) observe that part (iii) implies that the map

$$(\Delta \wedge 1)_*:\tilde{K}_1(\Sigma D_pS^1 \wedge D_pS^1;r) \to \tilde{K}_1(D_pS^2 \wedge D_pS^1;r)$$

is monic and that $(\Delta \wedge 1)_*(\Sigma v' \otimes v') = \Phi(\eta) \otimes v' + p\Phi(v) \otimes v'$. Hence it suffices to show that $(\Delta \wedge 1)_*(\Sigma\delta_*\Phi(\eta))$ is zero and that

$$(\Delta \wedge 1)_*\Sigma\delta_*\Phi(v) = \Phi(\eta) \times v' + p\Phi(v) \otimes v'.$$

Now let

$$h:S^1\ S^2 = S^1 \wedge (S^1 \wedge S^1) \approx (S^1 \wedge S^1) \wedge S^1 = S^2 \wedge S^1$$

be the associativity transformation and consider the diagram

$$\begin{array}{ccccc}
\Sigma D_pS^2 & & \xrightarrow{\quad\Sigma\delta\quad} & & \Sigma D_pS^1 \wedge D_pS^1\\
 & \searrow^{\Delta} & & & \\
\Sigma\delta\Big\downarrow & & D_p(S^1 \wedge S^2) \xrightarrow{D_ph} D_p(S^2 \wedge S^1) & & \Big\downarrow \Delta \wedge 1\\
 & & \delta & & \\
\Sigma D_pS^2 \wedge B\Sigma_p^+ & \xrightarrow{\cong} & D_pS^2 \wedge \Sigma B\Sigma_p^+ & \xrightarrow{1 \wedge \Delta} & D_pS^2 \wedge D_pS^1
\end{array}$$

The upper part clearly commutes, and the lower part also commutes since h is homotopic to the map switching the factors S^1 and S^2. Now

$$\delta_*:\tilde{K}_0(D_pS^2;r) \to \tilde{K}_0(D_pS^2 \wedge B\Sigma_p^+;r)$$

clearly takes $\Phi(\eta)$ to $\Phi(\eta)\otimes\eta$ and $\Phi(v)$ to

$$\Phi(\eta)\otimes v + \Phi(v)\otimes\eta + p\Phi(v)\otimes v.$$

Hence

$$(\Delta\wedge 1)_*(\Sigma\delta_*\Phi(\eta)) = (1\wedge\Delta)_*(\Phi(\eta)\otimes\Sigma\eta) = 0$$

by the diagram and part (ii), while

$$(\Delta\wedge 1)_*(\Sigma\delta_*\Phi(v)) = (1\wedge\Delta)_*[\Phi(\eta)\otimes\Sigma v + \Phi(v)\otimes\Sigma\eta + p\Phi(v)\otimes\Sigma v]$$

$$= \Phi(\eta)\otimes v' + p\Phi(v)\otimes v' .$$

6. Calculation of $\tilde{K}_*(D_pX;Z_p)$

In this section we define Q on $K_*(X;2)$ and prove Lemma 4.3. We work with K-theory on spectra in this section.

Our first result collects the information about $K_*(D_p\Sigma^nM_r;1)$ which will be used in this and later sections. We let i and j respectively denote the inclusion of the bottom cell of Σ^nM_r and the projection onto the top cell. Note that $j_*\Sigma^nu_r = \Sigma^nu$ and $i_*\Sigma^{n-1}u = \beta_r\Sigma^nu_r$, where u_r and u are the fundamental classes of M_r and S^0.

Lemma 6.1. (i) For any $n\in Z$ and $\alpha\in Z_2$, $K_\alpha(D_p\Sigma^nM_1;1)$ has dimension 1 over Z_p.

(ii) For any $n\in Z$, $\alpha\in Z_2$ and $r\geq 2$, $K_\alpha(D_p\Sigma^nM_r;1)$ has dimension 2 over Z_p.

(iii) $(D_pj)_*:K_0(D_pM_r;1)\to K_0(D_pS^0;1)$ is monic, and if $r\geq 2$ it is an isomorphism.

(iv) $(D_pj)_*\oplus\tau_*:K_1(D_p\Sigma M_r;1)\to K_1(D_pS^1;1)\oplus K_1((\Sigma M_r)^{(p)};1)^{\Sigma_p}$ is monic, and is an isomorphism if $r\geq 2$.

(v) $(D_pi)_*:K_0(D_pS^0;1)\to K_0(D_p\Sigma M_r;1)$ is onto. If $r = 1$ it has kernel generated by η and if $r\geq 2$ it is an isomorphism.

(vi) The sequence

$$K_1(D_pS^{-1};1)\xrightarrow{(D_pi)_*}K_1(D_pM_r;1)\xrightarrow{\tau_*}K_1((M_r)^{(p)};1)^{\Sigma_p}\longrightarrow 0$$

is exact, and if $r\geq 2$, $(D_pi)_*$ is a monomorphism.

In parts (iv) and (vi), $K_1((\Sigma^nM_r)^{(p)};1)^{\Sigma_p}$ denotes the subgroup invariant under the evident Σ_p-action; this subgroup can easily be calculated using 5.8(i). The proof of 6.1 is similar to that of 5.1 and is left to the reader.

We can now define elements $v_1 \in K_0(D_pM_2;1)$ and $v_1' \in K_1(D_p\Sigma M_2;1)$ by the equations $(D_pj)_*v_1 = v$, $(D_pj)_*v_1' = v'$, and $\tau_*v_1' = 0$. We use definition 1.6 to construct Q.

Definition 6.2. $Q:K_\alpha(X;2) \to K_\alpha(D_pX;1)$ is the generalized Dyer-Lashof operation Q_{v_1} if $\alpha = 0$ and $Q_{v_1'}$ if $\alpha = 1$.

Observe that $v_1 = Qu_2$ and $v_1' = Q\Sigma u_2$.

Next we turn to the proof of 4.3. We use the spectral sequence of I.2.4 with π equal to Z_p or Σ_p and $E = X$. This spectral sequence will be denoted by $E^r_{q,\alpha}(\pi;X)$; by Bott periodicity it is $Z \times Z_2$-graded, so that $\alpha \in Z_2$.

We can describe $E^2_{q,*}(\pi;X) = H_q(\pi;K_*(X;1)^{\otimes p})$ as follows. When $q = 0$ it is just the coinvariant quotient of $K_*(X;1)^{\otimes p}$. Let $\pi = Z_p$ with p odd. If $x \in K_\alpha(X;1)$ then x^p $K_*(X;1)^{\otimes p}$ generates a trivial π-submodule and we write $e_q \otimes x^p$ for the image of $e_q \in H_q(B\pi;1)$ under the inclusion of this submodule. Now $K_*(X;1)^{\otimes p}$ can be written as a direct sum of trivial π-modules of this kind and free π-modules generated by $x_1 \otimes \cdots \otimes x_p$ with not all x_i's equal. Hence the map

$$K_\alpha(X;1) \to E^2_{q,\alpha}(Z_p;X)$$

taking x to $e_q \otimes x^p$ is an isomorphism if $q > 0$ and p is odd. We continue to write $e_q \otimes x^p$ for the image of this element under the natural map

$$E^2_{q,\alpha}(Z_p;X) \to E^2_{q,\alpha}(\Sigma_p;X).$$

By [68,1.4] we see that this map is onto in all bidegrees, is an isomorphism when $q = (2i-\alpha)(p-1)$ or $(2i-\alpha)(p-1)-1$ for some $i \geq 1$, and is zero in all other bidegrees with $q > 0$. Finally, if $p = 2$ then by 3.2(vii)' the Z_2-action on $K_*(X;1)^{\otimes 2}$ is given by $x \otimes y \longmapsto y \otimes x + \beta y \otimes \beta x$; in particular, x^2 is invariant if and only if $\beta x = 0$. Using this it is easy to see that the map taking x to $e_q \otimes x^2$ induces an isomorphism from $\ker \beta/\operatorname{im} \beta$ to $E_{q,0}(Z_2;X)$ if $q > 0$, while $E_{q,1}(Z_2;X) = 0$ for $q > 0$.

Our next two results describe the groups $E^\infty_{q,\alpha}(\Sigma_p;X)$. Let A be a subbasis for X and let $\overline{A}_2 \subset K_*(X;2)$ be the set

$$\{\pi^{r-2}x \mid x \in A_r,\ 2 \leq r \leq \infty\} \cup \{\pi^{r-2}\beta_r x \mid x \in A_r,\ 2 \leq r < \infty\}.$$

Let $\overline{A}_{2,0}$ and $\overline{A}_{2,1}$ be the zero- and one-dimensional subsets of $\overline{A}_2$.

Proposition 6.3. (i) The kernel of the epimorphism $E^2_{0,*}(\Sigma_p;X) \to E^\infty_{0,*}(\Sigma_p;X)$ is generated by the set $\{(\beta x)^p \mid x \in K_1(X;1)\}$ if p is odd and by

$$\{(\pi\beta_2 x)^2 + (\pi x)^2 \mid x \in K_1(X;2)\} \text{ if } p = 2.$$

(ii) The terms $E^{\infty}_{q,\alpha}(\Sigma_p;X)$ with $q > 0$ are freely generated by the sets

$$\{e_{2p-2} \otimes (\pi x)^p | x \in \bar{A}_{2,0}\}$$

$$\{e_{p-1} \otimes (\pi x)^p \mid x \in \bar{A}_{2,1}\}$$

and, if p is odd,

$$\{e_{p-2} \otimes x^p \mid x \in A_{1,1}\}$$

<u>Proposition 6.4.</u> (i) If $x \in \bar{A}_{2,0}$ then Qx is represented in $E^{\infty}_{**}(\Sigma_p;X)$ by a nonzero multiple of $e_{2p-2} \times (\pi x)^p$.

(ii) If $x \in \bar{A}_{2,1}$ then Qx is represented by a nonzero multiple of $e_{p-1} \otimes (\pi x)^p$.

(iii) If $x \in A_{1,1}$ then $Q\beta_2 p_* x$ is represented by a nonzero multiple of $e_{p-2} \otimes x^p$.

Note that Lemma 4.3 is an immediate consequence of 6.3, 6.4 and the external versions of 3.3(iii), 3.3(v), and 3.6(iv).

When p is odd, Proposition 6.3 is Corollary 3.2 of [77]. We shall give a different proof, using the methods of Section 1, which also works for $p = 2$. First observe that there are two equivalent ways of constructing the spectral sequence $E^r_{**}(\pi;X)$; one can either apply mod p K-theory to the filtration of D_pX given in Section I.2 or one can apply mod p stable homotopy to the corresponding filtration of $K \wedge D_pX$. The latter procedure has the advantage that the map

$$\bar{D}_\pi f : D_\pi Y \to K \wedge D_\pi X$$

induced by any map $f : Y \to K \wedge X$ clearly gives rise to a homomorphism

$$(\bar{D}_\pi f)_{**} : E^r_{**}(\pi;Y) \to E^r_{**}(\pi;X)$$

of spectral sequences.

<u>Lemma 6.5.</u> If $\pi = Z_p$ or Σ_p and $y \in K_*(Y;1)$ (with $\beta y = 0$ if $p = 2$) then $(\bar{D}_\pi f)_{**}(e_q \otimes y^p) = e_q \otimes (f_{**}y)^p$.

<u>Proof of 6.5.</u> It suffices to consider the case $\pi = Z_p$. The composite

$$D_\pi X = D_\pi(X \wedge S^0) \xrightarrow{\delta} D_\pi X \wedge D_\pi S^0$$

induces a coproduct

$$\Psi : E^r_{**}(\pi;X) \to E^r_{**}(\pi;X) \otimes E^r_{**}(\pi;S^0)$$

and we have

$$\Psi \circ (\bar{D}_\pi f)_{**} = [(\bar{D}_\pi f)_{**} \otimes 1] \circ \Psi.$$

The lemma clearly holds for $q = 0$, and it follows for all q since the component of $\Psi(e_q \otimes y^p)$ in $E^2_{0*}(\pi;Y) \otimes E^2_{q*}(\pi;S^0)$ is $(e_0 \otimes y^p) \otimes e_q$.

Proof of 6.4. (i) Let x be represented by $f: M_2 \to K \wedge X$. Then $f_{**}u_2 = x$, $(\bar{D}_p f)_{**}Qu_2 = Qx$, and $(\bar{D}_p f)_{**}(e_{p-2} \otimes u_2^p) = e_{2p-2} \otimes x^p$. Hence we may assume that $X = M_2$ and $x = u_2$, and it suffices to show that $v_1 = Qu_2$ is not in the image of

$$K_0(M_2^{(p)};1) \to K_0(D_p M_2;1).$$

But this is clear since $(D_p j)_* v_1 = v$.

Part (ii) is similar. For part (iii) we may assume that $X = \Sigma M_1$ and $x = \Sigma u_1$. In this case it suffices to show that $Q\beta_2 p_* u_1$ is nonzero. But $\beta_2 p_* u_1 = i_* u$, where $u \in K_0(S^0;2)$ is the unit, and $Qu = v$. Hence $Q\beta_2 p_* u_1 = (D_p i)_* v$ is nonzero by 6.1(iii).

Proof of 6.3. First let $p = 2$. Since every element of $\ker \beta$ lifts to $K_*(X;2)$, Proposition 6.3 will be a consequence of the following facts.

(a) $d_2 = 0$

(b) $d_3(e_{2q-\alpha-1} \otimes (\pi x)^2) = e_{2q-\alpha-4} \otimes (\pi_{\beta_2} x)^2$

(c) $d_3(e_{2q-\alpha} \otimes (\pi x)^2) = e_{2q-\alpha-3} \otimes [(\pi x)^2 + (\pi\beta_2 x)^2]$.

Note that, when $\beta_2 x \neq 0$, formulas (b) and (c) differ from those given in [99, 3.8(a)(ii)].

First consider the case $X = S^0$. Then the spectral sequence of I.2.4 is isomorphic to the Atiyah-Hirzebruch spectral sequence, so that (a), (b) and (c) hold in this case by 5.1.

Next we need the coproduct Ψ defined in the proof of 6.5. this has the form

$$\Psi(e_q \otimes x^2) = \sum_{i=0}^{q} (e_i \otimes x^2) \otimes e_{q-i},$$

and it follows that if x and y satisfy

$$d_3(e_3 \otimes x^2) = e_0 \otimes y^2$$

then we also have

$$d_3(e_{2s+1} \otimes x^2) = e_{2s-2} \otimes y^2$$

and

$$d_3(e_{2s+2} \otimes x^2) = e_{2s-1} \otimes [y^2 + x^2]$$

for all $s \geq 1$.

Now let $X = S^1$. In this case $d_2 = 0$ for dimensional reasons, and there are only two possibilities for d_3 consistent with the coproduct, namely

$$d_3(e_{2q} \otimes (\Sigma u)^2) = e_{2q-3} \otimes (\Sigma u)^2$$

or
$$d_3(e_{2q-1} \otimes (\Sigma u)^2) = e_{2q-4} \otimes (\Sigma u)^2.$$

Only the second is consistent with 5.1, and hence (b) and (c) hold in this case.

Next observe that, by 6.5, d_2 vanishes in general if it does for M_2 and ΣM_2. In each of these cases, d_2 is zero for dimensional reasons except on $E^2_{2,0}$, and the only element that could be hit is $(\pi\Sigma^\alpha u_2)(\pi\beta_2\Sigma^\alpha u_2)$ in $E^2_{0,1}$. But the corresponding element of $K_1(D_2\Sigma^\alpha M_2;1)$ is nonzero since its transfer it nonzero in $K_1((\Sigma^\alpha M_2)^{(2)})$. Hence $d_2 = 0$.

Finally, (b) and (c) will hold for all x if they hold for $x = u_2$ and $x = \Sigma u_2$. First consider Σu_2. It suffices to show that

$$d_3(e_3 \otimes (\pi u_2)^2) = (\pi u_2)^2 + (\pi\beta_2 u_2)^2.$$

From inspection of the maps

$$E^3_{**}(Z_2;S^0) \to E^3_{**}(Z_2;\Sigma M_2)$$

and
$$E^3_{**}(Z_2;\Sigma M_2) \to E^3_{**}(Z_2;S^1)$$

we see that $d_3(e_3 \otimes (\pi\beta_2\Sigma u_2)^2)$ is zero and that $d_3(e_3 \otimes (\pi\Sigma u_2)^2)$ projects to $(\Sigma u)^2$ in $E^3_{0,0}(Z_2;S^1)$. Hence

$$d_3(e_3 \otimes (\pi\Sigma u_2)^2) = (\pi\Sigma u_2)^2 + \varepsilon(\pi\beta_2\Sigma u_2)^2$$

for some $\varepsilon \in Z_2$ and there are no further differentials. But by the external version of 3.3(x) we have $\iota_*(\pi\Sigma u_2)^{(2)} = \iota_*(\pi\beta_2\Sigma u_2)^{(2)}$ in $K_0(D_2\Sigma M_2;1)$, hence $\varepsilon = 1$ as required.

It remains to show that

$$d_3(e_3 \otimes (\pi u_2)^2) = (\pi\beta_2 u_2)^2.$$

For this we use the map

$$\Psi':E^r_{**}(Z_2;\Sigma M_2) \to E^r_{**}(Z_2;S^1) \otimes E^r_{**}(Z_2;M_2)$$

induced by

$$\delta:D_2\Sigma M_2 \to D_2S^1 \wedge D_2M_2 .$$

We have

$$\Psi'(e_q \otimes (\pi\Sigma u_2)^2) = \sum_{i=0}^{q} (e_i \otimes (\pi\Sigma u)^2) \otimes (e_{q-i} \times (\pi u_2)^2)$$

and therefore

$$d_3\Psi'(e_3 \otimes (\pi\Sigma u)^2) = (e_0 \otimes (\pi\Sigma u)^2) \otimes [d_3(e_3 \otimes (\pi u_2)^2) + e_0 \otimes (\pi u_2)^2]$$

while $\Psi' d_3(e_3 \otimes (\pi\Sigma u)^2) = (e_0 \otimes (\pi\Sigma u)^2) \otimes [e_0 \otimes (\pi u_2)^2 + e_0 \otimes (\pi\beta_2 u_2)^2]$

and the result follows.

Next let p be odd. We must show the following

(a) $d_i = 0$ for $i \leq p-2$

(b) $d_{p-1}(e_q \otimes x^p) = e_{q+1-p} \otimes (\beta x)^p$

(c) $d_i = 0$ for $p \leq i \leq 2p-2$

(d) $d_{2p-1}(e_q \otimes x^p) = e_{q+1-2p} \otimes x^p$

(e) $d_i = 0$ for $i \geq 2p$.

As before, when $X = S^0$ the spectral sequence is isomorphic to the Atiyah-Hirzebruch spectral sequence so that (a)-(e) hold for 5.1. They also hold for $X = S^1$ by 5.1 and the coproduct. Now 6.5 implies that (a) and (b) will hold for all X if they do for $X = M_1$ and $X = \Sigma M_1$. Inspection of the maps

$$E^r_{**}(\Sigma_p;S^{\alpha-1}) \to E^r_{**}(\Sigma_p;\Sigma^\alpha M_1)$$

and

$$E^r_{**}(\Sigma_p,\Sigma^\alpha M_1) \to E^r_{**}(\Sigma_p;S^\alpha)$$

and the coproduct shows in each case that <u>either</u> (a) and (b) hold <u>or</u> (a),(c),(d), and (e) hold with $d_{p-1} = 0$. Only the former gives an E_∞ term compatible with 6.1(i). Hence (a) and (b) hold for all x.

Now applying 6.5 again we see that (c), (d) and (e) will hold in general if they hold for M_2 and ΣM_2. But one can see that they do by inspection of the maps

$$E^r_{**}(\Sigma_p;S^{\alpha-1}) \to E^r_{**}(\Sigma_p;\Sigma^\alpha M_2)$$

and

$$E^r_{**}(\Sigma_p;\Sigma^\alpha M_2) \to E^r_{**}(\Sigma_p;S^\alpha),$$

and the proof is complete.

7. <u>Construction and properties of Q.</u>

In this section we complete the construction of Q and prove external and internal versions of Theorem 3.3.

As in section 6, we shall construct Q by specifying elements $v_{r-1} \in K_0(D_pM_r;r-1)$ and $v'_r \in K_1(D_p\Sigma M_r;r-1)$. In order to do this we need a stronger version of 6.1.

<u>Lemma 7.1.</u> Let $r \geq 2$. The maps

$$(D_pj)_* : K_0(D_pM_r;r-1) \to K_0(D_pS^0;r-1)$$

$$(D_pj)_* \oplus \tau_*: K_1(D_p\Sigma M_r;r-1) \to K_1(D_pS^1;r-1) \oplus K_1((\Sigma M_r)^{(p)};r-1)^{\Sigma_p}$$

and $$(D_pi)_*: K_0(D_pS^0;r-1) \to K_0(D_p\Sigma M_r;r-1)$$

are isomorphisms, and the sequence

$$0 \longrightarrow K_1(D_pS^1;r-1) \xrightarrow{(D_pi)_*} K_1(D_pM_r;r-1) \xrightarrow{\tau_*} K_1((M_r)^{(p)};r-1)^{\Sigma_p} \longrightarrow 0$$

is exact.

Note that the terms in 7.1 which involve iterated smash products may be calculated by using 5.8. Assuming 7.1 for the moment we may define v_{r-1} and v'_{r-1} by the equations $(D_pj)_*v_{r-1} = v$, $(D_pj)_*v'_{r-1} = v'$, and $\tau_*v'_{r-1} = 0$.

Definition 7.2. $Q: K_\alpha(X;r) \to K_\alpha(D_pX;r-1)$ is the operation $Q_{v_{r-1}}$ if $\alpha = 0$ and $Q_{v'_{r-1}}$ if $\alpha = 1$.

Observe that v_{r-1}, v, v'_{r-1} and v' are equal respectively to Qu_r, Qu, $Q\Sigma u_r$, and $Q\Sigma u$. From now on we shall always use the latter notations for these elements.

We shall prove 7.1 by showing that $E^1 = E^{r-1}$ in the K-theory BSS for $D_p\Sigma^nM_r$ when $r \geq 2$. For this we shall require a formula for the Bockstein of the external Q-operation, and this in turn depends on the other formulas collected in the following lemma.

Lemma 7.3. Let $x,y \in K_\alpha(X;r)$ with $r \geq 2$.

(i) $$\tau_*Qx = \begin{cases} 0 & \text{if } \alpha = 1 \\ -(p-1)!\,\pi x^{(p)} & \text{if } \alpha = 0 \text{ and } p \text{ is odd} \\ -\pi x^{(2)} + \omega 2^{r-2}\pi(\beta_r x)^{(2)} & \text{if } \alpha = 0 \text{ and } p = 2\,. \end{cases}$$

Here ω Z_2 is independent of x.

(ii) $\pi Qx = Q\pi x$ if $r \geq 3$.

(iii) $$Q(x+y) = \begin{cases} Qx + Qy - \pi\iota_*[\sum_{i=1}^{p-1} \frac{1}{p}\binom{p}{i} x^{(i)} \otimes y^{(p-i)}] & \text{if } \alpha = 0 \text{ and } p \text{ is odd} \\ Qx + Qy - \pi\iota_*(x \otimes y) + \omega 2^{r-2}\pi\iota_*[(\beta_r x) \otimes \beta_r y)] & \text{if } \alpha = 0 \text{ and } p = 2 \\ Qx + Qy & \text{if } \alpha = 1. \end{cases}$$

(iv) Let $k \in Z$. Then

$$Q(kx) = \begin{cases} kQx - \frac{1}{p}(k^p - k)\pi\iota_* x^{(p)} & \text{if } \alpha = 0 \\ kQx & \text{if } \alpha = 1. \end{cases}$$

(v) $$\Delta_*\Sigma Qx = \begin{cases} Q\Sigma x & \text{if } \alpha = 0 \\ \pi\iota_*(\Sigma x)^{(p)} + pQ\Sigma x & \text{if } \alpha = 1. \end{cases}$$

(vi) $$\beta_{r-1}Qx = \begin{cases} Q\beta_r x - p\pi\iota_*(x^{(p-1)} \otimes \beta_r x) & \text{if } \alpha = 0 \\ \pi\iota_*(\beta_r x)^{(p)} + pQ\beta_r x & \text{if } \alpha = 1. \end{cases}$$

The constant ω in parts (i) and (iii) will turn out to be 1, as required for 3.3(vi). In order to avoid circularity, we shall prove 7.1 and 7.3 by a simultaneous induction. More precisely, we shall assume that 7.1 holds for $r \leq r_0$ and that 7.3 holds for $r < r_0$ (vacuously if $r_0 = 2$) and then prove 7.3 for $r = r_0$ and 7.1 for $r = r_0 + 1$. Before beginning, we need two technical lemmas.

Lemma 7.4. Let $Y \xrightarrow{f} Z \xrightarrow{g} Cf \xrightarrow{h} \Sigma Y$ be a cofiber sequence in $\bar{h}\mathcal{S}$ and let $r \geq 2$. Suppose that β_{r-1} vanish on $K_1(Z;r-1)$. Let $y \in K_1(\Sigma Y;2r-2)$, $z \in K_0(Z;r-1)$ and $w \in K_1(Cf;r-1)$ be any elements satisfying $\pi^{r-1}y = h_*w$ and $p_*^{r-1}(\Sigma z) = f_*y$. Then $\beta_{r-1}w = g_*z$.

Proof Consider the following diagram in $\bar{h}\mathcal{S}$.

$$\begin{array}{ccccccc} K\wedge Cf & \xrightarrow{1\wedge h} & K\wedge\Sigma Y & \xrightarrow{1\wedge\Sigma f} & K\wedge\Sigma Z & \xrightarrow{1\wedge\Sigma g} & K\wedge\Sigma Cf \\ \uparrow w & & \uparrow \Sigma y & & \vdots\, \zeta & & \uparrow \Sigma w \\ \Sigma M_{r-1} & \longrightarrow & \Sigma M_{2r-2} & \longrightarrow & \Sigma M_{r-1} & \longrightarrow & \Sigma^2 M_r \end{array}$$

Here the bottom row is the evident cofiber sequence, with the first map induced by the inclusion $Z_{p^{r-1}} \subset Z_{p^{2r-2}}$ and the second by the projection $Z_{p^{2r-2}} \to Z_{p^{r-1}}$. Precomposition with the first, second, and third maps in this sequence induces the transformations π^{r-1}, p_*^{r-1} and (because of the suspension) $-\beta_{r+1}$, respectively. The left-hand square commutes up to homotopy since $\pi^{r-1}y = h_*w$. Hence there exists an element ζ making the other two squares commute, and we have $-\beta_{r-1}\Sigma w = (\Sigma g)_*\zeta$. Now the map

$$\Sigma z : \Sigma M_{r-1} \to K \wedge \Sigma Z$$

makes the middle square commute, hence $\zeta - \Sigma z$ restricts trivially to ΣM_{2r-2}. Thus $\zeta - \Sigma z$ extends to a map

$$\xi : \Sigma^2 M_r \to K \wedge \Sigma Z$$

with $\beta_{r-1}\xi = \zeta - \Sigma z$. Since β_{r-1} vanishes on $K_0(\Sigma Z;r-1)$ we have $\zeta = \Sigma z$. Thus $-\beta_{r-1}\Sigma w = \Sigma(g_* z)$ and the result follows.

Lemma 7.5. If $f:X \to K\wedge Y$ is any map then f_{**} commutes with π, β_r, p_* and Σ.

The proof of 7.5 is trivial. Before proceeding we use 7.5 to dispose of 3.2(iv).

Proof of 3.2(iv). For any $x \in K_*(X;r-1)$ and y $K_*(Y;r)$ there exist maps $f:\Sigma^{|x|}M_{r-1} \to K\wedge X$ and $g:\Sigma^{|y|}M_r \to K\wedge Y$ with $f_{**}\Sigma^{|x|}u_{r-1} = x$ and $g_{**}\Sigma^{|y|}u_r = y$. Thus by 7.5 and 1.3(ii) we may assume $X = \Sigma^{|x|}M_{r-1}$ and $Y = \Sigma^{|y|}M_r$ with $x = \Sigma^{|x|}u_{r-1}$ and $y = \Sigma^{|y|}u_r$. By 3.2(vi) we may assume $|x| = |y| = 0$. Clearly the set

$$\{u_{r-1} \otimes \pi u_r,\ u_{r-1} \otimes \pi\beta_r u_r\}$$

is a subbasis for $M_{r-1}\wedge M_r$. Hence by 3.9 we have

$$(1)\qquad (p_* u_{r-1}) \otimes u_r = a_1 p_*(u_{r-1} \otimes \pi u_r) + a_2\beta_r p_*(u_{r-1} \otimes \pi\beta_r u_r)$$

for some $a_1, a_2 \in Z_{p^{r-1}}$. Applying π to each side gives

$$\begin{aligned} pu_{r-1} \otimes \pi u_r &= a_1 pu_{r-1} \otimes \pi u_r + a_2\beta_{r-1}(u_{r-1} \otimes \pi\beta u_r) \\ &= a_1 pu_{r-1} \otimes \pi u_r + a_2\beta_{r-1}u_{r-1} \otimes \pi\beta_r u_r \,. \end{aligned}$$

Hence $a_2 = 0$. Now applying $(j\wedge j)_*$ to each side of equation (1) gives

$$p(u \otimes u) = a_1 p_*(u \otimes u) = a_1 p(u \otimes u)$$

in $K_0(D_pS\wedge D_pS;r) \cong Z_{p^r}$. Hence $a_1 = 1$ in $Z_{p^{r-1}}$.

Next we give the proof of 7.3 for $r = r_0$. The proof of each part will be quite similar to that just given for 3.2(iv). First we observe that by 1.3, 1.4, 1.5 and 7.5 we may assume in each part except (iii) that X is $\Sigma^\alpha M_r$ and that x is the fundamental class $\Sigma^\alpha u_r$.

(i). If $\alpha = 1$ the result holds by Definition 7.2. Suppose $\alpha = 0$ and consider the map

$$j_*^{(p)}:K_0(M_r^{(p)};r-1)^{\Sigma_p} \to K_0(S^0;r-1).$$

This is monic when p is odd and has kernel generated by $2^{r-2}\pi(\beta_r u_r)^{(2)}$ when $p = 2$. The result follows since $j_*^{(p)}u_r^{(p)} = u \in K_0(S^0;r)$ and

$$j_*^{(p)}\tau_* Qu_r = \tau_*(D_p j)_* Qu_r = \tau_* Qu = -(p-1)!u;$$

the last equality is 5.7(iv).

(ii). Let $\alpha = 1$. By 7.1 it suffices to show that

$$(D_p j)_*\pi Q\Sigma u_r = (D_p j)_* Q\pi\Sigma u_r$$

and that

$$\tau_*\pi Q\Sigma u_r = \tau_* Q\pi\Sigma u_r \ .$$

This second equation follows from part (i) and the first from 5.7(i). The case $\alpha = 0$ is similar.

(iii). Let $\alpha = 0$ with p odd. By 1.3, 1.4 and 7.5 we may assume that X is $M_r \vee M_r$ with x and y being the fundamental classes of the two summands. Let

$$F: \bigvee_{i=0}^{p} D_i M_r \wedge D_{p-i} M_r \to D_p(M_r \vee M_r)$$

be the equivalence of II.1.1 and let $f: M_r \to M_r \vee M_r$ be the pinch map. Then $(D_p f)_* Qu_r = Q(x + y)$, and it suffices to show that

$$F_*^{-1}(D_p f)_* Qu_r = Qu_r \otimes u + u \otimes Qu_r - \sum_{i=1}^{p-1} \frac{1}{p}\binom{p}{i}\pi\iota_* u_r^{(i)} \otimes \pi\,\iota_* u_r^{(p-i)}$$

since F_* applied to the right side of this equation clearly gives the right side of the desired formula. Now the projection of $F^{-1}\circ D_p f$ on the i-th wedge summand is the transfer

$$\tau_{i,p-i} : D_p M_r \to D_i M_r \wedge D_{p-i} M_r \ .$$

When i is 0 or p this transfer is the evident natural equivalence, hence it suffices to show

(2) $$(\tau_{i,p-i})_* Qu_r = -\frac{1}{p}\binom{p}{i}\pi\iota_* u_r^{(i)} \otimes \pi\iota_* u_r^{(p-i)}$$

for $0 < i < p$. Now the transfer

$$\tau'_{i,p-i}: D_i M_r \wedge D_{p-i} M_r \to M_r^{(p)}$$

induces a monomorphism since the order of $\Sigma_i \times \Sigma_{p-i}$ is prime to p for $0 < i < p$. We have

$$(\tau'_{i,p-i})_*(\tau_{i,p-i})_* Qu_r = \tau_* Qu_r = -(p-1)!u_r^{(p)}$$

by part (i) while

$$(\tau'_{i,p-i})_*[\pi\iota_* u_r^{(i)} \otimes \pi\iota_* u_r^{(p-i)}] = i!(p-i)!u_r^{(p)}$$

by the double coset formula. Equation (2) follows. The proof when $p = 2$ or $\alpha = 1$ is similar.

Part (iv) follows from (iii) by induction on k. When $p = 2$ and $\alpha = 0$ we need to know that $2^{r-2}\pi\iota_*(\beta_r x)^{(2)} = 0$. If $r > 2$ this is evident since $\iota_*(\beta_r x)^{(2)}$ has order 2 by 3.2(viii). If $r = 2$ then by 6.4(iii) we have

$$\iota_*(\pi\beta_2 x)^{(2)} = Q\beta_2 2_* \pi\beta_2 x = 0.$$

(v). Let $\alpha = 0$. By 7.1 is suffices to show

$$(\Sigma D_p j)_* \Delta_* \Sigma Q u_r = Q\Sigma u$$

and
$$\tau_* \Delta_* \Sigma Q u_r = 0.$$

The first equation is immediate from 7.2 and 5.7(ii). For the second, consider the diagram

$$\begin{array}{ccc} S^1 \wedge D_p M_r & \xrightarrow{\ \Delta\ } & D_p(S^1 \wedge M_r) \\ \Big\downarrow{\scriptstyle 1\wedge\tau} & & \Big\downarrow{\scriptstyle t} \\ S^1 \wedge M_r^{(p)} & \xrightarrow{\ \Delta'\ } & (S^1 \wedge M_r)^{(p)} \end{array} .$$

Here the map Δ' is induced by the diagonal of S^1. By definition, the map Δ is obtained by aplying the functor $E\Sigma_p^+ \wedge_{\Sigma_p} (\ \ )$ to the map of Σ_p-spectra

$$S_1 \wedge (M_r)^{(p)} \to (S_1 \wedge M_r)^{(p)}$$

induced by the diagonal of S^1. Hence the diagram commutes by naturality of τ. But the diagonal map of S^1 is nonequivariantly trivial, hence $\tau_*\Delta_* \Sigma Q u_r = 0$ as required. The proof when $\alpha = 1$ is similar.

(vi). Suppose first that $\alpha = 1$. Consider the following diagram

$$\begin{array}{ccccccc} D_pS & \xrightarrow{\ D_pf\ } & D_pS & \xrightarrow{\ g\ } & C & \xrightarrow{\ h\ } & \Sigma D_pS \\ & & & \searrow{\scriptstyle D_pi} & \Big\downarrow{\scriptstyle \gamma} & & \Big\downarrow{\scriptstyle \Delta} \\ & & & & D_p\Sigma M_r & \xrightarrow{\ D_pj\ } & D_pS^1 \end{array}$$

Here $f:S \to S$ has degree p^r and the top row is the cofiber sequence of D_pf. The map γ is that constructed in II.3.8, where it was called ψ, and the diagram commutes. For any $s \geq 1$ the map

$$(D_pf)_*:K_0(D_pS;s) \to K_0(D_pS;s)$$

is given by the formula $(D_pf)_*\eta = p^{pr}\eta$ and

$$(D_pf)_*Qu = Q(p^r u) = p^r Qu - (p^{p^{r-1}} - p^{r-1})\eta$$

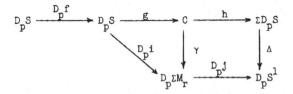

In particular, when $s = r-1$ the map $(D_pf)_*$ is zero, and since $K_1(D_ps;r-1) = 0$ we see that

$$h_*:K_1(C;r-1) \to K_1(\Sigma D_pS;r-1)$$

is an isomorphism. Thus there is a unique $w \in K_1(C;r-1)$ with $h_*w = \Sigma Qu$. Letting

$$y = \Sigma Qu \in K_1(*\Sigma D_pS;2r-2)$$

and $$z = pQu + \eta \in K_0(D_pS;r-1)$$

we have $\pi^{r-1}y = h_*w$ and $p_*^{r-1}\Sigma z = (D_pf)_*y$, hence by Lemma 7.4 we conclude that $\beta_{r-1}w = g_*z$ in $K_1(C;r-1)$.

Next we shall show that $\gamma_*w = Q\Sigma u_r$. Assuming this for the moment, we have

$$\beta_{r-1}Q\Sigma u_r = \gamma_*\beta_{r-1}w = \gamma_*g_*zx = (D_pi)_*z = pQ\beta_ru_r + \pi\iota_*(\beta_ru_r)^{(p)}$$

which gives (vi) when $\alpha = 1$. To show $\gamma_*w = Q\Sigma u_r$, we must show that $(D_pj)_*\gamma_*w = Q\Sigma u$ and $\tau_*\gamma_*w = 0$. The first equation is immediate from the diagram and part (v). For the second, we observe that D_pf and γ are obtained by applying $E\Sigma_p^+ \wedge_{\Sigma_p} (\ \ )$ to certain Σ_p-equivariant maps F and Γ, so that by naturality of τ we have the following commutative diagram of nonequivariant spectra.

$$\begin{array}{ccccc}
C & = & E\Sigma_p^+ \wedge_{\Sigma_p} CF & \xrightarrow{\tau} & CF \\
\downarrow{\scriptstyle\gamma} & & \downarrow{\scriptstyle E\Sigma_p^+ \wedge_{\Sigma_p}\Gamma} & & \downarrow{\scriptstyle\Gamma} \\
D_p\Sigma M_r & = & E\Sigma_p^+ \wedge_{\Sigma_p} (\Sigma M_r)^{(p)} & \xrightarrow{\tau} & (\Sigma M_r)^{(p)}
\end{array}$$

Thus it suffices to show $\Gamma_*\tau_* = 0$ on $K_1(C;r-1)$. As a nonequivariant map F is the map $S \to S$ of degree p^{pr}, hence the cofiber CF is nonequivariantly equivalent to ΣM_{pr}. The resulting Σ_p-action is clearly trivial on $K_0(\Sigma M_{pr};pr)$, hence also on $K_1(\Sigma M_{pr};pr)$ since the Bockstein β_{pr} is an isomorphism between these two groups. Thus

$$\Gamma_*:K_1(\Sigma M_{pr};pr) \to K_1((\Sigma M_r)^{(p)};pr)$$

lands in the Σ_p-invariant subgroup. We claim that this subgroup is generated by the element

$$p_*^{pr-r}[(\Sigma u_r) \otimes (\beta_r\Sigma u_r)^{p-1}]$$

when p is odd and by this element together with

$$2^{r-1}\beta_{2r}2_*^r[(\Sigma u_r) \otimes (\Sigma u_r)]$$

when $p = 2$. From this it will follow that π^{pr-r+1} vanishes on this subgroup and therefore that Γ_* vanishes on $K_1(\Sigma M_{pr};r-1)$, since π^{pr-r+1} maps onto the latter group; thus we will have shown $\Gamma_*\tau_* w = 0$ as required. To verify the claim we observe that the set

$$\{\Sigma u_r \otimes x_2 \otimes \cdots \otimes x_p \mid x_i = \Sigma u_r \text{ or } \beta_r \Sigma u_r\}$$

is a subbasis for $(\Sigma M_r)^{(p)}$. Using the basis for $K_1((\Sigma M_r)^{(p)};pr)$ given by 3.9, we see at once that the elements

$$z_1 = p_*^{pr-r}[(\Sigma u_r) \otimes (\beta_r \Sigma u_r)^{(p-1)}]$$

and
$$z_2 = \beta_{pr} p_*^{pr-r}[\Sigma u_r \otimes [\sum_{i=1}^{p-1} (\beta_r \Sigma u_r)^{(i)} \otimes \Sigma u_r \otimes (\beta_r \Sigma u_r)^{(p-i-1)}]]$$

are a basis for the $\Sigma_1 \times \Sigma_{p-1}$ invariant subgroup. Now if T is the map switching the first two factors of $(\Sigma M_r)^{(p)}$ we have $T_* z_1 = z_1$ and

$$T_* z_2 = z_2 - 2\beta_{pr} p_*^{pr-r}[(\Sigma u_r)^{(2)} \otimes (\beta_r \Sigma u_r)^{(p-2)}];$$

the claim follows.

Finally, we must prove part (vi) with $\alpha = 0$. By 7.1 we have

$$\beta_{r-1} Q u_r = a_1 Q \beta_r u_r + a_2 \pi \iota_*(u_r^{(p-1)} \otimes \beta_r u_r) \tag{3}$$

for some $a_1, a_2 \in Z_{p^{r-1}}$. Applying $\Delta_*\Sigma$ and using part (v) gives

$$\beta_{r-1} Q \Sigma u_r = a_1[\pi \iota_*(\beta_r \Sigma u_r)^{(p)} + pQ\beta_r \Sigma u_r].$$

Comparing this with the case $\alpha = 1$ of (vi) gives $a_1 = 1$. Now applying τ_* to (3) and using part (i) gives

$$-(p-1)!(\beta_{r-1}\pi(u_r^{(p)}) = a_2(p-1)!\pi[\sum_{i=0}^{p-1} u_r^{(i)} \otimes \beta_r u_r \otimes u_r^{(p-i-1)}].$$

But $\beta_{r-1}\pi(u_r^{(p)}) = p\pi\beta_r(u_r^{(p)})$ and it follows that $a_2 = -p$ as required.

This completes the case $r = r_0$ of 7.3. Next we must show 7.1 for $r = r_0+1 \geq 3$. It suffices to show that $E^1 = E^{r-1}$ in the K-theory BSS for $D_p M_r$ and $D_p \Sigma M_r$. We shall give the proof for $D_p M_r$, the other case being similar. Let x and y denote the elements πu_r and $\pi\beta_r u_r$. by 6.1,7.2 and 7.3(ii) we see that the set

$$\{\pi^{r-2}\iota_* x^{(p)}, \pi^{r-3}Qx, \pi^{r-2}\iota_*(x^{(p-1)} \otimes y), \pi^{r-3}Qy\}$$

is a basis for $K_*(D_p M_r;1)$. Since all elements of this basis lift to $K_*(D_p M_r;r-2)$ we have $E^1 = E^{r-2}$ in the BSS. The elements $\pi^{r-2}x^{(p)}$ and $\pi^{r-2}(x^{(p-1)} \otimes y)$ are $(r-2)$-

cycles since they clearly lift to $K_0(D_pM_r;r-1)$. Next we have

$$d_{r-2}\pi^{r-3}Qx = \pi^{r-3}\beta_{r-2}Qx = \pi^{r-3}Q\beta_{r-1}x = \pi^{r-3}Qpy = 0,$$

where the 2nd and 4th equalities follow from 7.3(vi) and 7.3(iv) respectively. Similarly,

$$d_{r-2}\pi^{r-3}Qy = \pi^{r-3}\beta_{r-2}Qy = \pi^{r-2}(\beta_{r-1}y)^{(p)} = 0.$$

This completes the inductive proof of 7.1 and 7.3.

Next we shall prove the external version of 3.3. Rather than write out the complete list of external properties, we give rules for changing the internal statements to their external analogs. All internal products and Dyer-Lashof operations are to be changed to external ones, with the map ι_* prefixed to any p-fold product which is to lie in $K_*(D_pX;r)$. The map δ_* is to be prefixed to the left-hand side of each Cartan formula. In the stability formulas, σ is to be changed to Σ and Δ_* prefixed to the left-hand side. These conventions give the correct external analog for each part of 3.3 except for part (ii) which has no external analog.

Proposition 7.6. The external Q-operation satisfies the external versions of each part of Theorem 3.3 except part (ii).

Before beginning the proof we need a lemma to deal with the prime 2. (See II.4.3 for another proof of this lemma.)

Lemma 7.7. Let X be any spectrum. The sequence

$$\Sigma D_2X \xrightarrow{\Delta} D_2\Sigma X \xrightarrow{\tau} \Sigma^2(X\wedge X) \xrightarrow{\Sigma^2\iota} \Sigma^2 D_2X$$

is a cofibering.

Proof. Consider the cofiber sequence

(4) $$S^1 \xrightarrow{\Delta} S^1\wedge S^1 \longrightarrow S^2\wedge S^2 \longrightarrow S^2$$

of Z_2-spaces. Here Z_2 acts trivially on the first and fourth terms and by switching factors (respectively, wedge summands) in the second and third terms. Now $S^1\wedge S^1$ is the one-point compactification S^V of the regular representation V of Z_2, and it is easy to see that the second map in the sequence (4) stabilizes to the transfer $S^V \to Z_2^+\wedge S^V$. The sequence of the lemma is obtained by applying the functor $EZ_2^+\wedge_{Z_2}(?\wedge X\wedge X)$ to the sequence (4).

Next we turn to the proof of 7.6. Part (i) is trivial and parts (iii), (v) and (viii) are contained in 7.3.

(iv). We may assume $X = \Sigma^{\alpha}M_r$, $x = \Sigma^{\alpha}u_r$. Suppose $\alpha = 1$. By 7.1 and 7.3(vi) we see that the set

$$\{Q\Sigma u_r,\ \iota_*[(\Sigma u_r)\otimes(\beta_r\Sigma u_r)^{(p-1)}], Q\beta_{r+1}p_*\Sigma u_r\}$$

is a subbasis of height r for $D_p\Sigma M_r$, hence the set

$$\{p_*Q\Sigma u_r,\ \iota_*[(\Sigma u_r)\otimes(\beta_r\Sigma u_r)^{(p-1)}]\}$$

is a basis for $K_1(D_p\Sigma M_r;r)$. It follows that the map

$$(D_pj)_*\oplus\tau_* : K_1(D_p\Sigma M_r;r)\longrightarrow K_1(D_pS^1;r) + K_1((\Sigma M_r)^{(p)};r)$$

is monic. Now

$$\begin{aligned}(D_pj)_*Qp_*\Sigma u_r &= Q(p_*j_*\Sigma u_r) = Q(p\Sigma u) = pQ\Sigma u\\ &= \begin{cases}0 & \text{if } r=1\\ (D_pj)_*p_*Q\Sigma u_r & \text{if } r\geq 2,\end{cases}\end{aligned}$$

and $\tau_*Qp_*\Sigma u_r = 0$ for all r. The result follows, and the case $\alpha = 0$ is similar.

Next we prove part (x). The proof is by induction on r. If r = 1 we have $\iota_*x^{(2)} = Q\beta_2 2_*x$ by 6.4(iii). Suppose $r \geq 2$. We may assume $x = \Sigma u_r$. The set

$$\{Q\Sigma u_r, \iota_*(\Sigma u_r\otimes\beta_r\Sigma u_r), Q\beta_{r+1}2_*\Sigma u_r\}$$

is a subbasis of height r for $D_2\Sigma M_r$, hence by 3.9 we have

$$\iota_*(\Sigma u_r)^{(2)} = a_1\beta_r 2_*Q\Sigma u_r + a_2 Q\beta_{r+1}2_*\Sigma u_r \tag{5}$$

with $a_1 \in Z_{2^{r-1}}$ and $a_2 \in Z_{2^r}$. Applying τ_* to (5) gives

$$0 = -a_2(\beta_r\Sigma u_r)^{(2)}$$

hence $a_2 = 0$. Now applying π to (5) gives

$$\iota_*(\pi\Sigma u_r)^{(2)} = a_1\beta_{r-1}Q\Sigma u_r. \tag{6}$$

If r = 2 the inductive hypothesis gives

$$\iota_*(\pi\Sigma u_{2'})^{(2)} = Q\beta_2 2_*(\pi\Sigma u_2) = Q(2\beta_2\Sigma u_2) = \pi\iota_*(\beta_2\Sigma u_2)^{(2)} = \beta Q\Sigma u_2$$

(where the third and fourth equalities follow from 7.3(iv) and 7.3(vi)) and we conclude that $a_1 = 1$ as required. If $r \geq 3$ the inductive hypothesis gives

$$\iota_*(\pi\Sigma u_r)^{(2)} = 2^{r-3}\beta_{r-1}2_*Q(\pi\Sigma u_r) = 2^{r-2}\beta_{r-1}Q\Sigma u_r$$

and comparing with (6) gives $a_1 = 2^{r-2}$ as required.

Next we show part (vi). This will follow immediately from 7.3(iii) and 7.3(iv) once we show that $\omega = 1$ in 7.3(i). Letting $X = \Sigma M_r$ in 7.7, we have

$$\begin{aligned} 0 &= (\Sigma^2\iota)_*\tau_*Q\Sigma^2 u_r \\ &= (\Sigma^2\iota)_*\Pi[-(\Sigma^2 u_r)^{(2)} + \omega 2^{r-2}(\beta_r\Sigma u_r)^{(2)}] \\ &= \Sigma^2\pi\iota_*[(\Sigma u_r)^{(2)} + \omega 2^{r-2}(\beta_r\Sigma u_r)^{(2)}]. \end{aligned}$$

By part (ix), we have

$$\pi\iota_*(\Sigma u_r)^{(2)} = 2^{r-2}\beta_{r-1}Q\Sigma u_r = 2^{r-2}\iota_*(\pi\beta_r\Sigma u_r)^{(2)} \neq 0.$$

Hence $\omega \neq 0$ as required.

(vii) Let $p = 2$; the odd primary case is similar and somewhat easier. First let $|x| = |y| = 1$. We may assume $x = \Sigma u_r$, $y = \Sigma u_r$. We assume by induction on r that we have chosen mod 2^s multiplications for $s < r$ such that the desired formula holds. We begin by giving a basis for

$$K_0(D_2\Sigma M_r \wedge D_2\Sigma M_r;r-1).$$

The set

$$\{\pi\iota_*(\Sigma u_r \otimes \beta_r\Sigma u_r), \pi\iota_*(\beta_r\Sigma u_r)^{(2)}, Q\Sigma u_r, Q\beta_r\Sigma u_r\}$$

is a subbasis of height r-1 for $D_2\Sigma M_r$ and in particular it is a basis for $K_*(D_2\Sigma M_r;r-1)$. By 5.8 we have

$$K_*(D_2\Sigma M_r \wedge D_2\Sigma M_r;r-1) \cong K_*(D_2\Sigma M_r;r-1) \otimes K_*(D_2\Sigma M_r;r-1)$$

with the tensor product taken in the Z_2-graded sense. We therefore obtain a basis for $K_*(D_2\Sigma M_r \wedge D_2\Sigma M_r;r-1)$ by taking all 16 external products of the elements in the set given above. It will be convenient to denote Σu_r by x in the first factor and by y in the second factor. Let $a_1,\ldots,a_8 \in Z_{2^{r-1}}$ be the coefficients of $\delta_*Q(x \otimes y)$ with respect to this basis, so that we have

$$\begin{aligned} (7)\qquad \delta_*Q(x\otimes y) &= a_1\pi\iota_*(x\otimes\beta_r x)\otimes\pi\iota_*(y\otimes\beta_r y) + a_2 Qx\otimes\pi\iota_*(y\otimes\beta_r y) \\ &+ a_3\pi\iota_*(x\otimes\beta_r x)\otimes Qy + a_4 Qx\otimes Qy + a_5\pi\iota_*(\beta_r x)^{(2)}\otimes\pi\iota_*(\beta_r y)^{(2)} \\ &+ a_6\pi\iota_*(\beta_r x)^{(2)}\otimes Q\beta_r y + a_7 Q\beta_r x\otimes\pi\iota_*(\beta_r y)^{(2)} + a_8 Q\beta_r x\otimes Q\beta_r y. \end{aligned}$$

We claim first that $2a_5 = 0$, so that a_5 is either 2^{r-2} or 0. When $r = 2$ this is trivial, while for $r \geq 3$ it follows from the inductive hypothesis and the equation $\pi Q(x \otimes y) = Q(\pi x \otimes \pi y)$. Now as in Remark 3.4(iv) we see that changing the choice of mod 2^r multiplication changes the value of a_5 without changing the other a_i. We can therefore choose the mod 2^r multiplication for which $a_5 = 0$. (When p is odd the commutativity of the multiplications gives $a_5 = 0$.)

It remains to determine the other coefficients in equation (7). If we apply the map $(D_2 j \wedge D_2 j)_*$ to this equation, the left side becomes $Q\Sigma u \otimes Q\Sigma u$ by 5.7(vii) while the right side becomes $a_4 Q\Sigma u \otimes Q\Sigma u$. Hence $a_4 = 1$. Next consider the following diagram

$$\begin{array}{ccccc} D_2(X \wedge Y) & & \xrightarrow{\quad\quad\delta\quad\quad} & & D_2X \wedge D_2Y \\ \downarrow \tau & & & & \downarrow \tau \wedge 1 \\ X \wedge Y \wedge X \wedge Y & \xrightarrow{1 \wedge T \wedge 1} & X \wedge X \wedge Y \wedge Y & \xrightarrow{1 \wedge \iota} & X \wedge X \wedge D_2Y \end{array}$$

The commutativity of this diagram will be proved in VI.3.10 of the sequel. With $X = Y = \Sigma M_r$ we obtain

$$(\tau \wedge 1)_* \delta_* Q(x \otimes y) = (1 \wedge \iota)_* (1 \wedge T \wedge 1)_* \tau_* Q(x \otimes y)$$

$$= (1 \wedge \iota)_* (1 \wedge T \wedge 1)_* \pi[-x \otimes y \otimes x \otimes y + 2^{r-2} \beta_r(x \otimes y) \otimes \beta_r(x \otimes y)]$$

$$= (1 \wedge \iota)_* \pi[x^{(2)} \otimes y^{(2)} + 2^{r-2} x^{(2)} \otimes (\beta_r y)^{(2)}$$

$$+ 2^{r-2} \beta_r x \otimes x \otimes y \otimes \beta_r y + 2^{r-2} x \otimes \beta_r x \otimes \beta_r y \otimes y + 2^{r-2} (\beta_r x)^{(2)} \otimes y^{(2)}]$$

$$= \pi x^{(2)} \otimes \pi \iota_* y^{(2)} + 2^{r-2} \pi x^{(2)} \otimes \pi \iota_* (\beta_r y)^{(2)}$$

$$+ 2^{r-2} \pi \tau_* \iota_* (x \otimes \beta_r x) \otimes \pi \iota_* (y \otimes \beta_r y) + 2^{r-2} \pi (\beta_r x)^{(2)} \otimes \pi \iota_* y^{(2)}$$

$$= 2^{r-2} \pi \tau_* \iota_* (x \otimes \beta_r x) \otimes \pi \iota_* (y \otimes \beta_r y) + 2^{2r-4} \pi (\beta_r x)^{(2)} \otimes \pi \iota_* (\beta_r y)^{(2)},$$

with the last equation following from part (x). Now applying $(\tau \wedge 1)_*$ to the right side of (7) and comparing coefficients gives $a_1 = 2^{r-2}$, $a_3 = 0$, $a_7 = 2^{2r-4}$ and $a_8 = 2a_6$. Similarly, applying $(1 \wedge \tau)_*$ to equation (7) gives $a_2 = 0$ and $a_6 = 2^{2r-4}$, whence $a_8 = 2a_6 = 0$. This completes the proof of part (vii) when $|x| = |y| = 1$.

Next let $|x| = 1$, $|y| = 0$. Consider the following commutative diagram

$$\begin{array}{ccc}
\Sigma D_2(X\wedge Y) & \xrightarrow{\ \Sigma\delta\ } & \Sigma D_2X\wedge D_2Y \\
\downarrow{\scriptstyle \Delta} & & \downarrow{\scriptstyle T\wedge 1} \\
D_2(\Sigma X\wedge Y) & & D_2X\wedge \Sigma D_2Y \\
\downarrow{\scriptstyle D_2(T\wedge 1)} & & \downarrow{\scriptstyle 1\wedge\Delta} \\
D_2(X\wedge \Sigma Y) & \xrightarrow{\ \delta\ } & D_2X\wedge D_2\Sigma Y
\end{array}$$

If we let $X = M_r$, $Y = \Sigma^{-1}M_r$ we obtain

(10) $$\delta_*[D_2(T\wedge 1)]_*\Delta_*\Sigma Q(-\Sigma u_r \otimes \Sigma^{-1}u_r) = (1\wedge\Delta)_*(T\wedge 1)_*(\Sigma\delta)_*Q(-\Sigma u_r \otimes \Sigma^{-1}u_r)$$

We can evaluate the left side of (10) using 7.3(v); the result is $\delta_*Q(\Sigma u_r \otimes u_r)$. On the other hand we can evaluate the right side of (10) by using 7.3(v) and the part of 7.6(vii) just shown; the result is

$$Q\Sigma u_r \otimes \pi\iota_* u_r^{(21)} + 2Q\Sigma u_r \otimes Qu_r + 2^{2r-4}\pi\iota_*(\beta_r\Sigma u_r)^{(2)} \otimes Q\beta_r u_r .$$

Thus equation (10) gives the desired formula when $x = \Sigma u_r$ and $y = u_r$, and therfore this formula holds in general.

Finally, let $|x| = |y| = 0$. We may assume $x = u_r$, $y = u_r$. The set

$$\{\pi\iota_* x^{(p)} \otimes \pi\iota_* y^{(p)}, Qx \otimes \pi\iota_* y^{(p)}, \pi\iota_* x^{(p)} \otimes Qy, Qx \otimes Qy,$$
$$\pi\iota_*(x \otimes \beta_r x) \otimes \pi\iota_*(y \otimes \beta_r y), Q\beta_r x \otimes \pi\iota_*(y \otimes \beta_r y),$$
$$\pi\iota_*(x \otimes \beta_r x) \otimes Q\beta_r y, Q\beta_r x \otimes Q\beta_r y\}$$

is a basis for $K_0(D_2M_r \wedge D_2M_r; r-1)$. Let $a_1,\ldots,a_8$ be the coefficients of $\delta_*Q(x\otimes y)$ in this basis. By 5.7(v) we have

$$(D_2j \wedge D_2j)_*\delta_*Q(x\otimes y) = \delta_*Q(u\otimes u) = Qu \otimes \eta + \eta \otimes Qu + pQu \otimes Qu,$$

hence $a_1 = 0$, $a_2 = a_3 = 1$ and $a_4 = 2$. Diagram (8) gives

$$(\tau\wedge 1)_*\delta_*Q(x\otimes y) = (1\wedge\iota)_*\delta_*\tau_*Q(x\otimes y)$$

and it follows that $a_5 = 2^{r-2}$ and $a_6 = 0$. Similarly,

$$(1\wedge\tau)_*\delta_*Q(x\otimes y) = (\iota\wedge 1)_*\delta_*\tau_*Q(x\otimes y)$$

and hence $a_7 = 0$. Thus we have

(11) $\delta_*Q(x \otimes y) = Qx \otimes \pi\iota_* y^{(2)} + \pi\iota_* x^{(2)} \otimes Qy + 2Qx \otimes Qy$

$$+ 2^{r-2}\pi\iota_*(x \otimes \beta_r x) \otimes \pi\iota_*(y \otimes \beta_r y) + a_8 Q\beta_r x \otimes Q\beta_r y$$

and it remains to determine a_8. Consider the following commutative diagram

(12)
$$\begin{array}{ccc} \Sigma D_2(X \wedge Y) & \xrightarrow{\Sigma\delta} & \Sigma D_2 X \wedge D_2 Y \\ \downarrow \Delta & & \downarrow \Delta\wedge 1 \\ D_2(\Sigma X \wedge Y) & \xrightarrow{\delta} & D_2\Sigma X \wedge D_2 Y \end{array}$$

With $X = Y = M_r$ we have

(13) $$(\Delta\wedge 1)_*\Sigma\delta_*Q(x \times y) = \delta_*\Delta_*\Sigma Q(x \otimes y).$$

We evaluate the left side of (13) using 7.3(v) and equation (11); the result is

$$Q\Sigma x \otimes \pi\iota_* y^{(2)} + 2Q\Sigma x \otimes Qy + a_8\pi\iota_*(\beta_r\Sigma x)^{(2)} \otimes Q\beta_r y + 2a_8 Q\beta_r\Sigma x \otimes Q\beta_r y.$$

Evaluating the right side of (13) using 7.34(v) and the part of 7.6(vii) already shown gives

$$Q\Sigma x \otimes \pi\iota_* y^{(2)} + 2Q\Sigma x \otimes Qy + 2^{2r-4}\pi\iota_*(\beta_r\Sigma x)^{(2)} \otimes Q\beta_r y.$$

Hence $a_8 = 2^{2r-4}$ as required.

(ix) We have seen in VIII.7.4 that ψ^k is an H_∞ ring map of $K_{(p)}$ for k prime to p. Hence we have

$$(\bar{D}_p f)_{**}\psi^k = \psi^k(\bar{D}_p f)_{**}: K_*(D_pY;r-1) \to K_*(D_pX;r-1)$$

for any map $f:Y \to K\wedge X$. Thus we may assume $x = \Sigma^\alpha u_r$ with $\alpha = 0$ or 1. First let $\alpha = 0$. Since the map

$$(D_pj)_*: K_0(D_pM_r;r-1) \to K_0(D_pS;r-1)$$

is monic and since $\psi^k u = u$, it suffices to show $\psi^k Qu = Qu$. Dually, it suffices to show that ψ^k is the identity on $K^0(B\Sigma_p;r-1)$. But this is immediate from 5.3 since ψ^k commutes with τ^*. Now, if $\alpha = 1$ we have

$$\psi^k Q\Sigma u_r = \psi^k\Delta_*\Sigma Qu_r = \Delta_*\Sigma\psi^k Qu_r = \Delta_*\Sigma Qu_r = Q\Sigma u_r.$$

This completes the proof of 7.6.

Next we must prove 3.3. Each part of this theorem is in fact an easy consequence of the corresponding external formula except for parts (ii) and (viii). For part (ii) we may clearly assume X = S, and it suffices to show that Qu goes to zero under the nontrivial map from $B\Sigma_p^+$ to S^0. But the induced map

$$\tilde{K}^0(S^0;r) \to \tilde{K}^0(B\Sigma_p;r)$$

takes 1 to 1, and $\langle 1,Qu\rangle = 0$ by Definition 5.6, whence the result follows.

The proof of part (viii) is more difficult. First recall that if X is any nondegenerately based space and $\lambda:X^+ \to X$ is the identity on X then the cofiber sequence

$$\Sigma^\infty S^0 \xrightarrow{\Sigma^\infty \eta} \Sigma^\infty X^+ \xrightarrow{\Sigma^\infty \lambda} \Sigma^\infty X$$

is naturally split by the evident retraction $\mu:X^+ \to S^0$. In particular, there is a natural transformation

$$\nu:\Sigma^\infty X \to \Sigma^\infty X^+$$

and the inclusion

$$\tilde{K}_*(X;r) \subset K_*(X;r)$$

can be identified with ν_*. Now let Y be an H_∞ space, let $Z = \Omega Y$, and let $\varepsilon:\Sigma Z \to Y$ be the counit. Then

$$\sigma:\tilde{K}_\alpha(\Omega Y;r) \to K_{\alpha+1}(Y;r)$$

is the composite $\nu_*\varepsilon_*\Sigma$.

Let $x \in \tilde{K}_0(\Omega Y;r)$; the case $|x| = 1$ is similar. First we must show that Qx is in $\tilde{K}_\alpha(\Omega Y;r-1)$, i.e., that $\mu_*Qx = 0$. But $\mu:\Sigma^\infty(\Omega Y)^+ \to \Sigma^\infty S^0$ is clearly an H_∞ ring map, and therefore $\mu_*Qx = Q\mu_*x = 0$. Next we state the required formula more precisely as follows:

(14) $\qquad \sigma\lambda_*Q\nu_*x = Q\sigma x.$

Since μ_* applied to each side of (14) gives zero, it suffices to show that λ_* makes the two sides of (14) equal, i.e., that

$$\varepsilon_*\Sigma\lambda_*Q\nu_*x = \lambda_*Q\nu_*\varepsilon_*\Sigma x.$$

This in turn follows at once from 7.3(v) and the commutativity of the following diagram in $h\bar{\mathcal{J}}$ (where we suppress Σ^∞ to simplify the notation).

(15)
$$\begin{array}{ccccc}
\Sigma D_pZ & & \xrightarrow{\Delta} & & D_p\Sigma Z \\
 & & & & \downarrow D_p\varepsilon \\
\downarrow \Sigma D_p\nu & & & & D_pY \\
 & & & & \downarrow D_p\nu \\
\Sigma D_pZ^+ & & & & D_p(Y^+) \\
 & & & & \downarrow \xi \\
\downarrow \Sigma\zeta & & & & Y^+ \\
 & & & & \downarrow \lambda \\
\Sigma(Z^+) & \xrightarrow{\Sigma\lambda} & \Sigma Z & \xrightarrow{\varepsilon} & Y
\end{array}$$

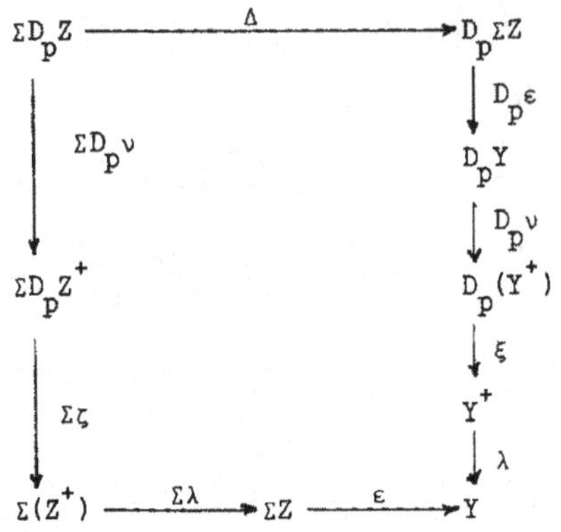

Here ζ and ξ are the H_∞ structural maps for Z^+ and Y^+ respectively. In order to see that (15) commutes we need two further diagrams. The first is the following in the catgory of spaces.

(16)
$$\begin{array}{ccccccc}
\Sigma D_p(Z^+) = \Sigma[(E\Sigma_p \times_{\Sigma_p} Z^p)^+] & & & \xrightarrow{\tilde{\Delta}} & & & E\Sigma_p \times_{\Sigma_p} (\Sigma Z)^p \\
 & & & & & & \| \\
\downarrow \Sigma\zeta & & & & & & D_p[(\Sigma Z)^+]/S^0 \\
 & & & & & & \downarrow D_p(\varepsilon^+) \\
\Sigma(Z^+) & \xrightarrow{\Sigma\lambda} & \Sigma Z & \xrightarrow{\varepsilon} & Y & \xleftarrow{\xi} & E\Sigma_p \times_{\Sigma_p} Y^p = D_p(Y^+)/S^0)
\end{array}$$

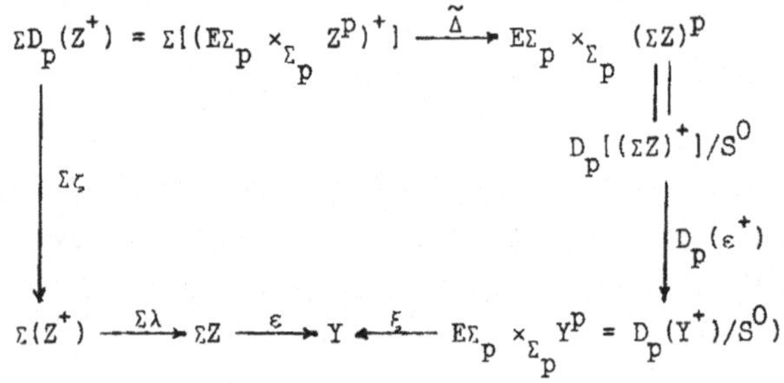

Here $\tilde{\Delta}$ is the evident diagonal map. This diagram commutes by definition of ζ; see [69, Lemma 1.5]. Next we have the following diagram in $\overline{h}\mathcal{S}$ (where we again suppress Σ^∞).

(17)
$$\begin{array}{ccccc}
W \wedge D_pZ & & \xrightarrow{\Delta} & & D_p(W \wedge Z) \\
 & \searrow^{\lambda\wedge 1} & & \swarrow^{D_p(\lambda\wedge 1)} & \\
 & \Sigma D_pZ & \xrightarrow{\Delta} & D_p\Sigma Z & \\
\downarrow 1\wedge D_p\nu & \downarrow \Sigma D_p\nu & & \downarrow D_p\nu \quad \text{(A)} & \downarrow D_p(1\wedge\nu) \\
 & & & D_p[(\Sigma Z)^+] & \\
 & & & \downarrow & \\
 & \Sigma D_pZ^+ & \xrightarrow{\tilde{\Delta}} & D_p[(\Sigma Z)^+]/S^0 & \\
 & \nearrow^{\lambda\wedge 1} & & \nwarrow \quad \nwarrow & \\
W \wedge D_p(Z^+) & & \xrightarrow{\Delta} & & D_p(W \wedge Z^+)
\end{array}$$

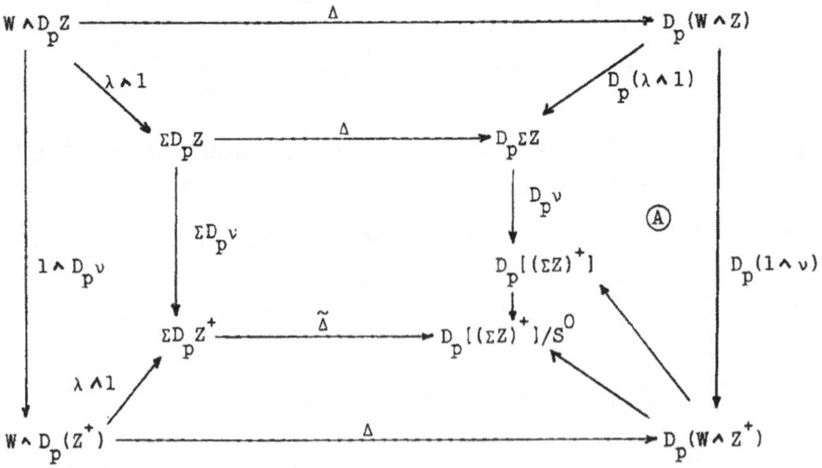

Here $W = (S^1)^+$ and the unlabeled arrows are the evident quotient maps. It suffices to show that the inner square of this diagram commutes, since combining it with diagram (16) gives diagram (15). Since

$$\lambda \wedge 1 : W \wedge D_p Z \to \Sigma D_p Z$$

is a split surjection, the commutativity of the inner square will be a consequence of the commutativity of the rest of the diagram. Each of the remaining parts clearly commutes except that marked Ⓐ. To show that Ⓐ commutes it suffices to show that the composites

$$W \wedge Z \xrightarrow{1 \wedge \nu} W \wedge Z^+ = (S^1 \times Z)^+ \longrightarrow (S^1 \wedge Z)^+$$

and

$$W \wedge Z \xrightarrow{\lambda \wedge 1} S^1 \wedge Z \xrightarrow{\nu} (S^1 \wedge Z)^+$$

are equal. But is is easy to see that these composites agree when composed with either of the maps $\lambda : (S^1 \wedge Z)^+ \to S^1 \wedge Z$ and $\mu : (S^1 \wedge Z)^+ \to S^0$; they are therefore equal since wedges are products in $\overline{h\mathcal{S}}$. This completes the proof of 3.3.

We conclude this section with the proof of 1.6. First we calculate

$$\beta_r p_* Q \Sigma u_r = \beta_r Q p_* \Sigma u_r = \iota_*(\beta_r \Sigma u_r)^{(p)} + pQ\beta_{r+1} p_* \Sigma u_r$$

in $K_0(D_p \Sigma M_r ; r)$. Multiplying by p^{r-1} gives

$$0 = p^{r-1} \beta_r p_* Q \Sigma u_r = p^{r-1} \iota_*(\beta_r \Sigma u_r)^{(p)},$$

hence $\iota_*(\beta_r \Sigma u_r)^{(p)}$ has order $\leq p^{r-1}$. Now suppose K_r has an H_∞ structure. Let $\overline{u} : S \to K_r$ be the unit map for this structure. Then $\overline{u} = cu \in K_0(S;r)$ for some c prime to p. Let f be the composite

$$\Sigma M_r = S \wedge \Sigma M_r \xrightarrow{u \wedge 1} K \wedge \Sigma M_r = K_r$$

and let F be the composite

$$K_0(D_p \Sigma M_r ; r) \xrightarrow{(D_p f)_*} K_0(D_p K_r ; r) \xrightarrow{\xi_*} K_0(K_r ; r) \longrightarrow K_0(S;r),$$

where the last map is induced by the product for K_r. We claim $c^{p+1} F \iota_*(\beta_r \Sigma u_r)^{(p)} = \overline{u}$, which contradicts the fact that $\iota_*(\beta_r \Sigma u_r)^{(p)}$ has order $\leq p^{r-1}$. The claim is a consequence of the commutativity of the following diagram

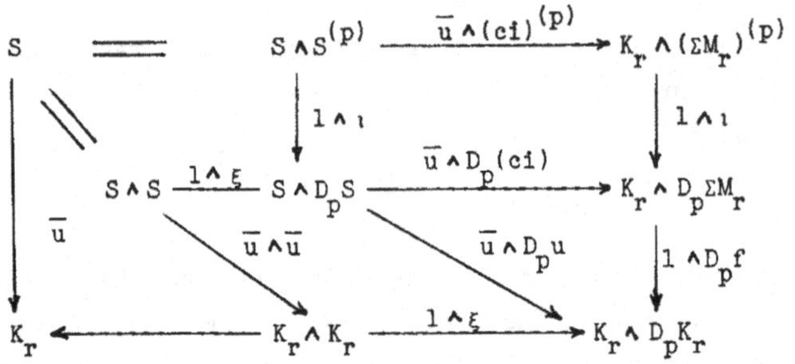

Here the composite $(1 \wedge \iota) \circ [\bar{u} \wedge (ci)^{(p)}]$ represents $c\iota_*(c\beta_r \Sigma u_r)^{(p)}$ and the diagram commutes since $\bar{u}$ is an H_∞ ring map.

8. Construction and properties of R and $\mathcal{Q}$.

In this section we construct R and $\mathcal{Q}$ and prove the external and internal versions of 3.6 and 3.7.

We begin with the construction.

Lemma 8.1. The map

$$\beta_{r+1}: K_1(D_p \Sigma M_r; r+1) \longrightarrow K_0(D_p \Sigma M_r; r+1)$$

is an isomorphism.

Lemma 8.2. The map

$$(D_p j)_*: K_0(D_p M_r; s) \longrightarrow K_0(D_p S; s)$$

is monic if $s = r$ or $s = r+1$, and $\eta \in K_0(D_p S; r+1)$ is in the image of $(D_p j)_*$.

Definition 8.3. Let $e \in K_1(D_p \Sigma M_r; r+1)$ be the unique element with $\beta_{r+1} e = Q\beta_{r+2} p_*^2 \Sigma u_r$. Let $e' \in K_0(D_p M_r; r+1)$ be the unique element with $(D_p j)_* e' = \eta$. Then

$$R: K_1(X; r) \to K_1(D_p X; r+1)$$

and

$$\mathcal{Q}: K_0(X; r) \to K_0(D_p X; r+1)$$

are the operations Q_e and $Q_{e'}$.

Note that e and e' are equal to $R\Sigma u_r$ and $\mathcal{Q} u_r$ respectively. We shall always use the latter notations for these elements. Also note that $\mathcal{Q} u = \eta$ in $K_0(B\Sigma_p; r+1)$.

Proof of 8.1. Let $r \geq 2$; the case $r = 1$ is similar. Consider the K-theory BSS for $D_p \Sigma M_r$. By 6.1 the set

$$\{\pi^{r-2}Q\Sigma u_r, \pi^{r-2}Q\beta_r\Sigma u_r, \pi^{r-1}\iota_*[\Sigma u_r \otimes (\beta_r\Sigma u_r)^{(p-1)}], \pi^{r-1}\iota_*(\beta_r\Sigma u_r)^{(p)}\}$$

is a basis for E^1. By 7.6(v) we have

(1) $$d_{r-1}\pi^{r-2}Q\Sigma u_r = \pi^{r-1}\iota_*(\beta_r\Sigma u_r)^{(p)},$$

while clearly $d_{r-1}\pi^{r-2}Q\beta_r\Sigma u_r = 0$ and

$$d_{r-1}\pi^{r-1}\iota_*[\Sigma u_r \otimes (\beta_r\Sigma u_r)^{(p-1)}] = 0;$$

hence the set

$$\{\pi^{r-2}Q\beta_r\Sigma u_r, \pi^{r-1}\iota_*[\Sigma u_r \otimes (\beta_r\Sigma u_r)^{p-1}]\}$$

is a basis for E^r. Now $d_r\pi^{r-2}Q\beta_r\Sigma u_r = 0$ by 7.6(v), and

$$d_r\pi^{r-1}\iota_*[\Sigma u_r \otimes (\beta_r u_r)^{(p-1)}] = \pi^{r-1}\iota_*(\beta_r\Sigma u_r)^{(p)} ,$$

which is zero in E^r. Thus there is an element x in $K_1(D_p\Sigma M_r;r+1)$ with

$$\pi^r x = \pi^{r-1}\iota_*[\Sigma u_r \otimes (\beta_r\Sigma u_r)^{(p-1)}],$$

and the set $\{Q\Sigma u_r, x, Q\beta_{r+2}p_*^2\Sigma u_r\}$ is a subbasis of height r+1 for $D_p\Sigma M_r$. In particular the group $K_\alpha(D_p\Sigma M_r;r+1)$ has the same order p^{2r} for $\alpha = 0$ and $\alpha = 1$. The lemma will follow if we show that $\beta_{r+1} \otimes Z_p$ maps onto $K_0(D_p\Sigma M_r;r+1) \otimes Z_p$. But the map

$$\pi^r \otimes Z_p : K_0(D_p\Sigma M_r;r+1) \otimes Z_p \to K_0(D_p\Sigma M_r;1) \otimes Z_p = K_0(D_p\Sigma M_r;1)$$

is an isomorphism, hence it suffices to show that $\pi^r\beta_{r+1}$ maps onto $K_0(D_p\Sigma M_r;1)$. Now equation (1) shows that $\pi^{r-1}\iota_*(\beta_r\Sigma u_r)^{(p)}$ is in the image of $\pi^r\beta_{r+1}$, and it remains to consider $\pi^{r-2}Q\beta_r\Sigma u_r$. By the exact sequence

$$K_1(D_p\Sigma M_r;r+1) \xrightarrow{\pi^r\beta_{r+1}} K_0(D_p\Sigma M_r;1) \xrightarrow{p_*^{r+1}} K_0(D_p\Sigma M_r;r+2)$$

it suffices to show $p_*^{r+1}\pi^{r-2}Q\beta_r\Sigma u_r = 0$. But 7.6(vi) gives

$$\begin{aligned} 0 = Qp^{r+1}\beta_{r+3}p_*^3\Sigma u_r &= p^{r+1}Q\beta_{r+3}p_*^3\Sigma u_r - (p^{pr+p-1} - p^r)\iota_*(\beta_{r+2}p_*^2\Sigma u_r)^{(p)} \\ &= p^{r+1}Q\beta_{r+3}p_*^3\Sigma u_r = p_*^{r+1}\pi^{r-2}Q\beta_r\Sigma u_r \end{aligned}$$

which completes the proof.

Proof of 8.2. It is easy to see that $\pi^{r-1}\beta_r\iota_*u_r^{(p)}$ and $\pi^{r-1}\beta_r\iota_*[u_r^{(p-1)} \otimes \beta_r u_r]$ are zero, hence by the exact sequence

$$K_\alpha(D_pM_r;r+1) \xrightarrow{\pi} K_\alpha(D_pM_r;r) \xrightarrow{\pi^{r-1}\beta_r} K_{\alpha-1}(D_pM_r;1)$$

there exist elements x and y with $\pi x = \iota_* u_r^{(p)}$ and $\pi y = \iota_*[u_r^{(p-1)} \otimes \beta_r u_r]$. Clearly the set $\{x,y,Qu_r\}$ is a subbasis of height r+1 for $D_p M_r$. In particular the set $\{x,p_*^2 Qu_r\}$ is a basis for $K_0(D_p M_r;r+1)$. Since $\{\eta,Qu\}$ is a basis for $K_0(D_p S;r+1)$ we have

$$(2) \qquad (D_p j)_* x = a_1\eta + a_2 Qu.$$

where $a_1, a_2 \; Z_{p^{r+1}}$. Applying π to both sides of (2) gives

$$\eta = (D_p j)_* \iota_* u_r^{(p)} = a_1\eta + a_2 Qu$$

in $K_0(D_p S;r)$, hence $a_1 = 1 + a_1' p^r$ and $a_2 = a_2' p^r$ for some $a_1, a_2 \in Z_p$. This fact, together with the equation $(D_p j)_* p_*^2 Qu_r = p^2 Qu$, shows that $(D_p j)_*$ is monic on $K_0(D_p M_r;r+1)$. A similar argument shows that $(D_p j)_*$ is monic on $K_0(D_p M_r;r)$. If $r \geq 2$ we have

$$(D_p j)_*[x - a_1' p^{r-1} p_* \iota_* u_r^{(p)} - a_2' p^{r-2} p_*^2 Qu_r] = \eta$$

so that $\eta \in K_0(D_p S;r+1)$ is in the image of $(D_p j)_*$ as required. If r = 1 we must show $a_2' = 0$. For this we need the map $j': M_1 \to M_2$ induced by the inclusion $Z_p \subset Z_{p^2}$. We have $j' \circ j = j : M_1 \to S$, hence

$$\begin{aligned}(D_p j)_*(D_p j')_*(x) &= (1 + a_1' p)\eta + a_2' pQu \\ &= (D_p j)_*[(1 + a_1' p)u_2^{(p)} + a_2' p_* Qu_2] .\end{aligned}$$

Since $(D_p j)_*$ is monic we conclude

$$(D_p j')_*(x) = (1 + pa_1')u_2^{(p)} + a_2' p_* Qu_2 .$$

Hence

$$(3) \qquad \pi\beta_2(D_p j')_*(x) = a_2'\beta Qu_2 = a_2' Q\beta_2 u_2 .$$

On the other hand, 6.1(vi) implies that $\iota_*[u_1^{(p-1)} \otimes \beta u_1]$ generates $K_1(D_p M_1;1)$, hence $\pi\beta_2 x = c\iota_*(u_1^{(p-1)} \otimes \beta u_1)$ for some $c \in Z_p$ and

$$(4) \qquad \pi\beta_2(D_p j')_*(x) = (D_p j')_*(\pi\beta_2 x) = c\iota_*[(j_*' u_1)^{(p-1)} \otimes j_*'\beta u_1] = 0$$

since $j_*'\beta u_1 = 0$. Comparing (3) and (4) gives $a_2' = 0$ and thus

$$(D_p j)_*[x - a_1' p_* \iota_* u_1^{(p)}] = \eta$$

which completes the proof.

Next we shall prove the external analogs of 3.6 and 3.7. The conventions preceding 7.6 give the correct external version of each statement except for 3.6(viii) and 3.7(ix). For 3.6(viii) we must prefix $(\beta_{p,p})_*$ to both sides, where

$\beta_{p,p}$ is the natural map $D_pD_pX \to D_{p^2}X$ defined in I.2, and for 3.7(ix) we prefix $(\beta_{p,p})_*$ to the left and $(\alpha_{p,p,\ldots,p})_*$ to the right.

Proposition 8.4. The operation

$$R:K_1(X;r) \longrightarrow K_1(D_pX;r+1)$$

satisfies the external analog of each part of 3.6.

Proposition 8.5. The operation

$$\mathcal{Q}:K_0(X;r) \to K_0(D_pX;r+1)$$

satisfies the external analog of each part of 3.7.

Theorems 3.6 and 3.7 will follow at once from 8.4 and 8.5 by the same proof given for 3.3. The rest of this section is devoted to the proofs of 8.4 and 8.5.

Proof of 8.4. Part (i) is trivial. In each of the remaining parts except (v) we may assume $X = \Sigma M_r$ with $x = \Sigma u_r$; part (iv) now follows from Definition 8.3. Observe that by the proof of 8.1 the set $\{Q\Sigma u_r, R\Sigma u_r\}$ is a subbasis for $D_p\Sigma M_r$ if $r \geq 2$ while $\{R\Sigma u_1\}$ is a subbasis for $D_p\Sigma M_1$.

(iii). The map

$$\pi\beta_{r+2}:K_1(D_p\Sigma M_r;r+2) \to K_0(D_p\Sigma M_r;r+1)$$

is an isomorphism since it takes the basis for the first group to that for the second. Now

$$\begin{aligned}\pi\beta_{r+2}Rp_*\Sigma u_r &= \pi Q\beta_{r+3}p_*^3\Sigma u_r = Q\beta_{r+2}p_*^2\Sigma u_r \\ &= \beta_{r+1}R\Sigma u_r = \pi\beta_{r+2}p_*R\Sigma u_r\end{aligned}$$

and the result follows.

(iv). The map

$$\beta_{r+1}p_*:K_1(D_p\Sigma M_r;r) \to K_0(D_p\Sigma M_r;r+1)$$

is monic since it takes the basis elements $\pi R\Sigma u_r$ and (when $r \geq 2$) $p_*Q\Sigma u_r$ to $p\beta_{r+1}R\Sigma u_r$ and $\beta_{r+1}p_*^2Q\Sigma u_r$ respectively. We have

$$\begin{aligned}\beta_{r+1}p_*\pi R\Sigma u_r &= p\beta_{r+1}R\Sigma u_r = pQ\beta_{r+2}p_*^2\Sigma u_r \\ &= \beta_{r+1}Qp_*^2\Sigma u_r - \iota_*(\beta_{r+1}p_*\Sigma u_r)^{(p)} \\ &= \beta_{r+1}p_*[Qp_*\Sigma u_r - \iota_*(\Sigma u_r \otimes (\beta_r\Sigma u_r)^{(p-1)})]\end{aligned}$$

which gives the first formula. For the second formula, we have

$$\begin{aligned}\beta_{r+1}p_*R\pi\Sigma u_r &= \beta_{r+1}Rp_*\pi\Sigma u_r = \beta_{r+1}Rp\Sigma u_r\\ &= Q\beta_{r+2}p_*^2(p\Sigma u_r) = Qp\beta_{r+2}p_*^2\Sigma u_r\\ &= pQ\beta_{r+2}p_*^2\Sigma u_r - (p^{p-1}-1)\iota_*(\beta_{r+1}p_*\Sigma u_r)^{(p)}\\ &= \beta_{r+1}Qp_*^2\Sigma u_r - p^{p-1}\iota_*(\beta_{r+1}p_*\Sigma u_r)^{(p)}\\ &= \beta_{r+1}p_*[Qp_*\Sigma u_r - p^{p-1}\iota_*(\Sigma u_r\otimes(\beta_r\Sigma u_r)^{(p-1)})]\end{aligned}$$

and the result follows.

(v). Let z denote Σu_r and fix i with $0 < i < p$. As in the proof of 7.3(iii) it suffices to show that the equation

$$(5)\qquad \begin{aligned}(\tau_{i,p-i})_*Rx &= a_1\iota_*[p_*z\otimes(\beta_{r+1}p_*z)^{(i-1)}]\otimes\iota_*(\beta_{r+1}p_*z)^{(p-i)}\\ &\quad + a_2\beta_{r+1}p_*[\iota_*(z\otimes(\beta_r z)^{(i-1)})\otimes\iota_*(z\otimes(\beta_r z)^{(p-i-1)})]\end{aligned}$$

holds in $K_1(D_i\Sigma M_r\wedge D_{p-i}\Sigma M_r;r+1)$ with $a_1 = -\frac{1}{p}\binom{p}{i}$ and $a_2 = \binom{p-1}{i}$. First observe that the group $K_*(D_i\Sigma M_r;1)$ is the Σ_i-coinvariant quotient of $K_*((\Sigma M_r)^{(i)};1) = K_*(\Sigma M_r;1)^{\otimes i}$, so that the set $\{\iota_*(z\otimes(\beta_r z)^{(i-1)}\}$ is a subbasis for $D_i\Sigma M_r$. Thus the set

$$\{\iota_*[z\otimes(\beta_r z)^{(i-1)}]\otimes\iota_*(\beta_r z)^{(p-i)},\iota_*[z\otimes(\beta_r z)^{(i-1)}]\otimes\iota_*[z\otimes(\beta_r z)^{(p-i-1)}]\}$$

is a subbasis for $D_i\Sigma M_r\wedge D_{p-i}\Sigma M_r$ and we see that equation (5) holds for <u>some</u> $a_1,a_2\in Z_{p^r}$. Now applying $(\tau'_{i,p-i})_*\beta_{r+1}$ to both sides of (5) gives

$$\tau_*\beta_{r+1}Rz = i!(p-i)!a_1(\beta_{r+1}p_*z)^{(p)}.$$

On the other hand we have

$$\tau_*\beta_{r+1}Rz = \tau_*Q\beta_{r+2}p_*^2z = -(p-1)!(\beta_{r+1}p_*z)^{(p)};$$

hence $a_1 = -\frac{(p-1)!}{i!(p-i)!} = -\frac{1}{p}\binom{p}{i}$. Next we apply π to (5) to get

$$(6)\qquad \begin{aligned}(\tau_{i,p-i})_*\pi Rz &= -\binom{p}{i}\iota_*[z\otimes(\beta_r z)^{(i-1)}]\otimes\iota_*(\beta_r z)^{(p-i)}\\ &\quad + a_2\iota_*(\beta_r z)^{(i)}\otimes\iota_*[z\otimes(\beta_r z)^{(p-i-1)}]\\ &\quad - a_2\iota_*[z\otimes(\beta_r z)^{(i-1)}]\otimes\iota_*(\beta_r z)^{(p-i)}.\end{aligned}$$

But we have

$$(\tau_{i,p-i})_*\pi Rz = (\tau_{i,p-i})_*[Qp_*z - \iota_*(z \otimes (\beta_r z)^{(p-1)}]$$

$$= -(\tau_{i,p-i})_*\iota_*(z \otimes (\beta_r z)^{(p-i)})$$

$$= -\binom{p-1}{i-1}\iota_*(z \otimes (\beta_r z)^{(i-1)}) \otimes \iota_*(\beta_r z)^{(p-i)}$$

$$- \binom{p-1}{i}\iota_*(\beta_r z)^{(i)} \otimes \iota_*(z \otimes (\beta_r z)^{(p-i-1)}),$$

where the last equality follows from the double-coset formula; comparing with (6) gives $a_2 = -\binom{p-1}{i}$ as required.

(vi). Let $r \geq 2$; the case $r = 1$ is similar. Let f be the composite

$$\Sigma^{-1}M_r = S^{-2} \wedge \Sigma M_r \xrightarrow{\Sigma^{-2}u \wedge 1} \Sigma^{-2}K \wedge \Sigma M_r \xrightarrow{B \wedge 1} K \wedge \Sigma M_r ,$$

where B is the Bott equivalence. We have $f_{**}\Sigma^{-1}u_r = \Sigma u_r$, hence it suffices to prove

$$\Delta_*\Sigma R(\Sigma^{-1}u_r) = p_*\iota_*u_r^{(p)} + p_*^2 Qu_r.$$

Now

$$(D_p j)_*\Delta_* R(\Sigma^{-1}u_r) = \Delta_*\Sigma R(\Sigma^{-1}u) = \Delta_*\Sigma R(\pi\Sigma^{-1}u)$$

$$= \Delta_*\Sigma Qp\Sigma^{-1}u = p\Delta_*\Sigma Q\Sigma^{-1}u$$

$$= p\iota_*u^{(p)} + p^2 Qu$$

$$= (D_p j)_*(p_*\iota_*u_r^{(p)} + p_*^2 Qu_r);$$

the result follows since $(D_p j)_*$ is monic by 8.2.

(vii) $$\beta_{r+1}\psi^k R\Sigma u_r = \psi^k\beta_{r+1}R\Sigma u_r = \psi^k Q\beta_{r+2}p_*^2\Sigma u_r$$

$$= Q\beta_{r+2}p_*^2\Sigma\psi^k u_r = \beta_{r+1}R\Sigma u_r ,$$

the last equality following from the fact that $\psi^k u_r = u_r$. The result now follows by 8.1.

(viii). Let z denote Σu_r, and abbreviate $(\beta_{p,p})_*$ by β_* and $(\alpha_{p,\ldots,p})_*$ by α_* (the reader is requested to remember that β_* is <u>not</u> a Bockstein). We must show

$$\beta_* QRx = \begin{cases} 0 & \text{if } r = 1 \\ \beta_* RQz & \text{if } r \geq 2 \end{cases}$$

in $K_1(D_{p^2}\,\Sigma M_r;r)$. We shall need the equation

(7) $$\delta_* Qx^{(n)} = \sum_{i=1}^{n} \binom{n}{i} p^{i-1}(\pi\iota_* x^{(p)})^{(n-i)} \otimes (Qx)^{(i)}$$

which holds in $K_0((D_pX)^{(n)};r-1)$ for each $x \in K_0(X;r)$ provided that p is odd (the proof is by induction on n from 7.6(ii)).

First let r = 1. The set {QRz, RRz} is a subbasis for $D_pD_p\Sigma M_1$, and it follows easily from Proposition 3.9 that the map

$$\beta_3 p_*^2 : K_1(D_pD_p\Sigma M_1;1) \longrightarrow K_0(D_pD_p\Sigma M_1;3)$$

is a monomorphism. Since $K_1(D_{p^2}\Sigma M_1;1)$ is imbedded in $K_1(D_pD_p\Sigma M_1;1)$ by the transfer we see that

$$\beta_3 p_*^2 : K_1(D_{p^2}\Sigma M_1;1) \longrightarrow K_0(D_{p^2}\Sigma M_1;3)$$

is a monomorphism. It therefore suffices to show that $\beta_*\beta_3 p_*^2 QRz$ is zero. We have

$$\begin{aligned}\beta_*\beta_3 p_*^2 QRz &= \beta_*\beta_3 Qp_*(Rp_*z) \quad \text{by 7.6(iv) and 8.4(iii)}\\ &= \beta_*\beta_3[R\pi Rp_*z + p^{p-1}\iota_*(Rp_*z \otimes (\beta_3 Rp_*z)^{(p-1)})]\\ &= \beta_*\beta_3 R[Qp_*^2z - \iota_*(p_*z \otimes (\beta_2 p_*z)^{(p-1)})] + p^{p-1}\beta_*\iota_*(\beta_3 Rp_*z)^{(p)},\end{aligned}$$

where the last two equalities follow from the second and first parts of 8.4(ii). Now $Qp_*^2z = 0$ by 7.6(iv), and

$$\begin{aligned}\beta_*\beta_3 R\iota_*(p_*z \otimes (\beta_2 p_*z)^{(p-1)}) &= \alpha_*\delta_*\beta_3 R(p_*z \otimes (\beta_2 p_*z)^{(p-1)}) \quad \text{by I.2.12}\\ &= \alpha_*\delta_* Q((\beta_4 p_*^3 z)^{(p)}) \quad \text{by 8.4(iv)}\\ &= p^{p-1}\alpha_*(Q\beta_4 p_*^3 z)^{(p)} \quad \text{by 7.6(vii) when } p = 2 \text{ and equation (7) when p is odd}\\ &= p^{p-1}\beta_*\iota_*(\beta_3 Rp_*z)^{(p)} \quad \text{by 8.4(iv) and I.2.11.}\end{aligned}$$

We conclude that $\beta_*\beta_3 p_*^2 QRz = 0$ as required, which concludes the case r = 1.

Next let r = 2. We have

$$\begin{aligned}\pi\beta_*(QRz - RQz) &= \beta_*[Q[Qp_*z - \iota_*(z \otimes (\beta_r z)^{(p-1)}]\\ &\quad - Qp_*Qz + \iota_*(Qz \otimes (\beta_{r-1}Qz)^{(p-1)})]\\ &= \beta_*[-Q\iota_*(z \otimes (\beta_r z)^{(p-1)}) + \iota_*(Qz \otimes (\beta_{r-1}Qz)^{(p-1)}]\\ &= \alpha_*[-\delta_*Q(z \otimes \beta_r z)^{(p-1)}) + Qz \otimes (\beta_{r-1}Qz)^{(p-1)}] \quad \text{by I.2.11 and I.2.12.}\\ &= \alpha_*[-Qz \otimes (\pi\iota_*(\beta_r z)^{(p)})^{p-1} - pQz \otimes \delta_*Q(\beta_r z)^{(p-1)}\\ &\quad + Qz \otimes (\iota_*(\pi\beta_r z)^{(p)} + pQ\beta_r z)^{p-1}].\end{aligned}$$

When p = 2 the last expression is clearly zero, while if p is odd it is zero by (7). Hence we have

(8) $$\pi\beta_*(QRz - RQz) = 0.$$

A similar calculation gives

(9) $$\beta_{r+2}P_*^2\beta_*(QRz - RQz) = 0.$$

To proceed further we need the case $k = p^2$ of 4.1. First we must check that the argument is not circular, since the present result is certainly used in the proof of 4.1. However, it enters only through the proof of 4.7, to be given in Section 9. An inspection of Section 9 will show that only the case r = 1 of the present result is used in proving the case $k = p^2$ of 4.7. Thus we may proceed. We suppose $r \geq 3$; the case r = 2 differs only slightly. By Remark 4.2 we obtain a subbasis

$$A = A_{r-2} \cup A_{r-1} \cup A_r \cup A_{r+2}$$

for $D_{p^2}\Sigma M_r$ with $A_{r-2} = \{\beta_*QQz\}$,

$$A_{r-1,1} = \{\alpha_*[Qz \otimes (\beta_{r-1}Qz)^{(i)} \otimes (\pi^2\beta_{r+1}Rz)^{(p-i-1)}] \mid 0 \leq i \leq p-2\},$$

$$A_{r-1,0} = \{\alpha_*[Qz \otimes \pi^2 Rz \otimes (\beta_{r-1}Qz)^{(i-1)} \otimes (\pi^2\beta_{r+1}Rz)^{(p-i-1)}] \mid 1 \leq i \leq p-2\},$$

$A_r = \{\beta_*RQz\}$ and $A_{r+2} = \{\beta_*RRz\}$. Therefore the set

$$\{\pi^{r-3}\beta_*QQz, \pi^{r-1}\beta_*RQz, \pi^{r+1}\beta_*RRz\} \cup \pi^{r-2}A_{r-1,1} \cup \pi^{r-2}\beta_{r-1}A_{r-1,0}$$

is a basis for $K_1(D_p\Sigma M_r;1)$, and the subset $\pi^{r-2}\beta_{r-1}A_{r-1,0}$ is a basis for the image of $\pi^{r-2}\beta_{r-1}$, hence for the kernel of P_*^{r-1}. By (8) we see that $\beta_*(QRz - RQz)$ is in the image of P_*^{r-1}, hence there exist constants $a,b,c,d_0,\ldots,d_{p-2}$ Z_p with

(10) $$\beta_*(QRz - RQz) = P_*^{r-1}[a\pi^{r-3}\beta_*QQz + b\pi^{r-1}\beta_*RQz$$
$$+ c\pi^{r+1}\beta_*RRz + \alpha_*\pi^{r-2}\sum_{i=0}^{p-2}(d_iQz \otimes (\beta_{r-1}Qz)^{(i)} \otimes (\pi^2\beta_{r+1}Rz)^{(p-i-1)})].$$

If we apply $\beta_{r+2}P_*^2$ to both sides of (10) then the left side becomes zero by (9), hence we have

$$0 = ap^{r-3}\beta_{r+2}P_*^4\beta_*QQz + bp^{r-1}\,P_*^2\beta_*RQz + cp^{r+1}\beta_{r+2}\beta_*RRz$$
$$+ \sum_{i=0}^{p-2} d_i p^{r-2}\beta_{r+2}P_*^3\alpha_*[Qz \otimes (\beta_{r-1}Qz)^{(i)} \otimes (\pi^2\beta_{r+1}Rz)^{(p-i-1)}].$$

Since the set A is a subbasis this gives $a = b = c = d_0 = \cdots d_{p-2} = 0$ as required. This completes the proof of 8.4.

Proof of 8.5. Part (i) is trivial.

(ii) We may assume $x = u_r$. We have

$$(D_pj)_*\pi\mathcal{Q}u_r = \eta = \iota_*u^{(p)} = (D_pj)_*\iota_*u_r^{(p)}$$

hence $\pi\mathcal{Q}u_r = \iota_*u_r^{(p)}$ by 8.2. If $r \geq 2$ then

$$(D_pj)_*\mathcal{Q}\pi u_r = \mathcal{Q}\pi u = \mathcal{Q}u = (D_pj)_*\iota_*u_r^{(p)} ,$$

hence $\mathcal{Q}\pi u_r = i_*u_r^{(p)}$ by 8.2.

(v) As in the proof of 7.3(iii) it suffices to show

$$\tau_*\mathcal{Q}u_r = \begin{cases} (p-1)!p_*u_r^{(p)} & \text{if } p \text{ is odd or } r \geq 2 \\ 2_*u_1^{(2)} + (\beta_2 2_*u_r)^{(2)} & \text{if } p = 2 \text{ and } r = 1. \end{cases}$$

We prove this when $p = 2$; the odd primary case is similar. The element $\tau_*\mathcal{Q}u_r$ is in the Σ_2-invariant subgroup of $K_0(M_r^{(2)};r+1)$, and this subgroup has a basis consisting of $2_*u_r^{(2)}$ with order 2^r and $2^{r-1}(\beta_{r+1}2_*u_r)^{(2)}$ with order 2. Thus we have

$$(11) \qquad \tau_*\mathcal{Q}u_r = a_1 2_*u_r^{(2)} + a_2 2^{r-1}(\beta_{r+1}2_*u_r)^{(2)}$$

with $a_1 \in Z_{2^r}$ and $a_2 \in Z_2$. Now

$$j_*^{(2)}\tau_*\mathcal{Q}u_r = \tau_*(D_2j)_*\mathcal{Q}u_r = \tau_*\eta = 2u;$$

thus applying $j_*^{(2)}$ to both sides of (11) gives $2u = 2a_1u$ in $K_0(S;r+1)$ so that $a_1 = 1$. Next we have

$$\pi\tau_*\mathcal{Q}u_r = \tau_*\iota_*u_r^{(2)} = \begin{cases} 2u_r^{(2)} & \text{if } r \geq 2 \\ (\beta u_1)^{(2)} & \text{if } r = 1, \end{cases}$$

hence applying π to (11) gives $a_2 = 0$ if $r \geq 2$ and $a_2 = 1$ if $r = 1$.

(iv) We may assume $x = u_r$. Let $r \geq 2$; the case $r = 1$ is similar. The set

$$\{Qu_r, \iota_*u_r^{(p)}, i_*(u_r^{(p-1)} \otimes b_ru_r)\}$$

is a subbasis of height r for D_pM_r, hence we have

$$(12) \qquad \pi\beta_{r+1}\mathcal{Q}u_r = a_1\iota_*(u_r^{(p-1)} \otimes \beta_ru_r) + a_2\beta_rp_*Qu_r$$

with $a_1 \in Z_{p^r}$, $a_2 \in Z_{p^{r-1}}$. Let $j':M_r \to M_{r+1}$ be the map induced by the inclusion $Z_{p^r} \subset Z_{p^{r+1}}$. Then $j \circ j' = j:M_r \to S$, hence $(j')_*u_r = \pi u_{r+1}$ and $(j')_*\beta_ru_r =$

$p\pi\beta_{r+1}u_{r+1}$. Thus

$$(D_pj')_*\pi\beta_{r+1}\mathcal{Q}u_r = \pi\beta_{r+1}\mathcal{Q}\pi u_{r+1} = \pi\beta_{r+1}\iota_*u_{r+1}^{(p)}$$
$$= p\pi\iota_*(u_{r+1}^{(p-1)} \otimes \beta_{r+1}u_{r+1})$$

and comparing with (12) gives $a_2 = 0$. Next we have

$$\tau_*\pi\beta_{r+1}\mathcal{Q}u_r = \pi\beta_{r+1}(p-1)!p_*u_r^{(p)} = (p-1)!\beta_r u_r^{(p)} = \tau_*\iota_*(u_r^{(p-1)} \otimes \beta_r u_r)$$

and comparing with (12) gives $a_1 = 1$.

(iii) By part (iv) we see that the set $\{Qu_r, \mathcal{Q}u_r\}$ is a subbasis for D_pM_r if $r \geq 2$, while $\{\mathcal{Q}u_r\}$ is a subbasis for D_pM_1. It follows that the map

$$(D_pj)_*: K_0(D_pM_r;r+2) \to K_0(D_pS;r+2)$$

is monic. But

$$(D_pj)_*\mathcal{Q}p_*u_r = \mathcal{Q}(pu) = \mathcal{Q}(\pi pu) = \iota_*(pu)^{(p)} = p^{p-1}p_*\eta = (D_pj)_*p^{p-1}p_*\mathcal{Q}u_r$$

and the result follows.

(vi). Let $p = 2$; the odd primary case is similar. First let $|x| = |y| = 0$ with $r \geq 2$. We may assume $x = u_r$, $y = u_r$. The set

$$\{\mathcal{Q}x \otimes \mathcal{Q}y, \pi\iota_*x^{(2)} \otimes Qy, Qx \otimes \pi\iota_*y^{(2)}, Qx \otimes Qy, \mathcal{Q}x \otimes \beta_{r+1}\mathcal{Q}y,$$
$$\mathcal{Q}x \otimes \beta_{r+1}4_*Qy, Qx \otimes \pi^2\beta_{r+1}\mathcal{Q}y, Qx \otimes \beta_{r-1}Qy\}$$

is a subbasis for $D_2M_r \wedge D_2M_r$, hence we have

$$\begin{aligned}(13) \qquad \delta_*\mathcal{Q}(x \otimes y) = {} & a_1\mathcal{Q}x \otimes \mathcal{Q}y + a_2\mathcal{Q}x \otimes 4_*Qy + a_3 4_*Qx \otimes \mathcal{Q}y \\ & + a_4 4_*(Qx \otimes Qy) + a_5\beta_{r+1}\mathcal{Q}x \otimes \beta_{r+1}\mathcal{Q}y \\ & + a_6\beta_{r+1}\mathcal{Q}x \otimes \beta_{r+1}4_*Qy + a_7\beta_{r+1}4_*Qx \otimes \beta_{r+1}\mathcal{Q}y \\ & + a_8\beta_{r+1}4_*Qx \otimes \beta_{r+1}4_*Qy\end{aligned}$$

with $a_1, a_5 \in Z_{2^{r+1}}$ and $a_2, a_3, a_4, a_6, a_7, a_8 \in Z_{2^{r-1}}$. Since

$$\pi\delta_*\mathcal{Q}(x \otimes y) = \delta_*\iota_*(x \otimes y)^{(2)} = \iota_*x^{(2)} \otimes \iota_*y^{(2)}$$

we have $a_6 = a_7 = a_8 = 0$. The equation

$$(D_2j \wedge D_2j)_*\delta_*\mathcal{Q}(x \otimes y) = \delta_*\mathcal{Q}u = \delta_*\eta = \eta \otimes \eta$$

implies $a_1 = 1$ and $a_2 = a_3 = a_4 = 0$. Hence we have

$$(14) \qquad \delta_* \mathcal{Q}(x \otimes y) = \mathcal{Q}x \otimes \mathcal{Q}y + a_5 \beta_{r+1} \mathcal{Q}x \otimes \beta_{r+1} \mathcal{Q}y$$

with a_5 depending on r. A similar argument shows that (14) holds also when $r = 1$. Now let T_1 and T_2 switch the factors of $M_r \wedge M_r$ and $D_2 M_r \wedge D_2 M_r$. Then

$$\delta_* \mathcal{Q}(T_{1*}(x \otimes y)) = T_{2*}\delta_* \mathcal{Q}(x \otimes y) = \mathcal{Q}y \otimes \mathcal{Q}x - a_5 \beta_{r+1} \mathcal{Q}y \otimes \beta_{r+1} \mathcal{Q}x.$$

On the other hand, if $r \geq 2$ then

$$\delta_* \mathcal{Q}(T_{1*}(x \otimes y)) = \delta_* \mathcal{Q}(y \otimes x) = \mathcal{Q}y \otimes \mathcal{Q}x + a_5 \beta_{r+1} \mathcal{Q}y \otimes \beta_{r+1} \mathcal{Q}x,$$

hence $2a_5 = 0$ as required. If $r = 1$ then

$$\begin{aligned} \delta_* \mathcal{Q}(T_{1*}(x \otimes y)) &= \delta_* \mathcal{Q}(y \otimes x + \beta y \otimes \beta x) \\ &= \delta_* \mathcal{Q}(y \otimes x) + 2\beta_2 \mathcal{Q}y \otimes \beta_2 \mathcal{Q}x. \end{aligned}$$

Hence in this case $-a_5 = a_5 + 2 \bmod 4$, so that $a_5 \equiv 1 \bmod 2$ as required.

Next let $|x| = 1$, $|y| = 0$ with $r \geq 2$ we may assume $x = \Sigma u_r$, $y = u_r$. Choosing a subbasis for $D_2 \Sigma M_r \wedge D_2 M_r$ as in the preceeding case, we see that

$$\begin{aligned} (15) \qquad \delta_* R(x \otimes y) = {} & a_1 Rx \otimes \mathcal{Q}y + a_2 Rx \otimes 4_* Qy + a_3 4_* Qx \otimes \mathcal{Q}y \\ & + a_4 4_*(Qx \otimes Qy) + a_5 \beta_{r+1} Rx \otimes \beta_{r+1} \mathcal{Q}y \\ & + a_6 \beta_{r+1} Rx \otimes \beta_{r+1} 4_* Qy + a_7 \beta_{r+1} 4_* Qx \otimes \beta_{r+1} \mathcal{Q}y \\ & + a_8 \beta_{r+1} 4_* Qx \otimes \beta_{r+1} 4_* Qy \end{aligned}$$

with $a_1, a_5 \in Z_{2^{r+1}}$ and the remaining a_i in $Z_{2^{r-1}}$. If f denotes the composite

$$D_2 \Sigma M_r \wedge D_2 M_r \xrightarrow{1 \wedge D_2 j} D_2 \Sigma M_r \wedge D_2 S^0 \xrightarrow{1 \wedge \xi} D_2 \Sigma M_r \wedge S^0 = D_2 \Sigma M_r$$

then the diagram

$$\begin{array}{ccc} D_1(\Sigma M_r \wedge M_r) & \xrightarrow{\delta} & D_2 \Sigma M_r \wedge D_2 M_r \\ \Big\downarrow D_2(1 \wedge j) & & \Big\downarrow f \\ D_2(\Sigma M_r \wedge S^0) & = & D_2 \Sigma M_r \end{array}$$

commutes. Applying f_* to (15) and using the equation $\xi_* Qu = 0$ (which was shown in the proof of 3.3(ii)) gives

$$Rx = a_1 Rx + a_3 4_* Qx,$$

hence $a_1 = 1$ and $a_3 = 0$. To determine a_2 and a_4 we calculate

$$\pi^2\beta_{r+1}\delta_*R(x\otimes y) = \delta_*Q\beta_r(x\otimes y) = \pi^2\beta_{r+1}[Rx\otimes \mathcal{Q}y + 4_*(Qx\otimes Qy)],$$

hence $a_2 = 0$ and $a_4 = 1$. Next we calculate

$$\pi\delta_*R(x\otimes y) = \delta_*\pi R(x\otimes y)$$

$$= \pi Rx\otimes \pi\mathcal{Q}y + \pi 4_*(Qx\otimes Qy) + 2^{r-2}\beta_r 2_*Qx\otimes \pi\beta_{r+1}\mathcal{Q}y$$

$$+ 2^{2r-3}\iota_*(\beta_r x)^{(2)}\otimes Q\beta_{r+1}\mathcal{Q}y.$$

Now the element $2^{2r-3}\iota_*(\beta_r x)^{(2)}$ is zero when $r \geq 3$ since $2r-3 \geq r$ while when $r = 2$ we have

$$0 = 2\beta_2 2_*Qx = 2\beta_2 Q2_*x = 2\iota_*(\beta_2 x)^{(2)}.$$

Thus applying π to both sides of (15) gives $2a_5 = a_6 = a_8 = 0$ and $a_7 = 2^{r-2}$. It remains to show $a_5 = 2^r\varepsilon_r$, where $\varepsilon_r \in Z_2$ is the constant in the formula for $\delta_*\mathcal{Q}(x\otimes y)$. But this follows from the equation

(16) $$(\delta\wedge 1)_*\delta_*R((\Sigma u_r\otimes u_r)\otimes u_r) = (1\wedge\delta)_*\delta_*R(\Sigma u_r\otimes(u_r\otimes u_r))$$

if we expand both sides using the formulas already shown.

Next let $x = \Sigma u_1$, $y = u_1$. A suitable choice of subbasis for $D_2\Sigma M_1\wedge D_2M_1$ gives

$$\delta_*R(x\otimes y) = a_1Rx\otimes\mathcal{Q}y + a_2\beta_2Rx\otimes\beta_2\mathcal{Q}y$$

and we see as before that $a_1 = 1$. Evaluating both sides of equation (16) in this case gives $a_2 = -(1+2\varepsilon_1)$. Finally, we have

$$\delta_*R(y\otimes x) = \delta_*R(T_{1*}(x\otimes y + \beta x\otimes\beta y))$$

$$= T_{2*}\delta_*R(x\otimes y + \beta x\otimes\beta y)$$

$$= \mathcal{Q}y\otimes Rx + (1+2\varepsilon_1)\beta_2\mathcal{Q}y\otimes\beta_2Rx$$

as required.

Now let $x = \Sigma u_r$ and $y = \Sigma u_r$, with $r \geq 2$. We have

(17) $$\delta_*\;(x\otimes y) = a_1Rx\otimes Ry + a_2Rx\otimes 4_*Qy + a_3 4_*Qx\otimes Ry$$

$$+ a_4 4_*(Qx\otimes Qy) + a_5\beta_{r+1}Rx\otimes\beta_{r+1}Ry + a_6\beta_{r+1}Rx\otimes\beta_{r+1}4_*Qy$$

$$+ a_7\beta_{r+1}4_*Qx\otimes\beta_{r+1}Ry + a_8\beta_{r+1}4_*Qx\otimes\beta_{r+1}4_*Qy$$

with $a_1, a_2 \in Z_{2^{r+1}}$ and the remaining a_i in $Z_{2^{r-1}}$. The equation

(18) $$\pi\delta_* \ (x \otimes y) = \delta_* \iota_* (x \otimes y)^{(2)} = \iota_* x^{(2)} \otimes \iota_* y^{(2)} = 2^{2r-4} \beta_r 2_* Qx \otimes \beta_r 2_* Qy$$

shows that $a_6 = a_7 = 0$, $a_8 = 2^{2r-4}$, and also that $a_1 \equiv 0 \bmod 2^r$ and that $a_2 \equiv a_3 \equiv a_4 = 0 \bmod 2^{r-2}$. Next we apply $(D_2 j \wedge D_2 j)_*$ to both sides of (17). The left side becomes

$$(D_2 j \wedge D_2 j)_* \delta_* \mathcal{Q}(x \otimes y) = \delta_* \mathcal{Q}(\Sigma u \otimes \Sigma u) = \pi\delta_* \mathcal{Q}(\Sigma u \otimes \Sigma u),$$

which is zero by (18). By 8.4(ii) we have

$$(D_2 j)_* R\Sigma u_r = R\Sigma u = R\pi\Sigma u = 2Q\Sigma u,$$

hence (since $8a_1 \equiv 8a_2 \equiv 8a_3 \equiv 0 \bmod 2^{r+1}$) the right side of (17) becomes $4a_4 Q\Sigma y \times Q\Sigma u$, so that $a_4 = 0$ in $Z_{2^{r-1}}$. Next we calculate

$$\pi\beta_{r+1} \delta_* \mathcal{Q}(x \otimes y) = 2^{r-2} \pi\beta_{r+1} [Rx \otimes 4_* Qy + 4_* Qx \otimes Ry],$$

hence $a_2 = a_3 = 2^{r-2}$. Finally, if we expand both sides of the equation

$$(\delta \wedge 1)_* \delta_* \mathcal{Q}((\Sigma u_r \otimes \Sigma u_r) \otimes u_r) = (1 \wedge \delta)_* \delta_* \mathcal{Q}(\Sigma u_r \otimes (\Sigma u_r \otimes u_r))$$

using the formulas already shown, it follows that $a_5 = 0$. The proof when $r = 1$ is similar.

(vii). We may assume $x = u_r$. Let $r \geq 2$; the case $r = 1$ is similar. Then

(19) $$\Delta_* \Sigma \mathcal{Q} u_r = a_1 R\Sigma u_r + a_2 p_*^2 Q\Sigma u_r$$

with $a_1 \in Z_{p^{r+1}}$ and $a_2 \in Z_{p^{r-1}}$. Applying π to (19) shows that $a_1 \equiv 0 \bmod p^r$, hence applying $(D_p j)_*$ to (19) gives $a_2 = 0$. It only remains to show that $\Delta_* \Sigma \mathcal{Q} u_r \neq 0$ when $p = 2$. But Lemma 7.7 gives the exact sequence

$$K_1(\Sigma M_r \wedge M_r; r+1) \xrightarrow{(\Sigma\iota)_*} K_1(\Sigma D_2 M_r; r+1) \xrightarrow{\Delta_*} K_1(D_2 \Sigma M_r; r+1).$$

Since $\Sigma \mathcal{Q} u_r$ has order 2^{r+1}, it cannot be in the image of $(\Sigma\iota)_*$ and the result follows.

(viii). We may assume $x = u_r$. We have

$$(D_p j)_* \psi^k \mathcal{Q} u_r = \psi^k \mathcal{Q} u = \psi^k \eta = \eta = (D_p j)_* \mathcal{Q} \psi^k u_r$$

since $\psi^k u_r = u_r$; the result follows by 8.2.

(ix) By equation (7) in the proof of 8.4(viii) and I.2.14 we have the following equation in $K_0(D_{p^2} X; r-1)$ when p is odd and $r \geq 2$.

(20) $$\beta_* Q\iota_* x^{(p)} = \iota_*\delta_* Qx^{(p)} = \iota_*[\sum_{i=1}^{p}\binom{p}{i}p^{i-1}(\pi\iota_* x^{(p)})^{(p-i)} \otimes (Qx)^{(i)}].$$

When p = 2 this equation follows from 7.6(vii) since $\iota_*(x \otimes \beta_r x \otimes x \otimes \beta_r x)$ and $\iota_*(Q\beta_r x \otimes Q\beta_r x)$ are zero by 7.6(x).

Let $r = 1$, $x = u_1$. The set $\{\mathfrak{J} u_1\}$ is a subbasis for D_pM_1, hence by 4.3 the set $\{Q\mathfrak{J}u_1, \iota_* u_1^{(p^2)}\}$ is a basis for $K_0(D_pD_pM_1;1)$. Lemma 4.3 also implies that the set

$$\{Q\mathfrak{J}u, \iota_* u^{(p^2)}\} \subset K_0(D_pD_pS;1)$$

is linearly independent. Hence $(D_pD_pj)_*$ is monic on $K_0(D_pD_pM_r;1)$. Since the transfer

$$\tau_*: K_0(D_{p^2}\, M_1;1) \to K_0(D_pD_pM_1;1)$$

is monic and $(D_pD_pj)_* \circ \tau = \tau_* \circ (D_{p^2}j)_*$, it follows that $(D_{p^2}\, j)_*$ is monic on $K_0(D_{p^2}\, M_1;1)$. But

$$(D_{p^2}\, j)_*\beta_* Q\mathfrak{J}u_1 = \beta_* Q\mathfrak{J}u = \beta_* Qi_* u^{(p)},$$

which is zero by (20), hence $\beta_* Q\ \ u_1 = 0$ as required.

Next let $r \geq 2$ and let y denote the element

$$\beta_* Q\mathfrak{J}u_r - \iota_* \sum_{i=1}^{p}\binom{p}{i}p^{i-2}[\iota_* u_r^{(p)}]^{(p-i)} \otimes p_*[(Qu_r)^{(i)}]$$

in $K_0(D_{p^2}\, M_r;r)$. Then (20) implies that $\pi y = 0$ and $(D_{p^2}\, j)_* y = 0$, and we must show $y = 0$. Since $\pi y = 0$ we see that y is in the image of p_*^{r-1}. To proceed further we need the case $k = p^2$ of 4.1; we may use this result without circularity since only the case r = 1 of the present result is used in proving it (see section 9). Now as in the proof of 8.4(viii) we see that the union of the sets

$$\{\iota_*[(\pi^{r-1}\iota_* u_r^{(p)}]^{(i)} \otimes (\pi^{r-2}Qu_r)^{(p-i)}] \mid 0 \leq i \leq p\}$$

$$\{\iota_*[(\pi^{r-1}\iota_* u_r^{(p)})^{(i-1)} \otimes \pi^{r-1}\iota_*(u_r^{(p-1)} \otimes \beta_r u_r)$$
$$\otimes (\pi^{r-2}Qu_r)^{(p-i-1)} \otimes \pi^{r-2}\beta_{r-1}Qu_r] \mid 1 \leq i \leq p-1\}$$

and, if $r \geq 3$, $\{\pi^{r-3}\beta_* QQu_r\}$, is a basis for $K_0(D_{p^2}M_r;1)$. The second of these sets generates the kernel of p_*^{r-1} and also the kernel of $(D_{p^2}\, j)_*$, and it follows that $(D_{p^2}\, j)_*$ is monic on the image of p_*^{r-1}. Since $(D_{p^2}\, j)_* y = 0$ we conclude $y = 0$ as required.

9. Cartan formulas

In this section we shall prove Lemma 4.7. As in the proof of 2.7, the basic idea is to "simplify" each expression in $C\mathcal{A}$ (respectively $C\mathcal{A}'$) to obtain an expression in C{x} (respectively C{y,z}). We shall refer to the simplified expression as a Cartan formula for the original one. Some explicit examples of such formulas will be given below. However, some of the formulas we need are too complicated to give explicitly, and instead we shall use an inductive argument to establish their existence.

In order to do so it is convenient to work in a suitable formal context. Let $\xi_1,\ldots,\xi_t$ be indeterminates and suppose that to each has been assigned a mod 2 dimension denoted $|\xi_i|$ and two positive integers called the height and filtration and denoted $\|\xi_i\|$ and $\nu\xi_i$. Intuitively, ξ_i should be thought of as an element of $K_{|\xi_i|}(D_{\nu\xi_i}X;\|\xi_i\|)$ for some spectrum X. We wish to consider certain finite formal combinations $E(\xi_1,\ldots,\xi_t)$ involving the ξ_i and the operations of section 3, namely those combinations which would represent elements in one of the groups $K_\alpha(D_jX;r)$ when interpreted "externally" as in section 4. More precisely, we define the allowable expressions $E(\xi_1,\ldots,\xi_t)$ and assign them dimensions, heights and filtration by induction on their length as follows.

Definition 9.1. (i) Each indeterminate ξ_i is an expression of length 1. For each $\alpha\in Z_2$, $r\geq 1$, $j\geq 1$ there is an expression $0_{\alpha,r,j}$ (called zero sub α,r,j) having length 1, dimension α, height r and filtration j. These are the only expressions of length 1.

(ii) Suppose that the expressions of length $\leq \ell$ have been defined and assigned dimensions, heights and filtrations. The expressions of length $\ell+1$ are the following, where E ranges over the expressions of length ℓ.

(a) p_*E. We define $|p_*E| = |E|$, $\|p_*E\| = \|E\|+1$ and $\nu(p_*E) = \nu E$.

(b) β_rE if $\|E\| = r$. We define $|\beta_rE| = |E|-1$, $\|\beta_rE\| = \|E\|$ and $\nu(\beta_rE) = \nu E$.

(c) πE if $2\leq \|E\|$. We define $|\pi E| = |E|$, $\|\pi E\| = \|E\|-1$ and $\nu(\pi E) = \nu E$.

(d) $E_1 + E_2$, where E_1 and E_2 are any expressions whose lengths add up to $\ell+1$ and which satisfy $|E_1| = |E_2|$, $\|E_1\| = \|E_2\|$, and $\nu E_1 = \nu E_2$. We define $|E_1+E_2| = |E_1|$, $\|E_1+E_2\| = \|E_1\|$ and $\nu(E_1+E_2) = \nu E_1$.

(e) $E_1\cdot E_2$ (the formal product) where E_1 and E_2 are any expressions whose lengths add up to $\ell+1$ and which satisfy $\|E_1\| = \|E_2\|$. We define $|E_1\cdot E_2| = |E_1| + |E_2|$, $\|E_1\cdot E_2\| = \|E_1\|$, and $\nu(E_1\cdot E_2) = \nu E_1 + \nu E_2$.

(f) QE if $2\leq \|E\|$. We define $|QE| = |E|$, $\|QE\| = \|E\|-1$ and $\nu QE = p\nu E$.

(g) $\mathcal{Q}E$ if $|E| = 0$. We define $|\mathcal{Q}E| = 0$, $\|\mathcal{Q}E\| = \|E\|+1$, and $\nu\mathcal{Q}E = p\nu E$.

(h) RE if $|E| = 1$. We define $|RE| = 1$, $\|RE\| = \|E\|+1$, and $\nu RE = p\nu E$.

Note that we have not required formal addition and multiplication to satisfy commutativity, associativity or other properties. However, in writing down particular expressions we shall often omit some of the necessary parentheses, since their precise position will usually be irrelevant. We shall also abbreviate $0_{\alpha,r,j}$ by 0.

We have given Definition 9.1 in complete detail as a pattern for other inductive definitions about which we will not be so scrupulous. For example, let E be an expression in the indeterminates $\xi_1,\ldots,\xi_t$. If $E_1,\ldots,E_t$ are expressions in another set of indeterminates $\eta_1,\ldots,\eta_s$ with $|E_i| = |\xi_i|$, $\|E_i\| = \|\xi_i\|$, and $\nu E_i = \nu u_i$ for $1 \leq i \leq t$ then we may (inductively) define the composite expression $E(E_1,\ldots,E_t)$ in $\eta_1,\ldots,\eta_s$. Again, if X is any spectrum and $x_i \in K_{|\xi_i|}(D_{\nu\xi_i}X;\|\xi_i\|)$ for $1 \leq i \leq t$ then we can define

$$E(x_1,\ldots,x_t) \in K_{|E|}(D_{\nu E}X;\|E\|)$$

as in section 4 by interpreting Q, $\mathcal{Q}$, R and the multiplication externally and applying α_* and β_* to formal products and composites.

<u>Definition 9.2</u>. Let $\xi_1,\ldots,\xi_t$ be a fixed set of indeterminates. <u>Equivalence</u>, denoted by $\sim$, is the smallest equivalence relation on the set of expressions in $\xi_1,\ldots,\xi_t$ which satisfies the following.

(1) $\sim$ is preserved by left composition with Q, $\mathcal{Q}$, R, π, p_* and β_r and by formal addition and multiplication.

(2) For each $r \geq 1$ the equivalence classes of expressions of height r, graded by dimension and filtration, form a $Z_2 \times Z$ graded ring (without unit) with the $0_{\alpha,r,j}$ as zero elements. The relation $E_1 \cdot E_2 = (-1)^{|E_1||E_2|}E_2 \cdot E_1$ is satisfied and left composition with π, β_r or p_* is additive.

(3) If x and y denote expressions E_1 and E_2 having height r and the required dimensions then the following hold with = replaced by $\sim$: 3.1; 3.2(iii),(iv) and (v); 3.3(iii), (iv), (v), (vi), (vii) and (x); 3.6(ii), (iii), (iv), (v) and (viii); 3.7(ii), (iii), (iv), (v), (vi) and (ix).

Roughly speaking, two expressions are equivalent if one can be transformed into the other by using the relations of Section 3.

It is easy to see that equivalent expressions must have the same dimension, height, and filtration but not necessarily the same length. An inductive argument shows that $E(E_1,\ldots,E_t)$ and $E'(E_1',\ldots,E_t')$ are equivalent if $E \sim E'$ and $E_i \sim E_i'$

for $1 \leq i \leq t$. A similar inductive argument using 3.1, 3.2, 7.6, 8.4 and 8.5 gives the following.

Lemma 9.3. Let E and E' be equivalent expressions in $\xi_1,\ldots,\xi_t$. Let X be any spectrum and let x_i be an element of $K_{|\xi_i|}(D_{\nu\xi_i}X;\|\xi_i\|)$, for $1 \leq i \leq t$. Then $E(x_1,\ldots,x_t) = E'(x_1,\ldots,x_t)$.

If $A = \{\xi_1,\ldots,\xi_t\}$ is any set of indeterminates we can define the filtered algebra CA and the subquotient groups D_jA with their standard bases exactly as in sections 3 and 4. If A' is another set of indeterminates and $f:A \to A' \cup \{0\}$ preserves degree, height and filtration we say that f is subbasic. Clearly, the constructions CA and D_jA are functorial with respect to subbasic maps. We can think of the elements of D_jA as expressions in $\xi_1,\ldots,\xi_t$ by inserting parentheses so that addition and multiplication are treated as binary operations. (Of course, up to equivalence it doesn't matter how the parentheses are inserted.) This identifies D_jA with a subset of the expressions of height 1 and filtration j in $\xi_1,\ldots,\xi_t$. By a Cartan formula for an expression E of height 1 we mean simply an equivalent expression in $D_{\nu E}A$. The next result, which will be proved later in this section, provides some examples which will be useful in the proof of 4.7. We say that two expressions E_1 and E_2 are equivalent mod p if there is an expression E' with $E_1 \sim E_2 + pE'$; in particular this implies $\pi^{\|E_1\|-1}E_1 \sim \pi^{\|E_1\|-1}E_2$.

Proposition 9.4. Let ξ_1, ξ_2, ξ_3, ξ_4 be indeterminates of height r with dimensions 0, 0, 1, 1 respectively. Let $1 \leq s < r$ and let $t \geq 1$.

(i) $\beta_{r-s}Q^s\xi_1 \sim Q^s\beta_r\xi_1$ mod p.

(ii) $\beta_{r-s}Q^s\xi_3 \sim (\pi^s\beta_r\xi_3)^{p^s}$ mod p.

(iii) $Q^s(\xi_1\xi_3) \sim (\pi^s\xi_1)^{p^s}Q^s\xi_3$ mod p if p is odd or $r \geq 3$.

(iv) $Q^s(\xi_3\xi_4)$ is equivalent to $(Q^s\xi_3)(Q^s\xi_4)$ if p is odd and to

$$(Q^s\xi_3)(Q^s\xi_4) + 2^{r-s-1}(\pi Q^{s-1}\xi_3)(\pi^s\beta_r\xi_3)^{2^{s-1}}(\pi Q^{s-1}\xi_4)(\pi^s\beta_r\xi_4)^{2^{s-1}}$$

if $p = 2$ and $r \geq 3$.

(v) $Q^s(\xi_1\xi_3\xi_4) \sim (\pi^s\xi_1)^{p^s}(Q^s\xi_3)(Q^s\xi_4)$ if p is odd.

(vi) If $1 \leq i \leq p-1$ then

$$\beta_{r-s}Q^s(\xi_1^i\xi_2^{p-i}) \sim i(\beta_{r-s}Q^s\xi_1)(\pi^s\xi_1)^{p^s(i-1)}(\pi^s\xi_2)^{p^s(p-i)}$$

$$- i(\pi^s\xi_1)^{ip^s}(\beta_{r-s}Q\xi_2)(\pi^s\xi_2)^{p^s(p-i-1)} \text{ mod } p$$

(vii) If $1 \leq i \leq p-1$ then $\pi^{r+t-1}\beta_{r+t}R^t[(\beta_r\xi_1)\xi_1^{i-1}\xi_2^{p-i}]$ is equivalent to

$$i(\pi^{r-1}\xi_1)^{(i-1)p^t}(\pi^{r-t-1}\beta_{r-t}Q^t\xi_1)(\pi^{r-1}\xi_2)^{(p-i-1)p^t}(\pi^{r-t-1}\beta_{r-t}Q^t\xi_2)$$

if $t < r$ and to zero otherwise.

(viii) $\beta Q^r\mathfrak{L}\,\xi_1 \sim 0$.

(ix) If $s \leq t$ then $Q^s p^t \xi_1$ is equivalent mod p^{t-s+2} to

$$p^{t-s+1}(\pi^{s-1}Q\xi_1)^{p^{s-1}} + c_1 p^{t-s}(\pi^s\xi_1)^{p^s},$$

where

$$c_1 = \begin{cases} 1 \text{ if } p \text{ is odd or } s < t \\ -1 \text{ if } p = 2 \text{ and } s = t. \end{cases}$$

$Q^s p^{s-1}\xi_1$ is equivalent mod p to

$$(\pi^{s-1}Q\xi_1)^{p^{s-1}} + c_2(\pi^s\xi_1)^{p^s},$$

where

$$c_2 = \begin{cases} 0 \text{ if } p \text{ is odd} \\ 1 \text{ if } p = 2. \end{cases}$$

There remain expressions, such that $Q^r\mathfrak{L}\,\xi_1$, for which the Cartan formula is too complicated to give explicitly. Our next result will guarantee the existence of such formulas. Let $A = \{\xi_1,\ldots,\xi_t\}$. We say that an element of D_jA is homogeneous if it is a sum of standard basis elements each of which involves every ξ_i. Note that such elements are in the kernel of D_jf whenever $f:A \to A' \cup \{0\}$ takes at least one ξ_i to 0.

Proposition 9.5. Any expression E of height 1 in $\xi_1,\ldots,\xi_t$ is equivalent to an expression in D_jA for some j. If the ξ_i have height r and degree 0 then the expression $\pi^{r-s-1}Q^s(\xi_1\cdots\xi_t)$ is equivalent to a homogeneous expression in D_jA for each $s < r$. If the ξ_i have height r and degree 1 then $\pi^{r+s-1}\beta_{r+s}R^s(\xi_1(\beta_r\xi_2)\cdots(\beta_r\xi_t))$ is equivalent to a homogeneous expression in D_jA for each $t \geq 0$.

The proof of 9.5 will be given at the end of this section. Unfortunately, there seems to be no direct algebraic proof that the Cartan formulas provided by 9.5 are unique, that is, that distinct elements of D_jA cannot be equivalent as expressions. If we had uniqueness in this sense then Lemma 4.7 would be an immediate consequence of 9.5. Instead we shall have to give a much more elaborate construction of γ_j and γ_j', making use of the explicit formulas of 9.4 in order to avoid appealing to uniqueness. (A similar difficulty in ordinary homology is implicit in our proof of 2.7). On the other hand, it is easy to see from 4.1 and 9.3 that uniqueness does hold, but of course such an argument cannot be used in proving 4.7. However, we can and shall use uniqueness in filtrations less than k in the following inductive proof of 4.7.

Proof of 4.7. We shall give the proof for $r < \infty$. The case $r = \infty$, which is similar and somewhat easier, requires some straightforward modifications in Definition 9.1 to allow for infinite heights; details are left to the reader.

First let $M = M_r$ with $r \geq 2$ (the $r = 1$ case is similar and easier). We define $\mathcal{A}$ to be $\{Qx, \mathcal{Q}x\}$. Let u_m and v_m respectively denote $y^m z^{p-m}$ and $(\beta_r y)y^{m-1}z^{p-m}$ for $1 \leq m \leq p-1$ and define $\mathcal{A}'$ to be

$$\{Qy, Qz, \mathcal{Q}y, \mathcal{Q}z\} \cup \{u_m \mid 1 \leq m \leq p-1\} \cup \{v_m \mid 1 \leq m \leq p-1\}.$$

Lemma 4.3 implies that $\mathcal{A}$ and $\mathcal{A}'$ are in fact subbases for D_pM_r and $D_p(M_r \vee M_r)$. Note that $(D_pg_1)_*$ takes Qy and $\mathcal{Q}y$ to Qx and $\mathcal{Q}x$ and takes all other elements of $\mathcal{A}'$ to zero. In particular $(D_pg_1)_*: \mathcal{A}' \to \mathcal{A} \cup \{0\}$ is a subbasic map and hence $F_1 = D_j(D_pg_1)_*$. Similarly, $F_2 = D_j(D_pg_2)_*$. On the other hand, $(D_pg_0)_*$ is not subbasic since it takes u_m to $\pi\mathcal{Q}x$ and v_m to $\pi\beta_{r+1}\mathcal{Q}x$, hence F_0 is not induced by functoriality from $(D_pg_0)_*$. It is determined by $(D_pg_0)_*$, however, in the following way. If

$$E(Qy, Qz, \mathcal{Q}y, \mathcal{Q}z, u_1, \ldots, u_{p-1}, v_1, \ldots, v_{p-1})$$

is any expression in $D_j\mathcal{A}'$ and E' is an expression in $D_j\mathcal{A}$ equivalent to

$$E(Qx, Qx, \mathcal{Q}x, \mathcal{Q}x, \pi\mathcal{Q}x, \ldots, \pi\mathcal{Q}x, \pi\beta_{r+1}\mathcal{Q}x, \ldots, \pi\beta_{r+1}\mathcal{Q}x)$$

then by 9.3 we have $\lambda_j(F_0(E)) = \lambda_j(E')$, hence $F_0E = E'$.

Next we shall construct γ_j and γ_j'. We assume inductively that γ_ℓ and γ_ℓ' with the required properties have been constructed for all $\ell < j$. By using the values of γ_ℓ and γ_ℓ' on indecomposables and extending multiplicatively, we can define γ_j and γ_j' on the decomposables of $D_j\mathcal{A}$ and $D_j\mathcal{A}'$ so that the diagram commutes when restricted to decomposables. It remains to define γ_j and γ_j' on the standard indecomposables of $D_j\mathcal{A}$ and $D_j\mathcal{A}'$. We may assume that $j = p^s$ for some s, since otherwise there are no indecomposables in filtration j.

Let $\xi_1, \ldots, \xi_p$ be indeterminates with dimension zero, height r, and filtration 1. If $s < r$ we use 9.5 to choose a homogeneous expression E in $D_k\{\xi_1, \ldots, \xi_p\}$ equivalent to $\pi^{r-s-1}Q^s(\xi_1 \cdots \xi_p)$. If $s = r$, let E be an expression in $D_k\{\xi_1, \ldots, \xi_p\}$ equivalent to $Q^r\mathcal{Q}\xi_1$. We define subbasic maps

$$f_m:\{\xi_1,\ldots,\xi_p\} \to A' \cup \{0\}$$

for $0 \le m \le p$ by

$$f_m(\xi_\ell) = \begin{cases} y & \text{for } \ell < m \\ z & \text{for } \ell > m\,. \end{cases}$$

Finally, we define $h:\{\xi_1,\ldots,\xi_p\} \to A$ by $h(\xi_\ell) = x$ for all ℓ. Note that $(g_0)_* \circ f_m = h$ for all m.

We define γ_j and γ_j' on indecomposables in table 1. The first column lists the standard indecomposables in $D_j\mathcal{A}'$, and the second column (we claim) gives the value of F_0 on each. The first four entries in column 2 are precisely the standard indecomposables in $D_j\mathcal{A}$, and the corresponding entries in column 3 define γ_j on each. The remaining entries in column 3 then give the resulting values of γ_j on the other entries of column 2. Finally, column 4 defines γ_j' on each entry in column 1. Note that we have denoted iterates of π in the table simply by π; the precise iterate intended can easily be determined since all entries in the table are to have height 1.

The values of F_0 claimed in column 2 are either obviously correct or follow easily from 9.4 or the formulas of section 3. For example, in line 10 we have

$$\pi^{r-s}\beta_{r-s+1}Q^s\pi\mathcal{Q}x \sim \pi^{r-s}\beta_{r-s+1}\pi Q^s\mathcal{Q}x \sim p\pi^{r-s+1}\beta_{r-s+2}Q^s\mathcal{Q}x \sim 0$$

and in line 12 we have

$$\pi^{r+s-1}\beta_{r+s}R^s\pi\beta_{r+1}\mathcal{Q}x \sim \pi^{r+s-1}Q^s\beta_{r+2s}P_*^{2s}\pi\beta_{r+1}\mathcal{Q}x \sim 0.$$

Table 1

		$\underline{F_0}$	$\underline{\gamma_j \circ F_0}$	$\underline{\gamma_j'}$
1.	$\pi Q^s(\mathcal{Q} y)$	$\pi Q^s(\mathcal{Q} x)$	$(D_k h)(E)$	$(D_k f_p)(E)$
2.	$\pi\beta_{r-s+1}Q^s(\mathcal{Q} y)$	$\pi\beta_{r-s+1}Q^s(\mathcal{Q} x)$	$\begin{cases} (\pi x)^{(p-m-1)j}\, \pi\beta_{r-s}Q^s x & \text{if } s < r \\ 0 & \text{if } s = r \end{cases}$	$\begin{cases} (\pi y)^{(p-m-1)j}\, \pi\beta_{r-s}Q^s y & \text{if } s < r \\ 0 & \text{if } s = r \end{cases}$
3.	$\pi Q^s(Qy)$	$\pi Q^s(Qx)$	$\pi Q^{s+1}x$	$\pi Q^{s+1}y$
4.	$\pi\beta_{r-s-1}Q^s(Qy)$	$\pi\beta_{r-s-1}Q^s(Qx)$	$\pi\beta_{r-s-1}Q^{s+1}x$	$\pi\beta_{r-s-1}Q^{s+1}y$
5.	$\pi Q^s(\mathcal{Q} z)$	$\pi Q^s(\mathcal{Q} x)$	$(D_k h)(E)$	$(D_k f_0)(E)$
6.	$\pi\beta_{r-s+1}Q^s(\mathcal{Q} z)$	$\pi\beta_{r-s+1}Q^s(\mathcal{Q} x)$	same as line 2	$\begin{cases} (\pi z)^{(p-m-1)j}\, \pi\beta_{r-s}Q^s z & \text{if } s < r \\ 0 & \text{if } s = r \end{cases}$
7.	$\pi Q^s(Qz)$	$\pi Q^s(Qx)$	$\pi Q^{s+1}x$	$\pi Q^{s+1}z$
8.	$\pi\beta_{r-s-1}Q^s(Qz)$	$\pi\beta_{r-s-1}Q^s(Qx)$	$\pi\beta_{r-s-1}Q^{s+1}x$	$\pi\beta_{r-s-1}Q^{s+1}z$
9.	$\pi Q^s u_m$	$\pi Q^s(\mathcal{Q} x)$	$(D_k h)(E)$	$(D_k f_m)(E)$
10.	$\pi\beta_{r-s}Q^s u_m$	0	0	$m(\beta_{r-s}Q^s y)(\pi y)^{(m-1)j}(\pi z)^{(p-m)j}$ $-m(\pi y)^{mj}(\beta_{r-s}Q^s z)(\pi z)^{(p-m-1)j}$
11.	$\pi Q^s v_m$	$\pi\beta_{r-s+1}Q^s(\mathcal{Q} x)$	$(\pi x)^{(p-m-1)j}\,\pi\beta_{r-s}Q^s x$	$(\pi y)^{(m-1)j}(\pi z)^{(p-m)j}\,\pi\beta_{r-s}Q^s y$
12.	$\pi\beta_{r+s}R^s v_m$	0	0	$m(\pi y)^{(m-1)j}(\pi z)^{(p-m-1)j}(\pi\beta_{r-s}Q^s y)(\pi\beta_{r-s}Q^s z)$ if $s < r$, 0 otherwise

The listed generators occur only for certain values of s. In lines 1,2,5 and 6 we require $s \le r$; in lines 9, 10, and 11, $s \le r-1$; and in lines 3,4,7 and 8, $s \le r-2$.

To complete the proof of 4.7 for $M = M_r$ it remains to show that diagram (*) of section 4 commutes for $i = 0, 1, 2$. In order to see that the inner square commutes it suffices, by Lemma 9.3, to show that the first four entries in columns 2 and 3 are equivalent as expressions in x. This is clear for lines 1, 3 and 4 and for line 2 if $s = r$ (by 9.4(viii)). If $s < r$ in line 2 we have

$$\pi^{r-s}\beta_{r-s+1}Q^s(\mathfrak{a}x) \sim \pi^{r-s-1}Q^s\pi\beta_{r+1}\mathfrak{a}x \sim \pi^{r-s-1}Q^s(x^{p-1}\beta_r x)$$

which is equivalent to the required formula by 9.4(iii).

To see that the outer square commutes, we must show that the entries in columns 1 and 4 are equivalent as expressions in y and z. The first eight cases follow as in the preceding paragraph. Line 9 follows from the definition of E, line 10 from 9.4(vi), line 11 from 9.4(iii), and line 12 from 9.4(vii).

For commutativity of the upper trapezoid when $i = 1$, we must show that $D_k(g_1)_*$ takes the first four entries in column 4 to the corresponding entries in column 3 (which is obvious) and takes the remaining entries in column 4 to zero. This follows in line 9 from the fact that E is homogeneous (since $(g_1)_* \circ f_m$ takes at least one ξ_ℓ to zero if $1 \leq m \leq p-1$) and the remaining cases are clear. Similarly, we see that the upper trapezoid commutes when $i = 2$. Finally, we observe that each entry of column 4 goes to the corresponding entry of column 3 under $D_k(g_0)_*$, and hence the upper trapezoid commutes when $i = 0$. This completes the proof of 4.7 for $M = M_r$.

Next suppose $M = \Sigma M_r$. We define $\mathfrak{a} = \{Rx\}$ when $r = 1$ and $\mathfrak{a} = \{Qx, Rx\}$ when $r \geq 2$. Let $u_m = y(\beta_r y)^{m-1}(\beta_r z)^{p-m}$ and $v_m = y(\beta_r y)^{m-1}z(\beta_r z)^{p-m-1}$ for $1 \leq m \leq p-1$. We define

$$\mathfrak{a}' = \{Ry, Rz\} \cup \{u_m | 1 \leq m \leq p-1\} \cup \{v_m | 1 \leq m \leq p-1\}$$

when $r = 1$ and

$$\mathfrak{a}' = \{Qy, Qz, Ry, Rz\} \cup \{u_m | 1 \leq m \leq p-1\} \cup \{v_m | 1 \leq m \leq p-1\}$$

when $r \geq 2$.

Then $(D_p g_1)_*$ and $(D_p g_2)_*$ induce subbasic maps from $\mathfrak{a}'$ to $\mathfrak{a}$ and we therefore have $F_i = D_j(D_p g_i)_*$ if $i = 1$ or 2. The map $(D_p g_0)_*$ takes u_m to $-\pi Rx$ when $r = 1$ and to $p_* Qx - \pi Rx$ when $r \geq 2$. it takes v_m to zero when p is odd. When $p = 2$, 3.3(x) implies

$$(D_p g_0)_* v_m = \begin{cases} Q\beta_2 2_* x & \text{uf } r = 1 \\ 2^{r-2}\beta_r 2_* Qx & \text{if } r \geq 2 \end{cases}$$

We begin with the case $r = 1$. We define γ_j and γ_j' on decomposables by inductive hypothesis as in the $M = M_r$ case. To define γ_j and γ_j' on indecomposables we use Table 2.

Table 2

		F_0	$\gamma_j \circ F_0$	γ_j'
1.	$Q(Ry)$	$Q(Rx)$	0	0
2.	$\pi\beta_{s+2}R^s(Ry)$	$\pi\beta_{s+2}R^s(Rx)$	$\pi\beta_{s+2}R^{s+1}x$	$\pi\beta_{s+2}R^{s+1}y$
3.	$Q(Rz)$	$Q(Rx)$	0	0
4.	$\pi\beta_{s+2}R^s(Rz)$	$\pi\beta_{s+2}R^s(Rx)$	$\pi\beta_{s+2}R^{s+1}x$	$\pi\beta_{s+2}R^{s+1}z$
5.	$\pi\beta_{s+1}R^s u_m$	$F_0(\pi\beta_{s+1}R^s u_m)$	0	0

Here the first column lists the indecomposables of $D_j\mathcal{Q}'$ and the second column (we claim) gives the value of F_0 each (note that lines 1 and 3 are relevant only when s = 1, i.e., when $k = p^2$). The first two entries in column 2 are the indecomposables of $D_j\mathcal{Q}$, and the corresponding entries in column 3 give our definition of γ_j on each, while the remaining entries in column 3 are claimed to be values of γ_j determined by the definition we have just given. The entries in column 4 define γ_j' on indecomposables. The necessary verifications are similar to those in the case $M = M_r$, and they are straightforward except in line 5. Here we must show that that $\gamma_j F_0(\pi^s\beta_{s+1}R^s u_m)$ is equal to zero and that $\pi^s\beta_{s+1}R^s(y(\beta y)^{m-1}(\beta z)^{p-m})$ is equivalent to zero as an expression in y and z. For simplicity we assume that p is odd -- the case p = 2 differs only slightly. First recall that to calculate $F_0(\pi^s\beta_{s+1}R^s u_m)$ we need only find an element of $D_j\mathcal{Q}$ which is equivalent to $-\pi\beta_{s+1}R^s\pi(Rx)$ as an expression in the indeterminate Rx. Now

$$-\pi^s\beta_{s+1}R^s\pi(Rx) \sim -\pi^s Q^s\beta_{2s+1}P_*^{2s}\pi(Rx) \qquad \text{by } 3.6(\text{iv})$$

$$\sim -Q^s p(\beta_{s+1}P_*^{s-1}(Rx)).$$

We see by induction on t using (3.3(vi) and 3.3(vii) that Q^t of a multiple of p is equivalent to a sum of terms each of which has either p or a p-th power as a factor. Hence $F_0(\pi^s\beta_{s+1}R^s u_m)$ is a sum of terms each of which has a p-th power factor, and the same is true for the element $\gamma_j F_0(\pi^s\beta_{s+1}R^s u_m)$ of $D_k\{x\}$. But by definition all p-th powers in $C\{x\}$ are zero when r = 1, so that $\gamma_j F_0(\pi^s\beta_{s+1}R^s u_m) = 0$ as required. The proof that $\pi^s\beta_{s+1}R^s(y(\beta y)^{m-1}(\beta z)^{p-m})$ is equivalent to zero is similar. We have

$$\pi^s\beta_{s+1}R^s(y(\beta y)^{m-1}(\beta z)^{p-m}) \sim \pi^s Q^s\beta_{2s+1}P_*^{2s}(y(\beta y)^{m-1}(\beta z)^{p-m})$$

$$\sim Q^s((\beta_{s+1}P_*^s y)^m(\beta_{s+1}P_*^s z)^{p-m}),$$

and 3.3(vi) and 3.3(vii) show that Q^t of a product of elements of degree zero is equivalent to a sum of terms each of which has either p or a p-th power as a factor. But again p-th powers in $C\{y,z\}$ are zero and we see that $\pi^s\beta_{s+1}R^s(y(\beta y)^{m-1}(\beta z)^{p-m}) \sim 0$ as required. This completes the proof of Lemma 4.7 for $M = \Sigma M_1$.

Next let $r \geq 2$. We can define γ_j and γ_j' on decomposables precisely as before. In defining γ_j and γ_j' on indecomposables when $r \geq 2$, it will be convenient to modify the standard basis we have been using as follows. Let η_1 and η_2 be indeterminates with dimension 1, filtration p and heights $\|\eta_1\| = r-1$, $\|\eta_2\| = r+1$. We use 9.5 to obtain an expression $E(\eta_1,\eta_2)$ in $D_j\{\eta_1,\eta_2\}$ equivalent to $\pi^{r+s-1}\beta_{r+s}R^s(p_*\eta_1 - \pi\eta_2)$. We claim that the coefficient of $\pi^{r+s-2}\beta_{r+s-1}R^s\eta_1$ in $E(\eta_1,\eta_2)$ is 1. To see this, write $E(\eta_1,\eta_2)$ as $E_1 + E_2$, where E_1 involves only η_1 and every standard basis element in E_2 involves η_2. If $f:\{\eta_1,\eta_2\} \to \{\eta_1\} \cup \{0\}$ takes η_1 to itself and η_2 to zero then $(D_jf)(E(\eta_1,\eta_2)) = E_1$. On the other hand,

$$(D_jf)(E(\eta_1,\eta_2)) \sim E(\eta_1,0) \sim \pi^{r+s-1}\beta_{r+s}R^s p_*\eta_1 \sim \pi^{r+s-2}\beta_{r+s-1}R^s\eta_1.$$

Since uniqueness holds (by inductive hypothesis) in filtration j we have

$$E_1 = \pi^{r+s-2}\beta_{r+s-1}R^s\eta_1 ,$$

proving the claim. We can therefore give new bases for the indecomposables of D_j and $D_j\mathcal{A}'$ when $r \geq 2$ by replacing $\pi^{r+s-2}\beta_{r+s-1}R^s(Qx)$, $\pi^{r+s-2}\beta_{r+s-1}R^s(Qy)$ and $\pi^{r+s-2}\beta_{r+s-1}R^s(Qz)$ in the standard bases by $E(Qx,Rx)$, $E(Qy,Ry)$ and $E(Qx,Rz)$ respectively.

Next let $\xi_1,\ldots,\xi_p$ be indeterminates with dimension 1, height r and filtration 1. We use 9.5 to choose a homogeneous expression $E'(\xi_1,\ldots,\xi_p)$ in $D_k\{\xi_1,\ldots,\xi_p\}$ equivalent to

$$\pi_{r+s-1}\beta_{r+s}R^s(\xi_1(\beta_r\xi_2)\cdots(\beta_r\xi_p)).$$

Finally, we define the subbasic maps f_m and h exactly as in the case $M = M_r$.

We can now define γ_j and γ_j' on indecomposables by means of Table 3. The first column lists the new basis for the indecomposables of $D_j\mathcal{A}'$. The second column (we claim) gives the values of F_0 on each basis element.

Table 3

		F_0	$\gamma_j \circ F_0$	γ_j'
1.	$\pi Q^s(Ry)$	$\pi Q^s(Rx)$	$\begin{cases} -(\pi Q^s x)(\pi\beta_r x)^{(p-1)j} & \text{if } s < r \\ 0 & \text{if } s = r \end{cases}$	$\begin{cases} -(\pi Q^s y)(\pi\beta_r y)^{(p-1)j} & \text{if } s < r \\ 0 & \text{if } s = r \end{cases}$
2.	$\pi\beta_{r+s+1}R^s(Ry)$	$\pi\beta_{r+s+1}R^s(Rx)$	$\pi\beta_{r+s+1}R^{s+1}x$	$\pi\beta_{r+s+1}R^{s+1}y$
3.	$\pi Q^s(Qy)$	$\pi Q^s(Qx)$	$\pi Q^{s+1}x$	$\pi Q^{s+1}z$
4.	$E(Qy,Ry)$	$E(Qx,Rx)$	$(D_k h)(E')$	$(D_k f_0)(E')$
5.	$\pi Q^s(Rz)$	$\pi Q^s(Rx)$	$\begin{cases} -(\pi Q^s x)(\pi\beta_r x)^{(p-1)j} & \text{if } s < r \\ 0 & \text{if } s = r \end{cases}$	$\begin{cases} -(\pi Q^s z)(\pi\beta_r z)^{(p-1)j} & \text{if } s < r \\ 0 & \text{if } s = r \end{cases}$
6.	$\pi\beta_{r+s+1}R^s(Rz)$	$\pi\beta_{r+s+1}R^s(Rx)$	$\pi\beta_{r+s+1}R^{s+1}x$	$\pi\beta_{r+s+1}R^{s+1}z$
7.	$\pi Q^s(Qz)$	$\pi Q^s(Qx)$	$\pi Q^{s+1}x$	$\pi Q^{s+1}z$
8.	$E(Qz,Rz)$	$E(Qx,Rx)$	$(D_k h)(E')$	$(D_k f_0)(E')$
9.	$\pi Q^s u_m$	$-\pi Q^s(Rx)$	$(\pi Q^s x)(\pi\beta_r x)^{(p-1)j}$	$(\pi Q^s y)(\pi\beta_r y)^{(m-1)j}(\pi\beta_r z)^{(p-m)j}$
10.	$\pi\beta_{r+s}R^s u_m$	$E(Qx,Rx)$	$(D_k h)(E')$	$(D_k f_m)(E')$
11.	$\pi Q^s v_m$	$(\pi\beta_{r-1}(Qx))^{2^s}$ if $p = 2,\ s = r-2$	$(\pi\beta_r x)^{2^{s+1}}$	
		$\begin{cases} (\pi\beta_r R(Qx))^{2^{s-1}} & \text{if } p = 2,\ s = r-1 \\ 0 & \text{otherwise} \\ 0 & \end{cases}$	$\begin{cases} (\pi\beta_{r+1}Rx)^{2^s} \\ 0 \\ 0 \end{cases}$	$(\pi Q^s y)(\pi\beta_r y)^{(m-1)j}(\pi Q^s z)(\pi\beta_r z)^{(p-m-1)j}$
12.	$\pi\beta_{r-s}Q^s v_m$			$(\pi\beta_r y)^{mj}(\pi Q^s z)(\pi\beta_r z)^{(p-m-1)j}$ $-(\pi Q^s y)(\pi\beta_r y)^{(m-1)j}(\pi\beta_r z)^{(p-m)j}$

The elements listed in lines 3 and 7 occur only when $s \leq r-2$. In lines 9, 11, and 12 they occur only when $s \leq r-1$, and in lines 1 and 5 only when $s \leq r$.

The first six entries in this column are the new basis for the indecomposables of $D_j\mathcal{A}$, and the first six entries in column 3 define γ_j, while the remaining entries in column 3 give the values of γ_j on the remaining entries in column 2. The entries in column 4 define γ_j'. The verifications necessary to prove 4.7 in this case are again similar to those in the case $M = M_r$. The less obvious ones are the following. If $s < r$ we have

$$\pi^{r-s}Q^sRx \sim \pi^{r-s-1}Q^s\pi Rx \sim \pi^{r-s-1}Q^{s+1}p_*x - \pi^{r-s-1}Q^s(x(\beta_r x)^{p-1})$$
$$\sim -(\pi^{r-s-1}Q^sx)(\pi^{r-1}\beta_r x)^{(p-1)j}$$

in lines 1,5 and 9 by 9.4(iii). (In particular we observe, as claimed in the proof of 8.4(viii), that the relation 3.6(viii) is not used in the present proof when $s = 1$ and $r \geq 2$.) If $s = r$ we have

$$Q^sRx \sim QRQ^{s-1}x \sim 0$$

in lines 1 and 5 by 3.6(viii). In line 11 with $p = 2$ we apply 9.4(ix) to show

$$F_0(\pi^{r-s-1}Q^sv_m) \sim \pi^{r-s-1}Q^s(2^{r-2}\beta_r 2_*Qx)$$

$$\sim \begin{cases} 0 & \text{if } s < r-2 \\ (\pi^{r-1}\beta_r 2_*Qx)^{2^{r-2}} & \text{if } s = r-2 \\ (\pi^{r-2}Q\beta_r 2_*Qx)^{2^{r-2}} + (\pi^{r-1}\beta_r 2_*Qx)^{2^{r-2}} & \text{if } s = r-1 \end{cases}$$

and the claimed values of F_0 follow from 3.1(ii), 3.5 and 3.6(iii) and (iv). This concludes the proof of 4.7.

Proof of 9.4. Let $\approx$ denote mod p equivalence. Parts (i), (ii), (iii), and (iv) follow easily by induction from 3.3(v) and 3.3(vii). For part (v) we have

$$Q^s((\xi_1\xi_2)\xi_4) \approx Q^s(\xi_1\xi_3)Q^s(\xi_4) \approx (\pi^s\xi_1)^{p^s}(Q^s\xi_3)(Q^s\xi_4)$$

by (iii) and (iv). For part (vi) we have

$$\begin{aligned}\beta_{r-s}Q^s(\xi_1^i\xi_2^{p-i}) &\approx Q^s\beta_r(\xi_1^i\xi_2^{p-i}) \\ &\approx Q^s[i(\beta_r\xi_1)\xi_1^{i-1}\xi_2^{p-i} = (p-i)\xi_1^i(\beta_r\xi_2)^{p-i-1}] \\ &\approx iQ^s[(\beta_r\xi_1)\xi_1^{i-1}\xi_2^{p-i}] - iQ^s[\xi_1^i(\beta_r\xi_2)\xi_2^{p-i-1}] \\ &\approx i(Q^s\beta_r\xi_1)(\pi^s\xi_1)^{(i-1)p^s}(\pi^s\xi_2)^{(p-i)p^s} \\ &\quad - i\,(\pi^s\xi_1)^{p^s}(Q^s\beta_r\xi_2)(\pi^s\xi_2)^{(p-i-1)p^s}\end{aligned}$$

and the result follows by part (i).

(vii) First we claim

(*) $$Q^r\beta_{r+1}p_*\xi_1 \sim 0.$$

This is true when r = 1 by 3.3(iv) and 3.3(v). If $r \geq 2$ we have

$$\begin{aligned} Q^r\beta_{r+1}p_*\xi_1 &\sim Q^{r-1}\beta_r Q p_*\xi_1 \\ &\sim Q^{r-1}\beta_r[p_*Q\xi_1 - (p^{p-1} - 1)\xi_1^p] \\ &\sim Q^{r-1}\beta_r p_* Q\xi_1 \end{aligned}$$

and the claim follows by induction on r.

Now we have

$$\begin{aligned} \pi^{r+t-1}\beta_{r+1}R^t[(\beta_r\xi_1)\xi_1^{i-1}\xi_2^{p-i}] &\sim \pi^{r+t-1}Q^t\beta_{r+2t}p_*^{2t}[(\beta_r\xi_1)\xi_1^{i-1}\xi_2^{p-i}] \\ &\sim \pi^{r+t-1}Q^t\beta_{r+2t}[(\beta_{r+2t}p_*^{2t}\xi_1)p_*^{2t}(\xi_1^{i-1}\xi_2^{p-i})] \\ &\sim -(\pi^{r+t-1}Q^t\beta_{r+2t}p_*^{2t}\xi_1)[\pi^{r+t-1}Q^t\beta_{r+2t}p_*^{2t}(\xi_1^{i-1}\xi_2^{p-i})] \end{aligned}$$

If $t \geq r$ then

$$\pi^{r+t-1}Q^t\beta_{r+2t}p_*^{2t}\xi_1 \sim Q^t\beta_{t+1}p_*(p_*^{t-r}\xi_1),$$

which is equivalent to 0 by (*). Otherwise we have

$$\begin{aligned} (\pi^{r+t-1}Q^t\beta_{r+2t}p_*^{2t}\xi_1)[\pi^{r+t-1}Q^t\beta_{r+2t}p_*^{2t}(\xi_1^i\xi_2^{p-i})] \\ \sim (\pi^{r-t-1}Q^t\beta_r\xi_1)[\pi^{r-t-1}Q^t\beta_r(\xi_1^i\xi_2^{p-i})] \end{aligned}$$

and the result follows from part (iii).

For (viii), we have

$$\begin{aligned} \beta Q^r\mathfrak{Z}\,\xi_1 &\sim Q^{r-1}\beta Q\mathfrak{Z}\,\xi_1 \\ &\sim \begin{cases} \frac{1}{p}\binom{p}{i}(\pi^{r-1}x)^{(p^2-p)(r-1)}Q^{r-1}\beta_r p_* Qx & \text{if } r \geq 2 \\ 0 \quad \text{if } r = 1, \end{cases} \end{aligned}$$

but the expression for $r \geq 2$ is also equivalent to zero by (*).

Finally, part (ix) follows from 3.3(vi) by induction on s.

It remains to prove 9.5. In order to keep track of when an element of $D_j\{\xi_1,\ldots,\xi_t\}$ is homogeneous, we make the following definition. Let S be a fixed set and suppose that we have assigned to each ξ_i a subset $h(\xi_i)$ of S called the homogeneity of ξ_i. Then we define the homogeneity of an arbitrary expression in $\xi_1,\ldots,\xi_t$ by requiring that $0_{\alpha,r,j}$ have homogeneity S, that $p_*,\beta_r,\pi,Q,\mathcal{Q}$ and R commute with h and that $h(E+E') = h(E)\cap h(E')$ and $h(E\cdot E') = h(E)\cup h(E')$. We say that an expression $E(\xi_1,\ldots,\xi_t)$ of height 1 is reducible with respect to h if there is an E' $D_j\{\xi_1,\ldots,\xi_t\}$ with $E' \sim E$ and $h(E') \supset h(E)$.

Proposition 9.6. If S is any set and $h(\xi_1),\ldots,h(\xi_t)$ are any subsets of S then every expression of height 1 in $\xi_1,\ldots,\xi_t$ is reducible with respect to h.

If $S = \{\xi_1,\ldots,\xi_t\}$ and $h(\xi_i) = \{\xi_i\}$ for $1 \le i \le t$ then the expressions listed in 9.5 have homogeneity S, while an expression in $D_j\{\xi_1,\ldots,\xi_t\}$ has homogeneity S if and only if it is homogeneous. Thus 9.5 follows from this case of 9.6. The extra generality allowed for S and h is technically useful in proving 9.6.

In the remainder of this section we prove 9.6. We fix a set S and assume from now on that any indeterminates mentioned have been assigned homogeneities contained in S as well as dimensions, heights and filtrations. It will be convenient to let ξ, η and θ denote indeterminates and to let E, F, G and H denote expressions. We say that two expressions (possibly involving different sets of indeterminates) match if they have the same dimension, height, filtration and homogeneity. We shall frequently use the fact that a sum or product of reducible expressions is reducible and that homogeneity is preserved by substitution, i.e., if F is any expression in $\eta_1,\ldots,\eta_s$ and $E_1,\ldots,E_s$ matching $\eta_1,\ldots,\eta_s$ respectively then $h(F(E_1,\ldots,E_s)) = h(F)$. Note, however, that equivalent expressions generally have different homogeneities; for example, $p\xi$ is equivalent to 0 if $\|\xi\| = 1$ but $h(\xi)$ is not necessarly equal to S.

For our next two results we fix a set $\{\eta_1,\ldots,\eta_s,\eta_1',\ldots,\eta_s',\eta_1'',\ldots,\eta_s''\}$ of indeterminates such that each η_i' matches $Q\eta_i$ and each η_i'' matches $R\eta_i$. Here and elsewhere we shall interpret $Q\eta_i$ as $0_{1,1,1}$ if $\|\eta_i\| = 1$ and $R\eta_i$ as $0_{1,1,1}$ if $|\eta_i| = 0$. We say that an expression is elementary if it does not involve Q or R.

Lemma 9.7. Let G be an elementary expression of length 2 in $\eta_1,\ldots,\eta_s$ and let θ match G.

(i) If F is $\pi^{\|\theta\|-1}\theta$ or $\pi^{\|\theta\|-1}\beta_{\|\theta\|}\theta$ then there is an elementary expression G' $D_{\nu G}\{\eta_1,\ldots,\eta_s\}$ with $G' \sim F(G)$ and $hG' \supset hF$.

(ii) If $F = Q\theta$ or $F = R\theta$ then there is an elementary expression $G'(\eta_1,\ldots,\eta_s,\eta_1',\ldots,\eta_s',\eta_1'',\ldots,\eta_s'')$ with $hG' \supset hF$ and

$$F(G) \sim G'(\eta_1,\ldots,\eta_s,Q\eta_1,\ldots,Q\eta_s,R\eta_1,\ldots,R\eta_s).$$

Proof. The possibilities for G are $\pi\eta_i, p_*\eta_i, \beta_r\eta_i, \eta_i+\eta_j, \eta_i\eta_j$ and η_i. The result can be checked in each case from the formulas of section 3.

Next we define the complexity $c(E)$ of a standard indecomposable E in $D_j\{\eta_1,\ldots,\eta_s\}$ to be the total number of Q's and R's that appear in it. We define $c(E)$ for an arbitrary expression E in $D_j\{\eta_1,\ldots,\eta_s\}$ to be the maximum of the complexities of the indecomposables that appear as factors in the terms of E.

Lemma 9.8. Let $H \in D_j\{\eta_1,\ldots,\eta_s,\eta_1',\ldots,\eta_s',\eta_1'',\ldots,\eta_s''\}$. Then there is an $H' \in D_j\{\eta_1,\ldots,\eta_s\}$ such that $h(H') \supset h(H)$, $c(H') \leq c(H)+1$ and H' is equivalent to

$$H(\eta_1,\ldots,\eta_s,Q\eta_1,\ldots,Q\eta_s,R\eta_1,\ldots,R\eta_s).$$

In particular, the latter expression is reducible.

Proof. We may assume that H is a standard indecomposable and hence that it involves only one of the indeterminates. If it involves one of the η_i the result is trivial. Otherwise H has one of the forms

$\pi^{\|\eta_i\|-t-2}Q^t\eta_i'$, $\pi^{\|\eta_i\|-t-2}\beta_{\|\eta_i\|-t-2}Q^t\eta_i'$, $\pi^{\|\eta_i\|-t}Q^t\eta_i''$, $\pi^{\|\eta_i\|+t-2}\beta_{\|\eta_i\|+t-1}R^t\eta_i'$, or $\pi^{\|\eta_i\|+t}\beta_{\|\eta_i\|+t+1}R^t\eta_i''$. In each case the result follows either trivially or from the formulas of section 3.

Lemma 9.9. Let $E_1,\ldots,E_r$ be elementary expressions in $\xi_1,\ldots,\xi_t$ and let $\theta_1,\ldots,\theta_r$ match $E_1,\ldots,E_r$ respectively. Let $F \in D_j\{\theta_1,\ldots,\theta_r\}$. Then there is an $H \in D_j\{\xi_1,\ldots,\xi_t\}$ such that $c(H) \leq c(F)$, $h(H) \supset h(F)$ and $H \sim F(E_1,\ldots,E_r)$. In particular, $F(E_1,\ldots,E_r)$ is reducible.

Proof. Let ℓ be the maximum of the lengths of the E_i. If $\ell = 1$ the result is trivial. We shall prove the result in general by induction on $c(F)$ with a subsidiary induction on ℓ. We may assume that F is a standard indecomposable, and hence that it involves only one of the θ_i, say θ_1. Now by Definition 9.1, E_1 can be written in the form $G(E_{11},E_{12})$, where $E_{11}(\xi_1,\ldots,\xi_t)$ and $E_{12}(\xi_1,\ldots,\xi_t)$ are elementary with lengths less than ℓ and $G(\eta_1,\eta_2)$ is elementary with length 2. If $c(F)=0$ then F has the form $\pi^{\|\theta_1\|-1}\theta_1$ or $\pi^{\|\theta_1\|-1}\beta_{\|\theta_1\|-1}\theta_1$ and the result follows by 9.7(i) and the subsidiary inductive hypothesis. Otherwise F has the form

$F'(F'')$, where $F'' = Q\theta_1$ or $R\theta_1$ and $c(F') = c(F) - 1$. Thus $F(E_1) = F'(F''(G(E_{11}, E_{12})))$. If $\eta_1', \eta_2', \eta_1'', \eta_2''$ are as in 9.7 then by 9.7(ii) there is an elementary expression $G'(\eta_1, \eta_2, \eta_1', \eta_2', \eta_1'', \eta_2'')$ such that $h(G')$ $h(F'')$ and $G'(\eta_1, \eta_2, Q\eta_1, Q\eta_2, R\eta_1, R\eta_2) \sim F''(G(\eta_1, \eta_2))$. Thus

$$F(G(\eta_1, \eta_2)) \sim F'(G'(\eta_1, \eta_2, Q\eta_1, Q\eta_2, R\eta_1, R\eta_2)).$$

Now since $c(F') < c(F)$ the inductive hypothesis gives an expression $H \in D_j\{\eta_1, \eta_2, \eta_1', \eta_2', \eta_1'', \eta_2''\}$ with $c(H) \leq c(F') < c(F)$, $h(H) \supset f(F') \supset h(F)$, and

$$H \sim F'(G'(\eta_1, \eta_2, \eta_1', \eta_2', \eta_1'', \eta_2''))$$

So that

$$F(G(\eta_1, \eta_2)) \sim H(\eta_1, \eta_2, Q\eta_1, Q\eta_2, R\eta_1, R\eta_2).$$

Now by Lemma 9.8 there is an expression $H' \in D_j\{\eta_1, \eta_2\}$ such that $c(H') \leq c(H) + 1 \leq c(F)$ and $h(H') \supset h(H) \supset h(F)$ with $H' \sim F(G(\eta_1, \eta_2))$. Hence $F(E_1) \sim H'(E_{11}, E_{12})$. Since E_{11} and E_{12} both have lengths less than ℓ, the result now follows by the subsidiary inductive hypothesis.

Finally, we complete the proof of 9.6. Let $G(\xi_1, \ldots, \xi_t)$ be any expression of height 1. The proof is by induction on the length of G, which we may assume is ≥ 2. It is easy to see from definition 9.1 (by another induction on the length of G) that G can be written in the form $G'(\xi_1, \ldots, \xi_t, E)$, where $G'(\xi_1, \ldots, , \xi_t, \eta)$ has length less than ℓ and E has length 2. Then G' has height 1 and $h(G') = h(G)$. By inductive hypothesis we may assume $G' \in D_{\nu G}\{\xi_1, \ldots, \xi_t, \eta\}$. If E is elementary the result now follows by 9.9, while if E is $Q\eta$ or $R\eta$ the result follows by 9.8. This concludes the proof.

Bibliography

1. J. F. Adams. Vector fields on spheres. Annals Math (2) 75(1962), 603-632.
2. J. F. Adams. On the structure and applications of the Steenrod algebra. Comm. Math. Helv. 32(1957-58), 180-214.
3. J. F. Adams. On the non-existence of elements of Hopf invariant one. Annals Math. 72(1960), 20-103.
4. J. F. Adams. A periodicity theorem in homological algebra. Proc. Camb. Phil. Soc. 62(1966), 365-377.
5. J. F. Adams. Lectures on generalized homology. Lecture Notes in Mathematics, Vol. 99, 1-138. Springer, 1969.
6. J. F. Adams. Stable Homotopy and Generalized Homology. The University of Chicago Press, 1974.
7. J. F. Adams. The Kahn-Priddy theorem. Proc. Camb. Phil. Soc. 73(1973), 45-55.
8. J. F. Adams. Infinite Loop Spaces. Annals of Mathematics Studies, Vol. 90. Princeton University Press, 1978.
9. J. F. Adams, J. H. Gunawardena, and H. R. Miller. The Segal conjecture for elementary Abelian p-groups. I. Topology. To appear.
10. D. W. Anderson. There are no phantom cohomology operations in K-theory. Pac. J. Math. 107(1983), 279-306.
11. S. Araki. Coefficients of MR-theory. Preprint.
12. S. Araki and T. Kudo. Topology of H_n-spaces and H-squaring operations. Mem. Fac. Sci. Kyusyu Univ. Ser. A. 10(1956), 85-120.
13. S. Araki and H. Toda. Multiplicative structures in mod q cohomology theories. I;II. Osaka J. Math. 2(1965), 71-115; 3(1966), 81-120.
14. S. Araki and Z-I Yosimura. Differential Hopf algebras modelled on K-theory mod p. I. Osaka J. Math. 8(1971), 151-206.
15. M. F. Atiyah. Thom complexes. Proc. London Math. Soc. 11(1961), 291-310.
16. M. F. Atiyah. Characters and cohomology of finite groups. IHES Publ. Math. No. 9 (1961), 23-64.
17. M. F. Atiyah. Power operations in K-theory. Quarterly J. Math. Oxford (2) 17(1966), 165-193.
18. M. F. Atiyah. K-Theory. Lecture notes by D. W. Anderson. W. A. Benjamin, 1967.
19. M. F. Atiyah, R. Bott, and A Shapiro. Clifford modules. Topology 3 Supp. 1(1964), 3-38.
20. M. G. Barratt, M. E. Mahowald, and M. C. Tangora. Some differentials in the Adams spectral sequence. II. Topology 8(1970), 309-316.
21. A. K. Bousfield. The localization of spectra with respect to homology. Topology 18(1979), 257-281.

22. W. Browder. Homology operations and loop spaces. Illinois J. Math. 4(1960), 347-357.

23. R. Bruner. The Adams Spectral Sequence of H_∞ Ring Spectra. Ph.D. Thesis. University of Chicago, 1977.

24. R. Bruner. An infinite family in $\pi_* S^0$ derived from Mahowald's η_j family. Proc. Amer. Math. Soc. 82(1981), 637-639.

25. R. Bruner. A new differential in the Adams spectral sequence. Topology 23(1984), 271-276.

26. S.R. Bullett and I. G. MacDonald. On the Adem relations. Topology 21(1982), 329-332.

27. J. Caruso, F. R. Cohen, J. P. May, and L. R. Taylor. James maps, Segal maps, and the Kahn-Priddy theorem. Trans. Amer. Math. Soc. 281(1984), 243-283.

28. F. R. Cohen, T. Lada, and J. P. May. The Homology of Iterated Loop Spaces. Lecture Notes in Mathematics, Vol. 533. Springer, 1976.

29. F. R. Cohen, J. P. May, and L. R. Taylor. James maps and E_n ring spaces. Trans. Amer. Math. Soc. 281(1984), 285-295.

30. C. Cooley. The Extended Power Construction and Cohomotopy. Ph.D. Thesis. University of Washington, 1979.

31. T. tom Dieck. Steenrod Operationen in Kobordismen-Theorien. Math. Z. 107(1968), 380-401.

32. T. tom Dieck. Orbittypen und äquivariante Homologie. I. Arch. Math. 23(1972), 307-317.

33. E. Dyer and R. K. Lashof. Homology of iterated loop spaces. Amer. J. Math. 84(1962), 35-88.

34. W. End. Über Adams-Operationen. I. Invent. Math. 9(1969/70), 45-60.

35. L. Evens. A generalization of the transfer map in the cohomology of groups. Trans. Amer. Math. Soc. 108(1963), 54-65.

36. B. Gray. Operations and two cell complexes. Conference on Algebraic Topology, 61-68. University of Illinois at Chicago Circle, 1968.

37. B. Gray. Homotopy Theory. An Introduction to Algebraic Topology. Academic Press, 1975.

38. J. H. Gunawardena. Segal's Conjecture for Cyclic Groups of (Odd) Prime Order. J. T. Knight Prize Essay. Cambridge, 1979.

39. J. H. Gunawardena. Cohomotopy of Some Classifying Spaces. Ph.D. Thesis. Cambridge University, 1981.

40. L. Hodgkin. On the K-theory of Lie groups. Topology 6(1967), 1-36.

41. L. Hodgkin. The K-theory of some well-known spaces. I. QS^0. Topology 11(1972), 371-375.

42. L. Hodgkin. Dyer-Lashof operations in K-theory. London Math. Soc. Lecture Notes Series No. 11, 1974, 27-32.

43. J. D. S. Jones. Root invariants and cup-r-products in stable homotopy theory. Preprint, 1983.

44. J. D. S. Jones and S. A. Wegmann. Limits of stable homotopy and cohomotopy groups. Math. Proc. Camb. Phil. Soc. 94(1983), 473-482.

45. D. S. Kahn. Cup-i products and the Adams spectral sequence. Topology 9(1970), 1-9.

46. D. S. Kahn and S. B. Priddy. Applications of the transfer to stable homotopy theory. Bull. Amer. Math. Soc. 78(1972), 981-987.

47. T. Kambe. The structure of K_Λ rings of the lens spaces and their applications. J. Japan Math. Soc. 18(1966), 135-146.

48. T. Kambe, H. Matsunaga, and H. Toda. A note on stunted lens space. J. Math. Kyoto Univ. 5(1966), 143-149.

49. P. S. Landweber. Cobordism and classifying spaces. Proc. Symp. Pure Math., Vol.22, 125-129. Amer. Math. Soc., 1971.

50. L. G. Lewis. The Stable Category and Generalized Thom Spectra. Ph.D. Thesis. University of Chicago, 1978.

51. L. G. Lewis, J. P. May, and M. Steinberger. Equivariant Stable Homotopy Theory. Lecture Notes in Mathematics. Springer. To appear.

52. H. H. Li and W. M. Singer. Resolution of modules over the Steenrod algebra and the classical theory of invariants. Math. Zeit. 181(1982), 269-286.

53. W. H. Lin. On conjectures of Mahowald, Segal, and Sullivan. Math. Proc. Camb. Phil. Soc. 87(1980), 449-458.

54. W. H. Lin, D. M. Davis, M. E. Mahowald, and J. F. Adams. Calculation of Lin's Ext groups. Math. Proc. Camb. Phil. Soc. 87(1980), 459-469.

55. A. Liulevicius. The Factorization of Cyclic Reduced Powers by Secondary Cohomology Operations. Memoirs Amer. Math. Soc. No. 42, 1962.

56. A. Liulevicius. Zeroes of the cohomology of the Steenrod algebra. Proc. Amer. Math. Soc. 14(1963), 972-976.

57. I. Madsen. Higher torsion in SG and BSG. Math. Zeit. 143 (1975), 55-80.

58. N. Mahammed. A propos de la K-théorie des espaces lenticulaires. C. R. Acad. Sci. Paris, 271(1970), A-639-A-642.

59. M. Mahowald. The Metastable Homotopy of S^n. Memoirs Amer. Math. Soc. 72, 1967.

60. M. Mahowald. A new infinite family in ${}_2\pi_*^s$. Topology 16(1977), 249-254.

61. M. Mahowald and M. Tangora. Some differentials in the Adams spectal sequence. Topology 6(1967), 349-369.

62. J. Mäkinen. Boundary formulae for reduced powers in the Adams spectral sequence. Ann. Acad. Sci. Fennicae Series A-I 562(1973).

63. A. F. Martin. Multiplications on cohomology theories with coefficients. Trans. Amer. Math. Soc. 253(1979), 91-120.

64. W. S. Massey. Products in exact couples. Annals math. 59(1954), 558-569.

65. C. R. F. Maunder. Some differentials in the Adams spectral sequence. Proc. Camb. Phil. Soc. Math. Phys. Sci. 60(1964), 409-420.

66. J. P. May. The Cohomology of Restricted Lie Algebras and of Hopf Algebras; Application to the Steenrod Algebra. Ph.D. Thesis. Princeton University, 1964.

67. J. P. May. Categories of spectra and infinite loop spaces. Lecture Notes in Mathematics, Vol. 99, 448-479. Springer, 1969.

68. J. P. May. A general algebraic approach to Steenrod operations. Lecture Notes in Mathematics, Vol. 168, 153-231. Springer, 1970.

69. J. P. May. The Geometry of Iterated Loop Spaces. Lecture Notes in Mathematics, Vol 271. Springer, 1972.

70. J. P. May. Classifying Spaces and Fibrations. Memoirs Amer. Math. Soc. No. 155, 1975.

71. J. P. May (with contributions by F. Quinn, N. Ray, and J. Törnehave). E_∞ Ring Spaces and E_∞ Ring Spectra. Lecture Notes in Mathematics Vol. 577. Springer, 1977.

72. J. P. May. H_∞ ring spectra and their applications. Proc. Symp. Pure Math. Vol. 32 part 2, 229-243. Amer. Math. Soc. 1978.

73. J. P. May. Multiplicative infinite loop space theory. J. Pure and Applied Algebra 26(1983), 1-69.

74. J. P. May and S. B. Priddy. The Segal conjecture for elementary Abelian p-groups. II. To appear.

75. J. E. McClure. H_∞^{d*} Ring Spectra and their Cohomology Operations. Ph.D. Thesis. University of Chicago, 1978.

76. J. E. McClure. Dyer-Lashof operations in K-theory. Bull. Amer. Math. Soc. 8(1983), 64-72.

77. J. E. McClure and V. P. Snaith. On the K-theory of the extended power construction. Math. Proc. Camb. Phil. Soc. 92(1982), 263-274.

78. R. J. Milgram. Steenrod squares and higher Massey products. Bol. Soc. Mat. Mex. (1968), 32-57.

79. R. J. Milgram. A construction for s-parallelizable manifolds and primary homotopy operations. Topology of Manifolds (Tokyo Conference), 483-489. Markham, 1970.

80. R. J. Milgram. Symmetries and operations in homotopy theory. Proc. Symp. Pure Math., Vol. 22, 203-210. Amer. Math. Soc., 1971.

81. R. J. Milgram. Group representations and the Adams spectral sequence. Pac. J. Math. 41(1972), 157-182.

82. H. R. Miller. An algebraic analogue of a conjecture of G. W. Whitehead. Proc. Amer. Math. Soc. 84(1982), 131-137.

83. H. R. Miller and V. P. Snaith. On the K-theory of the Kahn-Priddy map. J. London Mth. Soc. 20(1979), 339-342.

84. H. R. Miller and V. P. Snaith. On $K_*(QRP^n;Z_2)$. Can. Math. Soc. Conf. Proc. Vol. 2, part 1, 233-245. CMS-AMS, 1982.

85. J. W. Milnor. On the Whitehead homomorphism. J. Bull. Amer. Math. Soc. 69(1958), 79-82.

86. J. W. Milnor. The Steenrod algebra and its dual. Annals Math. 67(1958), 150-171.

87. J. W. Milnor and J. C. Moore. On the structure of Hopf algebras. Annals Math. 81(1965), 211-264.

88. S. Mukohda. The appliations of squaring operations for certain elements in the cohomology of the Steenrod algebra. The collected papers on natural science commemorating the 35th anniversary, 23-38. Fukuoka University, 1969.

89. G. Nishida. Cohomology operations in iterated loop spaces. Proc. Japan Acad. 44(1968), 104-109.

90. G. Nishida. The nilpotency of elements of the stable homotopy groups of spheres. J. Math. Soc. Japan 15(1973), 707-732.

91. S. P. Novikov. Cohomology of the Steenrod algebra. Dokl. Acad. Nauk. SSSR 128(1959), 893-895.

92. S. Oka and H. Toda. Nontriviality of an element in the stable homotopy groups of spheres. Hiroshima Math. J. 5(1975), 115-125.

93. D. G. Quillen. Elementary proofs of some results of cobordism theory using Steenrod operations. Advances in Math. 7(1971), 29-56.

94. G. B. Segal. Equivariant K-theory. IHES Publ. Math. 34(1968), 129-151.

95. G. B. Segal. Operations in stable homotopy theory. London Math. Soc. Lecture Notes Series No. 11, 1974, 105-110.

96. G. B. Segal. Power operations in stable homotopy theory. Preprint.

97. W. M. Singer. Squaring operations and the algebraic EHP sequence. J. Pure and Applied Algebra 6(1975), 13-29.

98. W. M. Singer. A new chain complex for the homology of the Steenrod algebra. Math. Proc. Camb. Phil. Soc. 90(1981), 279-282.

99. V. P. Snaith. Dyer-Lashof operations in K-theory. Lecture Notes in Mathematics, Vol. 496, 103-294. Springer, 1975.

100. N. E. Steenrod and D. B. A. Epstein. Cohomology Operations. Princeton University Press, 1962.

101. M. Steinberger. Homology Operations for H_∞ Ring Spectra. Ph. D. Thesis. University of Chicago, 1977.

102. R. Steiner. Homology operations and power series. Glasgow Math. J. 24(1983), 161-168.

103. M. C. Tangora. On the cohomology of the Steenrod algebra. Math. Zeit. 116(1970), 18-64.

104. H. Toda. Composition Methods in Homotopy Groups of Spheres. Princeton University Press, 1961.

105. H. Toda. An important relation in homotopy groups of spheres. Proc. Japan Acad. 43(1967), 839-842.

106. H. Toda. Extended pth powers of complexes and applications to homotopy theory. Proc. Japan Acad. 44(1968), 198-203.

107. A. Tsuchiya. Homology operations on ring spectrum of H^∞ type and their applications. J. Math. Soc. Japan 25(1973), 277-316.

108. G. W. Whitehead. Generalized homology theories. Trans. Amer. Math. Soc. 102(1962), 227-283.

INDEX

Index of Notations

Index of Notations - Greek, etc.